PALEOPALYNOLOGY

ALFRED TRAVERSE

*Department of Geosciences,
College of Earth and Mineral Sciences,
The Pennsylvania State University*

Boston
UNWIN HYMAN
London   Sydney   Wellington

89-0518

Allen & Unwin Inc.,
8 Winchester Place, Winchester, Mass. 01890, USA

Published by the Academic Division of
Unwin Hyman Ltd,
15/17 Broadwick Street, London W1V 1FP, UK

Allen & Unwin (Australia) Ltd,
8 Napier Street, North Sydney, NSW 2060, Australia

Allen & Unwin (New Zealand) Ltd in association with the
Port Nicholson Press Ltd,
60 Cambridge Terrace, Wellington, New Zealand

First published in 1988

Library of Congress Cataloging in Publication Data
Traverse, Alfred, 1925–
    Paleopalynology/Alfred Traverse.
      p. cm.
    Bibliography: p.
    Includes index.
    ISBN 0-04-561001-0 (alk. paper)
    ISBN 0-04-561002-9 (pbk.:alk. paper)
    1. Palynology
    I. Title
QE993.T67 1988
561′.13–dc19

British Library Cataloguing in Publication Data
Traverse, Alfred, 1925–
    Paleopalynology.
    1. Fossil microorganisms
    I. Title
    560
    ISBN 0-04-561001-0
    ISBN 0-04-561002-9 pbk

Typeset in 10 on 11 point Bembo by Oxford Print Associates Ltd
and printed in Great Britain by Biddles of Guildford

PALEOPALYNOLOGY

TITLES OF RELATED INTEREST

Preface

This book is intended to fulfill a need which I have recognized through nearly two decades of teaching paleopalynology. My approach in teaching the subject has always been laboratory-centered, and has emphasized learning by seeing and doing. This seems natural to me, as paleopalynology is not really a unified subject with an easily definable core of subject matter and a unified approach to its study. Rather, it is the application of a wide variety of techniques to the study of a hard-to-define, extremely diverse set of subjects. Inevitably a professor has favorite approaches and favorite aspects of a subject. Therefore my book obviously does not entirely cover all the possible areas of subject matter, nor does it cover the areas it does treat with an equal degree of thoroughness. Nevertheless, students from my courses have fairly often managed to commence practicing palynology with only my introductory, one-term course, followed by a term of rather independent laboratory work in my problems course. I believe that this book alone will enable even college teachers with little previous experience in the subject to present an adequate course in paleopalynology.

The shape of the book follows that of my course: definition and discussion of the subject matter of paleopalynology, followed by a stratigraphically based survey of palynofloras, and finally by a closer look at "non-pollen" palynomorphs such as dinoflagellate cysts, at "satellite" matters such as carbonization (= "maturation" or coalification) studies, and at some applicable techniques.

Bibliographies are not presented chapter-by-chapter as is often done in this sort of text, because I have frequently been annoyed by this. When one wants to find a certain reference, it is often maddening to have to figure out after which chapter to look. I really see no advantage to the piecemeal bibliography. The reader will find all references together in the back of the book.

Because mine is a laboratory-based approach to the subject, I present in the Appendix applicable laboratory techniques, a flowsheet for processing, and so forth. Ideally, my whole paleopalynology course should be two terms, or a year where the semester system exists. Ideally, also, students taking the course should have a basic understanding of both biology and geology. However, under the circumstances applying at our University if I were to attempt to realize all of the above-mentioned "ideals", there would not be enough qualified, interested students, with enough spare semester slots for such an elective course, to satisfy the minimum enrollment requirement in force here.

Acknowledgements

Obviously the author of a book such as this has been aided by a great array of people. This particular project has first of all been in many senses a joint effort shared by my wife, Elizabeth Insley Traverse. She has been research assistant, typist, word-processor operator, consultant, adviser, and much more. I am very grateful to her for all of this. The "official" reviewers selected by the publisher, W. G. Chaloner and A. C. Scott, have assisted me immeasurably with suggestions for changes, additions, deletions, and in finding errors. I have incorporated almost all of their recommended alterations. The same is true of C. W. Barnosky, who "unofficially" but very thoroughly reviewed the Cenozoic material and the subsequent chapters on pollen sedimentation and so forth, and the Appendix. Nevertheless, errors that remain are not to be laid to these reviewers' account, as most of them probably stem from either my failure to respond adequately to their criticism, or to material I have added since their reviews. I am grateful to Dr. K. J. Hsü for his encouragement of the project while I was at the Swiss Federal Technical Institute ("E.T.H."), Zürich, in 1980–81. Colleagues at E.T.H., R. Hantke and P. Hochuli, also have been helpful. Dozens of colleagues assisted by providing illustrative material and by reviewing the drafts of resulting figures. These persons are mentioned below in connection with credits for various figures. That sort of mention stands as admittedly far too meagre acknowledgement and expression of thanks for their inestimable help. All of my graduate students during this period have helped me in one way or another, especially by being sounding boards: D. K. Choi, V. S. Ediger, M. B. Farley, N. G. Johnson, R. J. Litwin, E. I. Robbins, D. J. Rue, and A. Schuyler. Former students, B. Cornet and D. J. Nichols, helped in a similar way. Dozens of people gave advice over the telephone or by letter (the file of such correspondence runs to many hundreds of items). It is not possible to acknowledge all such valuable help except by this general statement of hearty thanks! However, the patience and goodwill of Roger Jones, Director of Academic Publishing for Unwin Hyman, at all stages of this work, simply must be mentioned. Also, the assistance of D. G. Benson with some troublesome figures of phytoplankton, not one of my areas of expertise, is gratefully acknowledged, as is similar help with fungal spores by W. C. Elsik, and with sporopollenin and chitin diagenesis by K. J. Dorning. For the liberal use of the various excellent facilities in the Department of Geosciences at Penn State, I am also very thankful.

Specific Acknowledgements to Figures and Tables

Fig. 1.1 (a)–(z) reprinted by permission from Traverse (1974a); (aa) and (ac)–(ae) reprinted by permission of Rijks Geologische Dienst, Heerlen, The Netherlands, from Dijkstra & Piérart (1957); (ab) reprinted by permission from Piérart (1955).

Fig. 1.4 (a) reproduced from Sarjeant (1978) courtesy of the A.A.S.P. Foundation; (b) reprinted by permission from Erdtman (1954); (c) courtesy of W. R. Evitt.

Fig. 1.5 (a) reprinted by permission from Erdtman (1954).

Fig. 2.1 after DNAG Geologic Time Scale, *Geology*, September, 1983.

Fig. 2.2 redrafted from R. A. Peppers in Phillips & DiMichele (1981) and reprinted by permission from *Paleobotany, Paleoecology, and Evolution*, K. J. Niklas (ed.), copyright © 1981 Praeger Publishers.

Fig. 2.3 translated and adapted from Federova (1977).

Fig. 3.2 reproduced from Scott *et al.* (1985a) by permission of the author and the Royal Society of Edinburgh.

Table 3.1 from Hesse (1981), reproduced by permission from *Grana* **20**, 151.

Table 3.4 from Traverse (1965).

Table 3.5 from Havinga (1971), reproduced by permission of the author and Academic Press, London.

Fig. 5.2 reproduced from Chaloner (1976) by permission of Academic Press, London.

Fig. 5.5 reproduced by permission of Gustav Fischer Verlag from Beug (1961).

Fig. 5.8 (1)–(8) from Melville (1981) and (a)–(e) from Walker & Doyle (1975).

Fig. 5.12 reproduced from Solomon (1983) by permission of the author and the Botanical Society of America.

Fig. 5.13 electron micrographs courtesy of James W. and Audrey G. Walker.

Fig. 5.15 photos courtesy of J. W. Nowicke, and reproduced by permission of the authors from Nowicke & Skvarla (1977).

Fig. 5.18 from Stach (1964).

Table 5.3 modified and expanded from Faegri & Iversen (1975) by permission of the authors.

Fig. 6.2 reprinted by permission of *Geology* from Diver & Peat (1979).

Fig. 6.3 reproduced by permission of the authors from Vidal & Knoll (1983).

Fig. 6.4 photomicrograph (b) courtesy of Paul K. Strother; all others courtesy of Charles Downie.

Fig. 6.6 photos (a)–(c) are from Johnson (1984); (d)–(m) are courtesy of Charles Downie; (n)–(v) are courtesy of Gordon D. Wood; some of the latter group appeared originally in Wood (1984), and are reproduced by permission of the Board of Trustees of Southern Illinois University.

Fig. 6.7 all drawings are from Cramer & Diez (1979).

Fig. 6.8 all drawings are from Cramer & Diez (1979).

ACKNOWLEDGEMENTS

Fig. 6.9 photomicrographs courtesy of Dr. Karl Mädler; photos originally published in Mädler (1963).

Fig. 6.10 abridged and modified from Downie (1967), and used by permission of Elsevier Science Publishers.

Fig. 6.11 modified from Paris (1981).

Fig. 6.12 all S.E.M. pictures and photographs provided by Florentin Paris.

Fig. 6.13 reproduced by permission from Jansonius (1970).

Fig. 6.14 from Jansonius & Craig (1974), reproduced by permission.

Fig. 6.15 all photos courtesy of Jan Jansonius; (a)–(n) appeared in Jansonius & Craig (1974).

Fig. 7.1 after DNAG Geologic Time Scale, *Geology*, September, 1983.

Fig. 7.2 (a)–(e), (f), (g), and (j) are from Johnson (1984); (h), (i), and (k)–(p) are from Richardson & McGregor (1986), reproduced by permission of the Minister of Supply and Services, Canada; photos from this latter publication are by interference contrast microscopy.

Fig. 7.3 modified from Richardson & Ioannides (1973).

Fig. 8.1 (a) is reprinted from *Brittonia* **29**, 14–29 (1977) by permission of P. G. Gensel and the New York Botanical Garden; (b) is slightly modified from Scott (1984), reproduced by permission of the author and the Institute of Biology, London.

Fig. 8.2 from Chaloner (1967) reprinted by permission of the author and Elsevier Science Publishers.

Fig. 8.3 modified from Chaloner (1970a) (*Biol. Rev.* **45**, 353–77) by permission of Cambridge University Press.

Fig. 8.4 partly based on illustrations in Smith & Butterworth (1967) and used by permission of the authors.

Fig. 8.5 photomicrographs courtesy of D. C. McGregor and the Geological Survey of Canada, except (a), (b), (e), (f), and (w) courtesy of J. B. Richardson and the British Museum (Natural History), and (v) and (y) courtesy of W. Riegel.

Fig. 8.6 photographs courtesy of D. C. McGregor and the Geological Survey of Canada, except (c) and (k) courtesy of J. B. Richardson and the British Museum (Natural History).

Fig. 8.7 modified from Scott (1984) and reproduced by permission of the author and the Institute of Biology, London (originally based on a figure in Chaloner & Sheerin (1981)).

Fig. 8.8 (a)–(k) courtesy of A. C. Scott, originally appeared in Higgs & Scott (1982), and reproduced by permission; (n) and (o) courtesy of A. C. Scott, originally appeared in Scott & Meyer-Berthaud (1985), and reproduced by permission of the Royal Society of Edinburgh; and (m), (p), (q), and (s)–(u) courtesy of M. E. Collinson, originally appeared in Collinson *et al.* (1985), and reproduced by permission of the Geological Society (from *J. Geol. Soc. London* **142**, 375–95).

Fig. 8.9 modified from Chaloner (1970b) by permission of Louisiana State University.

Fig. 8.10 modified from Chaloner (1970b) by permission of Louisiana State University.

Fig. 8.11 information from D. C. McGregor – a previous version was published in McGregor (1979b).

ACKNOWLEDGEMENTS

Fig. 8.12 modified from McGregor (1981) by permission of Elsevier Science Publishers.

Fig. 8.13 photos (a)–(w) are courtesy of Maurice Streel and K. Higgs; most of them were published in Higgs & Streel (1984).

Fig. 8.14 modified from Streel & Traverse (1978) by permission of Elsevier Science Publishers.

Fig. 8.15 modified from Keegan (1981) by permission of Elsevier Science Publishers.

Fig. 9.1 courtesy of R. A. Peppers, Illinois Geological Survey, prepared 1984.

Fig. 9.3 all photomicrographs courtesy of M. S. Barss; they were published in Barss (1967) and are reproduced by permission of the author.

Fig. 9.4 photos (b), (c), (ac), (ad), (af), and (ag) courtesy of M. S. Barss and the Geological Survey of Canada; the photos appeared in Barss (1967); all the other photos are courtesy of R. A. Peppers and the Illinois Geological Survey.

Fig. 9.6 photos courtesy of M. S. Barss and the Geological Survey of Canada; originally published in Barss (1967).

Fig. 9.7 photos (b)–(d) are courtesy of M. S. Barss and the Geological Survey of Canada, from Barss (1967); photo (e) is courtesy of D. C. McGregor and the Geological Survey of Canada, from McGregor (1965).

Fig. 9.8 photos by and courtesy of K. M. Bartram.

Fig. 10.1 modified from Hart (1971) by permission.

Fig. 10.2 photos for (b) and (c) provided by M. S. Barss, Geological Survey of Canada, originally published in Barss (1967).

Fig. 10.3 photos courtesy of B. E. Balme; some photomicrographs first appeared in Balme (1964) and are reproduced by permission of The Palaeontological Association.

Fig. 10.4 the chart, prepared by W. A. Brugman (1983) for his informal publication *Permian–Triassic palynology*, was redrafted with his permission.

Fig. 10.5(a)–(o) are reproduced by permission of E. Schweizerbart'sche Verlagsbuchhandlung from Schaarschmidt (1963); (q)–(z) are from Sarjeant (1970) by permission of the University of Kansas; and (aa)–(aq) are from Sarjeant (1973) by permission of the Canadian Society of Petroleum Geologists.

Fig. 11.1 photomicrographs courtesy of T. Orłowska-Zwolińska, in whose publication (1979) they originally appeared; the more general ranges in parentheses were provided by W. A. Brugman.

Fig. 11.3 ranges and most of the spore/pollen diagrams are from Brugman (1983); a few of the diagrams are from Jansonius & Hills (1976–); all are reproduced by permission of the authors.

Fig. 11.4 redrawn from Dolby & Balme (1976) and used by permission of the authors and Elsevier Science Publishers; floral lists abstracted from same source.

Fig. 11.5 reproduced from Ash *et al.* (1982).

Fig. 11.6 redrawn from Olsen (1978).

Fig. 11.7 based on work of Cornet, Ediger, Robbins, Traverse, and others at Penn State.

Fig. 11.9 photos and explanations courtesy of W. A. Brugman (1983, personal communication).

ACKNOWLEDGEMENTS

Fig. 11.10 (h) is from Pettit & Chaloner (1964) and is reproduced by permission of the authors and *Pollen et Spores*; (i) is from Reyre (1970) and is reproduced by permission of the author and the Palaeontological Association.

Fig. 11.11 photos courtesy of Bruce Cornet, who has extensively studied such Triassic-Jurassic pollen.

Fig. 11.12 (a)–(g) are courtesy of T. Orłowska-Zwolińska, in whose publication (1979) they originally appeared; the line drawings are courtesy of B. Cornet, from unpublished work.

Fig. 12.1 photographs of acritarchs reproduced by permission from Orłowska-Zwolińska (1979); those of megaspores by permission from Marcinkiewicz (1979).

Fig. 12.2 modified from Pocock (1973).

Fig. 12.3 (b) photo courtesy of D. Wall, taken in 1978; (c) photo by J. G. Douglas, courtesy of M. E. Dettmann, taken in August, 1971, at the University of Queensland, at a symposium held in honor of I. C. Cookson.

Fig. 12.4 drawings from Evitt (1985), reproduced by permission of the author and the A.A.S.P. Foundation.

Fig. 12.5 drawings and explanations from Evitt (1985), used by permission of the author and the A.A.S.P. Foundation.

Fig. 12.6 upper and middle sets of drawings and explanations from Evitt (1985), used by permission of the author and the A.A.S.P. Foundation; the exploded projection at the bottom is from Lucas-Clark (1984) and is used with the author's permission.

Fig. 12.8 drawings from Evitt (1985), used by permission of the author and the A.A.S.P. Foundation.

Fig. 12.9 (a)–(c) are courtesy of V. D. Wiggins; (d) and (e) are from Harland *et al.* (1975), reproduced by permission of the authors and the Palaeontological Association; (f)–(j) are courtesy of D. Wall; (k)–(af) are courtesy of K. R. Pedersen, appeared originally in Lund & Pedersen (1985), and are reproduced here by permission of the authors and the Geological Society of Denmark.

Fig. 13.1 photomicrographs provided and identified by J. Medus, originally illustrated in Medus (1983), and reproduced here by permission.

Fig. 13.2 (a)–(d) rearranged from Kemp (1968), with permission; (e) rearranged from Doyle *et al.* (1977), with permission.

Fig. 13.3 all photomicrographs and electron micrographs courtesy of James W. and Audrey A. Walker, appeared originally in Walker & Walker (1984), and are reproduced here by permission.

Fig. 13.4 modified from Walker (1976) and used by permission.

Fig. 13.5 reprinted by permission from *Bot. Rev.* **43** (1), © 1977, L. J. Hickey & J. A. Doyle, and The New York Botanical Garden.

Fig. 13.6 (a) illustrations and caption reprinted by permission from Doyle & Hickey (1976, fig. 28, p. 178), © 1976, Columbia University Press (b) slightly revised from Doyle (1984), and reproduced by permission of VNU Science Press; (c) S.E.M. micrograph courtesy of J. A. Doyle, originally appeared in Doyle *et al.* (1977), and is reproduced here by permission of the authors and the Société Nationale Elf Aquitaine.

Fig. 13.7 reprinted from Muller (1970) by permission of Cambridge Philosophical Society and the Cambridge University Press.

ACKNOWLEDGEMENTS

Fig. 13.8 photographs courtesy of Chaitanya Singh, in whose publication (1983) they originally appeared.

Fig. 13.9 (a)–(d) courtesy of G. F. W. Herngreen, in whose paper (Herngreen & Chlonova 1981) they were originally published; (e), (f), and (s) all courtesy of D. J. Nichols; (g)–(j), (1), and (n) are reproduced with permission from Tschudy (1975), and (o) and (p) from Tschudy (1973); (k), (m), (q), and (t) are reproduced with permission from Góczán et al. (1967).

Fig. 13.10 photos and S.E.M. micrographs courtesy of D. J. Nichols.

Fig. 13.11 (b), (e)–(g), and (o)–(s) courtesy of D. J. Nichols, U.S.G.S., who also provided their identification and other information; (i)–(n) reproduced from Takahashi & Shimono (1982), and used with permission.

Fig. 13.12 the drawings and information are from Herngreen & Chlonova (1981) and are reproduced here by permission.

Fig. 13.13 photos courtesy of G. F. W. Herngreen & A. F. Chlonova, in whose paper (1981) they originally appeared.

Fig. 13.14 photos courtesy of G. F. W. Herngreen & A. F. Chlonova, in whose paper (1981) they originally appeared.

Fig. 13.15 photos courtesy of G. F. W. Herngreen & A. F. Chlonova, in whose paper (1981) they originally appeared.

Fig. 13.16 chart courtesy of G. F. W. Herngreen – a slightly different version appeared in Herngreen & Chlonova (1981).

Fig. 13.17 the photographs are courtesy of Dr. Chaitanya Singh, in whose publication (1983) they originally appeared.

Fig. 13.18 data compiled for the author by D. K. Choi.

Fig. 13.19 (a)–(c) are courtesy of D. J. Nichols and appeared previously in Nichols & Jacobson (1982); (d)–(x) are courtesy of Chaitanya Singh, and originally appeared in his publication (1983); all of the previously published photomicrographs are reproduced here by permission.

Fig. 13.20 photos (a), (c), (h), and (o) are courtesy of Chaitanya Singh, in whose publication (1983) they first appeared; (r) is courtesy of D. J. Nichols; all other photos are from the thesis of D. K. Choi (1983).

Fig. 14.2 forms identified as N.E.R. are courtesy of E. M. Truswell, in whose paper (Kemp & Harris 1977) the forms were originally described; forms identified as T.R.F.U. are courtesy of David Pocknall; forms identified as W. G. are courtesy of D. J. Nichols; all other photos are courtesy of N. O. Frederiksen, some of which appeared in Frederiksen (1980a) and are reproduced here by permission.

Fig. 14.3 all photos identified as N.E.R. are courtesy of E. M. Truswell, in whose paper (Kemp & Harris 1977) many of them originally appeared, and they are reproduced here by permission; all photos identified as M.E.S.D. are courtesy of N. O. Frederiksen, appeared originally in Frederiksen (1983), and are reproduced here by permission of the A.A.S.P. Foundation; most of the rest of the photos are from unpublished work of Frederiksen; (m), (q), (r), (y), and (ah) are courtesy of D. J. Nichols.

Fig. 14.4 all pictures and identifications courtesy of E. M. Truswell and M. Dettmann; (a), (b), and (o) appeared in Truswell (1983) and are reproduced here by permission.

ACKNOWLEDGEMENTS

Fig. 14.5 photos courtesy of M. C. Boulter; all but (y) appeared originally in Boulter & Craig (1979) and are reproduced here by permission of Elsevier Science Publishers.

Fig. 14.6 these diagrams are by W. C. Elsik and were distributed by him at a conference on fungal spores at Kent State University in 1979; some of them appeared in Elsik (1983) and are reproduced here by permission of the A.A.S.P. Foundation.

Fig. 14.8 photos (b), (c), (f), (g), (k), (m), (p), (s), (t), (x), and (y) are courtesy of W. C. Elsik; all other photos courtesy of E. M. Truswell, and are reprinted from her paper (Kemp 1978) by permission of the Director, Bureau of Mineral Resources, Geology and Geophysics, Canberra, Australia.

Fig. 14.9 (a) and (b) are slightly modified from figures in Hochuli (1984); (c) is from Jacobson & Nichols (1982); all are reproduced by permission.

Fig. 14.10 from Truswell & Harris (1982), reproduced by permission.

Fig. 14.11 these photomicrographs and S.E.M. micrographs are all courtesy of L. E. Edwards, U.S. Geological Survey; several of them have appeared previously in Edwards (1980, 1982b, 1984a) and in Edwards et al. (1984); all are reproduced here by permission.

Fig. 15.1 adapted from Reid (1920) and Barghoorn (1951).

Fig. 15.2 adapted from Axelrod (1958, fig. 9, p. 491) by permission of The New York Botanical Garden.

Fig. 15.3 all photos courtesy of M. C. Boulter, in whose publication (Boulter 1971a) they originally appeared; they are reproduced here by permission of the Trustees of the British Museum (Natural History).

Fig. 15.4 this appeared in Traverse (1982) and is based on data in Traverse (1978a, b) and in Hsü & Giovanoli (1979).

Fig. 15.5 data from Benda & Meulenkamp (1979) and Benda (1971); the figure originally appeared in Traverse (1982).

Fig. 15.6 from Bertolani Marchetti et al. (1979), reproduced by permission of Pollen et Spores.

Fig. 15.7 redrafted from Godwin (1975) and various other sources.

Fig. 15.8 (a) reproduced by permission of the Cambridge University Press from Godwin (1975); (b) is from Heusser (1977a), redrawn by permission of the University of Washington.

Fig. 15.9 diagram from Godwin (1975), reproduced by permission of the Cambridge University Press (data originally from West 1968).

Fig. 15.10 redrawn from Godwin (1975), reproduced by permission of the Cambridge University Press.

Fig. 15.13 from Brakenridge (1978), reproduced by permission of the University of Washington.

Fig. 15.14 much simplified from Van Zinderen Bakker (1976), by permission.

Fig. 15.15 slightly modified from Woillard (1978), used by permission of the University of Washington.

Table 15.1 data for northwest Europe from Van der Hammen et al. (1971); data for Germany from Von der Brelie in Boenigk et al. (1977).

Table 15.2 this material originally appeared in Traverse (1982); x-ray and isotope analyses by J. Pika, E.T.H., Zürich.

ACKNOWLEDGEMENTS

Fig. 16.1 (a) is from Faegri & Iversen (1975), and is reproduced by permission of the authors and Munksgaard International Publishers Ltd; (b) is from West (1977) and is reproduced by permission of the author and Longman Group Ltd; (c)–(e) are from Cohen & Spackman (1972) and are reproduced by permission of the authors.

Fig. 16.2 diagrams are from Anderson (1974, 1980) and are reproduced by permission.

Fig. 16.3 symbols are from Faegri & Iversen (1975), reproduced by permission; the pollen diagram is by Burga, as presented in Hantke (1983), reproduced by permission.

Fig. 16.4 modified slightly from Janssen (1974) and reproduced by permission.

Fig. 16.5 both diagrams are modified from Straka (1975), and are reproduced here by permission.

Fig. 16.6 from Straka (1975), reproduced by permission.

Fig. 16.7 (a) is reproduced from McDowell et al. (1971) with permission; (b) is from Anderson (1980), reproduced with permission.

Fig. 16.8 reproduced by permission from Watts (1979).

Fig. 16.9 diagrams from Hall (1985), reproduced by permission.

Fig. 16.10 (a) is from Anderson et al. (1985), reproduced by permission; (b) is from Short (1985), reproduced by permission.

Fig. 16.11 (a) is from Holloway & Bryant (1985) – redrawn from Van Zant (1979) – and is reproduced by permission; (b) is from Baker & Waln (1985), reproduced by permission.

Fig. 16.12 (a) and (b) are from Ritchie (1985) and are reproduced by permission; (c) is from Ager & Brubaker (1985) and is reproduced by permission.

Fig. 17.1 reproduced from Leuschner & Boehm (1981) *Grana* **20** (3), 161–7 by permission.

Fig. 17.3 (a) reproduced by permission from Tauber (1967); (b) reproduced by permission from Jacobson & Bradshaw (1981).

Fig. 17.4 reproduced from Straka (1975) by permission.

Fig. 17.5 from Bernabo & Webb (1977), reproduced by permission of the authors and the University of Washington.

Fig. 17.6 (a)–(e) are from Delcourt & Delcourt (1985a) by permission; (f) and (g) are from Delcourt (1979) by permission.

Fig. 17.7 information reproduced from Adam (1974) by permission.

Fig. 17.8 reproduced by permission of the National Research Council of Canada from Mudie (1982).

Fig. 17.9 reproduced from Stover & Williams (1982) by permission.

Fig. 17.10 slightly modified from Muller (1959) and reproduced by permission.

Fig. 17.11 figures reproduced from Traverse & Ginsburg (1966) by permission of Elsevier Science Publishers.

Fig. 17.12 reproduced from Cross et al. (1966) by permission of the authors and Elsevier Science Publishers.

Fig. 17.13 reproduced from Heusser (1983) with permission of the author and Elsevier Science Publishers.

Fig. 17.14 from Williams (1971), reproduced by permission of the Cambridge University Press.

Fig. 17.15 (a) is reproduced from Chaloner & Muir (1968) by permission of the authors;

ACKNOWLEDGEMENTS

(b) is from Chaloner (1968b), reproduced by permission of the author and the Yale University Press.

Fig. 17.16 modified slightly from Wilson (1976) by permission of the author and the Shreveport Geological Society.

Fig. 17.18 (a) is reproduced from Masran & Pocock (1981) by permission of the authors and Academic Press (London); (b) is from Habib (1982b), reproduced by permission of the author and Academic Press (New York).

Fig. 17.19 photos and explanations courtesy of D. J. Batten.

Fig. 17.22 (a) and (b) from Traverse (1974b); (c) reproduced from Richardson & Rasul (1978), by permission of the Geological Association.

Table 17.1 data abstracted from Birks & Birks (1980) and Wijmstra (1978), but originally from Pohl (1937a,b).

Table 17.2 data from Straka (1975) and Firbas (1949).

Fig. 18.1 (a) is redrawn by permission from Potonié & Kremp (1955); (b) is redrawn by permission from an unpublished poster accompanying Dorning (1984), to which the information about fungal spores has been added – Dorning (1986) gives additional explanations.

Fig. 18.2 photos courtesy of E.I. Robbins, from Robbins (1982).

Fig. 18.3 simplified from Clayton et al. (1977) by permission.

Fig. 18.4 from Brugman (1983), used by permission.

Table 18.1 from Batten (1980).

Fig. A.3 reproduced with permission from Ediger (1986b).

Fig. A.4 reproduced with permission from Ediger (1986b).

Fig. A.11 after Rittenhouse (1940).

Plate 1 left side is from Pearson (1984), reproduced by permission; coal ranks are from various sources; fluorescence data are from Van Gijzel (1981); Munsell color standards are available from Munsell Color, Macbeth Division, Kollmorgen Corp., 2441 N. Calvert St., Baltimore, MD 21218, U.S.A.

Contents

CONTENTS

CONTENTS

CONTENTS

List of tables

CHAPTER ONE

What paleopalynology is and is not

DEFINITION OF THE SUBJECT

There are those who would insist that palynology, and hence also paleo-palynology, applies only to pollen and spores or, more specifically even, only to pollen and the spores of embryo-producing (embryophytic) plants. And it is true that Hyde & Williams (1944) had that in mind when they coined the term "palynology", a word from the Greek παλυνω ("I sprinkle"), suggestive of ("fine meal"), which is cognate with the Latin *pollen* ("fine flour", "dust"). However, that is not the way most paleopalynologists use the term. Instead, they use a pragmatically based working definition that in effect says that paleopalynology consists of the study of the organic microfossils that are found in our maceration preparations of sedimentary rocks, i.e., "What my net catches is a fish." This means that palynomorphs, the microfossils which are the subject matter of this study, consist of very resistant organic molecules, usually sporopollenin, chitin or pseudochitin (there are a few exceptions). Palynomorphs are also by common consent in the approximately 5–500 μm (= micrometer = micron = μ) size range. Many megaspores are larger, and some "seed" megaspores of the late Paleozoic are much larger. Species of *Tuberculatisporites* are reported by Potonié & Kremp (1955) to be 3,000 μm (= 3 mm!). From the pragmatic point of view, such huge spores might be viewed as beyond the pale of palynology – in megafossil paleobotany. However, it is more logical to bend the definition a little and include them as palynomorphs. On the other hand, nannofossils are not palynomorphs on two scores. First, they are calcium carbonate ($CaCO_3$) and hence are destroyed by the dilute hydrochloric acid (HCl) we usually employ as a first treatment in palynological maceration of sediments. Secondly, they are also too small, prevailingly less than 5 μm. Diatoms are not palynomorphs, because they are usually siliceous and destroyed by the hydrofluoric acid (HF) that is the major weapon in the paleopalynological armory. A curious oddity is the report of spores and pollen that are apparently permineralized and lack residual sporopollenin (Srivastava & Binda 1984). According to my definition, such fossils, though spores and pollen, are not palynomorphs!

The word "palynology" should be pronounced pal-ih-nol-o-jee. The first "a" is pronounced as in "map". Avoid "pahl . . ." or "pohl . . .", as if directly taken from the word "pollen". Avoid also "pail . . .", suggesting by the beginning of the word that it comes from "paleo-", as paleontology.

1

Figure 1.1 illustrates the common sorts of palynomorphs, and Figure 1.2 presents the range in time of the various categories of fossils. The categories of things included are as follows, more or less in order of stratigraphic appearance.

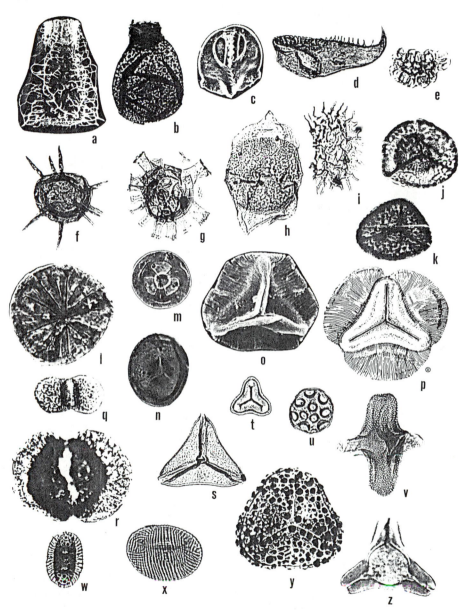

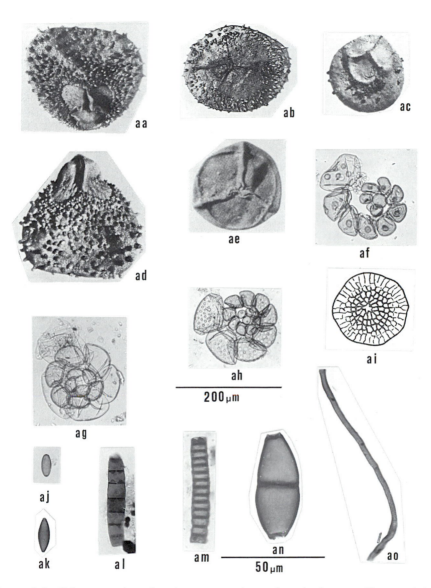

Figure 1.1 Palynomorphs of various categories and geologic ages, illustrated by photomicrographs (P.M.) (= transmitted light), microphotographs (M.P.) (= reflected light, relatively low power), scanning electron micrographs (S.E.M.), and line drawings. All major categories of palynomorphs are included: "microforams", chitinozoans, scolecodonts, colonial algae, acritarchs, dinoflagellates, isospores (or microspores), megaspores, pollen, and fungal spores. Because the fossils come from many sources and represent specimens in many different size ranges, the magnification

3

varies considerably. Approximate size of each specimen is indicated in micrometers, except that a bar is provided under (ah) for the "microforams" (af)–(ah), and a bar is given under (an) for fungal spores (ai)–(ao). (a) Chitinozoan: *Herochitina* sp., Upper Ordovician, England. S.E.M. by W. A. M. Jenkins, length 200 μm. (b) Chitinozoan: *Kalochitina multispinata* Jansonius, Upper Ordovician, Oklahoma. P.M. by R. W. Hedlund, length 150 μm. (c) Scolecodont: *Xanthoprion albertensis* Jansonius & Craig. Dorsal view of partial apparatus. S.E.M. by J. Jansonius, length 350 μm. (d) Scolecodont: *Arabellites* sp., Upper Ordovician, Oklahoma. P.M. by R. W. Hedlund, length 150 μm. (e) Colonial alga: *Botryococcus* sp., Oligo-Miocene, New South Wales, Australia. P.M. length 40 μm. (f) Acritarch: *Baltisphaeridium* sp., Upper Ordovician, Oklahoma. P.M. by R. W. Hedlund, diameter 100 μm. (g) Dinoflagellate cyst: *Hystrichokolpoma unispinum* Williams & Downie, Lower Cretaceous, Ellef Ringnes Island, Arctic Canada. P.M. by C. J. Felix, diameter 85 μm. (h) Dinoflagellate cyst: *Deflandrea granulifera* Manum, Upper Cretaceous, Ellef Ringnes Island. P.M. by C. J. Felix, length 120 μm. (i) Colonial alga: *Pediastrum* sp., Upper Pleistocene, Black Sea. P.M., length 50 μm. (j) Spore: *Retispora lepidophyta* (Kedo) Playford (= *Spelaeotriletes lepidophytus* (Kedo) Keegan), uppermost Devonian, Pennsylvania. P.M., diameter 70 μm. (k) Spore: *Rugospora flexuosa* (Jushko) Streel, uppermost Devonian, Pennsylvania. P.M., diameter 70 μm. (l) Spore: *Emphanisporites robustus* McGregor, Upper Devonian. P.M. by R. W. Hedlund, diameter 65 μm. (m) Spore: *Knoxisporites stephanephorus* Love, Upper Mississippian, Oklahoma. P.M. by C. J. Felix, diameter 55 μm. (n) Spore: *Calamospora* sp., uppermost Devonian, Pennsylvania. P.M., diameter 70 μm. (o) Spore: *Reinschospora speciosa* (Loose) Schopf, Wilson, & Bentall, Upper Mississippian, Iowa. S.E.M. by J. B. Urban, diameter 65 μm. (p) Spore: as (o), Carboniferous. Line drawing by R. Potonié (Potonié & Kremp 1954, p. 139) adopted as trademark of *Catalog of Fossil Spores and Pollen*. (q) Pollen grain: *Vitreisporites pallidus* (Reissinger) Nilsson (al. *Pityopollenites* and *Caytonipollenites*), Upper Triassic, Texas. P.M., length 30 μm. (r) Pollen grain: *Platysaccus nitidus* Pautsch, Upper Triassic, Texas. P.M., length 55 μm. (s) Pollen grain: *Expressipollis ocliferius* Khlonova, Upper Cretaceous, Ellef Ringnes Island, Arctic Canada. P.M. by C. J. Felix, diameter 65 μm. (t) Pollen grain: *Expressipollis accuratus* Khlonova, Upper Cretaceous, Ellef Ringnes Island. Compare with (s)–forms referred to the same form-genus of dispersed spores are often very heterogeneous. P.M. by C. J. Felix, diameter 30 μm. (u) Pollen grain: *Kuylisporites lunaris* Cookson & Dettman, Middle Cretaceous, Lougheed Island, Arctic Canada. P.M. by C. J. Felix, diameter 40 μm. (v) Pollen grain: *Aquilapollenites trialatus* Rouse, Upper Cretaceous, Alaska. P.M. by C. J. Felix, length 95 μm. (w) Pollen grain: *Wodehouseia spinata* Stanley, Upper Cretaceous, Alaska. P.M. by C. J. Felix, length 50 μm. (x) Spore: *Schizaeoisporites* sp., Middle Cretaceous, Oklahoma. P.M. by R. W. Hedlund, length 50 μm. (y) Spore: *Trilobosporites sphaerulentus* Phillips & Felix, Lower Cretaceous, Louisiana. P.M. by C. J. Felix, diameter 80 μm. (z) Pollen grain: *Nudopollis* sp., Paleocene, Gulf Coast, U.S.A. P.M. by R. W. Hedlund, diameter 30 μm. (aa)–(ae) M.P. of megaspores; sizes, given in micrometers, are approximate. (aa) *Lagenicula acuminata* (Dijkstra) Dybová-Jachowicz *et al.*, Lower Carboniferous, Moscow Basin, U.S.S.R., 1,800 μm. Proximal view showing prominent contact structure or gula. Although the parts of the gula look as though they could represent aborted spores, they are actually expanded contact faces. (ab) *Tuberculatisporites mammilarius* (Bartlett) Potonié & Kremp, Carboniferous (Westphalian), Belgium, 1,800 μm. Proximal view showing laesura and contact faces. (ac) *Triletes grandispinosus* Dijkstra, Lower Carboniferous, Moscow Basin, 1,100 μm. Proximal view, showing trilete laesura with curvaturae perfectae and contact faces. (ad) *Triletes acuminata* (Dijkstra) Dybová-Jachowicz *et al.*, Lower Carboniferous, Moscow Basin, 1,800 μm. Lateral view for comparison with (aa) – note prominent proximal gula (see comment under (aa)). (ae) *Triletes patulus* Dijkstra, Lower Carboniferous, Moscow Basin, 800 μm. Proximal view showing trilete laesura and contact faces. Raised figure of

Acritarchs   (Fig. 1.1f. Range: Proterozoic to present.) The term means "of doubtful origin" (Greek ακριτοσ-αρχη). It was introduced by Evitt (1963) as part of a package to replace an older grab-bag term, "hystrichosphaerid" ("spiny sphere"; also by analogy to the genus, *Hystrichosphaera* Wetzel, a synonym for *Spiniferites* Mantell). Many of the former hystrichosphaerids were shown by Evitt and by Wall to be dinoflagellate cysts. The "hystricho-sphaerids" which could not be shown to be dinoflagellate cysts were then left as "acritarchs". However, the term now includes a very large range of presumed algal bodies, mostly marine (but there are many brackish-water or freshwater forms including probable green algal akinetes and hypnozygotes), in the palynomorph size range, mostly less than 100 µm. The wall is sporopolleninous. They may be more or less psilate, scabrate, spiny, reticulate, and a nearly bewildering array of other sculpturing types. They first appear in the Proterozoic as simple more or less spherical, more or less featureless monads about 30 µm in diameter. Some palynologists follow Diver & Peat (1979) in separating off such bodies lacking spines, plates or other features suggesting algal affinity as *cryptarchs* (see Glossary and later discussion). However, it is as yet more common to include the types in the one category, *acritarch*. Acritarchs range to the present time, but their greatest significance and high point of abundance was reached in the early Paleozoic. Tappan (1980) shows that, after the early Jurassic, spiny acritarchs are rare and unimportant, and though leiospheres and tasmanatids continue to the present, acritarchs as a whole are not an important factor among extant phytoplankton. The statement in Pflug & Reitz (1985) that acritarchs ". . . finally disappear in the Pleistocene" is, however, too extreme. There are even some acanthomorphs in modern sediments.

Chitinozoans   (Figs. 1.1a, b. Range: Cambrian to Permian.) These are pseudochitinous palynomorphs of disputed origin, possibly produced by graptolites (see Jenkins 1970), for which relationship the arguments are circumstantial but interesting (chemical similarity, frequent association, more or less identical range). Others have made various proposals regarding the source, e.g., the Fungi (Loquin 1981). They first appear in Cambrian rocks, but are most abundant in the Ordovician. They tail off by Devonian time and are entirely gone by the end of the Permian. They are always found in marine rocks, unless reworked. They sometimes occur in chains, but usually as single "individuals". Because of their thick, more or less opaque walls, they are

---

laesura has characteristic flaps near center which are sometimes called "tecta". (af)–(ah) P.M. of the chitinous inner tests of spiral foraminifera, recent sediment, Great Bahama Bank. Magnification indicated by bar under (ah). (ai)–(ao) P.M. of chitinous palynomorphs of fungal origin. Magnification as shown under (an). (ai) Drawing of characteristic flattened, multichambered, fruiting body (ascostroma) of a sort produced by some ascomycetes. *Asterothyrites* sp., Cenozoic (from Elsik 1979). (aj) Small, non-aperturate fungal spore, recent sediment, Gulf of Mexico. (ak) Small diporate fungal spore, Pleistocene, Black Sea. (al) Chain of fungal spore units, with characteristic thickenings on septa, Pleistocene, Black Sea. (am) Chain of fungal spore units, with very thick septa, recent sediment, Gulf of Mexico. (an) Diporate, two-celled (septate) fungal spore body, Pleistocene, Black Sea. (ao) Fungal hypha, Pleistocene, Black Sea.

usually best studied by scanning electron microscopy (S.E.M.). Because they are not present in the abundance often seen for spores/pollen and acritarchs (about 10^2/g range instead of 10^3–10^5/g range), larger samples must be processed. For this reason and because large specimens are easily broken, somewhat different processing techniques must be employed.

Scolecodonts (Figs. 1.1c, d. Range: Cambrian to present.) These are chitinous mouthparts of marine annelid worms. Although they range from Cambrian to the present, they have been mostly studied in Paleozoic rocks. As is true of chitinozoans, the scolecodonts mostly occur in the range up to 10^2/g and require different processing techniques from those for spores/pollen, though fragments are frequently encountered in slides from conventional macerations. As pointed out by Jansonius & Craig (1971), the mouth-lining parts from one worm may exist united or dispersed, and the dispersed parts may not all be alike. The taxonomy is therefore difficult.

Microscopic colonial algae (Figs. 1.1e, i. Range: Ordovician to present.) *Botryococcus* (Fig. 1.1e) is a colonial alga occurring in a wide range of freshwater to brackish aquatic environments. The walls of the colonies that are preserved apparently consist partly of hydrocarbons; the hydrocarbon "mineral", coorongite, consists largely of *Botryococcus* colonies. Although I once described a "new species", I now regard all *Botryococcus* colonies as representing *B. braunii* Kutzing, which would be, if this is correct, a candidate for the oldest surviving species of Planta, ranging from Ordovician to the present. Niklas (1976) has shown *Botryococcus*, both extant and fossil, to possess an extraordinarily diverse suite of organic compounds. Stratigraphically and paleoecologically almost worthless, *Botryococcus* must nevertheless be treated here, as it so commonly is found in palynological preparations. The other principal "colonial" alga occurring as a palynomorph is *Pediastrum* spp. (Fig. 1.1i), the various species of which range from early Cretaceous to present. Although it is multicellular, it is more precise and technically correct to call it a green algal coenobium, as the number of cells is fixed at the origin of the organism. The resistance to biodegradation of *Pediastrum* is one of the puzzles of palynology. The wall seems to be fairly delicate and cellulosic and should be hydrolyzed quickly. Obviously the walls must be impregnated with something additional – sporopollenin? The various species of *Pediastrum* are all freshwater forms.

Embryophyte spores (Figs. 1.1j–p, x, y. Range: Lower Silurian to present.) Embryophytic plant spores actually include pollen, in the sense that the exines of pollen we study as sporomorphs are the microspore walls of seed plants (see "Pollen" below, and presentation of life-cycles in Chapter 4). However, for palynologists, "spore" as usually employed refers to sporopolleninous microspores and homospores (= isospores) of embryophytes. Embryophyte pollen, megaspores, and fungal spores are thought of as "different", despite the fact that pollen and free megaspores are sporopolleninous and clearly are part of the same category as "spores". Fungal spores are equally clearly another story.

Pollen (Figs. 1.1q–w, z. Range: latest Devonian to present.) The definition of pollen is not morphological but functional: the microspore wall of seed plants, plus the microgametophyte that develops within the wall. Only the outer microspore wall survives as a fossil. The earliest pollen grains were not at all different morphologically from homospores or microspores, and such pollen is hence often called "prepollen" (Chaloner 1970b). We would not realize that they are pollen except for the fact that they are known from paleobotanical investigations to be the fecundating element of seed plants. Later gymnosperm, and especially angiosperm, pollen differ markedly in morphology from spores. Chemically, the walls are apparently the same, that is, sporopolleninous. Biologically, a whole mature pollen grain represents a plant generation and is therefore, in a sense, a plant. In fact, whole (haploid) plants can be grown in culture from a single pollen grain. Nitsch & Nitsch (1969) have produced tobacco plants by culture from single pollen grains. The plants mature normally and flower but, as they are haploid, cannot set seed.

Dinoflagellates (Figs. 1.1g, h. Range: (Silurian?)–Triassic to present.) Sporopolleninous cysts of dinoflagellates are common from Triassic rocks to present, mostly in marine environments, but also in sediments deposited in fresh and brackish water. The range problem indicated above has to do with the difficulty of proving that a given cyst is a dinoflagellate. This requires certification of the presence of dinoflagellate-type archeopyle–operculum and/or dinoflagellate plates and related morphological features. Many Paleozoic cysts may be of dinoflagellates, but the required proofs are not present in sufficiently convincing manner. On the other hand, many brackish-water or freshwater dinoflagellate cysts are more or less featureless "bags" that require examination of thousands of specimens to prove that a dinoflagellate made them, e.g., in Black Sea Deep-Sea Drilling Project (D.S.D.P.) cores (see Traverse 1978a).

Chitinous fungal spores and other fungal bodies (Figs. 1.1ai–ao. Range: Jurassic(?) to present.) Although the kingdom Fungi ranges from Proterozoic to present, these organisms did not until the Jurassic (Elsik personal communication 1981) commonly produce chitinous walls in hyphae or spores, permitting preservation as maceration-resistant fossils. (There are some Permo-Triassic exceptions, but they are not common.) This curiously coincides with the rise of the angiosperms, and it is tempting to think that the development of abundant chitinous walls by fungi is somehow related to their exploitation of the flowering plants. "Spore" in the fungi is a far different concept from "spore" in embryophytic plants. There are many kinds of fungal spores: conidiospores, ascospores, basidiospores, etc. Some are sexually produced, others asexually. They may be single-"celled" or multi-"celled". Many of the things loosely called by paleopalynologists "fungal spores" are actually not strictly spores, e.g., ascocarps and ascomata. Also, pieces of chitinous-walled vegetative tissues of fungi occur as palynomorphs: hyphae and mycelia (see Fig. 1.1ao). It would be good to check the hypothesis that resistant-walled fungal parts are really always chitinous. The statement that they are is based mostly on the knowledge that chitin does occur in the Fungi and is a resistant substance.

Microforaminiferal inner tests (= "microforaminifera") (Figs. 1.1af–ah. Range: Lower Cretaceous(?) to present.) These frequently occur in paleopalynological preparations of marine rock, especially in Cenozoic sediments. They represent the chitinous inner tests of foraminifera, almost always of planispiral forms. The size range is very much less than that of the foraminifera from which they are presumed to have come. The jury is still out on the question of how they are produced and how they should be treated taxonomically: Are they in general referable to particular existing foraminifera taxa (some certainly are), etc? (See Tappan & Loeblich 1965, Traverse & Ginsburg 1966, Cohen & Guber 1968.) The assertion that they are chitinous is based on the fact that the substance behaves and looks like chitin.

Megaspores (Figs. 1.1aa–ae. Range: Lower Devonian to present.) Megaspores are the spores of heterosporous embryophytes, inside of the walls of which the megagametophytes develop. The common practice in paleopalynology is to follow Guennel (1952) in setting an arbitrary lower size limit at 200 μm. The first such large spores occur in the Emsian stage of the latest Lower Devonian. They represent an evolutionary stage or "experiment" largely superseded by development of seeds, and reached the peak of their development in Carboniferous time. There have remained some heterosporous lycopsids and ferns ever since, and hence fossil megaspores have been preserved in sediments. After the Cretaceous they are not abundant, but they can be common and paleoecologically useful. Chemical constitution is apparently sporopollenin.

Varia (Figs. 17.17–17.19. Range: Proterozoic to present.) Palynological preparations always contain more or less non-palynomorph organic "junk". Coal petrologists and palynologists have made a virtue of this by studying the color and/or reflectivity of pellets of such things (plus palynomorphs) to determine the geothermal history of a sedimentary rock. Four categories of such particles are especially common and are occasionally useful to paleopalynologists: (1) wood (tracheids, wood fibers, vessel elements), (2) cuticular fragments, (3) ubisch bodies (orbicules), and (4) various degraded algal and other plant tissues. Wood fragments in palynological preparations can seldom be identified, almost never closer than to a class, e.g., "conifer tracheids". Their abundant presence in a shale, however, usually indicates lagoonal or deltaic environment. Cuticular fragments, especially if well-preserved stomata are present, on the other hand, can often be identified. However, their identification is a complex matter, a field of its own, and few palynologists do more than report presence. Very abundant presence in shale usually indicates lacustrine or fluvio-lacustrine deposition. Some palynologists use relative amounts of cuticles, wood fragments, sporomorphs, dinoflagellate cysts, etc., for "palynofacies" studies: classification of the organic residue in a sedimentary rock (see discussion in Ch. 17). Ubisch bodies, also called orbicules, are tiny bits of sporopollenin about 1–5 μm, which seem to represent sporopollenin not used by the tapetum in laying down the exine of spores and pollen. They are left over as surplus building material and occasionally are abundant in paleopalynological preparations. I am not aware that they are at present regarded by anybody as indicating much paleoenvironmentally. Degraded algal and other plant tissues can sometimes indicate probable marine deposition.

The general stratigraphic range of palynomorphs is shown in Figure 1.2. A liberal view of what is a palynomorph is assumed, in order to be as complete as possible.

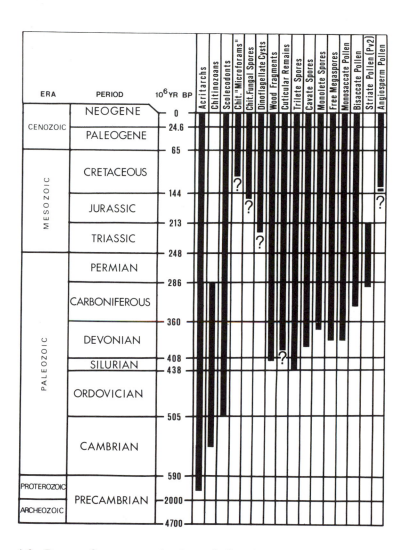

Figure 1.2 Range of occurrence in time of all major categories of palynomorphs, derived from the current literature. "Cuticular remains", column 8, refers to plant cuticular coverings. Depending on interpretation of problematic structures, such structures may go back to early Silurian. "Pv2" after "striate pollen" means bisaccate. Striate (= taeniate) pollen other than bisaccate ranges to the present.

HISTORICAL MATTERS

Wodehouse (1935) presents a marvelous history of the study of extant pollen grains and spores, and this book should be consulted by those who need more information about the early history of palynology proper. A more concise, modern summary of the history of pollen studies is presented by Ducker & Knox (1985). The development of pollen research follows the history of plant anatomy and morphology in general and is dependent to a large degree on the development of microscopes. Nehemiah Grew first observed pollen microscopically in Britain about 1640. Malpighi noted differences in size and color of pollen about the same time. Various people later studied the biology of pollen and spores, especially with reference to the function of pollen in fertilization of ovules, in the 18th century. Camerarius usually gets credit for proving the maleness and fertilizing function of pollen in the late 17th century. Curiously, ancient peoples knew what pollen was for, and such aboriginal people as American Indians have understood the maleness, and the precise function, of pollen, apparently for thousands of years: pollen played a prominent role in some Indian puberty ceremonies. Southwestern Indians seem also to have understood the dietary advantage of eating pollen, long before health-food stores began promoting it (see Fig. 1.3). The Indians perhaps noticed that various insects, especially beetles and hymenopterids, use pollen as a major food staple.

Western Apache Writing Symbol:

"He who is decorated with and enriched by pollen."

Figure 1.3 The vital, dynamic nature of pollen has long been recognized by humans in many cultures. The symbol shown above, from Silas John's Western Apache writing system (Basso & Anderson 1973), bespeaks the ceremonial use of pollen by American Indians, symbolizing fertility, among other things. The pollen used by Silas John (a shaman) was apparently *Typha*, although the symbol would seem to represent a *Zea* tassel, the pollen of which was and is also used ceremonially by Indians. The jars above illustrate the widespread modern use of pollen as a dietary supplement. Such pollen is sometimes harvested by human vacuum cleaners, sometimes taken from domesticated bees, which use pollen for hive nutrition. The human use is based on the vitamin, mineral, and nutritive (lipids, carbohydrates, amino acids) content of pollen. The Graystone apiary near Penn State University sells untreated bee leg loads stripped from worker bees by a device at the hive entrance – hence the variously colored blobs. "Vibrant Health Bee Pollen" is compressed into tablets, for a considerable elevation in price per gram.

E. J. LENNART VON POST

a

b

c

Figure 1.4 Important founding figures of paleopalynology: (a) Christian Gottfried Ehrenberg, 1795–1876; (b) E. J. Lennart Von Post, 1884–1951; (c) left, Gunnar Erdtman, 1897–1973; right, William S. Hoffmeister, 1901–80.

From Saxony in Germany, Ehrenberg was originally a mycologist, but in 1837 he presented a paper to the Berlin Academy of Science, in which most of the major categories of what we now call palynomorphs were described.

The science of palynology is usually reckoned as commencing in 1916 with the introduction by Von Post of analytical pollen diagrams for post-glacial sediments. Von

11

In the 19th century, with the coming of much improved microscopes, the anatomy of pollen and spores was carefully studied and catalogued by German scientists, e.g., von Mohl (d. 1872), Fritsche (d. 1871), and Fischer (pollen work published in 1889). Robert Brown noted in 1809 that pollen could be used to advantage for systematic studies of seed plants, and Brown's illustrator, F. Bauer, described 175 species of pollen for this purpose (see Graham & Barker 1981). Pollen morphologists of today continue the work of Brown, Fritsche, and Fischer, employing better optical microscopes, and especially scanning electron microscopes (S.E.M.) and ultramicrotome techniques, in conjunction with transmission electron microscopes (T.E.M.). The study of spore/pollen morphology has its impact on paleopalynology, of course, but the history of paleopalynology is in practice a separate matter.

The first person to describe fossil spores/pollen, and figure them in line drawing, was apparently Goeppert in Germany in 1838. *Alnus*-like and *Betula*-like pollen are easily recognizable in Goeppert's plates. Ehrenberg (Fig. 1.4a), who pioneered almost everything in micropaleontology, certainly also saw them, and described what we later were to discover as acritarchs and dinoflagellate cysts. By 1867, Schenck was illustrating with good line drawings the *in situ* fossil spores he removed from fossil fern compressions. Reinsch in 1884 published the first photomicrograph of a fossil spore. It was of Carboniferous age, and the genus to which it belonged was long afterwards named, in his honor, *Reinschospora* by Schopf *et al.* (1944). Bennie & Kidston published in 1886 descriptions of megaspores from the Carboniferous of Scotland. Actually, that is about where the paleopalynological matter ended for many decades, with somewhat peripheral exceptions, such as studies in the early 1900s by Thiessen of spores seen in Pennsylvanian coal thin sections, and unpublished studies by Wodehouse of about the same time of pollen and spores in thin sections of the Paleogene Green River Oil Shale.

However, the investigation of possible use of fossil (as some would say, "subfossil") spores/pollen in investigation of "Holocene" or "post-glacial" ("present interglacial" would be a preferable term for the last approximately 10,000 years) sediments went off rather independently, more or less unheeded by paleobotanists, about 1900. Details of this story are related by Erdtman (1954). A Swedish botanist, Lagerheim, realized that the pollen in, e.g., peats of Sweden, told the story of the vegetation in the vicinity of the peat deposition, and he published some brief notes on preliminary studies based on

---

Post, a Swede, was developing ideas actually pioneered earlier by Gustav Lagerheim and others in Scandinavia. Von Post did not publicize pollen analysis much outside of Scandinavia, partly for linguistic reasons.

Erdtman, a younger one-time associate of Von Post's in Sweden, was urbane, fluent in various languages, and traveled very widely, evangelizing for the new scientific method. To Erdtman should go most of the credit for putting fossil spores/pollen studies on the map, over much of the world. He was very conscious of his contribution and would not have liked appearing beneath Von Post here, as Erdtman felt Von Post never sufficiently recognized his work.

Hoffmeister, a paleontologist for Esso (Exxon), more than any other person in industry was responsible for recognition that palynomorphs were a critically important group of microfossils that could be used for practical correlation, where other microfossils failed or were less satisfactory.

Figure 1.5 A few guiding spirits, and one poltergeist, of paleopalynology.

(a) Gunnar Erdtman's self-portrait as a wood gnome. One must not deduce from this bit of whimsy that Erdtman was jovial and light-hearted. He could excoriate a younger palynologist for such a self-defined infraction as publishing a trilete spore photo with no radius of the laesura pointing up.

rather primitive pollen spectra. Lagerheim himself depended on previous work based on geological and paleobotanical studies of Blytt, Sernander, and others, showing that vegetational changes marked the climatic history of the latest Neogene (Faegri 1974, 1981). It remained, however, for a Swedish assistant and student of Sernander, a protégé of Lagerheim, Lennart Von Post (see Fig. 1.4b), to put the subject on a sound footing with thorough studies of a number of cored sequences of Flandrian peat in Sweden. The publication of Von Post's dissertation work in 1916 is usually accepted as the beginning of pollen analysis or pollen statistics, as such studies came to be known. Von Post was apparently not as versatile in languages as some Scandinavians (he published only in Swedish and German), and did not widely popularize the new subject outside of Scandinavia. That task fell to the one man who has probably most influenced the subject, Von Post's doctoral student, Gunnar Erdtman (Figs. 1.4c & 1.5a). A very gifted musician (flutist) and surrealist artist (see frontispiece "self-portrait" sketch in Figure 1.5a of pollen analyst with flute!), Erdtman was urbane and very skilled at languages. (He was, in fact, so good at English and so confident of his talents that he would argue nuances of the language even with native speakers of English such as this writer! He was often, but not always, right!) He traveled widely, indeed he loved to travel, and wherever he went he practiced and "sold" palynology, e.g., trapping pollen grains with a vacuum cleaner on a transatlantic cruise. He was responsible for a great expansion of pollen analytical/statistical studies in many parts of the world in the 1920s, 1930s, and 1940s. The terminology he developed for pollen morphology came to be dominant, partly because of the pre-existing partial vacuum, partly because of his talent for coining new terminology, sometimes even in anticipation of the discovery of features not yet found! Also, his prolific production of publications served to popularize and establish his ideas. In his later years he became rather intolerant of what he regarded as deviant, i.e., non-erdtmanian, practices in spore/pollen work and frequently wrote letters to errant authors, dressing them down for their sorry

---

(b) Knut Faegri, a keen student of pollination mechanisms, has also been the leading Norwegian Quaternary palynologist for decades. Though not primarily a paleopalynologist, he has especially contributed to paleopalynology by his (with J. Iversen) *Textbook of Pollen Analysis.*

(c) Leonard R. Wilson, one of the first American palynologists, is still active, although he (usually called "Dick" or "Doc") published papers on what we would now call paleopalynology soon after Erdtman's visits to North America in the 1920s.

(d) James M. Schopf (1911–78), American paleobotanist, coal petrologist, and paleopalynologist, made very important early contributions to establishing the systematic study of palynomorphs on a sound basis. For example, Schopf *et al.* (1944), largely his work, became a model for such studies.

(e) Robert Potonié (1899–1974), son of the well-known German paleobotanist, Henri Potonié, was one of the first people to recognize and apply the stratigraphic possibilities of paleopalynology, especially in the German coal fields and, through students and associates, elsewhere. After World War II, Potonié's encyclopedic studies of fossil spores/pollen, including introduction of his suprageneric "turmal" system of classification, had great importance in emphasizing the potential for systematic work in the field.

ways. (I have such a letter scolding me for orienting spores/pollen upside down: the laesura ray of a trilete laesura must point up at 90°; the long axis of bisaccate pollen must be parallel to the bottom of the page with the distal side up; etc.) Nevertheless, it must be acknowledged that Erdtman's impact on palynology is unequaled, and we all owe his memory a great debt.

After Von Post and Erdtman came many other Holocene pollen analysts in Scandinavia, such as Iversen, Jessen, and Faegri (see Fig. 1.5b), but also Godwin in Britain, Sears and Potzger in America, Neustadt in the U.S.S.R., Firbas in Germany, and a long list of others. But from the time of Erdtman and Von Post, this subject, linked to plant geography, ecology generally, paleoclimatology, and archeology–anthropology, tended to go its own way, as it still does today. Most of the practitioners are botanically rather than geologically oriented, and the field and laboratory techniques for taking cores of the relatively shallow, usually unindurated, post-glacial sediments and processing them to obtain the usually well-preserved palynofloras are more or less special to Holocene palynologists. The pollen/spore types studied are 100% extant species. Unfortunately, because of all these factors, and other, more personal ones, there tends to be little contact between pollen analysts and paleopalynologists. This was not always the case, and some pioneer paleopalynologists, such as A. Raistrick in Britain, L. R. Wilson (Fig. 1.5c) in the U.S.A., and others, worked in both Pleistocene (including Holocene = or present interglacial) pollen analysis and with palynomorphs as old as Paleozoic.

As already noted, coal petrologists studied spores/pollen seen in coal thin sections long ago, and the American coal petrologist, Thiessen, early in this century even suggested the use of spores for coal-bed stratigraphy. Raistrick, in Britain in the 1920s, was a pioneer in the use of spores for this important task (see Chaloner 1968a). Unfortunately, coal beds are very difficult to correlate by spores because coal is primarily derived from woody swamp peat. The spore/pollen flora of such a sediment is notoriously local in derivation, e.g., as compared to deltaic silts, which have a rich, fluvially derived pollen flora representing a large area. Also the palynofloras of coals of an area usually represent a persistent biofacies that tends to recur mostly in response to the environment. Thus, within a given time frame, it is not always possible to correlate coal beds by correlating the facies, as almost identical palynofloras may occur in widely separated horizons. Nevertheless, because the "original" palynologists were the Holocene pollen analysts who preferred to work with post-glacial peats, the idea persisted for decades that paleopalynologists should be looking at fossil peats, that is, coals. Even now it is hard to convince some field geologists that it is usually better to collect the associated shales than coals for palynology, and that presence of plant megafossils in a shale does not necessarily correlate with presence of palynomorphs! It should be noted that some palynologists (Smith & Butterworth 1967, Kosanke 1950) have successfully correlated coal beds by their spore content, despite the attendant difficulties.

In the late 1920s and early 1930s, Robert Potonié (Fig. 1.5e), son of the paleobotanist, Henri Potonié, began to study spores/pollen from German coals and associated sediments, especially at first the Cenozoic lignitic coals, but later the Carboniferous coals as well. Potonié and his students and coworkers made a very significant contribution to paleopalynology with systematic and biostratigraphic studies. Just before the outbreak of World War II, in the late 1930s, Potonié was engaged by the Royal Dutch/Shell petroleum interests to investigate the possibility of using palynology as a biostratigraphic tool. The war ended that, but soon after the war Shell remembered Potonié's work and began palynological research in earnest, employing especially Dutch palynologists such as Waterbolk, whose ultimate palynological roots were in Holocene pollen analysis. Other, more geologically oriented, persons such as Kuyl were soon involved, however, and even such botanists as Jan Muller were geologically adept enough to assure a sound geological approach. By the early 1950s, Shell had employed paleopalynology very successfully in the Maracaibo Basin in Venezuela, where marine micropaleontology (= study of foraminifera and ostracodes mostly) was not fully satisfactory because of the extensive non-marine sections (see Kuyl *et al.* 1955). In North America, Esso (Standard Oil of New Jersey = Exxon) also began looking at palynology as a biostratigraphic tool quite early. By the late 1940s, they also had a laboratory in Venezuela under R. H. Tschudy, but the center of their palynological operation soon shifted to Oklahoma, where W. S. Hoffmeister (Fig. 1.4c), a micropaleontologist working for Esso's research subsidiary, Carter Oil Co., and L. R. Wilson (Fig. 1.5c), an academic palynologist working as a consultant, together developed Esso's palynostratigraphic program. American paleopalynology owes a great debt also to J. M. Schopf (Fig. 1.5d), who spent his whole career with the Bureau of Mines and the U.S. Geological Survey and recognized in the early 1930s the great potential of spores/pollen studies for solution of geologic problems. The classic work of Schopf *et al.* (1944) was one flowering of Schopf's pioneer efforts. By 1955 when I joined Shell Development Company (a research subsidiary of the Royal Dutch/Shell group of companies) as a palynologist, Shell's palynological operations were worldwide, from Nigeria to western Canada. About the same time, nearly all the other oil companies in the world of any size introduced palynology at least into their research programs. The 1950s were the time of greatest expansion of the subject.

A treatment such as this is not intended to be complete, and I clearly should discuss at length the important and early contributions of the many Soviet paleopalynologists: Naumova, Bolkhovitina, Zaklinskaya, and many others. In the English-speaking world, probably no other one institution has had the impact of the University of Sheffield, under L. R. Moore, Charles Downie, and their many students, now practicing the profession all over the world (see Sarjeant 1984).

As far as I am aware, my Ph.D. dissertation on Paleogene spores/pollen, in 1951, was the first paleopalynological thesis in North America; it certainly was one of the first, but I never had a course specifically in palynology because there were none at Harvard or Cambridge Universities. However, Sir Harry Godwin's ecology course at Cambridge, England, and E. S. Barghoorn's paleobotany course at Harvard, both of which I took, contained some of what is now called paleopalynology. By the 1960s, Ph.D.s in palynology were common, and by 1981, the American Association of Stratigraphic Palynologists had over 800 members. There were about 3,000 professional paleopalynologists in the world at that time (see Traverse 1974a, for a history of palynology to 1972). When I was working on my doctoral dissertation in the late 1940s, there was no journal specifically for palynology, and the relatively manageable number of publications in the subject appeared in a wide variety of journals. There was an informally mimeographed newsletter about pollen/spore research in the 1930s and 1940s called the *Pollen and Spore Circular*, edited and distributed by a Holocene pollen analyst, P. B. Sears. It was chiefly intended to inform "friends of pollen" about new developments, but it has achieved lasting recognition, mostly because of the fortuitous circumstance that the word "palynology" was coined by Hyde & Williams in No. 8, October, 1944. Erdtman's *Grana Palynologica* (now *Grana*) began in 1954, and the French *Pollen et Spores* in 1959. There are quite a number of other journals worldwide (see Annotated bibliography at the end of this chapter). It is now impossible to "stay on top" of the whole palynological literature, even with the aid of the many society newsletters and bibliographies (see above-mentioned bibliography).

The work of Evitt, Wall, Williams, Norris, and many others, beginning in the early 1960s, put a new face on paleopalynology by showing that dinoflagellates (apparently almost always their cysts) can be used biostratigraphically. Evitt (Fig. 12.3a) has had an especially important impact on this work. Dinoflagellate cysts are in the same size range as spores/pollen, and apparently possess a resistant sporopollenin framework in their walls, as do spores/pollen. Most (but by no means all) of the cyst-producing dinoflagellates are marine, and their study has greatly expanded the usefulness of paleopalynology by providing more control for marine rock sequences where spores/pollen may be rare or absent, and because dinoflagellate cysts are often far better chronostratigraphic indicators. The pioneer in this field was G. Deflandre (Ehrenberg, as noted above, had seen dinoflagellate cysts in the 1840s), when palynology was in its infancy in the 1930s. However, it was the proof that many of what had been known as "hystrichosphaerids" are dinoflagellates that opened up this field. Dinoflagellate palynologists mostly started as spore/pollen palynologists, and perhaps this and the fact that spores/pollen and dinoflagellate cysts occur in the same preparations is responsible for keeping spore/pollen and dinoflagellate people closely allied. The "hystrichosphaerids" which were not transferred to the dinoflagellates were then recognized as acritarchs ("unknown" origin), following a proposal of W. R. Evitt. The study of these, mostly algal cysts or reproductive bodies, was pioneered especially by Alfred Eisenack (1891–1982) in Germany, by Charles Downie (see Fig. 6.5a) in England, and by F. H. Cramer and M. d. Carmen R. Diez (Fig. 6.5b), working primarily in Spain and the U.S.A.

Pleistocene (including Holocene, as used here) pollen analysis researchers tend to go their own way, now as before. Their orientation has always been ecological, paleoclimatological, and archeological, and still is. Paleopalynologists have been oriented to biostratigraphy and especially to its economic application. Paleopalynology owes its origins to present interglacial ("postglacial" = Holocene) pollen analysis, and even the very name "palynology" to exponents thereof, but very few paleopalynologists have managed to remain closely associated with Pleistocene pollen analysis as well.

17

ANNOTATED GENERAL BIBLIOGRAPHY OF READILY AVAILABLE PUBLICATIONS

The following bibliography is intended to help the student get "established in the neighborhood", by helping him/her find the way around in the basic literature. It is aimed primarily at North Americans and others whose language is English. The principal weakness this occasions is that the coverage of the Soviet literature is practically nil, despite the great volume of Soviet palynological publication. Most Soviet publications are, of course, in Russian, and English-speaking students will, in my experience, mostly not attempt to read them.

Textbooks and other general works

Andrews, H. N. 1961. *Studies in paleobotany*. New York: Wiley. Good short text for paleobotany. Has a chapter by C. Felix on palynology which, despite the date, is worth reading.

Artzner, D., *et al.* 1979. *A systematic illustrated guide to fossil organic-walled dinoflagellate genera*. Royal Ontario Mus. Life Sci. Misc. Publ. Classification and brief introductory text, but mostly consists of hundreds of interpretative line drawings of dinoflagellates. Should be carefully studied in conjunction with photographs.

Beck, C. B. (ed.) 1976. *Origin and early evolution of angiosperms*. New York: Columbia University Press. A really excellent series of papers on a very important subject. Palynologically speaking, the most important contributions are by Brenner, Doyle & Hickey, and J. W. Walker.

Berglund, B. E. (ed.) 1986. *Handbook of Holocene palaeoecology and palaeohydrology*. Chichester: Wiley. This bulky collection of papers includes many that have direct bearing on paleopalynology, especially on Pleistocene studies. Contributions by Birks (general background and statistical methods), Aaby & Digerfeldt (coring and other sampling techniques), Berglund & Ralska-Jasiewiczowa (pollen analysis and pollen diagrams), Prentice (forest-composition calibration of pollen data), and other chapters will all be useful to the student.

Birks, H. J. B. & H. H. Birks 1980. *Quaternary palaeoecology*. London: Edward Arnold. This is absolutely up to date as to palynological methods for all latest Cenozoic studies, from laboratory techniques to multivariate analysis. However, many of the explanations are very terse, and will send you scurrying for other references, of which they have many.

Brooks, J. (ed.) 1981. *Organic maturation studies and fossil fuel exploration*. London: Academic Press. Papers presented at a symposium at the 5th Int. Palynol. Conf., Cambridge, U.K., 1980. Deals with "carbonization" studies of organic matter in sediments, a subject that grew out of palynology, is still more or less satellite to it and coal petrology, and has considerable importance in organic geochemistry. Level of "carbonization" (= maturation or thermal alteration) is a useful geothermometer.

Brooks, J., *et al.* (eds.) 1971. *Sporopollenin*. London: Academic Press. Collection of papers presented at a symposium in September, 1970, at Imperial College, London. After 718 pages we still do not know for sure of what exines consist, but it is an interesting trip.

Cramer, F. H. & M. d. C. R. Diez 1979. Lower Paleozoic acritarchs. *Palinologia* **1**, 17–160, 227–304 (glossary of morphological terms). This is a very useful compendium of the acritarchs found in Lower Paleozoic rocks, rocks that of course contain no pollen or dinoflagellates, and no spores until the Ordovician/Silurian boundary (see also Downie 1973, Downie *et al.* 1974).

Cross, A. T. (ed.) 1964. *Palynology in oil exploration*. Soc. Econ. Paleont. Mineral. Spec. Publ., no. 11. Though now dated, an excellent group of general papers on petroleum-oriented palynology.

Cushing, E. J. & H. E. Wright, Jr. (eds.) 1967. *Quaternary paleoecology*. Proc. VII Congr. Int. Assoc. Quat. Res., Boulder–Denver, Colorado, Vol. 7. New Haven, Conn.: Yale University Press. Collection of papers on various topics but with considerable emphasis on palynology.

Dimbleby, G. W. 1985. *The palynology of archaeological sites*. London: Academic Press. A rather brief summary of the methods of, problems encountered in, and potential application of, palynology in the science of archaeology.

Downie, C. 1973. Observations on the nature of the acritarchs. *Palaeontology* **16**:2, 239–59. Excellent, short illustrated article explaining the general morphology and natural history of acritarchs. Good bibliography.

Downie, C., S. Jardiné & H. Visscher (eds.) 1974. *Acritarchs*. Spec. Issue of *Rev. Palaeobot. Palynol.* **18**, 1–186. Series of papers of rather uneven quality on acritarchs, but it is important to use this to get a start on acritarch studies.

Dragastan, O., J. Petrescu & L. Olaru 1980. *Palinologie: cu applicati in geologie*. Bucharest: Editura Didactica si Pedagogica. This seems so good a text that, were it in English, we could perhaps use it in lieu of this book. Nicely illustrated, it covers all or most of the subtopics I do, some in much more detail.

Erdtman, G. 1952–71. *Pollen morphology and plant taxonomy: angiosperms – an introduction to palynology*, Vol. 1. Stockholm: Almqvist and Wiksell. (Vol. 2, 1957; Vol. 3, 1965; Vol. 4, 1971.) An attempt to present a complete catalog of flowering plant pollen, plus an introduction to pollen morphology is given in Vol. 1. Erdtman's trouble was diarrhea of terminology, but the pictures are very good, and the book as a whole is indispensable to anybody studying pollen. Vols. 2 and 3 treat in much briefer fashion than Vol. 1 (for angiosperms) the conifers, the bryophytes, and the lower vascular plants. Vol. 2 is mostly composed of illustrations. Vol. 3 mostly consists of text covering the illustrations in Vol. 2. Vol. 4 has descriptions and illustrations (S.E.M. pictures and T.E.M. pictures of sections, but most palynologists use light microscopes, and such pictures have limited use for identification) of spores of Lycopsida, Psilopsida, Sphenopsida, and ferns (mostly of the latter).

Erdtman, G. 1954 (1943). *An introduction to pollen analysis*. Waltham, Mass.: Chronica Botanica. The first real palynology text. Still very useful for getting an overall picture of Quaternary palynology (= pollen analysis).

Erdtman, G. 1969. *Handbook of palynology: morphology-taxonomy-ecology: an introduction to the study of pollen grains and spores*. New York: Hafner. Poorly organized and meandering, but contains much useful stuff, including information about non-spore/pollen palynomorphs. Paleopalynology is spotty and scattered. Laboratory techniques not completely or well presented.

Evitt, W. R. 1985. *Sporopollenin dinoflagellate cysts: their morphology and interpretation*. Am. Assoc. Strat. Palynol. Found. Presents in book form and in great detail the information from Evitt's justly famous Stanford short courses on dinoflagellates. This book is a must for any palynological library.

Faegri, K. & J. Iversen 1975. *Textbook of pollen analysis*, 3rd edn. New York: Hafner. Short and sweet. This is, page for page, the most useful book on palynology going, despite its one-sided concentration on Pleistocene pollen. Terminology clear, simple, and easy to understand and apply.

Faegri, K. & L. Van der Pijl 1979. *The principles of pollination ecology*, 3rd edn. Oxford: Pergamon. For background information about pollen as it operates in nature, you can do no better than to read this. Of course, the pollen of most significance to us (wind-propelled) is not much considered!

Ferguson, I. K. & J. Muller (eds.) 1976. *The evolutionary significance of the exine*. Linnean Soc. London Symp. Ser., no. 1. As is true of most symposium collections, this volume is a very mixed bag. Chaloner on adaptive features of spores/pollen and Walker on angiosperm pollen evolution are important to our concerns.

Gothan, W. & H. Weyland 1964. *Lehrbuch der Paläobotanik*, 2nd edn. Berlin: Akademie Verlag. If you read German, this is a good, though dated, introductory book on fossil plants.

Haq, B. V. & A. Boersma (eds.) 1978. *Introduction to marine micropaleontology*. New York: Elsevier. Contains several articles of interest to palynologists, viz., "Dinoflagellates, acritarchs and tasmanitids", by G. L. Williams; "Spores and pollen", by L. Heusser; "Chitinozoa", by J. Jansonius & W. A. M. Jenkins. Each article is beautifully illustrated, informative, and useful.

Jardiné, S. (ed.) 1971. *Les acritarches*. C.I.M.P. Microfossiles Organiques du Paléozoique, Part 3, Centre Nat. Rech. Sci., Paris. Well-illustrated, detailed information on one genus, *Polyedryxium* (acritarchs), and a very good annotated bibliography on tasmanatids (one group of acritarchs).

Kapp, R. O. 1969. *How to know pollen and spores*. Dubuque, Iowa: Wm. C. Brown. This will assist you to "know" only a relatively few of the more common modern pollen and spores and a few other recent and fossil palynomorphs (fungal spores, diatoms, etc.). Mostly notable as the only "pictured key" to spores and pollen, and spiral bound, too!

Kedves, M. 1986. *Introduction to the palynology of pre-Quaternary deposits*, Parts I & II. Budapest: Akadémiai Kiadó. An English language two-volume paperback textbook, covering much of the same ground as this book, with many useful diagrams and literature lists. Kedves' text will provide an interesting companion piece to this book, whenever students are able to obtain it.

Kremp, G. O. W. 1965. *Morphologic encyclopedia of palynology*. University Arizona Press. Kremp's glossary or dictionary of palynologic terms is useful in trying to read palynological papers in which a variety of terminology is employed. Its usefulness is limited by its lack of critical comment on the terms listed and by incompleteness; one term used in the foreword to the book is not in the *Encyclopedia*!

Kummel, B. & D. Raup (eds.) 1965. *Handbook of paleontological techniques*. San Francisco: W. H. Freeman. You know what they say about the camel having been designed by a committee. This book is of very uneven quality, but students of palynology will find parts of it quite useful, especially the summary of palynological techniques by J. Gray.

Lewis, W. H., P. Vinay & V. E. Zenger 1983. *Airborne and allergenic pollen of North America*. Baltimore: Johns Hopkins University Press. Sumptuously illustrated treatment of pollen found in the air of North America and of their natural history. Unfortunately for paleopalynologists, the illustrations are almost exclusively S.E.M. of pollen surfaces and photomicrographs of "raw", whole (untreated), pollen. This means that *Sassafras*, for example, is illustrated, despite its lack of sporopollenin in the exine, and that the appearance of most forms is quite different from (and is less revealing than) that of acetolyzed or fossilized pollen, which lack protoplasmic contents.

Moore, P. D. & J. A. Webb 1978. *An illustrated guide to pollen analysis*. New York: Wiley. The only way this is better than Faegri & Iversen (*q.v.*), which it resembles, is that it has many photomicrographs and extensive keys to modern pollen types – of Europe. For paleopalynology it is of hardly any use.

Muir, M. D. & W. A. S. Sarjeant (eds.) 1977. *Palynology*, Parts I (spores and pollen) and II (dinoflagellates, acritarchs and other microfossils): *Benchmark papers in geology*. New York: Academic Press. Compilation of several dozen "classics" of palynology. Speaking as one of the original authors, I can say that there is no way this is as good as having the original publications, but it is handy if you do not, and there is some interesting introductory material by the editors.

Muller, J. 1981. *Fossil pollen records of extant angiosperms*. Spec. Issue of *Bot. Rev.* **47**:1. Though preliminary, biologically oriented, and not mistake-free, this little book is an essential aid for palynostratigraphic work in the Cenozoic.

Nair, P. K. K. 1966. *Essentials of palynology*. Bombay: Asia Publ. House. This thin little volume has an overly optimistic title. There is much of interest in it, but it hardly answered to the need for a general text in palynology in English.

Nelson, R. 1975. *Pollen guide for allergy*. Hollister-Stier, Div. of Cutter Labs., Inc. Short, handy summary, with maps, of pollen allergy producing vegetation of the U.S.A. Obtainable from Hollister-Stier, P.O. Box 3145, T.A., Spokane, WA 99220, U.S.A.

Ogden, E. C., *et al.* 1974. *Manual for sampling airborne pollen*. New York: Hafner. A beautifully illustrated book about how the "airborne pollen people" operate. Compared with paleopalynology it is a different world, but very interesting.

Pokrovskaya, I. M. (ed.) 1966. *Paleopalynologia*, 3 vols. Trudy VSEGEI, Leningrad, N.S. no. 141 (in Russian). If this work were in English and generally available we would have used it as the text for our Penn State course. Vol. 1 gives descriptions and synonyms for the more significant forms. Vol. 2 is a systematic explanation of the complexes of spores and pollen found for each significant time segment, Precambrian through Pliocene, plus a discussion of the stratigraphic application of palynology through the Pleistocene. Vol. 3 consists entirely of illustrations of the forms. Treats only spores/pollen.

Sarjeant, W. A. S. 1974. *Fossil and living dinoflagellates*. New York: Academic Press. Sarjeant is a top man on dinoflagellates, both fossil and living, but Evitt (1985) is now a more up-to-date book.

Scagel, R. F., G. E. Rouse, *et al.* 1965. *An evolutionary survey of the plant kingdom*. Belmont, Cal.: Wadsworth. A stimulating general look at plant phylogeny that is helpful to a student of palynology.

Seward, A. C. 1963. *Fossil plants*, 4 vols., reprinted by Hafner Press, New York (originally published as part of Cambridge Biological Series, Cambridge University Press, Vol. 1, 1898, and regularly thereafter). Regarded as a pivotal work in paleobotany for the past three-quarters of a century.

Smith, A. H. V. & M. A. Butterworth 1967. *Miospores in the coal seams of the Carboniferous of Great Britain*. Palaeont. Assoc. London Spec. Pap. Palaeont., no. 1. Already recognized as the "Bible" of Carboniferous palynology, this book is a must for work with the spores of Mississippian–Pennsylvanian age.

Stanley, R. G. & H. F. Linksens 1974. *Pollen*. New York: Springer. Although of little direct importance to paleopalynologists, this book presents much very interesting and potentially important information about the biology of pollen and should be read for background information.

Staplin, F. L. *et al.* 1982. *How to assess maturation and paleotemperatures*. Soc. Econ. Paleont. Mineral. Short Course, no. 7. Contains a series of papers on study of organic matter in sedimentary rocks for determination of thermal maturation level. Examples: Staplin paper deals with the study of color of spores/pollen exines for this purpose. Dow & O'Connor treat the use of reflectance microscopy of kerogen (= organic residue), and van Gijzel the application of fluorescence microscopy and other techniques.

Stewart, W. N. 1983. *Paleobotany and the evolution of plants*. Cambridge: Cambridge University Press. Basic paleobotany text and hence competitive with Taylor (1981). However, it is very different in including much information, some of it speculative, about paths of plant evolution. Very little palynology.

Straka, H. 1975. *Pollen- und Sporenkunde*. Stuttgart: Gustav Fischer Verlag. Introductory general text for palynology, but emphasizes Holocene (present interglacial) work. In German.

Tappan, H. 1980. *The paleobiology of plant protists*. San Francisco: W. H. Freeman. This is an invaluable reference book for the paleopalynologist because we deal constantly with dinoflagellates and acritarchs and occasionally with diatoms and other "teenies" in Tappan's book. Helen Tappan's husband, Alfred Loeblich, is an equally famous micropaleontologist whose help per the preface approaches co-authorship.

Taylor, T. N. 1981. *Paleobotany: an introduction to fossil plant biology*. New York: McGraw-Hill. This book supersedes the older English-language paleobotany texts (Arnold, Andrews, the lot). It is well-written, thorough, and includes considerable information about paleopalynology. The information about fossil plant groups is up to date, trimerophytes and all.

Tschudy, R. H. & R. A. Scott (eds.) 1969. *Aspects of palynology*. New York: Wiley-Interscience. Although the book suffers from being done by a panel, with the usual problems of uneven quality, it is the most comprehensive textbook in English available. One of its principal failings is the lack of a section on laboratory methods. It is now slightly dated and out of print.

Visscher, H. & G. Warrington (eds.) 1974. *Permian and Triassic palynology*. Spec. Issue of *Rev. Palaeobot. Palynol.* **17**, 1–246. Usual problems of a multi-author publication, but a number of very good papers in this area of work.

Wodehouse, R. P. 1935. *Pollen grains*. New York: McGraw-Hill. The first modern book on pollen morphology and its history, plus information on the state 50 years ago of what we would now call palynology. For pollen morphogenesis, this is still the best reading there is.

Catalogs, atlases, keys, etc.

Adams, R. J. & J. K. Morton 1972. *An atlas of pollen of the trees and shrubs of eastern Canada and the adjacent United States*, Part I, *Gymnospermae to Fagaceae*. Dept. Biology, Univ. Waterloo, Waterloo, Ont. (privately printed and available from them). When complete, this series will be very useful in conjunction with McAndrews *et al.* (1973). Only Part I is so far available. The illustrations are great, but all are S.E.M., which are not very useful to the average palynologist unless he also has photomicrographs for comparison.

Andrew, R. 1980. *A practical pollen file of the British flora*. Sub-Dept. Quaternary Res., Botany School, Cambridge University. This informally "published" (photocopied) little catalog provides thumbnail word sketches of practically all pollen/spores found in the British flora. It would not be of much use without a pollen collection, but with one, the classification can be quite helpful.

Andrews, H. N., Jr. 1970. *Index of generic names of fossil plants, 1820–1965*. U.S. Geol. Surv. Bull., no. 1300. A comprehensive list of genera of megafossil plants, useful to palynologists for checking names of fossil plants associated with sporomorphs. Unfortunately now a little out of date.

Bassett, I. J., C. W. Crompton & J. A. Parmalee 1978. *An atlas of airborne pollen grains and common fungus spores of Canada*. Can. Dept. Agric., Res. Branch, Monograph, no. 18. Profusely illustrated (light, S.E.M., light interference contrast pictures), provided with keys. Useful for Pleistocene studies of temperate North American and Eurasian sediments.

Boros, Á. & M. Járai-Komlódi 1975. *An atlas of recent European moss spores*. Budapest: Akadémiai Kiadó. Despite the title, includes also hornworts and liverworts, as well as mosses, *sensu stricta*. Very interesting presentation of spores of bryophytes, profusely illustrated, though the S.E.M. and light pictures average only fair.

Charpin, J., R. Surinyach & A. W. Frankland 1974. *Atlas of European allergenic pollens*. Paris: Sandoz Editions. Never mind the error ("pollens") in the title. This is a good reference if you wish to know what kinds of pollen to expect in the air at a certain time and place in Europe.

Combaz, A., *et al.* (ed.) 1967. *Les chitinozaires*. C.I.M.P. Microfossiles Organiques du Paléozoique, Parts 1 and 2, Centre Nat. Rech. Sci., Paris. Well-illustrated and handsomely printed catalog description of chitinozoan morphology (in French).

Eisenack, A. 1964–. *Katalog der fossilen Dinoflagellaten, Hystrichosphären und verwandten Mikrofossilien*, Vol. I, *Dinoflagellates* (3 parts), Vol. II, *Dinoflagellates*, cont. (2 parts),

Vols. III–VI, *Acritarchs* (4 parts). Stuttgart: E. Schweizerbart'sche Verlag. This fine catalog is intended to cover fossil dinoflagellates and "hystrichosphaerids" (many of latter now called acritarchs). Original descriptions in originally published languages.

Jansonius, J. & L. V. Hills 1976 *et seq. Genera file of fossil spores and pollen.* Dept. Geol., Univ. Calgary, Alberta, Spec. Publ. A set of (two longish boxes) "cards" (= papersheets), usually one per genus of fossil spores and pollen, with line drawings and "diagnosis" (description) along with bibliographic reference to type species, and comments. Absolutely invaluable for paleopalynological taxonomy. Supplemental cards are being issued from time to time. I recommend organizing the sheets alphabetically in ring notebooks to prevent misplacement.

Kremp, G. O. W., *et al.* 1957–. *Catalog of fossil spores and pollen*, 43 vols. as of 1984. Pennsylvania State University. Intended as a systematic compendium of original (specific) descriptions and illustrations for the specialist, and very incomplete, but nevertheless the best way for a starting student to get an impression of the sorts of fossil pollen to be found at various stratigraphic levels. If possible it is good to have two sets, one of cards filed stratigraphically and (secondarily) morphologically (the "strat file") in a filing cabinet, and the other, a set of bound shelf volumes.

Lamotte, R. S. 1952. *Catalog of the Cenozoic plants of North America through 1950.* Geol. Soc. Am. Mem., no. 51. For Tertiary plants, an indispensable systematic survey of the literature.

Lentin, J. K. & G. L. Williams 1985. *Fossil dinoflagellates: index to genera and species*, 1985 edn. Canadian Tech. Rep. Hydrogr. Ocean Sci. no. 60. Invaluable tool for work with fossil dinoflagellate cysts, listing all genera and species and their nomenclatural history, synonyms, etc.

McAndrews, J. H., *et al.* 1973. *Key to the Quaternary pollen and spores of the Great Lakes region.* Royal Ontario Mus. Misc. Publ. For identification of modern pollen and spores from peats, etc., of northeastern North America, this key is an indispensable aid. Good photomicrographic illustrations. Available from the Royal Ontario Museum, Toronto, Canada.

Nilsson, S. (ed.) 1973–. *World pollen and spore flora.* Stockholm: Almqvist and Wiksell. Treats pollen and spores of all sorts of plants, rather randomly. Eventually the set will be an important research tool, if it continues to grow. As of late 1986, no. 12, on the Onagraceae, is in hand, so only about one thin monograph per year has so far appeared. It can be purchased separately. If you get *Grana* (*q.v.*, journals), you will get this as part of the subscription price.

Nilsson, S., J. Praglowski & L. Nilsson 1977. *Atlas of airborne pollen grains and spores of northern Europe.* Stockholm: Natur och Kultur. Light pictures made of whole (= untreated) spores/pollen and therefore of limited use to paleopalynologists. S.E.M.–T.E.M. pictures sensational. Is moderately useful for learning morphology of many of dominant spores/pollen forms of northern Eurasia and northern North America.

Potonié, R. 1956–75. *Synopsis der Gattungen der Sporae Dispersae*, 7 vols: I–VI, Beihefte zum geologischen Jahrbuch 23 (1956), 32 (1958), 39 (1960), 72 (1966), 87 (1970), and 94 (1970); VII, Fortschr. Geol. Rheinld. Westf. 25 (1975). These monographs outline Potonié's system of spore classification (Turmas, etc.), and list with commentary (in German) most of the genera of fossil spores. The classification is mostly of use for Paleozoic and (to a much lesser extent) for Mesozoic spores, and has been somewhat revised by M. E. Dettmann (1963) ("Upper Mesozoic microfloras from South-eastern Australia", *Proc. R. Soc. Victoria*, **77**:1), among others. Related works that should be consulted are Potonié & Kremp (1954, 1955, 1956a, 1956b) (see Bibliography).

Punt, W. (ed.) 1976. *The northwest European pollen flora I.* Amsterdam: Elsevier. The first volume of a projected series provided with keys, illustrations, and descriptions. Will be especially handy for Pleistocene studies, especially of Old World materials.

Stover, L. E. & W. R. Evitt 1978. *Analyses of pre-Pleistocene organic-walled dinoflagellates.* Stanford, Cal.: Stanford University Press. A very useful catalog of the genera and species of fossil dinoflagellates, along with a very limited amount of descriptive material about them.

Bibliographies, computer-based databanks

Bibliography and index of geology (continuing). Washington, DC: Am. Geol. Inst. Monthly in magazine form, published annually in hardback, as bibliographic and index volumes. Available in most good geological libraries, it is as useful for literature searches in palynology as in any branch of geology. Now worldwide in coverage and prepared by computer. Formerly published (until Vol. 43, 1979) by Geol. Soc. Am. Incorporates the former U.S. Geological Survey *Bibliography of North American geology*, and the former Geol. Soc. Am. *Bibliography of geology outside of North America.* (Pre-1966, you must search both of these!)

Boureau, E., *et al.* (eds.) 1956–73. *World report on paleobotany.* Int. Org. Palaeobot. (an arm of Int. Assoc. Plant Taxonomy). This was sent automatically to all paleobotanists who were members of I.A.P.T. Covered the waterfront – general paleobotany to paleopalynology – but covered it rather inadequately. There is at the moment much debate about how to have a truly up-to-date international bibliography for paleobotany and paleopalynology.

Cousminer, H. J., J. Golden, *et al.* (eds.) 1972–. *Bibliography and index of micropaleontology.* New York: Am. Geol. Inst. and Am. Mus. Nat. Hist. Covers palynomorphs as well as most other microfossils. Rather incomplete and behind, but worth having and using.

Downie, C. & W. A. S. Sarjeant 1964. *Bibliography and index of fossil dinoflagellates and acritarchs.* Geol. Soc. Am. Mem., no. 94. A useful list of older works dealing with acritarchs such as *Veryhachium*, algal forms such as *Tasmanites*, and dinoflagellates, including what used to be known as the "hystrichosphaerids". Also lists the names and stratigraphic position of the taxa which have been described in these groups.

Erdtman, G. 1927–60. *Literature on palynology.* Stockholm. Mostly published in Geol. Fören. I Stockholm Förhand. This goes back to the beginning of palynology. If you have access to a complete set, you can find everything (well, practically everything) ever published on a certain aspect of palynology up to 1960. For a time, this was continued in the journal *Grana Palynologica* (*q.v.*). This bibliography was eventually abandoned by Erdtman because of the *Pollen et Spores* bibliographies (*q.v.*).

Hulshof, O. K. & A. A. Manten 1971. *Bibliography of actuopalynology, 1871–1966.* Spec. Issue of *Rev. Palaeobot. Palynol.* **12**, 1–243. Does same thing for extant spores/pollen that Manten (1969) does for fossil sporomorphae.

Jongmans, W. J. 1910–13. *Die paleobotanische Literatur*: Vol. I (1910), before 1908; Vol. II (1911), 1909 and Nachträge für 1908; Vol. III (1913), 1910–1912 and Nachträge für 1909. Jena: Gustav Fischer. Includes the few palynological things there were at the time. Written in German. World War I ended this effort at an international bibliography for paleobotany.

Kremp, G. O. W. 1977–. *Paleo data banks.* Tucson, Ariz.: Palynodata. Palynological computerization. Project begun by Kremp with cooperation of many oil companies. Separate bibliographies for various time segments now available. For example, no. 13 is "Jurassic palynological literature . . .". Photocopies of almost all papers in palynology are also available at or near cost. Dr. Kremp is now retired, but still has copies of most of the books. Associated with the publication effort has been a computer-based databank of stratigraphic ranges, and nomenclatural–taxonomic information, on most taxa of palynomorphs. This service is now available to the public for an annual fee. One can connect his own personal computer to the database by suitable modem. As of July, 1986, contact R. A. Morgan, Chevron Inc., 935 Gravier St., New Orleans, Louisiana 70112, U.S.A., for more information about this unique service.

Lejal-Nicol, A. 1977. *Rapport sur la Paléobotanique et la Paléopalynologie, France, Belgique, Suisse, 1972–1976.* Paris: Lab. Paléobot., Univ. Pierre et Marie Curie. Example of the many small bibliographies available for various regions, linguistic groups or subjects. It is impossible to be on top of all of these.

Manten, A. A. (ed.) 1969. *Bibliography of palaeopalynology 1836–1966.* Spec. Issue of *Rev. Palaeobot. Palynol.* **8**:1–4, 1–572. A very handy bibliographic tool. If you have this, and the *Pollen et Spores* (*q.v.*) bibliographies, you are fairly complete (see also Hulshof & Manten), but you should also use *Bibliography and index of geology* (*q.v.*).

Paleobotanical section, Botanical Society of America: bibliographies (continuing). These are sent out free to members of the section, and deal only with North America. They are very complete and up to date.

Pollen et Spores: bibliographies (continuing). Published once a year as part of the journal, *Pollen et Spores* (*q.v.*). These overlapped Erdtman's bibliographies at first but were more nearly current. Now they are alone in the field as a comprehensive, exclusively palynological bibliography.

Tralau, H. 1974. *Bibliography and index to paleobotany and palynology, 1950–1970.* Vol. I = author and title; Vol. II = index. The index goes by "keyword", all computer-handled. Only the first few letters of the author names count, so Traverso, Travers and Traverse are mixed together! After Dr. Tralau's death, B. Lundblad brought out a supplement covering 1970–75.

N.B. Local linguistic or subject matter paleobotanical–palynological societies also produce bibliographies, often excellent, an example of such a society is: Arbeitskreis für Paläobotanik und Palynologie (c/o Dr. F. Schaarschmidt, Forschungsinstitut-Natur Museum Senckenberg, Senckenberganlage 25, 6000 Frankfurt am Main 1, West Germany). For a small fee, one can usually get copies.

Journals
(Palynological articles appear in such unlikely places as the *J. Sed. Petrol.* and *Am. Midland Naturalist*, but the following are worth special note.)

Advances in Pollen–Spore Research. An Indian periodical with occasional important articles.

Grana (formerly *Grana Palynologica*). Originally published by G. Erdtman's laboratory in Stockholm, now by a board of editors in Scandinavia. Printed by Almqvist and Wiksell, Stockholm. Mostly on extant pollen. This journal is now included with membership in I.A.A. (see societies).

(Indian) Journal of Palynology. Lucknow, India: Nat. Bot. Gardens. A less formal publication, *Palynological Communications*, is included in the deal. Semi-annual publ. of Palynol. Soc., India. So far, not very productive.

Journal of Micropalaeontology. London: Br. Micropalaeont. Soc. Publishes many palynological papers and has had several special all-palynology issues.

Micropaleontology. New York: Am. Mus. Nat. Hist. Averages 6–8 palynological articles per year. News report sometimes useful for finding out who is doing what.

Oklahoma Geology Notes. Norman: University Oklahoma. This was for some years used as a house organ by L. R. Wilson and associates and was, hence, palynologically important. *Oklahoma Geology Notes* now does not publish many palynological papers.

(The) Palaeobotanist. Lucknow, India: Birbal Sahni Inst. Palaeobot. Has numerous palynological papers. Unfortunately quite irregular in publication.

Palaeontographica, Abt. B, *Paläobotanik*. This private German publication (Stuttgart) is too expensive for individual subscribers. Contains very important survey papers in palynology.

Palaeontology. British publ. that publishes many of the papers of Chaloner, Hughes, and other British palynologists. Should be followed closely for titles.

Palinologia. León, Spain: Inst. Palinologico. A journal started at the above-mentioned institute by Fritz Cramer. The future both of the journal and the institute is now unclear.

Palynology. (see proceedings of A.A.S.P.)

Pollen et Spores. Paris: Madame Van Campo-Duplan's laboratory. Very productive and important. Unfortunately, too expensive for individuals.

(proceedings of) A.A.S.P. Publ. formerly in *Geoscience and Man* (Baton Rouge, Louisiana: Louisiana State University), A.A.S.P. Vol. I is *Geoscience and Man* Vol. I, A.A.S.P. Vol. II is *Geoscience and Man* Vol. III, A.A.S.P. Vol. III is *Geoscience and Man* Vol. IV, etc. (confusing, eh?). Contains most of the papers presented at the annual A.A.S.P. meetings (see societies). After 1976, volumes in this series were published under the new title, *Palynology.* Vol. I in this new series appeared in 1977 and represents proceedings of the 1975 A.A.S.P. meeting, etc. Subsequent volumes have appeared annually and continue to do so.

Review of Palaeobotany and Palynology. Amsterdam: Elsevier. The first five volumes consist entirely of papers presented at the 1966 2nd Int. Conf. on Palynology in Utrecht. The magazine is nicely printed and edited and is a must to have around, but it is expensive.

N.B. Newsletters – Most societies (see next section) have newsletters. Many of them are very useful and important. Also, some working groups and some individual laboratories produce valuable newsletters. However, newsletters, except those for societies, tend to come and go, and it is not practicable to list them here. For example, there is at the time of writing a chitinozoan newsletter (produced by C.I.M.P. Subcommission on Chitinozoa – see section following), an acritarch newsletter (produced by C.I.M.P. Subcommission on Acritarchs – see next section), a very interesting newsletter called *Stuifmail* by the Palaeobotanical Palynological Society of Utrecht, more or less coterminous with the Laboratory of Palaeobotany and Palynology, University of Utrecht, Netherlands.

Societies (U.S.A.)

American Association of Stratigraphic Palynologists (A.A.S.P.). This society almost got the better (?) name, Society of North American Palynologists (S.N.A.P.). Emphasizes pre-Pleistocene, applied palynology, but really covers just about the whole field. Puts out a *Newsletter*, proceedings volumes (*Palynology*), a *Contributions* series, and special publications. To join, see information in *Palynology.*

Coal Geology Division, Geological Society of America. The annual meetings of this group, held with G.S.A., almost always include palynological papers dealing with coal-oriented problems. Join by first joining G.S.A. and then informing the current Secretary (name available from G.S.A., Boulder, Colorado, U.S.A.).

Paleobotanical Section, Botanical Society of America. At one time this group included both megafossil and microfossil people, but now it is dominated by those who study megafossil plants. Nevertheless, palynologists are paleobotanists and find it significant to be associated with P.S.B.S.A. An annual bibliography is provided to members. Members first join the Botanical Society of America at the regular fee, or one may be an associate member of the Section only by paying a single fee. In either case, to join, write to the current Secretary (listed in *American Journal of Botany*).

N.B. There are about 25 paleobotanical–palynological societies worldwide. Some are national, e.g., the Canadian Association of Palynologists (C.A.P.), and do not particularly encourage non-nationals to join. (C.A.P. will send you all its newsletters and other notices if you affiliate, but you cannot join as a full member unless you are a

Canadian.) Others are linguistic, e.g., Association des Palynologes de Langue Française (A.P.L.F.). A.P.L.F. has a number of members in Quebec, Canada, for example. Still others are subject-matter oriented, e.g., the important Commission Internationale de Microflore du Paleozoique (C.I.M.P.), devoted to Paleozoic studies. One can join by contacting the current Secretary-General. The International Association for Aerobiology (I.A.A.) has mostly to do with airborne pollen and spores, and especially their impact on human health. Nevertheless, the *I.A.A. Newsletter* contains much of interest and importance to paleopalynologists. Palynologists mostly interested in the Pleistocene tend to belong to such organizations as the American Quaternary Association (AMQUA) or the International Quaternary Association (INQUA), and their papers appear in journals such as *Quaternary Research* or *Ecology*. The newsletters are often of paleopalynological interest. If you want a list of all societies now affiliated with the International Federation of Palynological Societies (I.F.P.S.), write to the current Secretary-Treasurer (name and address from I.U.B.S. Secretariat, Paris), requesting a copy of the current Federation *Newsletter*. (As for I.F.P.S. and most of the other societies mentioned here, current information can be obtained from the Secretariat of the International Union of Biological Sciences (I.U.B.S.), or of the International Union of Geological Sciences (I.U.G.S.), both quartered in Paris.) Some societies that are not affiliated with I.F.P.S. as palynological societies nevertheless have a palynological section and/or a microplankton (dinoflagellates and acritarchs) section. The British Micropalaeontological Society, for example, has both. At this writing the I.F.P.S. newsletter is *Palynos* (Vol. 9, no. 2, was published in Dec. 1986, edited by J. E. Canright, Arizona State University, Tempe, AZ 75287).

CHAPTER TWO

Why one "does" paleopalynology and why it works

PURPOSES

The primary reasons for doing paleopalynology are as follows.

Geochronology Palynomorphs represent parts of the life cycles of various plants and animals that have at times evolved quite rapidly, with the result that the palynomorphs are characteristic of a fairly narrow time range and hence are useful for age dating (geochronology). (See also biostratigraphic correlation of strata, below.) Fairly often each year an academic palynologist is given a piece of sandstone, siltstone or claystone which is "unfossiliferous" – meaning no megafossils – with the request that he/she use palynological methods to determine the relative age. Sometimes, before palynological study was available, the geologist concerned did not know, even within a period or two, what the age of the rock was. Paleopalynology has now been practiced on countless thousands of samples of sedimentary rocks from late Precambrian to present (about one billion years; see Fig. 2.1), so that well-established, well-dated (by radioactive and paleontological methods) reference palynofloras are readily available. In our laboratory, for example, the Mississippian/Devonian boundary in Centre County, Pennsylvania, was shown from well-preserved palynofloras to lie within the local Pocono Formation, which is otherwise not very fossiliferous. We also showed from well-preserved palynofloras that the Newark Supergroup rocks in the Hartford–Springfield Basin of Connecticut and Massachusetts range over a very large timespan from late Triassic to well up into the Jurassic, whereas formerly most geologists regarded these rocks as Triassic only, and practically all geologists thought the timespan of deposition was much narrower. Each student in my beginning course in paleopalynology is issued an "unknown" sample of sedimentary rock near the beginning of the course. In the final week the student hands in a report, in which he/she establishes from work in the laboratory and library the age of the rock, among other things (e.g., environment of deposition). The students almost invariably succeed in determining the age to the correct stage of the correct period (one student issued a Cretaceous coal even succeeded in identifying the coal bed from which the sample originated, but he used some non-palynological, albeit scientific, sleuthing in the final stages of his investigation!).

Biostratigraphy Paleopalynology has become economically important mostly because palynofloras can be used, beginning with about one-billion-year-old rocks (acritarch palynofloras), to show correlation of a section of rocks from

28

one place with another section of rocks of perhaps quite different thickness and quite different lithology. This work of biostratigraphic (in this case, palynostratigraphic) correlation is what oil company palynologists mostly do. The sections to be correlated may be hundreds of kilometers apart. More often they are not widely separated, and may be in the same oil field, where it is very important to know at what level one is drilling, not in meters of depth, but with reference to known gas or oil production levels. Sometimes this sort of stratigraphy is attempted across a whole continent, but this is risky, and intercontinental correlation is only possible with great caution and with appropriate concern for the paleolocation of the continents concerned. Paleopalynology is particularly well-suited for correlation because palynomorphs of one kind or another are found in sedimentary rocks of all ages from about one billion years ago to present, and in all sorts of environments, from freshwater lacustrine deposits to deep-marine sediments.

A curiosity in the history of paleopalynology is that the earliest pre-Pleistocene use of palynostratigraphy was the attempt to correlate coal beds. It is probable that the success of present interglacial (= "post-glacial") pollen analysis of peats encouraged the belief that coals were the things to look at. Furthermore, there was economic incentive, because both coal and peat were mined in Europe and North America. As mentioned earlier, however, coals are nearly all autochthonous deposits of woody swamps. As such, the palynoflora reflects almost entirely the very local swamp environment, and correlation of coals by palynology is therefore very tricky – variations in spore/pollen content of coal beds are likely to be palynofacies depending on the local condition of the swamp.

Paleoecology For a variety of reasons it may be of importance to know as much as possible about various sorts of environments represented by a sedimentary rock. Palynology can help here in several ways. Palynomorphs can be sensitive indicators of the processes of sedimentation and the source of sediments. Some palynomorphs, e.g., dinoflagellates, are marine organisms, and may be indicators of the biological environment of the organisms when alive. Spores/pollen occurring as sporomorphs originate almost exclusively on the continents. They indicate therefore the presence of source vegetation. Because plants are sensitive indicators of continental environments (mostly climates), spores/pollen have much to tell us about climatic paleoenvironments. This is, of course, the reason for the original successes of palynology/pollen analysis in post-glacial vegetational analysis. On a microscale, the same sort of reflection in the palynoflorules of climatically caused vegetational change as was first noted by Von Post *et al.* in the Flandrian peats can be seen, e.g., in Carboniferous coals. Figure 2.2 shows such an example for Pennsylvanian coal in Illinois. *Lycospora granulata* is the spore of *Lepidophloios*, and *Cappasporites* that of *Lepidodendron*, both lycopsid trees, whereas *Thymospora* was produced by *Psaronius* tree ferns; a plant succession depending on conditions in the swamp is probably reflected, though one must always be cognizant of a spore succession depending on paleosedimentological factors. Pollen in human and and animal coprolites have been used (see Waldman & Hopkins 1970) to deduce the climate and other aspects of the environment in which the producing animals lived!

Era	Period	Epoch	Beginning of interval, 10^6 yr B.P.	Initiation of life forms	Plant-based "eras"
Cenozoic	Neogene	Holocene [Quaternary]	0.01	humans	"Cenophytic"
		Pleistocene	1.6		
		Pliocene	5.3		
	Paleogene	Miocene [Tertiary]	23.7	herbaceous angiosperms	
		Oligocene	36.6	anthropods	
		Eocene	57.8	horses	
		Paleocene	66.4	primates	
Mesozoic	Cretaceous		144	angiosperms	
	Jurassic		208	birds, chitinous walled fungi, mammals	"Mesophytic"
	Triassic		245	cycadophytes, dinosaurs, dinoflagellates	
Paleozoic	Permian		286	conifers	
	Carboniferous { Pennsylvanian		320	reptiles, seed ferns, scale trees	"Paleophytic"
	{ Mississippian		360		
	Devonian		408	amphibians, seed plants, ferns	
	Silurian		438	land plants, fish, spore plants	
	Ordovician		505		
	Cambrian		570	metazoans	
Proterozoic			ca. 1000	higher algae, sporopolleninous acritarchs	"Proterophytic"
Archeozoic			2500		
			4500	monerans	"Archeophytic"

Phanerozoic

Cryptozoic = Precambrian

Figure 2.1 General geological timescale. (B.P. = before present; My = million years; Precambrian = "Cryptozoic"; all time since the Cryptozoic = "Phanerozoic".) Time not to scale. Source of dates: DNAG Geologic Time Scale, *Geology*, September, 1983. Plant-based "eras" (= "phytic" eras) are discussed at various places in the text – see also Figure 6.1. The last epoch of the Cenozoic is labeled Holocene, as it appears in the DNAG Scale. However, in the text the Western European expressions "Flandrian", or "present interglacial", are sometimes used for the last approximately 10,000 years because these terms do not imply things that "Holocene", "recent", and "post-glacial" do. Note that the Cenozoic era is divided into major subdivisions in two different ways: (1) Tertiary (about 65 My) plus Quaternary (about 2 My), which is the traditional way; and (2) Paleogene (about 43 My) plus Neogene (about 24 My). The Neogene does not as usually defined include the Pleistocene and Holocene (sometimes collectively called the "Anthropogene"), but in this book Neogene is treated as including all the time from the end of the Oligocene to present, as indicated by the dotted bracket.

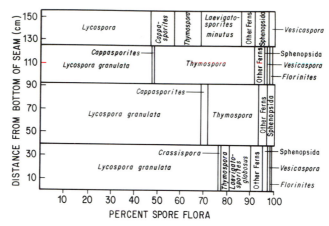

Figure 2.2 Spore succession in Herrin (no. 6) coal, Illinois (Pennsylvanian).

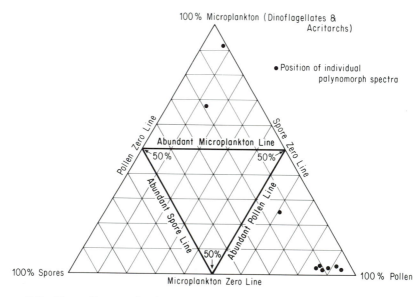

Figure 2.3 Phase diagram for showing paleoecological relationships for palynofloras. Dots represent position of nine Berriasian–Valanginian (Lower Cretaceous) palynomorph assemblages from the Pre-Caspian Depression, U.S.S.R.

Federova (1977) has pointed out that, in large sequences of sediments deposited over a considerable timespan, the complexes of associated palynomorphs reveal much about sedimentation regimes and the coordinated climatic conditions. She has shown that these broad-scale relationships can be plotted as triangular diagrams (see Fig. 2.3), which succinctly demonstrate the coordination of sedimentation and general environmental conditions on land (per spores/pollen) and in the marine environments (per phytoplankton, especially dinoflagellates). Federova studied the Lower Cretaceous of the

U.S.S.R., but similar sorts of relationships can be demonstrated for other time and space frames.

WHY PALEOPALYNOLOGY WORKS

Ubiquity of palynomorphs   Beginning with Precambrian acritarchs up to 1.4 billion years old, sporopolleninous, chitinous or "pseudochitinous" palynomorphs occur in sedimentary rocks of all ages and from many different sedimentary and biological environments (but there are problems – see Disadvantages and limitations, below). They originate both on land (spores/pollen = sporomorphs) and in fresh water (*Botryococcus* and other algae, some dinoflagellates and a few sporomorphs) and salt water (most dinoflagellates, the extinct chitinozoans, scolecodonts, "microforaminifera", a few sporomorphs).

Abundance and durability of palynomorphs   Most palynomorphs, especially spores/pollen and dinoflagellates (when present at all), tend to be much more abundant than most other fossils, even than other categories of microfossils. I have worked on a sample of silt containing five million dinoflagellates per gram and one containing four million pollen grains per gram. That is exceptional, but 10,000–100,000 palynomorphs per gram is common for siltstones. It is routine for palynological strew slides to contain 5,000 specimens. A 100-place slide box often contains therefore 500,000 specimens, more than the specimen catalog of most museums! This provides possibilities for statistics and population studies nearly unique in paleontology.

Further, sporopollenin and chitin, the principal components of the walls of palynomorphs, are among the most, probably *the* most, chemically inert of the major naturally occurring organic compounds. Palynomorphs thus not only tend to be preserved despite chemical vicissitudes during and subsequent to deposition, but also can be readily separated by relatively easy laboratory procedures from the enclosing sedimentary rock.

Fast evolution   Palynomorphs of various sorts represent preserved parts of the life cycles of various organisms that during one or more segments of Earth history were comparatively fast evolving, as well as abundant and easy to collect. This is a "must" for biostratigraphy and is true of acritarchs for the latest Precambrian, Cambrian, and Ordovician; scolecodonts and chitinozoans in marine sediments of the Ordovician and Silurian; spores in the Devonian; spores/pollen in all periods since the Devonian; and dinoflagellates in marine rocks since the Triassic. (However, it should be noted that, although paleopalynology for the first time brought into the field a paleontological discipline in which a single slide of a single preparation may contain easily 5,000 specimens, paleopalynology has not as yet provided compelling evidence for gradual evolution. Though centimeter-by-centimeter sampling is ·possible, taxa caught in the act of gradually turning into other taxa have not been widely reported. On the other hand, some palynologists, e.g., de Jersey (1982) for the genus *Aratrisporites*, deduce evolutionary change in quanta, but this is as yet very controversial.)

DISADVANTAGES AND LIMITATIONS

It is not fair to oversell! Paleopalynology has disadvantages and limitations!

Palynomorphs are silt-sized and are therefore sparse or absent in well-sorted sandstones and claystones.

Palynomorphs are sensitive to oxidation and to high alkalinity and therefore are not usually recoverable from red bed deposits, "clean" limestones generally, evaporitic deposits generally, or weathered rocks, although there are exceptions.

Palynomorphs are sensitive to high temperatures and pressures and apparently to recrystallization processes in rocks, and are therefore not usually studied with profit from metamorphosed rock, e.g., slate, or rocks in which the organic matter has been carbonized by heat and/or pressure (anthracite or very low-volatile bituminous coals; shales near a lava flow) or secondarily cemented or recrystallized rock such as dolomite.

Pollen and spores of modern plants can ordinarily be determined with certainty only to genus of producing plant, sometimes, e.g., Chenopodiaceae, Gramineae, and Cyperaceae, only to family. Almost certainly, the same situation applied in, say, the Permian. Obviously this is a handicap to both biostratigraphy and paleoecology because different species of the same genus can mean very different things ecologically. Also, only certain species of a genus may become extinct, and the extinction would be useful stratigraphically if it could be observed. Actually, in a limited number of special cases, species of pollen can be distinguished. For example, Ammann (1977) did it for two species of *Pinus*, based on nodules under the sacci; Holloway & Bryant (1984) have separated some species of *Picea* on sculpturing differences; and others have done the same sort of thing with size, or the ratio of one measurement to another, but this kind of procedure is seldom practicable. Dinoflagellate cysts are usually not subject to this limitation and are therefore intrinsically better suited to biostratigraphic use than spores/pollen. Unfortunately dinoflagellate cysts are not regularly found in non-marine sediments.

CHAPTER THREE

The natural history of palynomorphs

Palynomorphs are derived from four of the five kingdoms of organisms now recognized: Protista, Planta, Fungi, and Animalia. Only the kingdom Monera produces no palynomorphs. Representatives of all four of these kingdoms have some part of the life cycle that produces a cell, tissue or organ with some sort of "wall" that is highly resistant (= inert) to organic decay or inorganic degradation, e.g., attack by enzymes, oxygen, or high or low pH. *Botryococcus* (Fig. 1.1e) colonies are preserved apparently mostly because the "skeletons" contain waxy hydrocarbons, though they may also contain some sporopollenin, and certainly contain a complex mix of other compounds (Niklas 1976). *Pediastrum* (Fig. 1.1i) colonies (coenobia) occur commonly as fossils even though they seem to be cellulosic; they must be composed of something other than pure cellulose, which easily hydrolyzes and is very sensitive to attack by microorganisms. Chitinozoans (Figs. 1.1a and b) occur as well-preserved palynomorphs in marine early Paleozoic sediments. Their relationship to the kingdom Animalia is fairly certain, though some consider them fungal. In any event, their walls are generally agreed to consist not of "true" chitin but of another complex C–H–N–O compound, "pseudochitin".

However, the walls of the "main line" of palynomorphs consist of two compounds, or at least two families of compounds: chitin and sporopollenin.

CHITIN

Chitinous palynomorphs include: spores, mycelia and other organs of certain fungi, scolecodonts, arthropod organs such as insect mouthparts, and certain foraminiferal inner tests ("microforaminifera").

Chitin is a close "relative" of cellulose. This is puzzling, because cellulose is normally quite easily degraded in nature to simple sugars such as glucose, by microorganisms, by oxidation, by enzymes of certain animals, and by hydrolysis. For example, lye (KOH) is quite destructive of cellulose. On the other hand, various celluloses are quite disparate in resistance to chemical and microbiological attack. Household bleach (5.25% NaOCl, sodium hypochlorite) attacks linen, but not so readily cotton. The cellulosic primary wall of some fossil wood may persist even after the lignified secondary wall has been destroyed.

The structural formula of cellulose is:

Cellulose

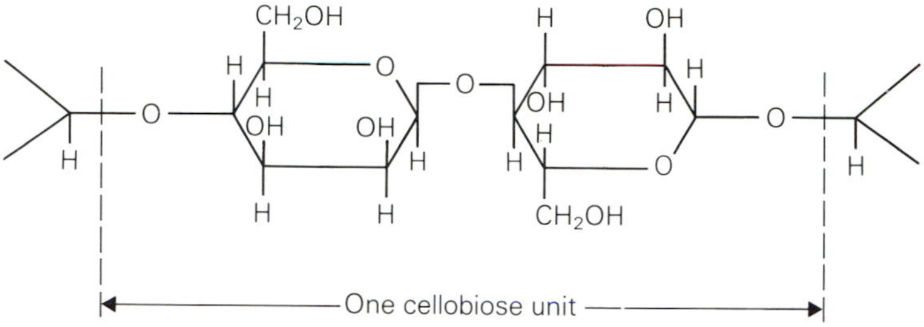

Molecule has 2,000–3,000 cellobiose units.
Empirical formula for cellobiose: $C_{12}H_{22}O_{11}$
(Glucose: $CH_2OH–CHOH–CHOH–CHOH–CHOH–CHO = C_6H_{12}O_6$)

The structure of chitin is very similar to cellulose, differing only in the presence of side acetamide groups:

Chitin

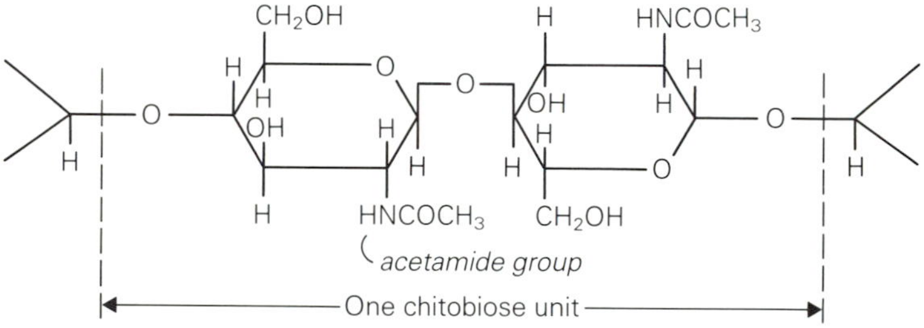

Chitin and sporopollenin behave very similarly in sediments and in laboratory procedures. In unconsolidated sediments sporopollenin is a very pale yellow color, whereas chitin can be almost clear, but often is darker, brown or orange-brown to grayish brown. Other properties seem similar to sporopollenin, although I have noticed some tendency for some chitinous palynomorphs to be a bit more resistant to both chemical and biological attack and to color change on carbonization (= "thermal maturation", or coalification).

SPOROPOLLENIN

Sporopollenin forms the basic structure of the resistant wall of most palynomorphs: dinoflagellates, acritarchs, spores (isospores, megaspores, microspores), and pollen. It is also present in a number of algae, e.g., in the walls of *Phycopeltis epiphyton*, a green alga (see Good & Chapman 1978). It is a very interesting substance, probably the most inert organic compound known. It resists acetolysis, but is degraded by strong oxidants such as H_2O_2

or CrO_3, and exhibits secondary fluorescence when stained with primuline (Good & Chapman 1978). It was first observed and named (as "sporonin") by John in 1814, and characterized by Berzelius in the 1830s (see Zetsche & Huggler 1928). Much later, Zetsche and coworkers (Zetsche & Kalin 1931, Zetsche & Vicari 1931) were able to establish an approximate (and arbitrary) empirical formula of $C_{90}H_{142}O_{27}$ for *Lycopodium* and similar numbers for *Picea*, *Pinus*, and *Corylus* sporopollenin. Characterization of the structural formula is very difficult because procedures which break down the substance (oxidation, "solution" in monoethanolamine) produce simple sugars and other compounds that do not prove the structure of the original molecule. Further, preparation techniques used to remove the other constituents of spores and pollen (KOH lysis, acetolysis) also change the sporopollenin, as it has a marked tendency to pick up halogens, metallic ions, and other groups when treated. In a series of analyses I did long ago, sporopollenin of *Beta vulgaris* pollen, a very resistant exine type, was prepared by prolonged KOH lyzing. Infrared absorption spectra of the sporopollenin so prepared matched quite well with the K salt of glucuronic acid:

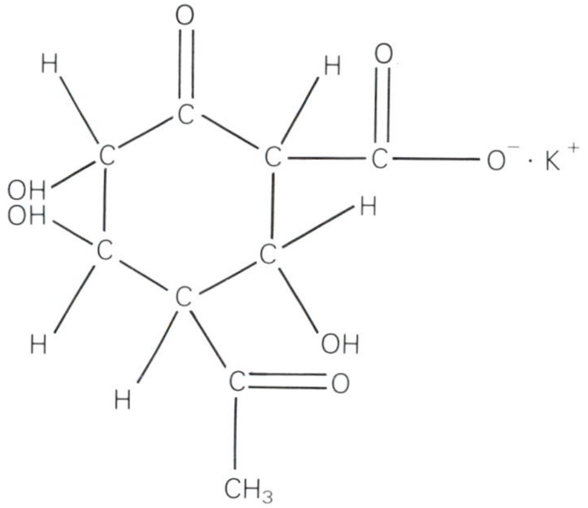

Inasmuch as I also was able to show that the breakdown of beet sporopollenin by oxidation (by H_2O_2) produced hydroxycarboxylic acids (see flowsheet, Fig. 3.1), I thought at the time that sporopollenin was probably a condensed carbohydrate of some sort. (That chitin is a carbohydrate derivative is certainly interesting in this connection.) In the meantime many others have used more sophisticated tools on this substance. Shaw and coworkers, Heslop-Harrison, Brooks, and others (see Brooks *et al.* 1971) have published research done by following the buildup of compounds in the tapetum in anthers of flowers at the time that sporopollenin is seen to be deposited in the exines of pollen. Based on these studies, it has been suggested that sporopollenin is a copolymer of β-carotene, a xanthophyll such as antheraxanthin, and fatty acids. If true, the substance should have repeating units of an isoprenoid sort. Given (1984) notes that the predominance of straight chains over isoprenoid structure in fossil spore walls makes it difficult to accept the Brooks *et al.* theory. Given *et al.* (1985) have also noted the presence of aliphatic chains with little or no

37

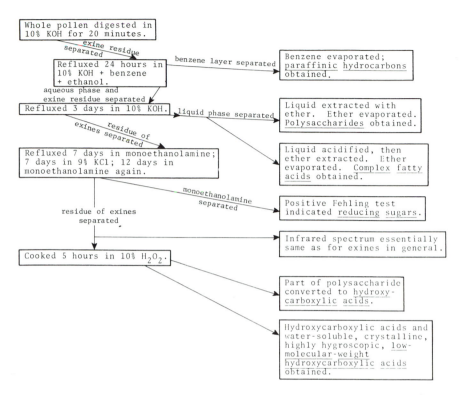

TREATMENTS   RESULTS

Figure 3.1 Flowsheet for chemical investigation of *Beta vulgaris* L. exines.

branching in samples of modern sporopollenin, as well as considerable structural variability in samples from different plant groups. Potonié & Rehnelt (1971), for example, suggested the structural unit:

This would make sporopollenin a sort of "cousin" of rubber. They share much in common: elasticity, sensitivity to oxidation and alkalinity, general "durability". However, rubber's structure contains no oxygen, and the analogy is really only a caricature. The structure of rubber is:

(Very large number of
units in molecule)

The structural manner in which sporopollenin occurs in spores/pollen is very diverse. In general, it is limited to the outer wall, the exine. Most fern spores and some gymnosperm pollen have an additional sporopollenin-bearing wall, the perine or perispore (seldom seen in fossils). Some, mostly aquatic, flowering plants have abandoned the production of sporopollenin altogether, and others, e.g., the family Lauraceae, manage to operate with very little sporopollenin in their exines, meaning that these exines seldom are preserved as fossils. Other sporopollenin-containing elements associated with spores/pollen are the following:

(a) Viscin threads (Hesse 1981), ektexinous extensions, producing a sticky (but not viscous) webbing, possessed by various, entomophilous (insect-pollinated), ornithophilous (bird-pollinated), and chiropterophilous (bat-pollinated) angiospermous pollen. These threads cause adherence to insect appendages. Viscin threads were described for a fossil onagraceous pollen by Traverse (1955). Note that anemophilous pollen and spores generally are not sticky. The stickiness of animal-dispersed pollen is usually provided by pollenkitt, a tapetal derivative (see Table 3.1).
(b) Elaters, perispore-like bands attached to *Equisetum* spores, probably used for dispersal by air currents.
(c) Ubisch bodies or orbicules, small (2–5 μm) pieces of sporopollenin apparently produced by the tapetum as a "throw away" after completion of the exine proper. Ubisch bodies are often loosely attached to the surface of fossil spores/pollen exines and also occur as free bodies in palynological macerations (see Taylor 1976a, b).

Measurement of the amount of sporopollenin in spores and pollen is usually achieved by acetolyzing them and assuming (as microscopic examination shows to be generally true) that only exine is left. Table 3.2 shows a representative list for a variety of spores/pollen producers. The range is very considerable. In general, the more sporopollenin, the more resistant to decay, oxidation, etc. However, the distribution of the sporopollenin also makes a difference: the more sporopollenin concentrated in the outer part of the exine (ektexine or sexine), the more durable the exine. Elemental analysis of the exinal material remaining after acetolysis or KOH lysis is interesting in several respects (see Table 3.3). Acetolysis obviously destroys some carbonaceous matter in the exines that KOH treatment leaves intact. KOH treatment introduces some

Table 3.1 Origin and characteristics of pollenkitt and viscin threads.

Characters	Pollenkitt	Viscin threads
occurrence in angiosperms	in all families investigated up to now	only in a few distinctly entomophilous families
origin	synthesis by plastids of the anther tapetum, mostly during the tetrad stage and later on	like the material of the ektexine, long before the pollenkitt synthesis
chemistry	a complex mixture of lipid substances	similar or equal to the ektexine, containing sporopollenin
state of aggregation	fluid, viscous, sticky, partially with crystalline inclusions	firm, elastic, not fluid, not sticky without crystalline inclusions
sculpture	no sculpturing	often smooth surfaces, but partially species-specific knobs or furrows
fine structure	opaque, but with varying homogeneity and electron density, partially with lamellae	the same homogeneity and electron density as the ektexine, without lamellae
distribution on the pollen surface	in entomophilous angiosperms mostly as an electron-dense, homogeneous film on the pollen surface; in anemophilous angiosperms usually as small and heterogeneous lumps in the cavities of the exine	the threads often originate near the apertures, but otherwise they are not connected with the exine
mode of pollen attachment	the pollen grains (or tetrads) adhere only by their stickiness	the pollen grains (or tetrads) become entangled by the threads, pollen attachment by friction or adhesion

Table 3.2 Proportion of various species of pollen and spores that consist of sporopollenin, as calculated by residue from acetolysis.

Species	Percent of weight of air-dried pollen lost in acetolysis	Percent sporopollenin
Spores		
moss (mixed sample of commercial origin)	69.0	31.0
Lycopodium sp.	69.5	30.5
Equisetum sp.	95.0	5.0
fern (mixed sample of commercial origin)	87.0	13.0
Pollen		
Pinus silvestris	74.1	25.9
Poa pratensis	85.6	14.4
Alnus incana	82.2	17.8
Populus trema	94.4	5.6
Ulmus scabra	84.7	15.3
Ulmus scabra (2 h maceration with 10% KOH)	93.1	6.9
Trifolium pratense	90.8	9.2
Beta vulgaris	82.6	17.4
Brassica napus	90.9	9.1

Table 3.3 Elemental analysis (%) of spores/pollen exines obtained by various treatments of whole spores/pollen.

Species	Treatment	Ash	C	H	O	S	N	K
Calvertia gigantea (puffball fungus spores)	acetolysis	0.4	45.3	6.3	35.9	6.5	0.9	★
	KOH	0.2	43.3	6.6	33.3	0.8	5.6	★
Lycopodium sp. (club-moss spores)	acetolysis	3.0★★	57.8	7.4	28.0	3.3	0	★
Juniperus communis (juniper pollen)	acetolysis	11.5	46.7	6.8	29.5	3.1	0	★
Pinus silvestris (Scots pine pollen)	KOH	2.3	55.2	7.6	27.2	0.1	0	★
Typha latifolia (cat-tail pollen)	KOH	0.7	62.1	8.7	24.6	0	0	★
Poa pratensis (bluegrass pollen)	KOH	6.1	51.5	7.4	28.8	0	0	★
Ulmus scabra (elm pollen)	acetolysis	6.1	50.5	6.6	31.8	5.9	0.4	★
	KOH; H$_2$O wash	8.4	55.3	9.1	25.7	0	0.2	5.0
	KOH; HC1, H$_2$O washes	0.5	60.0	9.4	30.8	0	0.5	0.04
Quercus robur (oak pollen)	acetolysis	3.1	51.7	7.5	30.7	4.2	0	★
	KOH; H$_2$O wash	11.8	51.3	7.8	28.5	0	0.5	0.5
	KOH; HC1, H$_2$O washes	2.2	63.5	9.6	27.6	0	0.4	0.1
Artemisia vulgaris (mugwort pollen)	KOH	23.5	44.9	6.4	19.1	0	0	★
Beta vulgaris (beet pollen)	acetolysis	10.9	47.7	7.0	32.0	5.1	★	★
	KOH; HC1, H$_2$O washes refluxed 1 week in benzene–alcohol–KOH, 1 week in HC1, 3 weeks in monoethanolamine	11.3	48.2	4.8	34.0	0.2	★	★
		11.6	62.6	8.2	13.0	★	★	★

★ Not analyzed.   ★★ Estimated.

Table 3.4 Degradation of pollen/spore exines on micro-slides, 1951–61 (= "autoxidation").

Amount of swelling (%)	Number of species
0–1	13
1–5	7
6–10	11
20	16
>20	1

potassium to the exine, whereas acetolysis introduces sulfur. Some sorts of spores/pollen, e.g., *Juniperus*, contain much more non-combustible matter than others. Chitinous fungal spore walls, e.g., *Calvertia*, contain much more nitrogen than sporopolleninous exines, as might have been expected from the acetamide content of chitin. Acetolysis, however, largely strips the acetamide groups away.

Sporopollenin's propensity to pick up oxygen is so great that it will "autoxidize" on microscope slides. The observable effect on exines is swelling, eventually grotesque swelling and dissolution. (N.B. Some have suggested that this effect is due to hydrolysis, but my observation is that it does not occur if O_2 is really excluded.) Some years ago I remeasured the modern spores/pollen illustrated in Traverse (1955) after 10 additional years on microscope slides in glycerin jelly. The results are shown in Table 3.4. Havinga (1971) showed that exines of spores/pollen buried in soils of various sorts were attacked progressively, presumably by bacteria and natural hydrolysis. Table 3.5 shows the results of some of Havinga's experiments. Havinga (1984) has also shown that spores/pollen of different taxa have conspicuously different rates of corrosion (usually observable as thinning). *Lycopodium* spores are the most resistant, followed by conifer pollen and various angiosperm pollen. Generalization is difficult as illustrated by the fact that one of the most sensitive forms to corrosion is *Polypodium*, a fern spore.

The evolutionary history of sporopollenin is very long (Fig. 1.2). The oldest sporopolleninous acritarchs occur in Precambrian rocks 1.2–1.4 billion years old. In these organisms sporopollenin probably played the role of protector of protoplasm against ultraviolet radiation. The green algae are presumably responsible for the development of sporopollenin and its introduction into the armament of most higher green plants, where its principal function is protection against oxidation and desiccation (while sporopollenin can be oxidized, it is more resistant to O_2 than are most other organic compounds).

I would now like to summarize the salient features. Sporopollenin is a highly inert C–H–O compound, probably of the carotenoid–terpenoid sort (though in my opinion, some sort of condensed carbohydrate structure has not been altogether excluded). Its natural color is pale yellow, but with thermal maturation, as the sporopollenin increases in "rank", with loss of O and H and increasing percentage of C (parallel to the coalification series of other organic matter), the color deepens through dark yellow, orange, reddish brown,

Table 3.5 Unaffected (u), intermediately affected (i) and severely affected (s) pollen grains and spores in various soil types for different lengths of time. (November 1964 is eight months after placement of the pollen in the soils.)

	Percentages													
	Sphagnum peat			*Carex* peat			Podsolized sand soil			River clay soil			Leaf mould in greenhouse		
	u	i	s	u	i	s	u	i	s	u	i	s	u	i	s
November 1964	92	8	0	84	16	0	63	37	0	26	61	13	23	60	17
March 1965	78	21	1	48	45	7	45	51	4	–	–	–	21	31	48
November 1965	–	–	–	–	–	–	26	61	13	17	36	47	16	25	59
November 1969	68	31	1	23	60	17	24	58	18	4	26	70	4	27	69

finally to black. During this series, the reflectance increases. The agents of thermal maturation are temperature elevation plus time. Time alone does little. Pressure alone does little. Gutjahr (1966) showed that *Quercus* pollen darkens in the manner described slowly at 100°C, and more rapidly at 150 and 200°C (so dark that 90% of transmitted light is absorbed by exines treated less than 3 days at the latter temperature). Other experiments show that 150°C over very long periods will eventually blacken exines, and that the geothermal gradient alone is sufficient to do the job completely at depths of 5,000 m or so. The specific gravity is about 1.4, close to that of solid wood substance (because of air spaces, whole wood is commonly less than 1.0). The index of refraction is 1.48. Sporopollenin is sensitive to oxidation, though not as much so as most organic matter in sediments. It is also sensitive to high pH over prolonged periods of time. Enzymes mostly do not affect sporopollenin, so pollen and spore exines pass through most animal guts (including human) unchanged, though the contents of the grains are digested (Scott *et al.* 1985a). However, I once heard the noted paleobotanist, Tom Harris (1974b), relate that he fed pollen from his beehives to his geese, through which the pollen passed unscathed, but that by the third "pass" through the recycling ruminant stomachs of his goats, the exines were attacked, and by the sixth day they were destroyed by enzymes. Chaloner (1976, 1984) has shown that fern spores can even *germinate* in fair numbers after passing through the gut of a locust (American: "grasshopper"; see Fig. 3.2). The poor preservation of pollen in reptilian coprolites reported by Waldman & Hopkins (1970) was not likely due to corrosion of the sporopollenin by the animals' enzymes, as supposed by the authors. Harris's experiment with exines of pollen passing through goats' digestive systems showed that it takes the multiple cycles of a goat's ruminating digestive system plus a goat's powerful enzymes to destroy sporopollenin over a six-day period.

Elsik (1966) and Srivastava (1976a) have shown that some fungi can digest sporopollenin and can attack spores/pollen, apparently after deposition in sediments. (*In situ* fossil spores/pollen showing fungal attack of the spore *contents* have been described by Stubblefield & Taylor (1984), but the

VIABILITY OF FERN SPORES EATEN BY LOCUSTS

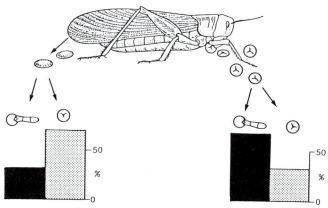

Figure 3.2 The sporopollenin-rich exine of some fern spores is so tough and protects the sporoplast so well that, even after passing through an insect's digestive system, many spores can still germinate. About half as many spores germinate in the feces (left) as germinate from fresh spores (right). Pattern designated by trilete spore represents ungerminated spores, and that designated by a sporeling represents germinated spores.

sporopolleninous exines were apparently not the fungal target.) Havinga (1967) has demonstrated that some soil bacteria can also accomplish this difficult trick. Various people, e.g., Southworth (1974), have noted that sporopollenin "dissolves" in 1-ethanolamine (monoethanolamine), especially if the exines treated are pretreated by acetolysis (Denizot 1977). However, my studies showed that what really happens is breakdown of the sporopollenin, producing, among other things, sugars (see Fig. 3.1). Loewus *et al.* (1985) report that 4-methylmorpholine *N*-oxide monohydrate (MMNO·H$_2$O), a patent polysaccharide solvent, "dissolves" sporopollenin, a technique which is used for freeing pollen sporoplasts by removing the exine.

From a paleopalynological point of view, the significance of the properties of sporopollenin is that sporopollenin-bearing palynomorphs, once delivered to a sediment, tend to stay there, though the contents, and other wall layers, of the palynomorphs are quickly lost. That small (1–3 μm) sporopollenin bodies produced by plants as a byproduct of exine formation are frequently preserved and are indeed often quite abundant in sediments further demonstrates the resistance of sporopollenin. These bodies are called ubisch bodies or orbicules. In formation of a sediment, the following original and post-depositional factors can destroy the sporopollenin:

(a) oxidizing environment,
(b) highly alkaline environment,
(c) carbonization (= coalification, thermal maturation, as a result of relatively low temperature elevation over a long time),
(d) high temperature (over a relatively short time, as a result, e.g., of volcanic intrusion),

(e) recrystallization of the minerals in the sediments.

The effect of recrystallization on sporopollenin has not to my knowledge been directly studied, but I have observed from study of several dozens of dolomite samples from most continents that dolomites never contain palynomorphs. Limestones are not as a group very good sources of palynomorphs, presumably in large part because of formation in an alkaline environment. (In studies of recent sediments of the Bahamas, I showed progressive loss of palynomorphs in lime muds with time.) But limestones are not as a class barren of palynomorphs. In fact, some limestones contain excellently preserved palynomorphs (Blome & Albert 1985). Therefore, as dolomites *are* barren, I conclude that the $CaCO_3$ to $CaMgCO_3$ (limestone to dolomite) conversion process destroys palynomorphs.

PALYNOMORPHS IN PETROLEUM

A number of studies have turned up the interesting fact that spores/pollen are capable of being swept along with migrating petroleum as it moves through porous sedimentary rocks. De Jersey (1965), for example, was able to show in Australia that petroleum in the producing sands of the Moonie oil field in Queensland contained a characteristically Jurassic palynoflora, whether the reservoir rocks were Permian, Jurassic or Triassic. A Jurassic source for the petroleum was therefore assumed, though previous information had favored a Permian source. Jiang (1984), in the Peoples' Republic of China, has described a taxonomically diverse palynoflora of Jurassic age in crude oils from reservoir rocks of Jurassic, Cretaceous, and Cenozoic age.

GENERAL OCCURRENCE OF PALYNOMORPHS IN TIME

As we have seen, the palynomorphs commonly studied include representatives of four of the five usually recognized "kingdoms" of organisms (all but the Monera). This information is very briefly presented in Table 3.6. This book will concentrate on embryophytic (= archegoniate, i.e., Bryophyta and Tracheophyta) spores/pollen ("sporomorphs"), although dinoflagellates and acritarchs, chitinozoans, scolecodonts, fungal spores, and miscellaneous other forms will also be treated. One of the strong selling points for paleopalynology is that from about one billion years ago (late Precambrian) to present, sporopolleninous or chitinous palynomorphs of one group or another are always present in appropriate sedimentary rocks, and they are biostratigraphically useful (that is, rapidly evolving, ubiquitous, abundant). The range of the main groups is shown in Figure 1.2.

The oldest sporopollenin-containing palynomorphs are sphaeromorph acritarchs over one billion years old from the Soviet Union and from other parts of the world. The oldest chitinous and pseudochitinous palynomorphs are Cambrian scolecodonts and chitinozoans, respectively. (Chitinous fungal spores appear much later. With probable exceptions in the Permian and

Table 3.6 A classification of organisms showing palynomorphs occurring as fossils. The makeup of the kingdoms is as per Margulis, 1981.

Palynomorphs produced	Organisms
	KINGDOM MONERA (Prokaryotes)
No palynomorphs produced	bacteria
	blue–green "algae"
	KINGDOM PROTOCTISTA
	Subkingdom Protozoa
Foraminiferal chitinous inner tests	
	Subkingdom Thallophyta
	Pyrrophyta
Dinoflagellate cysts	
Frustules of SiO_2 (diatoms) or platelets of $CaCO_3$ (coccoliths and discoasters = nannofossils). These are not here accepted as palynomorphs	Chrysophyta (diatoms, coccoliths, probably discoasters)
No palynomorphs produced	Phaeophyta (brown algae)
$CaCO_3$ deposited, but no palynomorphs produced	Rhodophyta (red algae)
Desmid zygospores, e.g., *Staurastrum*, *Pediastrum* colonies (cellulose, ?sporopollenin). *Botryococus* colonies (fatty–waxy). Characeae oögonia (cellulosic, often mineralized, not usually considered "sporomorphs"). Probably other zygospores, e.g., "*Circulisporites*" (?), and many acritarchs	Chlorophyta (green algae)
No palynomorphs produced	Myxomycophyta (slime molds)

Table 3.6 *contd.*

Palynomorphs produced	Organisms
	KINGDOM FUNGI Eumycophyta
Fossil fungi range from Precambrian to present. Chitinous-walled fungal "spores" and mycelia occur from Jurassic to present, with a few earlier exceptions. A variety of fossil "spores" occur, small unicellular to giant multicellular "sclerotia"	
	KINGDOM ANIMALIA Porifera (sponges)
Spicules (SiO_2) and other skeletal parts sometimes seen in preparations, not usually regarded as palynomorphs	
No palynomorphs produced	Coelenterata
No palynomorphs produced (bryozoan statoblasts sometimes seen in preparations, not considered palynomorphs)	Ctenophora, Platyhelminthes, Nematohelminthes, Bryozoa, Brachiopoda
No palynomorphs produced	Mollusca
Chitinous mouth-lining parts often occur as palynomorph fossils from Lower Paleozoic to present	Annelida (class Polychaeta)
Chitinous exoskeleton parts, from carapaces of cladocerans to butterfly scales, frequently occur in preparations; none is as yet considered a palynomorph	Arthropoda
Possibly the pseudochitinous Chitinoza from the Lower Paleozoic belong here (if graptolites)	Chordata

Atracheophyta (non-vascular plants)
Bryophyta (mosses and hepatics)

Isospores: trilete in many hepatics and some mosses; mostly alete or monolete and very small and thin-walled. These not likely to be preserved

Tracheophyta (vascular plants)
Primitive, extinct mid-Paleozoic plants: rhyniophytes, trimerophytes, etc.

Trilete and monolete(?) isospores

Psilotales (two living genera, no megafossil record)

Monolete and trilete isospores

Lycopsida
Lycopodiales
Selaginellales

Trilete isospores
Trilete microspores and megaspores

Lepidodendrales (scale trees – extinct)

Trilete microspores and megaspores (approaches seed habit)

Pleuromeiales (extinct, reduced lepidodendrid)

Presumably as just above

Isöetales (quillworts)
Sphenopsida

Monolete microspores, trilete megaspores

Hyeniales (extinct)

Trilete isospores(?)

Sphenophyllales (extinct)

Trilete isospores

Equisetales (living: Equisetum – herbaceous; extinct: Calamitaceae – woody)

Equisetum: apparently alete isospores with elaters and perine. Calamitaceae: apparently alete to trilete, isospores or microspores and megaspores. Some approach seed habit

Table 3.6 *contd.*

Palynomorphs produced	Organisms
	Pteropsida
	Filicineae (ferns)
Mostly monolete and trilete isospores, many with perine, some (e.g., *Stauropteris*) with megaspores and microspores	Coenopteridales (extinct)
Isospores, trilete	Ophioglossales (extinct and living)
Isospores, trilete or monolete	Marattiales (some living ferns, some extinct tree ferns)
Mostly isospores (Marsileaceae and Salviniaceae heterosporous!), trilete to monolete, often with perine	Filicales (most modern ferns)
Trilete microspores and megaspores	Progymnospermopsida Archaeopteridales: *Archaeopteris*
	Gymnospermae
"Prepollen", monolete or trilete (and leaf cuticles)	Cycadofilicales ("seed ferns" – extinct)
Pollen, monosulcate (and leaf cuticles)	Bennettitales (extinct, cycad-like)
Pollen, monosulcate (and leaf cuticles)	Cycadales (modern cycads and fossil relatives)
Pollen, monosulcate to vesiculate (which is monosulcate also) (and leaf cuticles)	Cordaitales (extinct progenitors of conifers)
Pollen, monosulcate (and leaf cuticles)	Ginkgoales (living *Ginkgo* and extinct relatives)
Pollen: vesiculate (*Pinus et al.*), inaperturate (*Juniperus et al.*), "monosulcate" (*Larix et al.*), "monoporate" (*Sequoia et al.*) (and leaf cuticles)	Coniferales (pine, juniper, etc.)

Pollen: mostly polyplicate; *Gnetum* (others?): inaperturate. Some fossils probably polyplicate–vesiculate (and leaf cuticles)

Gnetales (living *Ephedra* and associates, living and fossil)

Pollen: monosulcate, trichotomosulcate, inaperturate (and leaf cuticles)

Angiospermae (flowering–fruiting plants)
Monocotyledonae (one seed leaf, parallel venation, floral parts usually in threes or derivative)

Pollen: monosulcate (primitive Ranales), tricolpate, triporate, periporate, syncolpate, etc. (and leaf cuticles)

Dicotyledonae (two seed leaves, net venation, floral parts usually in fives or derivative)

Notes

1 Homospore = isospore, i.e., spores all of one type. Contrasts with types which have microspores and megaspores, of which pollen–seed plants are the extreme example. Pollen consists of a microspore coat, inside of which a male gametophyte develops.
2 Vesiculate pollen is also prevailingly monosulcate, though the sulcus is often obscure, especially in Cenozoic forms.
3 Comparison of larger taxa, in classical and five kingdoms views, is as follows (from Margulis 1981):

Classical view

Two kingdoms	Major members
Planta	bacteria, fungi, algae, green plants
Animalia	protozoans, metazoans

Five kingdoms view

Five kingdoms	Major members
Monera (prokaryotes)	bacteria, blue–green algae
Protoctista (lower eukaryotes)	protozoans, protists, nucleate algae, slime molds
Fungi	mushrooms, molds, yeasts, lichens
Planta	green plants (bryophytes, tracheophytes)
Animalia	metazoans

Triassic, they do not occur regularly until the Jurassic and are not abundant until mid-Cretaceous). Precambrian sphaeromorph, sporopolleninous acritarchs were joined by hordes of acritarchs with processes and other modifications in the Cambrian and Ordovician. Even non-marine sediments began to contain sporopolleninous fossils in latest Ordovician, and the earliest unquestioned embryophyte spores are Llandovery–Wenlock trilete spores.

In referring palynomorphs to their appropriate stratigraphic level we use, of course, the general geological timescale presented in Figure 2.1. However, it is very helpful to one's understanding of the plant history involved to speak (very informally!) of plant-based "eras", shown on the right side of Figure 2.1, based on events in plant evolution. The "Archeophytic" extends from about 3.5 billion years (and the earliest known fossils) to the level of the first robust-walled acritarchs and the eucaryotes, at about 1.2 billion years ago. The "Archeophytic" had only monerans, which produced non-robust fossils. With the first sporopolleninous acritarchs begins the "Proterophytic", extending up to the level of the first spores or spore-like tetrads of the late Ordovician–early Silurian. This marks the commencement of the "Paleophytic" which is typified by ancient sorts of vascular plants and persists until the middle of the Permian, when conifers, cycadophytes, and other advanced gymnosperms came to dominate the land flora, and the "Mesophytic" began. The "Mesophytic" gave way to the present "Cenophytic" in the very early Cretaceous, with the appearance of angiosperms. It is interesting that the "Mesophytic" and "Cenophytic" each began well before the "-zoic" eras with the same prefix. Remy, e.g., in Gothan & Remy (1957), presented practically the same "phytic" classification many years ago.

CHAPTER FOUR
Spores/pollen basic biology

Embryophytic spores and pollen grains (which *are* spores, but more than *just* spores) are the main subject matter of this book. It is unavoidable at this point to explain what they are in some detail. This is best done by examination of life cycles of embryophytic plants (Figs. 4.1 & 4.2).

BRYOPHYTE LIFE CYCLES

The Bryophyta comprise the mosses and liverworts (and hornworts, too). They are non-vascular green plants, apparently related to the green algae on the one hand, and to the vascular green plants on the other. Their general need for a wet environment to effect reproduction invites comparison with the amphibians, and it is tempting to regard them as survivors of the transition from aquatic green algae to land plants in Ordovician–Silurian time, although there is no proof of this. That some of the first certain sporopolleninous embryophytic spores of the Silurian resemble certain liverwort spores is fascinating, but the bryophytes have practically no fossil record, and what, if anything, they have to do with land plant evolution is as yet uncertain and controversial. Some even regard them as degenerate tracheophytes! Nevertheless, the Bryophyta are "right" for a primitive, non-vascular land plant. (Some green algae produce sporopolleninous zygospores and some have alternating generations with a prevailing haploid life cycle, as a bryophyte ancestor "should"!)

The typical moss shown diagramatically in Figure 4.1 is a small, haploid (1N = gametophytic) plant with no true roots or vascular system. It thus is bound closely to water, both because it cannot easily obtain, store or transport it, and because its reproduction depends on water. This plant has female and male sex organs, which produce eggs and spermatozoids, respectively. The fertilized egg (zygote) develops into a small, diploid plantlet which remains attached to the parent plant and is generally not as green as the parent because it is in a sense a parasite – its only function is to produce a capsule (sporangium). In the capsule tetrads of spores are differentiated through meiosis (two-stage reduction division). The spores are haploid and are isospores (or homospores, a synonym), meaning that all the spores are alike. Liverwort spores are often but not always trilete (having a Y-shaped contact scar from the contact in the original tetrad). This is the primitive or original structure for spores. Some liverwort spores, however, are monolete. Moss spores are typically small, are sometimes trilete (*Sphagnum, Andreaea*), but are more often alete or, in a few cases, monolete. Most moss spores contain rather

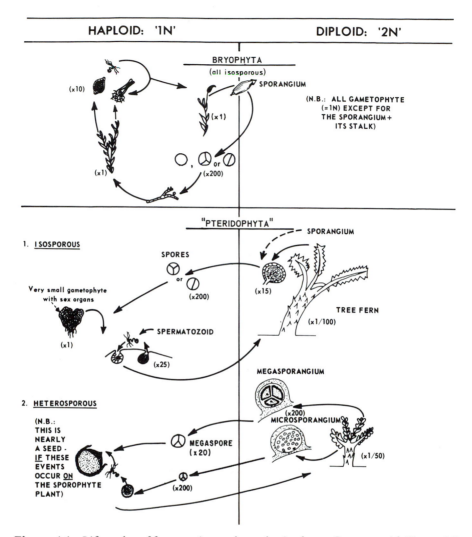

Figure 4.1 Life cycles of free-sporing embryophytic plants. Contrast with Figure 4.2. Note that the spore-bearing organ (= sporangium) of the bryophyte is usually called a capsule by bryologists.

little sporopollenin and are not, therefore, good candidates for fossilization. *Sphagnum* is an exception, and this sort of moss spore has a fossil record reaching back at least 150 million years, as the genus *Stereisporites*. Moss spores can produce perispores (Brown & Lemmon 1984), and sporopollenin deposition in moss spores in exine and perine during sporogenesis is not unlike that of vascular cryptograms. Bryophyte spores are ejected from or fall from the capsule (sporangium). In order to germinate, they must land in a moist place or a place that will later be moist. When they germinate, a new moss or liverwort plant may eventually result. Water remains important to the plant, as the spermatozoids must swim in "casual" water to the egg in order to bring about fertilization.

PTERIDOPHYTE LIFE CYCLES

"Pteridophyta" is not a natural division in plant classification. Nevertheless, for paleopalynological palynology it is a convenient category, including all vascular embryophytes that do not produce true seeds.

Isosporous pteridophytes

In Figure 4.1 I have used the example of a cyatheaceous tree fern, but the basic plan is good also for other homosporous ferns and for what used to be called the homosporous "fern allies": psilopsids, lycopsids, and sphenopsids. (These are rather insignificant in modern vegetation, but apparently at least collaterally ancestral forms were much more important in the Paleozoic.) Here the conspicuous plant that comprises more than 99.9% of the life cycle is diploid (2N = sporophytic). It produces sporangia, 2N of course, as in the case of the bryophytes, but the sporangium is an organ of the main plant, not a 2N "parasite"! In the sporangia, spore mother cells divide meiotically to produce ultimately tetrads of haploid (1N) spores. Again they are isospores: all alike, trilete most commonly, but in some instances monolete or alete. Actually, there may be a considerable range of size and other morphological features in the spores from one sporangium, but the critical matter for the spores to be "iso-" is that, when they germinate in a moist place, a tiny, separate gametophytic plant is produced which produces both male and female sex organs. Again, as in bryophytes, a little water is necessary at this stage in order for the spermatozoids to swim to the egg-bearing organ for fertilization. From the fertilized egg grows a 2N adult spermatophyte. Most non-botanists never see gametophytes, as they are very tiny, but gardeners are usually aware that one may "plant" pteridophyte spores on moist soil or a wet filter paper and get gametophytes and eventually adult ferns.

It should be noted that, as fossils, isospores (= homospores) are indistinguishable from the microspores of heterosporous plants (see below). Paleopalynologists should always use the term "miospore" (explained later in this chapter) or "small spores", *not* "microspore", for small dispersed spores whose life cycle function cannot be determined. Regrettably this dictum is often disobeyed, and the student will find small spores in the literature described as "microspores", when in fact they are more probably isospores.

Heterosporous pteridophytes

Some living pteridophytes, such as the fern, *Marsilea*, and the lycopod, *Selaginella*, are heterosporous. This suggests that the spores are of two different types, which is correct. Usually, as shown in Figure 4.1, there is a conspicuous size difference between the two spore types, the microspore (male) being smaller, and the megaspore (female) being larger. However, the critical matter is that the microspores (usually numerous, smaller) produce on germination a male gametophyte, which makes in its sex organs spermatozoids. The megaspores (usually very reduced in number, larger), on the other hand, produce a female gametophyte, which makes eggs in its sex organs. The germination of the megaspores occurs on moist ground, and produces a very tiny megagametophyte, confined, or mostly confined, to the exine of the germinated spore. The microgametophyte also germinates on the ground, producing a very tiny male gametophyte which makes spermatozoids. These swim in films of water to the eggs and cause fertilization. From the zygote grows an adult (2N) sporophyte.

SEED PLANT LIFE CYCLES

Gymnosperms

Suppose that the megaspores in the sporangium of the heterosporous pteridophyte in Figure 4.1 were reduced in number to one, and that the germination of this one spore to produce a reduced mega- (female) gametophyte occurred in the megasporangium while this is attached to the "parent" plant. Then suppose that the egg of the female gametophyte is fertilized by a spermatozoid from a reduced micro- (male) gametophyte that landed, still in its microspore coat (exine), near the opening of the megasporangium. (The spermatozoid swims in a liquid drop provided by the adult plant at the mouth of the megasporangium.) Further suppose that the embryo resulting from the fertilization develops before the megasporangium and its supporting "maternal" tissues are shed from the "parent" plant. What is produced is a seed: the megasporangium and its protective maternal structure, and remains of the megagametophyte, and the embryo (see Fig. 4.2). The microspore exine with its included, very reduced (ultimately in angiosperms to three nuclei; more in gymnosperms) microgametophyte is then called a pollen grain! This is a functional, embryological, not a morphological, definition! Actually, a pollen grain occurring as a fossil is only the microspore exine. Hence, it really is not wrong to call it a microspore. Note however that it is wrong, though frequently done, to call an isospore a "microspore", because it is best to reserve "microspore" exclusively for the male spores of heterosporous plants! The gymnosperms are a huge group, possibly but not certainly monophyletic, ranging from the first seed and pollen producers of the uppermost Devonian through the various Paleozoic and Mesozoic seed fern groups (Pteridospermae or Cycadofilicales: Paleozoic medullosans, Mesozoic caytonialeans, corystospermaleans, and others), cycadeoids and true cycads, ginkgoaleans, gnetaleans, and, more familiar to most people, coniferaleans. The pollen is usually sulcate, but there are many variants. In the Paleozoic, the pollen of primitive gymnosperms often resembles morphologically the microspores of heterosporous plants, including trilete laesurae. Such primitive pollen with spore-like morphology is usually called "prepollen". Later gymnosperm pollen is usually monosaccate, bisaccate, multisaccate, monosulcate, trichotomosulcate, inaperturate, or a variant of one of these

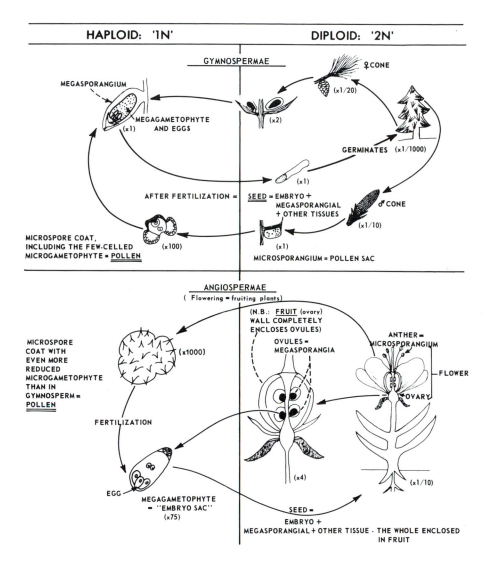

Figure 4.2 Life cycles of pollen-bearing (= seed-bearing) plants. Note that fossil pollen consists only of the exine, which is homologous with the exine (exospore) of a heterosporous "pteridophyte" (see Fig. 4.1). Although not shown in the angiosperm diagram, it should be emphasized that, when the pollen arrives at the stigma of the flower (below the letter "s" of microsporangium) it germinates, if accepted by the stigma, on the basis of recognition compounds, and a pollen tube grows through the style (directly beneath the stigma) to reach the embryo sac, where the nuclear fusions leading to zygote and triploid endosperm occur.

morphological types (see Ch. 5). In some extinct gymnosperms, such as the cycadeoids, the megasporangia and microsporangia were produced in a single strobilus, but in living gymnosperms, such as conifers, there are separate male and female strobili ("male cones" and "female cones"). Sometimes, e.g., in *Ginkgo*, these are even produced by separate, male and female, trees. In primitive gymnosperms the spermatozoids were presumably flagellate and motile, because they are in living cycads and *Ginkgo*. However, in living conifers and gnetaleans, a pollen tube is produced which penetrates into the megagametophyte of female cones and introduces non–motile fertilizing nuclei into the vicinity of the egg. (Pollen tubes have naturally not often been preserved as fossils, but Rothwell (1972) described one from a saccate gymnosperm grain of Pennsylvanian age.)

Angiosperms

This group of plants dominates the modern vegetation in terms of number of species as no animal group (except possibly insects) dominates the modern fauna. (Conifers cover some vast areas of the world but are two orders of magnitude less diverse.) Angiosperms apparently first appear in the fossil record in the Early Cretaceous, and already dominated the world's flora by early Late Cretaceous time. The group is often called the "flowering plants", but the angiosperm flower is not really unique. Bisexual cycadeoid (gymnosperm) strobili of the Mesozoic are really flowers, too. The carpellate fruit is a more distinctively angiosperm structure, in which the seeds are enclosed by carpel walls, but some gymnosperms approximate this, and a few apparently primitive angiosperms have incompletely closed carpels. The ovules or megasporangia (see Fig. 4.2), plus the enclosing carpel walls, in a flower comprise the ovary, and the ovary plus accessory "maternal" tissues make up the fruit. The megasporangium produces by reduction–division in a several-celled megagametophyte or embryo sac, one of the cells of which is the egg (see Fig. 4.2). The embryo sac is the really distinctive angiosperm feature. We really should call the angiosperms the embryo-sac plants, or a Latin equivalent! The flowers (or sometimes separate male flowers) also have microsporangia (anthers), which with their stalks are called stamens. In them, meiosis produces from spore mother cells the tetrads of microspores, within which 3-nucleate, fully enclosed microgametophytes develop. The microspore wall, plus the enclosed gametophyte, is the pollen grain. This agent of fertilization is carried by various vectors – wind, water or animal – to an extension of the ovary (stigma) of the same or different flower. Some angiosperms have elaborate biochemical and structural features to prevent self-fertilization. Others are even cleistogamous, with flowers that do not open but pollinate themselves. Still others are open to normal pollination in good times but self-pollinate in poor times. Some make pollen, and the flowers are visited by insects who take the pollen to other flowers, where it germinates, but the whole act is a farce: the ovules are fertilized apomictically, and the apparently normal pollen has nothing to do with it! Angiosperms also are wind- or animal-(usually insect) pollinated in very variable patterns. Some species are even partly wind-pollinated, partly insect-pollinated. (Meeuse & Morris (1984) present a good and entertaining discussion of these matters.)

On the stigma, the pollen germinates, producing a pollen tube. The pollen tube penetrates the stigma and grows down inside the style to the ovule, where two of the three nuclei of the pollen grain enter the embryo sac. Another uniquely (well, almost uniquely – the Gnetales have an analogous feature) angiospermous event then occurs: double fertilization, in which one nucleus fertilizes the egg to produce the 2N zygote, and the other nucleus combines with two other nuclei of the embryo sac to produce the triploid nucleus from which a 3N tissue, the endosperm, develops. This in many angiosperms is later the major source of nutrients to the embryo during germination. (In the form of cereal grass endosperm, it also produces the principal source of nourishment for human beings.) All angiosperm pollen probably derives from monosulcate–trichotomosulcate forms. From that base, however, practically all imaginable structural variants of pollen derive, except saccate pollen, which seem to be confined to gymnosperms.

Pollination biology ("anthecology") is a fascinating field in itself, with much to tell paleopalynologists. One of the nestors of paleopalynology, K. Faegri, is also co-author of a fundamental book in this field (Faegri & Van der Pijl 1979). The subject has a hoary history within biological history generally (Baker 1983), and the coevolution of angiosperms with their pollination vectors (Crepet 1983) is a very important aspect of the evolution of the plant kingdom.

SPORES, POLLEN, "MIOSPORES", AND OTHER TERMINOLOGICAL TROUBLES

The reader has already noticed that we have some difficulty with what inclusive term to use for those palynomorphs that are spores/pollen of embryophytic plants. I have already stressed that various such things with sporopolleninous exines that occur as fossils may be homospores (= isospores), microspores, megaspores, prepollen, or pollen. I will use "spores/pollen" because there are difficulties with other terms. Grayson once proposed "polospore", but it never caught on, perhaps because the suggested sport is not very common. Guennel's (1952) proposal of "miospore" was more popular: this makes all pollen or spores less than 200 μm "miospores", those more than 200 μm "megaspores". There are a few problems with the term "miospore", though in general it works well. First, it is pronounced the same as "meiospore", a term mostly used for meiotically produced algal spores such as zoospores and aplanospores (Scagel *et al.* 1965). Secondly, the 200 μm boundary, while reasonably good as a biological division between functional microspores and functional megaspores, is not completely dependable for that purpose. Some functional megaspores are less than 200 μm, some extant and fossil pollen grains are occasionally more than 200 μm, e.g., *Cucurbita* spp., *Oenothera* spp., and *Schopfipollenites* prepollen. "Miospores" include pollen, isospores, microspores, and some (only a few) small megaspores. It is essentially equivalent to the older expression "small spore". As was stressed earlier, however, it is not correct to use "microspore" in this sense. The arbitrary 200 μm boundary was selected by Guennel because of the use of ca. 200 μm screens in sedimentological screening for particle size.

CHAPTER FIVE

Spores/pollen morphology

As we noted earlier, paleopalynology's subject matter includes far more than embryophytic spores/pollen. Nevertheless, spores/pollen are the "heart" of the subject and are the major emphasis of this book. I have found in over 20 years of teaching the subject that beginning the study of paleopalynology with both laboratory and lecture exposure to extant spores/pollen is the "correct" approach, even though it seems to be putting the cart before the horse to study Holocene things before Paleozoic!

Figure 5.1 provides a very general glance, without exact time, of the evolution (presumably monophyletic) of spores/pollen types, beginning with early Silurian trilete spores. Most, but perhaps not all, of the types listed are still extant. (At the moment I cannot think of an extant plant that produces pseudosaccate spores or pollen grains.) The appearance in time of the major morphological novelties in spores/pollen is depicted in Figure 5.2.

MORPHOLOGICAL TYPES

The basic morphological types encountered in extant embryophytic spores/ pollen are shown diagrammatically in Figures 5.3 and 5.4. The fossil spore/pollen is basically a hollow, tough, variously ornamented and grooved bag, ball, or case, from which the contents (inner wall layers and protoplasm) have been removed through biodegradation by bacteria, by fungi, and possibly also by non-biological lysis. (Figure 5.4 shows schematically what a section of the whole grain, before removal of contents, looks like.) As a fossil, the bag or ball is usually found squashed flat and variously contorted. Students first encountering fossil sporomorphs in an uncovered preparation, e.g., in attempting to pick them up for single-grain mounts, are usually startled to discover that these fossils are not spheres as idealistically drawn but wafers that twist and turn in liquid mountant, like snowflakes or, better, tiny pancakes. The bag consists of a wall (exine) of sporopollenin which retains considerable resilience until much of the hydrogen and oxygen has been removed during increase of rank ("coalification" = "carbonization" = "maturation"). Therefore, during palynological processing of the enclosing rock, low-rank palynomorphs will often re-expand to some extent. Preparations of modern spores/pollen are usually produced by acetolysis (see Appendix), which removes the contents and inner walls in much the same way as fossilization, and also slightly carbonizes the wall, therefore changing the color from pale yellow to a darker yellow or orange. However, unless the wall was very thin (as in Juncaceae or *Populus*), the prepared grains are not much collapsed or folded. In any event,

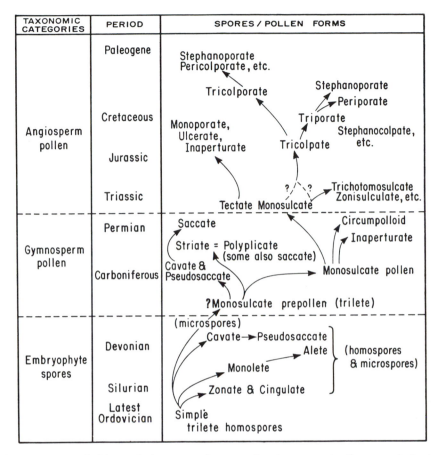

TAXONOMIC CATEGORIES	PERIOD	SPORES / POLLEN FORMS
Angiosperm pollen	Paleogene Cretaceous Jurassic Triassic	
Gymnosperm pollen	Permian Carboniferous	
Embryophyte spores	Devonian Silurian Latest Ordovician	

Figure 5.1 Probable evolutionary pathways of major spores/pollen morphological types. Time periods not to scale. Spores/pollen of course do not evolve independently of the taxa producing them, but, for example, logic and the fossil record indicate that plants producing monolete homospores were derived from plants producing trilete spores, etc. There is inevitably oversimplification in a diagram such as this. For example, at the interface between spores and pollen, prepollen represents the stage at which microspores still germinated proximally, as defined by Chaloner (1970b). However, some trilete forms representing this stage also seem to be sulcate, suggesting that haustorial pollen tubes probably formed.

the diagrams of major types presented in Figure 5.3 are of hypothetical, non-collapsed forms. Study of these morphological types is best accomplished with a set of accompanying slides representing a majority of the types.

The basic morphological categories shown in Figure 5.3 depend on obvious differences in external organization. Of these differences, the most common are scars representing former contact with other members of the original tetrad (laesurae), or thin places or openings in the wall, which are usually the site of

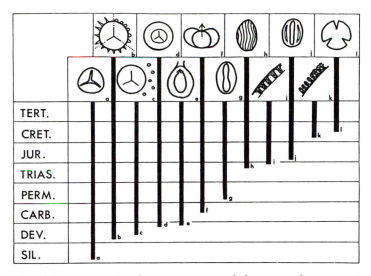

Figure 5.2 Major events in the appearance of features of spore exines. The bottom of each bar represents the time of first appearance of each item. Not all of these exine features (e.g., (e), (i), and (j)) persist to the present time. (a) Spores with evident triradiate aperture (now known from at least earliest Silurian). (b) Triradiate spores, showing diverse sculpture of the exine (now known from late Silurian). (c) Size differentiation indicative of heterospory. (d) Spores with markedly cavate exine. (e) Megaspore tetrad with three members aborted, retained in seed-like structure (*Archaeosperma*). (f) Bisaccate pollen (of seed ferns) with distal germinal area. (g) Monosulcate pollen of unknown (presumably gymnospermous) origin. (h) Polyplicate pollen, comparable to that of modern *Ephedra*. Early records of this type of pollen have been attributed to a conifer, but later (Tertiary) occurrences probably represent Gnetales. (i) Pollen exine with inwardly directed columellae simulating angiospermous tectate structure (as seen in the conifer *Hirmeriella*). (j) Asymmetrically tricolpate pollen (*Eucommiidites*) of gymnospermous origin. (k) Incompletely tectate angiospermous monosulcate pollen. (l) Symmetrically tricolpate (presumed angiospermous) pollen. Names of the geological periods abbreviated from Silurian, Devonian, Carboniferous, Permian, Triassic, Jurassic, Cretaceous, and Tertiary.

exit of pollen tube or other germinal material (pores, sulcus, colpus). As Chaloner (1984) has noted, various authors have stressed different functions of the laesurae, sulci, pores, etc. For some, e.g., Potonié, the "exitus" function is the most significant, and it is true that even the laesura normally serves this function. For others, e.g., Wodehouse, the accordion-pleat action of the colpi to accommodate for moisture-related harmomegathic expansion and contraction is more significant. For still others, e.g., Heslop-Harrison, the colpi of at least the angiosperms are pathways for chemical recognition signals. In any event, these features, plus a few other major morphological characteristics such as possession of sacci (the function of which is still uncertain) provide a convenient means for classifying spores and pollen.

This is a good point to emphasize that spores/pollen, unlike dinoflagellate

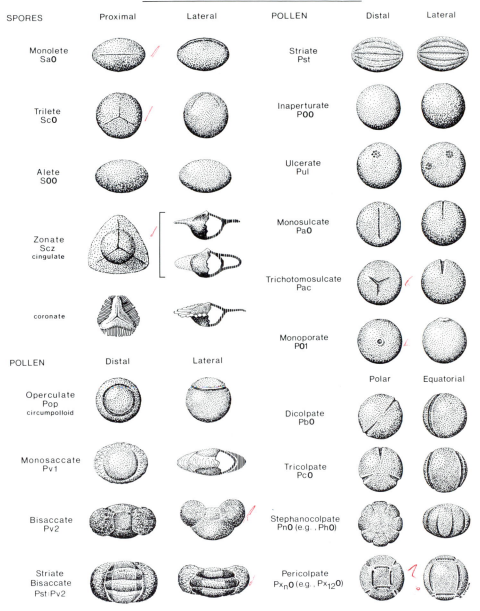

MORPHOLOGIC TYPE AND CODE DESIGNATION

SPORES — Proximal — Lateral — POLLEN — Distal — Lateral

Monolete Sa0
Trilete Sc0
Alete S00
Zonate Scz cingulate
coronate

POLLEN — Distal — Lateral

Operculate Pop circumpolloid
Monosaccate Pv1
Bisaccate Pv2
Striate Bisaccate Pst:Pv2

Striate Pst
Inaperturate P00
Ulcerate Pul
Monosulcate Pa0
Trichotomosulcate Pac
Monoporate P01

Polar — Equatorial

Dicolpate Pb0
Tricolpate Pc0
Stephanocolpate Pn0 (e.g. , Ph0)
Pericolpate Px_n0 (e.g., $Px_{12}0$)

Figure 5.3 General morphological types of embryophytic spores and pollen, plus "Shell code" designations. These are letters-and-numbers symbols for the various types, as explained in the text. The classification depends to some extent on Iversen & Troels-Smith (1950) and Faegri & Iversen (1975). Note under zonate (Scz) the lower

63

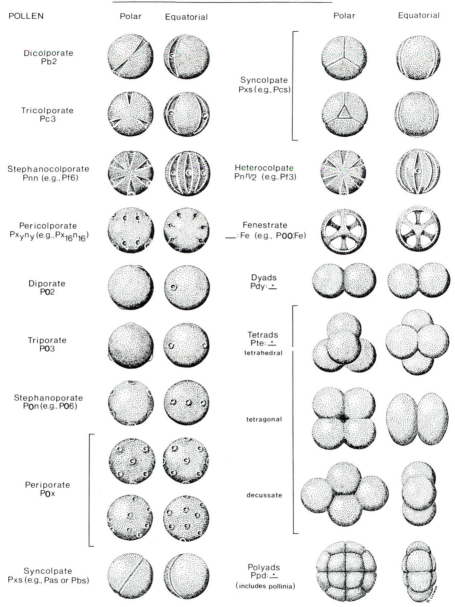

* Code for individual grains

form shown in diagrammatic section would be considered by some as camerate–cavate (see glossary), but such forms are included as zonate in this classification. The sectional view of monosaccate pollen (Pv1) is intended to demonstrate that the saccus of saccate pollen is not completely hollow but contains a more or less webby inner lining.

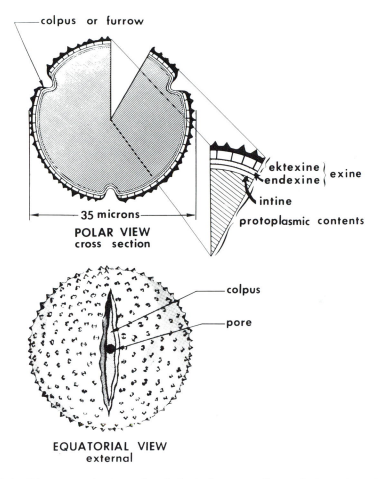

colpus or furrow

ektexine
endexine } exine

intine

protoplasmic contents

35 microns

POLAR VIEW
cross section

colpus

pore

EQUATORIAL VIEW
external

Figure 5.4 Diagrammatic view of typical tricolporate pollen grain in section, above, and external view, below.

cysts, do not unfortunately have anything like perfect integrity as to morphological type per species. They are very variable as to size, and even as to morphological type in a single sporangium, or anther; in exceptional cases, monolete and trilete spores are found in one sporangium, or triporate, tetraporate, and pentaporate pollen in one anther, or (very commonly) pollen of very different size in one species. Furthermore, it is well-documented that many angiosperms are regularly ("intentionally") dimorphic or even polymorphic as to size and morphological type, presumably as an anti-self-pollinating mechanism. Chinappa & Warner (1982) found 13 species and varieties of *Coffea* to be di-, tri-, tetra-, penta-, and polymorphic! *Coffea arabica*, the source of commercial coffee, for example, was found to have six pollen types, ranging from 3- zonocolporate to spiraperturate, with irregularly disposed ora and colpal-like apertures. Ferguson (1984) reports just as great "lack of integrity" among the palms. The genus *Daemonorops*, for example, has diporate, disulcate, and monosulcate pollen, with intectate echinate or tectate psilate, rugulate, foveolate or reticulate sculpture! Teratological monsters occur and are found as fossils. Nevertheless, it has been recognized almost since Linnaeus' time that pollen morphology can be used to show indication of generic and family relationships (Graham & Barker 1981), though plant systematists have only irregularly availed themselves of this opportunity. Many of the morphological features of spores/pollen probably have some sort of adaptive significance, but Muller (1984) and others have pointed out that some characters are determined by the mathematical–physical constraints imposed by development in the anther. Nevertheless, it is fair to assume that the sacci of saccate pollen do serve a flotation function for some reason and that various sorts of pleats (sulci, colpi, taeniae) serve an accordion–like expansion–contraction function (harmomegathy) somehow related to moisture loss and gain.

Table 5.1 presents data on the numbers of representatives of various morphological categories and sculpturing types in my reference core collection of modern spores/pollen. The collection has been assembled since 1947 in connection with various Cenozoic research projects and is broadly representative of the common plants of the North American flora, with some cultivated forms and many forms from other continents. (The total collection is over 5,000 species, but only the core collection was tallied.) Table 5.2 gives size information for the same core collection. (The total number of species under "size" is a bit less, because size was not measured for a few forms at the time of study.) Note that the "super-typical" spore/pollen in the modern flora is tricolporate (27%), reticulate (26%), and 21–40 µm (52%) in maximum dimension.

Chaloner (1984) has noted that, whereas the evolutionarily basic homospore is often sculptured, gymnosperm pollen seldom has much sculpture, and angiosperm pollen varies enormously, from strongly sculptured to smooth, with all sorts of intermediate stages. The subject of pollination mechanisms and the adaptive significance of the size, shape, and other features of pollen is mostly beyond the scope of this book. The investigator should be aware that features he/she observes mostly if not always had significance for the producing plant. Large and/or ornately sculptured pollen ordinarily means insect pollination, though *Picea* and *Zea* pollen are both examples of wind-pollinated pollen that can be near or over 100 µm in one dimension. Smooth and/or small pollen is ordinarily wind-pollinated, though, as Basinger & Dilcher (1984) point out, the very tiny pollen (as small as 8 µm!) of a Cretaceous flower they describe was likely not wind-pollinated, because particles that small do not have impact velocity enough to adhere to a stigma. Some smooth pollen, such as *Prosopis* sp., is clearly insect-pollinated. Nevertheless, despite exceptions, it is pretty safe to assume that ornately sculptured fossil pollen, or pollen greater than 40 µm in size (or less than 10 µm?), was insect-pollinated, and that relatively smooth pollen in the 15–35 µm range was wind-pollinated. Chaloner (1984) has pointed out that the sculpture of angiosperm pollen in some instances is related to static electricity: the negative charge of pollen, in part given it by insect vectors, means that pollen is attracted to the (induced) positive charge of the stigma, and protruding sculptural elements such as spines protect the charge from premature grounding. Wind-pollinated forms seem even to have evolved features on the female side adapted to reception of pollen: aerodynamically tuned structures of female cones or inflorescences to deflect air flow containing pollen in such a way as to maximize pollination success (Niklas 1985).

It should also be emphasized that spores/pollen of extant plants are characteristically determinable only to genus, not to species. In a few unfortunate cases (Gramineae, Chenopodiaceae, and others) only the family can be determined in routine analysis, though S.E.M and T.E.M studies show differences on which separation would in many cases be possible. On the other hand, Tiwari (1984) and de Jersey (1982) both emphasize that, for fossil sporomorphs, even such a seemingly minor feature as sculpture may be of

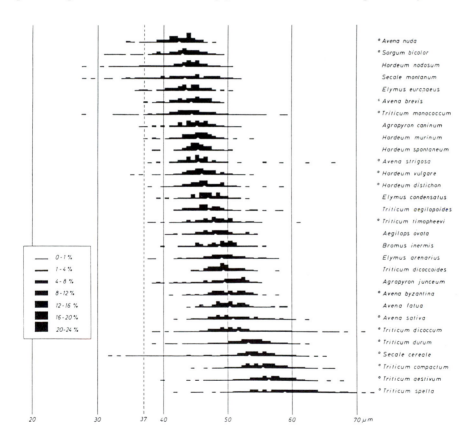

Figure 5.5 Pollen sizes of selected grass taxa: cultivated cereals are marked with a ·small circle. These are larger grass pollen. The original diagram includes many forms in the 20–40 μm range.

Table 5.1 Numbers of representatives (species) of morphological categories and sculpturing types in Traverse's reference core collection of extant spores/pollen.

Morphological category	Total species for class	Sculpturing type													
		Psilate	Micropitted	Foveolate	Fossulate	Scabrate	Gemmate	Clavate	Verrucate	Baculate	Echinate	Rugulate (incl. retirugulate)	Striate	Reticulate	Mixed and other
Trilete	142 (=6%)	38	4	2	2	21		2	28	2	6	16		21	6
Monolete (for spores, sculpture sometimes of exospore, sometimes of perispore)	136	72			3	15			20	2	3	10	1	4	6
Alete	52	4	3			31	2	3	3		3	1		5	6
Bisaccate (= prevailing sculpture of corpus) + > bisaccate	54			20		1			3					27	3
Polyplicate	15	15													
Inaperturate	138	15	14	1	2	39	1		4	2	21	1		29	9
Monosulcate (includes Pac)	219 (=9%)	45	34	13	2	17	6	3	6	2		14	4	66	9
Ulcerate	3		2												1
Monoporate	63	5	19	2		16			4		2	1		11	3
Dicolpate	7	2												3	2
Tricolpate	282 (=12%)	21	24	19	31			10	5	2	19		9	129	13

Type														
Stephanocolpate (Pd0–P__0)	43	11	6			4								13	9
Pericolpate	15	1		2		2						3			7
Dicolporate	2			1		1									
Tricolporate	614 (=27%)	117	33	44	1	104	1	13	3	73	6	14		190	15
Stephanocolporate (Pd4–P__)	58	21	5	2		7	3	2		1		1			17
Pericolporate	2							1		1					
Diporate	20	5	4	1		4								6	6
Triporate	165 (=7%)	35	40	9	1	35		7	1	7	2	3		19	6
Stephanoporate (P04–P0__)	40	11	2	1	1	9		2	1	3	3			8	
Periporate	97	9	19	4		15	2	2		24	1	1		15	6
Syncolpate (Pbs–P__s)	42	11	5		1	5		1		4		1		12	2
Heterocolpate	5	3				1				1					
Fenestrate	11	1						2		4				4	
Dyads	1							1		1				1	
Tetrads	63	12	1	1	1	31	1	4	1	4	4			8	
Polyads (includes massulae)	17	9	1	2	2	2								3	
Totals	2306	462 (=20%)	217	120	46	360 (=16%)	12	20	105	16	177	56	33	592 (=26%)	90

Table 5.2 Numbers of species in various size groups in Traverse's reference core collection of extant spores/pollen.

Morphological category	Maximum dimension															Total species
	<11 μm	11–15 μm	16–20 μm	21–25 μm	26–30 μm	31–35 μm	36–40 μm	41–45 μm	46–50 μm	51–60 μm	61–70 μm	71–80 μm	81–90 μm	91–100 μm	>100 μm	
Trilete			1	3	5	13	23	17	20	22	17	8	6	1	3	139
Monolete					4	13	12	24	14	28	18	8	2	3	5	131
Alete	5	11	13	3	1	2	1	3	2	3	3	1	1	1		50
Bisaccate + >bisaccate						1	1		2	3	3	6	8	8	22	54
Polyplicate						1	2	5	1	5	1					15
Inaperturate	1	7	12	12	20	23	12	7	4	11	6	5	8	3	2	133
Monosulcate (includes Pac)	1	1	7	10	30	30	29	29	11	27	12	7	7	3	10	214
Ulcerate				1		1										2
Monoporate			2	5	15	13	11	9	5	6		1			1	68
Dicolpate					2	2	2									6
Tricolpate	1	5	28	44	51	39	26	18	9	13	14	7	6	2	3	266
Stephanocolpate (Pd0–P_0)			1	10	14	4	3	4		2	1	2				41
Pericolpate			1	1		2	3			3	2			1	2	15
Dicolporate									1	1						2
Tricolporate		30	62	132	120	69	74	33	36	31	9	5	5	1	5	612
Stephanocolporate (Pd4–P___)		1	6	11	8	4	14	1	2	3	2	1	2	1		56
Pericolporate											1			1	1	3
Diporate		2	4	3		4	3	1	1	3		1				22
Triporate		1	9	12	32	28	19	10	9	12	3	8	6	4	4	157

																Total
Stephanoporate (P04–P0___)		2	1	4	14	11	3	1	1							37
Periporate	2	7	9	14	11	9	5	8	6	4	3	2	1		15	96
Syncolpate																
Pbs–P___s	4	4	8	8	5	2	2	1	1	2	1	1				39
Heterocolpate		1	3	1												5
Fenestrate		2	1		1	2	1	2								9
Dyads				1												1
Tetrads	1	4	4	15	6	9	4	3	3	2	5	3	1	1		61
Polyads (includes massulae)	1	2	3	1	3	1	3	1	1					1		17
Totals	10	67	169	273	346	284	272	177	135	186	101	69	56	31	75	2251
				(=12%)	(=15%)	(=13%)	(=12%)									

(=52%)

great significance in separating genera from each other. Nevertheless, it is clear that the lack of distinctiveness at the specific level implies that fossil spores/pollen species seldom really represent plant species.

Figure 5.5 shows part of a plot of size ranges of common grass pollen. Note that the overlap is too great to permit use of size alone for identification. Most grass pollen is smaller than the species shown. All grass pollen is monoporate (P01). Most of the largest grass pollen classes are of cultivated cereals, presumably reflecting polyploidy. Use of S.E.M. and other sophisticated techniques to separate species or species groups of large genera such as *Pinus* or *Quercus* is possible, though it is doubtful that this will be practicable on a routine basis in the near future (see Fig. 5.12). Efforts have been made to separate pollen of species of various genera on the basis of size or other measurement ratios, but they are seldom really reliable, just because of the great plasticity that characterizes pollen and spore morphology.

"SHELL CODE"

The diagrams in Figure 5.3 are accompanied by the name of each morphological type, plus a letter-and-number shorthand "code" designation for each type. Such diagrams were originally developed by Iversen & Troels-Smith (1950), and have appeared since the first edition (1950) in Faegri & Iversen's (1975) famous *Textbook of pollen analysis*. The Royal Dutch/Shell group of oil companies has used this basic chart and the code designations in its practical paleopalynology since the mid-1950s (Hopping 1967). For easy reference, Shell developed the three-symbol code for designation of the morphological types. When I was a Shell palynologist I learned this "code" and have employed it ever since. The principle is very simple.

Spores

1st symbol: "S" for spore
2nd symbol: laesura type
   "c" for trilete (the basic type)
   "a" for monolete
   "b" for dilete, or "chevron" (but this is very rare)
   "0" if no laesura (Note: here and elsewhere 0 is zero, not o, though it is usually spoken "oh")
3rd symbol: special features, if any
   "0" if none
   "z" for zonate

Thus, Sc0 is for ordinary trilete spores, S00 for alete spores, Sa0 for monolete spores, Scz for zonate, trilete spores. "Sc0", "Scz", etc., were used by Shell palynologists 20 years ago as if they were super-genera. A number following the code designation was the locally used "species": thus Sc0-12, Scz-29, etc.

Pollen

1st symbol: "P" for pollen
2nd symbol: usually refers to colpi (or sulci)
  "a" for monocolpate (monosulcate)
  "c" for tricolpate, etc.
  "0" for none
3rd symbol: usually refers to pores
  "1" for one pore
  "3" for three pores, etc.

Thus, a monocolpate pollen grain is Pa0, a tricolporate pollen grain is Pc3, a triporate pollen grain is P03.

Saccate pollen, a special problem

1st symbol: "P" for pollen
2nd symbol: "v" for vesiculate (= saccate)
3rd symbol: number of sacci

Thus, a bisaccate is Pv2, a monosaccate is Pv1.

Other features of the shorthand code are more easily understood by referring to the figures than by more textual explanation.

Paleozoic and Mesozoic spores/pollen present many features that are difficult to handle with the code. In its basic form the code does not distinguish between pseudosaccate and saccate, for example, nor does it provide for the possibility that a pollen grain may have laesurae, as prepollen do. Nevertheless, the code is a handy shorthand way of referring to morphological types, both fossil and extant, and a palynologist will easily invent his own extra symbols for forms of importance to him/her. In my own laboratory when working up a new palynoflora I usually use the Shell code designations, plus an abbreviation for sculpture, plus a number for unnamed forms, e.g., P0x-ret-4 (for a certain reticulate periporate pollen grain form.) Erdtman and others have provided far more complicated classifications which are useful for detailed morphological studies, but the one presented here has worked well for practical purposes in our laboratory for over 30 years.

MORPHOLOGICAL TYPES IN DETAIL

The reader should consult Figures 5.3–5.7 when reading this section.

Spores

(See Figs. 5.3, 5.6, & 5.7.)

Trilete (Sc0) This is the basic spore type, first appearing in the early Silurian. It owes its trilete laesura to the contact between it and the other three members of the tetrad of spores produced by meiosis from a spore mother cell

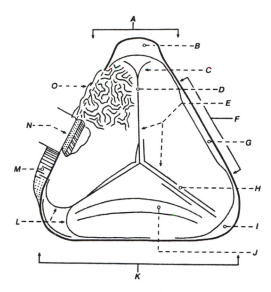

Figure 5.6 Schematic representation of principal morphological features of trilete spores (monolete spore terminology is basically the same): A, radial region (area); B, auricle (radial thickening; a limited zona); C, curvatura imperfecta (does not join other radii); D, commissure (center of the suture of "dehiscence mark"); E, radii (arms) of laesura; F, interradial region (area); G, interradial thickening; H, labrum (lip of suture); I, valva (slight to moderate radial thickening); J, torus (≃ "kyrtome", "margo"; often a fold feature); K, equatorial diameter (= spore size); L, curvatura perfecta (joins other radii); M, cingulum or zona (equatorial thickening or flange); N, exospore in cross section (≃ exine; often double-layered); O, perispore (≃ perine).

in the sporangium. The trilete laesura is also called a "Y mark". Trilete is an adjective, not a noun. Thus, a spore has a trilete laesura, not "a trilete". Laesurae are haptotypic features, resulting from position in the spore tetrad. Extant examples: *Lycopodium, Botrychium, Aneimia* (Fig. 5.7 a–g).

Monolete (Sa0)   Appears much later in the fossil record, seemingly derived from the trilete condition by change of tetrad form so that the three-pronged laesura is not formed. (An intermediate, dilete form (Sb0) with a V-shaped laesura exists but is rare.) Examples: *Marattia, Lorinseria* (Fig. 5.7h–j).

Alete (S00)   Apparently derived from monolete or trilete condition by non-formation of haptotypic marks. Example: many moss spores (Fig. 5.7l,m).

Zonate (Scz, rarely Saz)   Trilete (rarely monolete) spores with an equatorial extension. More common in Paleozoic than since. Extant example: *Gymnogramme* (Polypodiaceae).
    This general category includes variants, such as the following:
    Cingulate, in which the zone is thick, more of a flange than a thin zone. Extant example: *Lophosoria*.

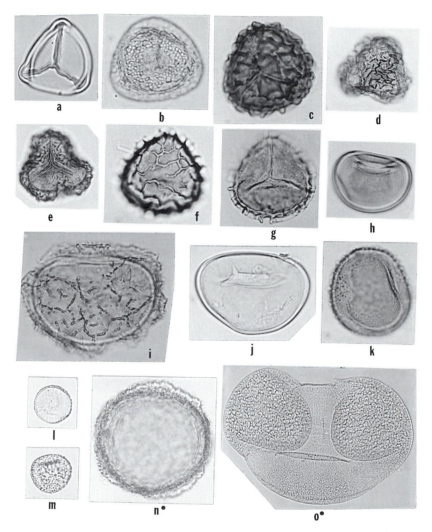

Figure 5.7 Photomicrographs of principal spores/pollen morphological types. All from acetolyzed preparations. Magnification indicated by bar under (ah). Items with a dot after the letter are 0.6 × as magnified as those with no dot. Item (aw) is at very low magnification. For it, the reference bar represents 440 μm. Erdtman suggested conventions for illustrating spores and pollen, and he felt quite strongly that they should be followed. Some consistency *is* a good idea, and I have followed Erdtman, e.g., in orienting trilete spores with one radius of the laesura always pointing up, and in placing bisaccates in lateral view with the distal side up, as well as in orienting equatorial views of colp(or)ate grains with the colpi pointing up. However, I have oriented polar views of triporate and tricolp(or)ate grains with one colpus or pore at the top, whereas Erdtman oriented them with a colpus or pore pointing *down*. I only recently realized I was being non-erdtmanian in this respect. Regarding monolete and monosulcate forms, Erdtman was uncharacteristically inconsistent, except that the

75

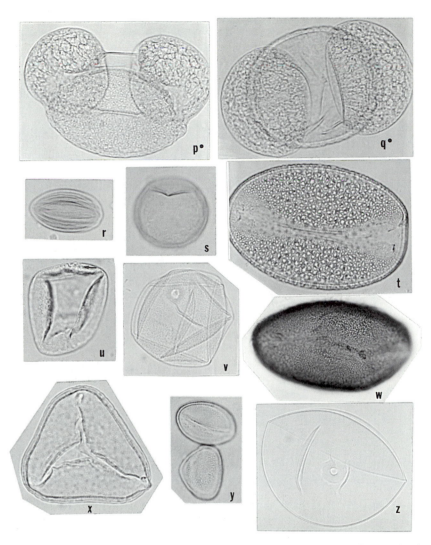

laesura of a monolete form is never portrayed on the up side, but is either on the down side and parallel to the bottom of the page, or pointing up at exactly 90°. For illustrating monoletes with laesura on the up side and triporates–tricolp(or)ates upside down, one would hear from Professor Erdtman were he still alive. (a) Trilete (Sc0): *Hemionitis palmata* L., Guatemala. Psilate exospore in proximal view. Auriculae developed at ends of radii of laesurae. (b) Same taxon as (a). Perispore in proximal view with rugulate–reticulate sculpture enclosing exospore. (c) Trilete (Sc0), trending to zonate (Scz): *Pityrogramma triangularis* (Kaulf.) Maxon., California. Rugulate sculpture presumably on perispore, proximal view. (d) Trilete (Sc0): *Botrychium lunaria* (L.) S.W., Pennsylvania. Perispore, distal view, rugulate–reticulate sculpture. (e) Same taxon as (d). Proximal view showing trilete laesura and outline of exospore within the perispore. (f) Trilete (Sc0): *Lycopodium annotinum* L., Quebec. Loosely reticulate

76

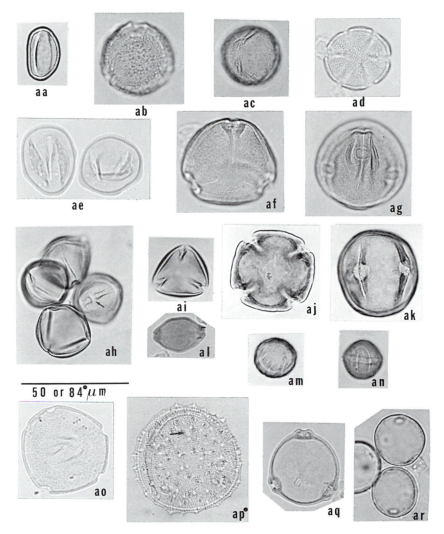

50 or 84°μm

sculpture of exospore, distal view, mid-focus. (g) Same taxon as (f). Proximal view showing trilete laesura. High focus. (h) Monolete (Sa0): *Dryopteris hexagonoptera* (Michx.) C. Chr., Pennsylvania. Psilate exospore, lateral view, but showing monolete laesura. (i) Monolete (Sa0): *Thelypteris gongylodes* (Schkuhr) Kuntze, Florida. Beady rugulate perispore, partly abrading away, showing exospore. Lateral view. High focus. (j) Same taxon as (i). Exospore without perispore, psilate sculpture. Proximo-lateral view showing that this "monolete" laesura is really more or less trilete, indicating origin of the monolete condition. (k) Operculate (P0p) (= zonisulcate), a variant of monosulcate (Pa0): *Nymphaea candida* Presl. (cult.). Single encircling sulcus (= sulculus) seen in distal view. High focus. Sculpture scabrate and baculate. (l) Alete (S00): *Bryum bimúm* Schreb., Colorado. Scabrate and verrucate sculpture. (m) Alete (S00): *Anoectangium anomalum* Bartr., New Guinea. Scabrate and verrucate sculpture. As (l), a

77

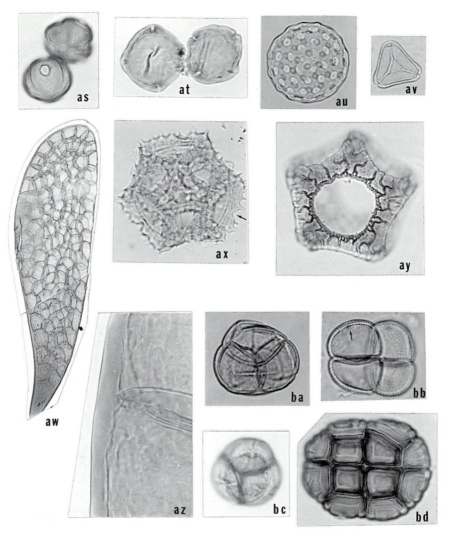

moss spore. (n) Monosaccate (Pv1) or inaperturate (P00): *Tsuga heterophylla* (Raf.) Sarg., Washington. Mid-focus. Sporopolleninous envelope of internally alveolate structure is interpretable as a single saccus with rugulate sculpture. (o) Bisaccate (Pv2): *Picea likiangensis* Pritzel var. *purpurea* Dallimore & Jackson, China. Lateral view. Sacci reticulate sculpture, corpus psilate on distal surface (= cappula), reticulate on proximal surface (= cappa). (p) Bisaccate (Pv2): *Pinus resinosa* Ait., New Hampshire. Lateral view. High focus. Sculpture information as for (o). (q) Same taxon as (p). Distal view. High focus. (r) Polyplicate (= striate) (Pst): *Ephedra foliata* C.A.M., India. If the plicae are interpreted as morphological features, the sculpture is psilate. Lateral or proximal–distal view. (s) Inaperturate (P00): *Populus* sp., Pennsylvania. See (n). Sculpture scabrate. (t) Monosulcate (Pa0): *Hippeastrum vittatum* Herb. (cult.). Distal view. Mid–high focus. Sculpture reticulate. Columellate structure observable in section around the edge and in small light dots on the surface. (u) Ulcerate (Pu1): *Carex*

78

variabilis Bailey, Colorado. Lateral view. Sculpture scabrate. Ulcus (arrow) is a place where the exine is incomplete; it serves as a pore. (v) Monoporate (P01): *Arundinaria tecta* (Walt.) Muhl., Louisiana. Distal view. Note thickening (annulus) around the pore, and the small block of ektexine (operculum) in the center of the pore membrane. Sculpture scabrate. (w) Monosulcate (Pa0): *Yucca* aff. *louisianensis* Trel., Texas. Distal view. High focus, very thick exine, hence darkness. Sculpture reticulate. (x) Trichotomosulcate (Pac): *Cocos nucifera* L., Yucatan, Mexico. Distal view. High focus. The single sulcus is tricornered, sometimes closely simulating a trilete laesura. Sculpture micropitted. (y) Monosulcate (Pa0): *Xerophyllum tenax* (Pursh) Nutt., Montana. Two specimens, upper one is distal view showing sulcus, lower one a proximal view. Sculpture reticulate. (z) Monoporate (P01): *Zea mays* L. (cult.). See comments for (v). Cereal grass pollen is usually larger than non-cereal grass pollen. Sculpture counts as psilate, is actually microscabrate. (aa) Dicolpate (Pb0): *Pontederia cordata* L., Florida. Sculpture psilate. (ab) Tricolpate (Pc0): *Quercus phellos* L., District of Columbia. Polar view. High focus. Sculpture scabrate and verrucate. (ac) Same taxon as (ab). Equatorial view. Note that colpus is somewhat modified equatorially, a condition called tricolporoidate, if more pronounced, as in (ah) and (ai). (ad) 6-Stephanocolpate (Pf6): *Mentha rotundifolia* (L.) Huds., North Carolina. Polar view. Microreticulate sculpture. (ae) Pericolpate (Px0): *Batis maritima* L., Yucatan, Mexico. Equatorial views, showing that colpi are not located on lines connecting the poles. Psilate sculpture. Note knobby thickenings of ektexine. (af) Tricolporate (Pc3): *Nyssa ogeche* Marsh., Georgia. Polar view. Sculpture counts as psilate per Faegri classification, although very small pits and granae are visible. LO analysis at 1,000 × reveals columellate structure. (ag) Same taxon as (af). Equatorial view. Note costae (thickenings) on edges of colpi, and the complex equatorial structure with an inner aperture (= os) with a thickened exinous rim (see those out of focus, both here and in (ag)). (ah) Tricolpate–tricolporoidate (Pc0): *Phoradendron serotinum* (Raf.) M. C. Johnston, Texas. Variety of equatorial views, various focal levels. Note (arrow) equatorially modified colpus. Psilate sculpture. (ai) Same taxon as (ah). Polar view. Note that from polar view alone, even with critical focusing, it is hard to prove whether it is Pc0, Pc0 or Pc3. (aj) 4-Stephanocolporate (Pd4): *Melia azedarach* L., Texas. Polar view. Sculpture counts as psilate per Faegri classification, though a slightly verrucate texture is observable. See (ak). (ak) Same taxon as (aj). Equatorial view. Note costate thickenings along the colpi and the large equatorial pore structure. (al) Diporate (P02): *Itea virginica* L., Alabama. Polar or equatorial view (in P02s, hard to distinguish). Sculpture psilate. (am) *n*-Stephanocolporate (Pnn): *Spermacoce glabra* Michx., Texas. Obliquely polar view. Sculpture psilate. (an) Same taxon as (am). Note that the equatorial "pore" structure is actually an equatorial band of thinning (= "transverse colpus") (Faegri & Iversen 1975: colpus transversalis). In some stephanocolporate pollen the transverse colpus is separate for each longitudinal (meridional) colpus, so that the two present a cruciate appearance. The colpi in (an) are not syncolpate, though from this photo one might draw that conclusion. (ao) 5-Stephanoporate (P05): *Pterocarya stenoptera* D.C., China. Polar view. Sculpture scabrate. (ap) Pericolporate (Pxx): *Malvastrum spicatum* (L.) Gray, Texas. Sculpture echinate plus microreticulate. The colpi and associated rectangular "pores" are small and scattered over the surface (see arrow). Pxx is a relatively rare condition. (aq) Triporate (P03): *Betula nigra* L., Tennessee. Polar view. Sculpture psilate. Ektexine and endexine separate at pores producing a vestibulum. Erdtman called the outer, ektexinous pore the "pore", the inner endexinous one the "os", and he would call *Betula* pollen "pororate". (ar) Triporate (P03): *Maclura pomifera* (Raf.) Schneid. (cult.). Oblique, mostly equatorial views. Sculpture psilate. Pores annulate (surrounded by thickened rim). (as) Heterocolpate (Pf3): *Lythrum salicaria* L., Quebec. Polar (top) and equatorial views. Sculpture psilate. Equatorial pores occur in only half the colpi. (at) 4-Stephanocolpate (Pd0): *Haloragis*

79

Coronate, in which the zona is feathery or broken up to a tattered fringe. The best examples are all fossils, such as, for example, *Reinschospora*, a Carboniferous spore shown in S.E.M. picture and diagrammatically in Figure 1.1o,p.

Grebe (1971) has produced an excellent compendium of standard terms for, and methods of, spore description, which the student should consult for more information. Unfortunately, the publication is by no means available in every library.

Pollen

(See Figs. 5.3, 5.4, & 5.7.)

Operculate (P0p) Has an encircling sulcus or pseudosulcus. The circum-polloid Mesozoic pollen group including *Corollina* (= *Classopollis*) could conceivably be put here, as can pollen of the extant pond lily, *Nymphaea*. The encircling feature in the latter is a true sulcus, and that of *Corollina* not (Fig. 5.7k), but sometimes *Corollina* pollen does break apart at the rimula.

Saccate (Pv2, etc.) Having at least one saccus (= "vesicle", "bladder", "wing"). Strictly speaking, a true saccus is not a hollow sack; it has an internal spongy or "webby" lining of varying thickness and density. If a similar vesicle is present but lacking the internal "webbing", the grain is pseudosaccate, a common form in the Paleozoic. In practice, I include pseudosaccate in Pv1,

---

erecta (Murr.) Schindler, New Zealand. Polar views. Sculpture psilate. This is a good example of form with very short colpi often described as "pores". (au) Periporate (P0x): *Salicornia virginica* L., New Jersey. Sculpture psilate (micropitted). (av) 3-Syncolporate (Pcs): *Melaleuca quinquenervia* (Cav.) S. T. Blake (cult.). Polar view. Sculpture psilate. Colpi unite to an "island" comprising the polar area. "Pcs" includes both the syncolpate and syncolporate forms. (aw) Pollinium (Ppd:P00): *Asclepias tuberosa* L., Georgia. Composite of two photomicrographs at very low magnification – bar under (ah) represents 440 μm. All of the pollen of one chamber of a stamen is shed as a single mass. This is thus a special sort of polyad. The individual grains are P00. Sculpture psilate. See (az). (ax) Fenestrate (Pfe:Pc0): *Sonchus arvensis* L., Pennsylvania. Polar view. Echinate sculpture. The large "windows" (hence, fenestrate) dominate, but the grain is also tricolpate, though the colpi are hard to see (arrow). (ay) Fenestrate (Pfe): *Passiflora incarnata* L., Texas. Sculpture loosely reticulate. The "windows" are left by opercula dropping out. These are sporopolleninous and also appear in pollen preparations. (az) Same pollinium as in (aw), 1,000 ×. Outer edge of pollinium showing that outer exines of individual grains tend to fuse with each other. (ba) Tetrahedral tetrad, tricolporate (Pte:Pc3): *Phyllodoce caerula* (L.) Bab., Finland. Lateral view. High focus of tetrad, showing colpi running toward each other from adjoining grains (Fischer's rule). Sculpture verrucate. See (bc). (bb) Tetragonal tetrad, monoporate (Pte:P01): *Typha latifolia* L., Texas. "Top" view of tetrad. Mid-focus in which out-of-focus pore shows for only one grain (arrow). Sculpture reticulate. (bc) Tetrahedral tetrad, tricolporate (Pte:Pc3): same taxon as (ba). See "top" view of tetrad, high focus. (bd) Polyad, syncolpate (Ppd:Pcs?): *Acacia* sp. (cult.). Sculpture psilate. Sixteen grains comprise the polyad. Each grain is undoubtedly syncolpate, but that they are basically 3-syncolpate is a guess.

Pv2, etc. The function of sacci is still debated, but the "idea" was already on the go in the Devonian and this was a very dominant type in the Mesozoic. The function of the spongy lining of the true saccus would seem clearly to be at least partly to provide firmness to the saccus, to prevent its collapse, and the saccus obviously makes a saccate grain float better in water, perhaps even significantly in air, because of the reduction in total specific gravity. Extant examples: *Pinus, Picea, Podocarpus* (Fig. 5.7n–q).

Striate (= polyplicate) (Pst) Pollen grains with multiple grooves (striae), plicae ("pleats"), or straps (taeniae) dominating the surface. Such grains in the Paleozoic and Mesozoic are often also saccate, and one can use a double code reference, e.g., Pst–Pv2. The plicae, etc., may have a harmomegathic (expansion–contraction) function. Extant examples: *Ephedra, Welwitschia* (Fig. 5.7r).

Inaperturate (P00) No haptotypic features. Unfortunately not easy to distinguish from S00 or even from "baggy" acritarchs and dinoflagellate cysts. Extant example: *Populus* (Fig. 5.7s).

Monosulcate (≃ monocolpate) (Pa0) Having a single germinal furrow or colpus or sulcus. Technically, a sulcus is such a furrow when located on the distal surface, usually with the distal pole as its center, whereas a colpus is a longitudinal furrow on a "meridional line" crossing the equator. In day-to-day work palynologists do not usually trouble with this erdtmanian distinction. For example, some of the "colpi" of pericolpate grains (Fig. 5.3) are not technically colpi but sulculi (see Glossary). In any event, the code designation is Pa0. Extant example: *Lilium* (Fig. 5.7t,w,y).

Ulcerate (Pul) Grains with no well-organized pores or colpi, but having one or more areas with thinned or partially broken exine. Extant example: Cyperaceae (Fig. 5.7u).

Trichotomosulcate (-colpate) (Pac) A variant of Pa0, in which the germinal furrow is drawn out to a three-pronged shape, like the outline of a tricorn hat, but resembling sometimes a trilete laesura. As such a furrow is always a distal feature, some would insist on the use of trichotomosulcate not trichotomocolpate, but in our laboratory we regard the distinction as of little practical importance. Extant examples: many palms, e.g., *Cocos*, which sometimes has Pac and Pa0 in the same anther (Fig. 5.7x).

Monoporate (P01) Obviously a modification of Pa0. P01 grains are especially a feature of Gramineae, in which they invariably occur. Grass pollen always has a thickened rim of exine around the pore, the annulus, and a small pad of ektexine, the operculum, on the pore membrane (Fig. 5.7z).

Dicolpate (Pb0) Uncommon type of angiosperm pollen. Extant example: *Pontederia* (Fig. 5.7aa).

81

Tricolpate (Pc0) and tricolporoidate (Pc0)   The basic type of dicot angiosperm pollen, with three meridional colpi, 120° apart as viewed from the pole. Extant examples: certain *Quercus* spp., *Ilex* (Fig. 5.7ab,ac). (Many *Quercus* spp. and *Ilex* are actually tricolporoidate, meaning that the colpal membrane is narrowed and/or somewhat modified at the equator. See comment under Tricolporate.)

Stephanocolpate (Pn0: Pd0, Pe0, Pf0, etc.)   Colpi arranged on meridians connecting the poles of the grains, perpendicular to the equator, evenly spaced, and greater than three in number. Extant example: Labiatae (e.g., *Mentha* is usually Pf0) (Fig. 5.7ad,at).

Pericolpate (Px$_n$0: Px$_4$0, etc.)   Colpate pollen in which the colpi are not located on "meridians" but are in quite other positions or are skewed with reference to lines connecting the poles. (According to Erdtman, such "colpi" should be called either sulculi or colpoids depending on orientation.) Extant example: *Batis* (Fig. 5.7ae).

Dicolporate (Pb2)   Dicolporate pollen (with colpi and pores), an uncommon pollen type. Extant example: some Acanthaceae.

Tricolporate (Pc3)   One of the fundamental dicot angiosperm pollen types, with three equally spaced colpi on "meridians" (see Pc0), each colpus also provided with an equatorial pore, os, ulcus, or other membranal modification. There are literally thousands of extant examples, e.g., *Nyssa* (Fig. 5.7af,ag).
   Note that many species make pollen grains with colpal configuration intermediate between tricolporate and tricolpate. Erdtman and others call these *tricolporoidate*. Such -colporoidate grains lack a well-defined equatorial pore or even ulcus, but a thinning of some sort of the membrane, or a shallowing or other modification of the colpus, is apparent. I normally include these with -colpate, but underline the 0. Thus, a tricolporoidate grain is a Pc0. Extant example: some species of *Quercus* (see Fig. 5.7ac, ah, ai).

Stephanocolporate (Pnn: Pd4, Pd5, etc.)   As Pc3, but more than three equally spaced meridional colpi, each with equatorial pores or other membranal modifications. Extant example: *Citrus* (Fig. 5.7aj,ak,am,an).

Pericolporate (Px$_y$n$_y$: Px$_4$n$_4$, etc.)   Colporate pollen in which the colpi (Erdtman: colpoids) are not arranged meridionally but otherwise, or at least are skew to lines connecting the poles (see Px$_n$0). Extant example: *Malvastrum* (Fig. 5.7ap).

Diporate (P02)   Pollen with two more or less isodiametric germinal apertures, including both pores and ulci (if the latter are regularly arranged – otherwise these are Pu1). Extant examples: *Itea*, some *Ficus* (Fig. 5.7al).

Triporate (P03)   One of the most common dicot angiosperm pollen types, having three equatorial, more or less isodiametric germinal apertures, including pores and ulci (if the latter are arranged as described – otherwise these are Pu1), including forms with complex apertures, with an outer pore, an inner opening (os), and an intervening chamber (vestibulum) between them. (Erdtman calls the latter "-pororate".) Extant examples: *Urtica, Maclura, Celtis* (simple pores), *Betula* ("vestibulate triporate") (Fig. 5.7aq,ar).

Stephanoporate (P0n: P04, P05, etc.)   Similar to P03, but with more than three equatorial, equally spaced pores or ulci. Extant examples: *Alnus* (sometimes P03, usually P04–5), *Pterocarya* (usually P06 or P07) (Fig. 5.7ao).

Periporate (P0x)   With pores arranged other than on the equator, characteristically all over the surface, as in Chenopodiaceae, but also included here are grains such as those of *Juglans*, in which there are a few off-equator pores, as well as a number arranged in the stephanoporate manner. *Carya* has three pores 120° apart. They are slightly off the equator in one hemisphere. Nevertheless, we exercise some license and consider *Carya* as P03! (Fig. 5.7au).

Syncolpate (Pas, Pbs, Pcs, Pxs)   Pollen with anastomosing colpi. In practice Pas and Pbs are not separable; one continuous colpus girdles the grain. The most common syncolpate is Pcs with a triangular polar colpal connection delimiting the polar area. Extant examples: *Nymphaea* (Pas or Pbs, also called zonisulculate), *Syzgium* and other Myrtaceae (Pcs) (Fig. 5.7av illustrates a syncolporate form; see also Fig. 5.7bd).

Heterocolpate (Pfs, etc.)   Grains that are partly -colpate, partly -colporate; that is, only some of the colpi have pores. Extant examples: *Combretum, Lythrum* (Fig. 5.7as).

Fenestrate (P03-Fe, etc.)   The suffix in the code designation signifies that this is an exceptional case. The basic morphological type can be P03, Pc0 or Pc3, but this is so dominated by very large "windows" in the ektexine that the germinal apertures (which the windows are not) are often hard to observe. This morphological type is common in the tribe Cichorieae of the Compositae, e.g., *Sonchus. Another example: Passiflora* (Fig. 5.7ax,ay). In *Passiflora* the "windows" completely dominate, and there is no sign of other morphological features. The code designation therefore is only Pfe.

Dyads (Pdy:_____)   The double code designation means that the pollen grains are shed from the anthers as doubles, pairs of grains united. This represents presumably incomplete breakup into individual grains or monads. This is not a common condition. The code designation Pdy is followed by the code for the individual grains, e.g., Pdy:P00 for the extant genus *Scheuchzeria*.

Tetrads (Pte:_____)   Grains shed in fours presumably are the unseparated product of meiosis. Although most pollen are shed as single grains ("monads"), tetrads are a common pollen type. The double code designation means that the code for the individual grain follows, e.g., Pte:Pc3 for many Ericaceae such as *Vaccinium*, Pte:P01 for *Typha*, Pte:P03 for *Gardenia*. A further complication is that there are a number of types of tetrads (see

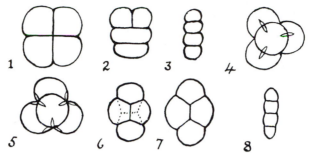

Types of pollen tetrads. 1-3, Rigid uniplanar tetrads due to early wall formation. 1, tetragonal ; 2, T-shaped 3-linear ; 4-8, mobile tetrads determined by surface tension. 4, Tetrahedral tetrad with six pairs of inter-radial colpi according to Fischer's rule ; 5, Tetrahedral tetrad with four groups of three radial colpi according to Garside's rule. 6-8, Tetrad configurations determined by surface tension and increasing pressure. 6, Decussate ; 7, Rhomboidal ; 8, Linear.

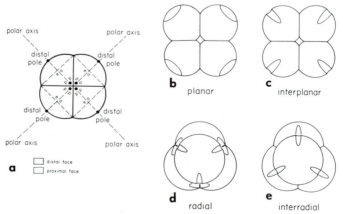

Relationships of pollen tetrads and aperture positions.—a. Diagram of a tetragonal pollen tetrad showing the polar axis, one equatorial axis (e.a.) of the equatorial plane, the distal face, the proximal face, the distal pole, and the proximal pole (p.p.) of each pollen grain in the tetrad.—b–c. Tetragonal tetrads showing placement of distal-polar, furrow-like apertures in planar and interplanar positions.—d–e. Tetrahedral tetrads showing placement of equatorial, furrow-like apertures in radial and interradial positions; top pollen grain shown in polar view with its polar axis perpendicular to the plane of the figure.

Figure 5.8 (1)–(8) Diagrammatic illustrations of kinds of pollen tetrad types (from Melville 1981); (a)–(e) aperture positions in some tetrad types (from Walker & Doyle 1975). Note that "d" below is equal to "5" above and "e" below to "4" above. That is to say, "Fischer's rule" refers to interradially and "Garside's rule" to radially arranged colpi in the tetrad.

Fig. 5.8). Only two types are common as fossils, tetragonal (e.g., *Typha*) and tetrahedral (e.g., *Vaccinium* and most other members of the family Ericaceae). Most tetrahedral tetrads have interradially placed apertures ("Fischer's rule": apertures located as though halfway between the ends of an imaginary triradiate mark). However, radially placed apertures also occur ("Garside's rule": apertures sited as though at the ends of radii of an imaginary triradiate mark) (Fig. 5.7ba–bc).

Obligate ("permanent") tetrads, those normally released from sporangia (anthers) in the tetrad condition, have a very long and interesting history. The very first spore-like bodies of latest Ordovician–early Silurian time are typically such tetrads. Obligate tetrads reappear regularly throughout the subsequent history of land plants. Krutzsch (1970) has summarized some of the literature on the subject of fossil tetrads.

Polyads (Ppd:_____) Grains shed in united groups, usually including multiples of four (16, 32, 64). Quite common in the mimusoid section of the Leguminosae. The code designation for individual grains follows that for polyad, but often the individual grains are so small and so closely packed that this is very difficult to determine. Feuer *et al.* (1985) demonstrate that in some legume polyads the ektexines are fused so that individual grains are not separately discernible. Extant example: (Ppd:Pbs?) *Acacia* (Fig. 5.7bd).

Note that, in a broad sense, the polyad designation includes *massulae* consisting of an irregular, large number of grains, and *pollinia*, in which the whole contents of an anther or anther locule may be shed as one united mass of pollen. Extant example: *Asclepias* (Fig. 5.7aw,az).

SUPPLEMENTAL NOTES ON MORPHOLOGY

Colpus This is really a furrow or "pleat" of the wall exine, part of which furrow may be membranous (thinned). The colpus thus serves as an accordion pleat to accommodate swelling of the grain (a "harmomegathic" function). It may also serve as a site for germination of the grain. Technically, a colpus is supposed to be located on a line connecting the poles and to cross the equator of the grain (see sulcus), but most palynologists use the term more loosely than this. Erdtman would call many of the "colpi" of a pericolpate grain *sulculi*, as they are not meridional, do not cross the equator, but are more or less parallel to it, or *colpoids*, if otherwise located.

Sulcus As colpus (see above), but distal and not crossing the equator of the grain. Typically one pole is located in the center of a sulcus. Most Pa0 grains are therefore technically monosulcate, not monocolpate, as the sulcus is a distal (polar) feature. However, "-colpate" and "-sulcate" are really used more or less interchangeably in practical palynology.

Tricolpate (Pc0), tricolporate (Pc3), stephanocolpate (Pn0) and stephano-colporate (Pnn) vs. pericolpate (Px$_n$0) and pericolporate (Px$_y$n$_y$) Tricolpate means, technically, having three colpi which must be separated in polar view by more or less 120°. Also, the colpi must be more or less on "meridians" of the grain, determined by theoretically projecting the surface of the grain onto a perfect sphere. The colpi must also be more or less bisected by the equator. If the colpi are not so arranged (are at angles to the equator other than 90°), or not bisected by the equator (are therefore mostly in one "hemisphere"), or are not separated as viewed from a "pole" by more or less 120°, the grains are pericolpate (Px_n0) or pericolporate (Px_yn_y). Stephanocolpate (Pn0) and stephanocolporate (Pnn) grains have colpi arranged regularly just as in tricolp(or)ate, but the number is four or more (which number is the "n" in the code formula). 6-Stephanocolpate (see illustration) and 6–stephanocolporate grains, for example, have the colpi on meridians 60° apart.

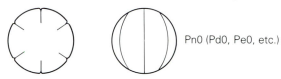

Pn0 (Pd0, Pe0, etc.)

Stephanocolpate

–colpate vs. –colporate vs. –colporoidate   –Colporate grains also have, in addition to the colpus, a further modification of the exine, usually a thinning, usually in the colpus, usually more or less in the equatorial region. This modification is often an ulcus or pore, but may be an additional, transverse colpus. Sometimes the main colpus and the equatorial colpus form a cross, i.e., are cruciate. –Colporoidate grains have a modification of the colpus, usually more or less equatorial, which is not a true pore or transverse colpus. Sometimes it is an ulcus, or it may be just a roughening, wrinkling, thinning, or other modification of the colpal membrane. (The fossil record supports the idea that –colporoidate is transitional from –colpate to –colporate.) Rarely, some of the colpi of a pollen grain are –colporate and others are –colpate. This rather rare, mixed condition is *heterocolpate* (Pf3, etc.).

Ulcus   (See Ulcerate – Pu1.) An irregularly thinned area of a pollen grain, apparently functioning as a pore. Cyperaceous pollen, for example, are characteristically ulcerate. There are often multiple ulci on ulcerate grains.

Pore

Pore vs. ulcus and colpus   The distinction between a pore and an ulcus is that pores are more or less uniform in size, shape, and distribution, and they have a membrane (usually thinner than the rest of the exine) that is a regular thickness. Ulci are irregular in size, shape, and distribution, and they consist of irregular patchy thinnings rather than having membranes of uniform thickness. The dimensions of a pore are such that no axis is greater than twice that of another axis:

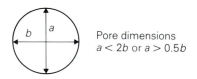

Pore dimensions
$a < 2b$ or $a > 0.5b$

Colpus dimensions
$a > 2b$ or $a < 0.5b$

Annulus and operculum   Pores may have a thickened rim, an annulus (= annular thickening). On the pore membrane may be a disk that thickens the membrane, called an operculum. The operculum is usually ektexinous, the rest of the membrane often endexinous.

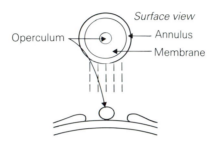

Operculum — Surface view

Annulus

Membrane

Vestibulum and os   (Plural, ora; adjectival form, -orate.) Sometimes an external, ektexinous pore is associated with an internal complex structure involving the endexine. If the internal structure involves an additional opening, this is an os. The space between the (external) pore and the back of the internal structure is the vestibulum. Some grains are triporate (P03, e.g., *Celtis*), others tripo(or)ate (also P03, e.g., *Betula*).

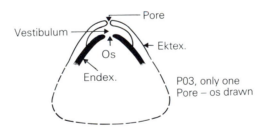

Vestibulum — Pore

Os

Ektex.

Endex.

P03, only one
Pore – os drawn

Triporate (P03) and stephanoporate (P0n) vs. periporate (P0x)

Triporate grains are presumably homologous to tricolpate grains. Therefore, the pores must be 120° apart as viewed from a pole. Also, the pores must be bisected by, i.e., lie upon, or nearly on, the equator. Stephanoporate grains are multipored versions of the same plan. P06 (stephanoporate) grains therefore have the pores on the equator and more or less exactly 60° apart. Arrangements of more than three pores off the equator are periporate (P0x). Characteristically, periporate grains have pores over much of the surface (example: Chenopodiaceae). However, *Juglans*, with only one or two pores off the equator, is technically P0x, as is even *Carya*, with three pores 120° apart but shifted into one hemisphere.

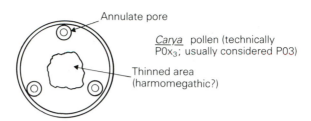

Annulate pore

Carya pollen (technically
P0x3; usually considered P03)

Thinned area
(harmomegathic?)

Margos Modified exinal margins of pores or colpi, usually thinnings or thickenings, are margos. Those associated with pores are usually called annuli.

Costa(e) and arci When there is a rib-like thickening in the exine underneath the edge of a colpus, this is called a costa. The thickening bands in the exine of *Alnus*, running from pore to pore, are not costae. They are arci.

Fenestrate This is a condition especially typical of certain Compositae pollen, e.g., *Sonchus*, in which there are large "windows" in the ektexine. The "windows" are neither a sculpturing type, nor a colpal type. Thus a fenestrate pollen grain can be Pc3 or P03, and echinate in sculpture, as well as fenestrate.

Polar area The degree to which the colpae of tricolp(or)ate or stephano-colp(or)ate grains encroach on the proximal or distal polar areas has diagnostic value. The size of the non-encroached-upon polar area is sometimes measured by a polar area index (*PAI*).

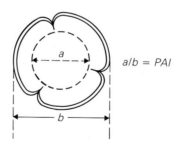

$a/b = PAI$

Polar vs. equatorial: You will find it helpful in understanding this to look at pollen of forms that are shed as tetrads, e.g., *Rhododendron, Vaccinium,* and *Gardenia.* P_1 and P_2 are poles. In the tetrad it is possible to tell that P_1 on the outside is the distal and P_2 the proximal (= toward center) of the tetrad, but this is not possible as a rule when the pollen occurs (as is usually the case) as monads. For this reason, "polar" and "equatorial" are used instead of distal–proximal and lateral for Pc0 and derived forms of dicot angiosperm pollen.

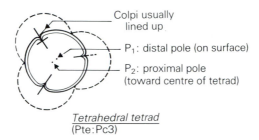

Colpi usually lined up

P_1: distal pole (on surface)

P_2: proximal pole (toward centre of tetrad)

Tetrahedral tetrad
(Pte:Pc3)

Tetrads and polyads Morphological analysis of these should, wherever possible, include analysis of the individual grains of the tetrad, e.g., Pte:Pc3, Ppd:Pc0. (See Fig. 5.8 for explanation of the various sorts of tetrads.)

Trichotomosulcate (loosely, "trichotomocolpate") (Pac) This is a variant of monosulcate in which the colpus is triangular instead of keel-shaped. It may even sometimes very closely approximate a trilete laesura in appearance.

or

Pac

Pac can usually be differentiated from Sc0, however, by the greater regularity of length and orientation of the radii of Sc0, and by the fact that the colpal margins of Pac are often more or less ragged or irregular compared to the edges of an open commissure of a spore's laesura.

Perine (= perisporium in spores) This is a sporopolleninous envelope outside the exine, rather loosely organized compared to the exine, and therefore not usually persisting in fossil sporomorphs. Perines do occur on some gymnospermous pollen grains, e.g., *Taxodium*, but are much more common on spores. Often, acetolyzed preparations of perisporate spores show scraps of the perine still adhering to the exine, and these are easily misinterpretable as exinous sculpture.

Amb (= "limb") It is often necessary to describe the "outline" of a spore or pollen grain. However, I see no reason to provide a chart of such terms, as triangular, oval, squarish, etc., all have well-understood meanings! You also should consult the shape classes chart (see Fig. 5.17).

Of course, the "outline" can be the outline seen in either equatorial (or lateral) view or polar (or proximal–distal) view. The outline seen in equatorial view is the one used in determining whether a grain is more or less oblate, more or less prolate, etc. In other words, the shape classes are of the shape around the polar axis. The shape seen in polar view is the amb/=outline of the equator or "limb". For example, this P03 grain is of peroblate shape with a triangular amb ($a/b=0.3$; according to Erdtman, $a/b< 0.5$ is peroblate). When not otherwise specified, *amb* refers to what is seen from a pole.

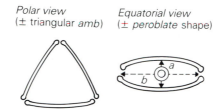

Polar view
(± triangular *amb*)

Equatorial view
(± *peroblate* shape)

Laesura   This is the "scar" which shows the contact of spores (or of some, mostly fossil, pollen grains) with their neighbors in the original tetrad from which they have separated. Laesurae can be (most common) trilete (Sc0 = "Y mark"), monolete (Sa0), dilete (Sb0; quite rare), or they can be absent (= alete, S00). (Yes, it is very difficult in some cases to distinguish S00 from P00, or even from some relatively featureless acritarchs or dinoflagellate cysts!) Note that trilete and monolete are adjectives. It is never correct to speak of "the trilete" when "the trilete laesura" is meant. Some palynologists use laesura to mean each arm or radius of a trilete laesura. They therefore use laesurae (plural) for a single contact figure (= laesura in this book).

A laesura has a center suture or commissure, which usually serves the purpose of providing a zone of weakness for rupture upon germination of the spore. There may also be modifications of the spore wall adjacent to the laesura, such as thickenings, usually called lips (labiae). However, there are many terms for other marginal modifications next to the laesura, such as kyrtomes, interradial thickenings, etc. The separate arms of a trilete laesura are called radii. The terminal ends of radii may be forked; these extensions are called curvaturae. If these connect with neighboring curvaturae to surround contact areas, they are curvaturae perfectae. Otherwise, they are curvaturae imperfectae. Curvaturae are much more evident in Paleozoic spores than in extant spores.

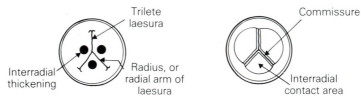

Trilete
laesura

Commissure

Interradial
thickening

Radius, or
radial arm of
laesura

Interradial
contact area

Sc0 with curvatura imperfecta   *Sc0 with curvatura perfecta*

Saccate pollen   Monosaccate (Pv1) and trisaccate (Pv3), as well as pseudo-saccate and cavate or camerate grains simulating saccates, are common as Mesozoic and Paleozoic fossils and they are not all coniferous pollen. However, all extant saccates are coniferous, and practically all are Pv2, e.g., *Pinus*. It is necessary to measure and describe the morphology and sculpture of the corpus (= "body") and sacci (= "vesicles" or "bladders") separately, as well as to measure the grains as a whole. (See Fig. 5.9 for further explanation).

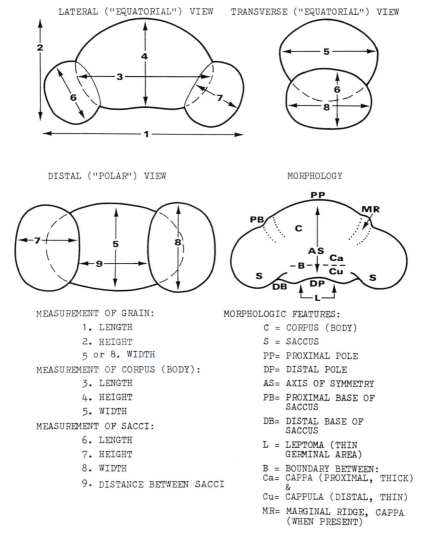

LATERAL ("EQUATORIAL") VIEW   TRANSVERSE ("EQUATORIAL") VIEW

DISTAL ("POLAR") VIEW   MORPHOLOGY

MEASUREMENT OF GRAIN:
    1. LENGTH
    2. HEIGHT
    5 or 8. WIDTH
MEASUREMENT OF CORPUS (BODY):
    3. LENGTH
    4. HEIGHT
    5. WIDTH
MEASUREMENT OF SACCI:
    6. LENGTH
    7. HEIGHT
    8. WIDTH
    9. DISTANCE BETWEEN SACCI

MORPHOLOGIC FEATURES:
    C = CORPUS (BODY)
    S = SACCUS
    PP= PROXIMAL POLE
    DP= DISTAL POLE
    AS= AXIS OF SYMMETRY
    PB= PROXIMAL BASE OF
        SACCUS
    DB= DISTAL BASE OF
        SACCUS
    L = LEPTOMA (THIN
        GERMINAL AREA)
    B = BOUNDARY BETWEEN:
    Ca= CAPPA (PROXIMAL, THICK)
        &
    Cu= CAPPULA (DISTAL, THIN)
    MR= MARGINAL RIDGE, CAPPA
        (WHEN PRESENT)

Figure 5.9 Bisaccate pollen present special problems in measurement because of their tripartite structure. This figure presents the solution to the problems suggested by the Soviet palynologist, Zauer (1977).

EXINE TEXTURE, SURFACE, AND SUBSURFACE: SCULPTURE AND STRUCTURE

Figure 5.4 shows diagrammatically the major features of an average dicot angiosperm pollen grain (monocot pollen differs in morphological features but is similar in wall construction). Note that the intine (cellulosic–pectic) layer of the wall or integument of the pollen grain is not preserved in fossils and, along with the protoplasmic contents, is also completely removed by acetolysis or by boiling in solutions of KOH in the laboratory. The exine consists of a framework of sporopollenin (plus other compounds in the interstices and on the surface, which are also lost in fossilization) and constitutes the shell or exine that is found as a palynomorph. The exine is most complex in angiosperms, but gymnosperm pollen exines have a similar makeup. Various pieces of evidence suggest that the exine is deposited on the basic cell of the spore under control of the "parental", not the spore, genome. For example, Rogers & Harris (1969) showed that aberrant *Linum* microspores, lacking essential chromosomal parts, turned into miniature pollen grains with a normal exine. (Chunks of sporopollenin produced by the tapetum of the anther or sporangium but not "used" in making exines often are shed as "ubisch bodies" or "orbicules" (Rowley 1963), which may even occur as Paleozoic fossils (Taylor 1976), showing that the basic mechanism of sporopollenin production and deposition in spores/pollen was well established over 300 million years ago.)

The exine of pollen usually has two layers. For the moment, we shall refer to these as ektexine and endexine, or sexine and nexine, without further comment, although we must deal with some complications later on. The inner layer (endexine or nexine) is relatively amorphous, and usually only its presence or absence is noted. It is frequently missing, is very thin, granular or may be present only in the vicinity of the apertures. The outer layer (ektexine or sexine), however, may be complex either internally or externally or both. Surface complexity, such as verrucae (warts), spines, etc., is referred to as sculpture. This is sometimes called "ornamentation". I dislike this, as implying an esthetic value or purpose, but "sculpture" perhaps offends also by suggesting a sculptor. There is further the problem that some see ontogenetic implications for *both* terms, sculpture being the result of removal, ornamentation the result of addition, of sporopollenin! I follow Faegri in distinguishing internal ektexine features as structure. With a light microscope, 100 × objective (thus about 1,000 × magnification), sculpture can be readily studied, but structure is hard to analyze in this manner and cannot be established or described with certainty without thin sections and transmission electron microscopy.

Sculpture: LO analysis and edge analysis

Table 5.3 is the classification, modified and expanded from Faegri & Iversen (1975), which I have always used in our laboratory. It is far simpler than the sculpture classifications used by others, and yet has served us well, with few problems requiring other special terms. One major exception to this statement is that Faegri's classification apparently regards 1 µm as the practical limit of accurate observation of sculpture. Therefore, e.g., if the surface of a pollen grain or spore is more or less smooth, but has holes less than 1 µm, it is still regarded as psilate (= laevigate). I have no trouble recognizing holes of 0.5 µm diameter as holes, and therefore designate this kind of sculpture as micropitted. Similarly, Faegri's classification excludes reticulate sculpture from that category if the elements of the muri of the reticulum are less than 1 µm in length, and echinate is not so recognized if the spines are less than 1 µm in length. Sometimes I can recognize such features quite well under oil immersion, and I get around the problem of the Faegri classification by calling these features microreticulate (this is a quite common type) and microechinate.

Erdtman, Faegri, and others have long ago described the possibilities of microscopic manipulation in understanding sculpture of exines. The problem is complicated, especially for beginners, by the fact that the sculpture is ordinarily present on a curved surface, and there is a background created by the underlying internal ektexine structure, the endexine, and the entire exine of the opposite side of the grain as well. (For aerobiologists, who insist on studying *whole* pollen, including the protoplasm, surface lipids and oils, and pollenkitt, the situation is desperate.) "LO analysis" (an erdtmanian term) refers to the possibility of distinguishing sculptural nuances by taking advantage of the fact that, as one focuses up and down in ordinary bright-field microscopy (90–100 × objective, oil immersion) on an exine surface, the "brightness" or "darkness" of features varies. LO analysis was introduced by Erdtman and the initials are from the Latin words *lux* (light) and *obscurus* (dark). For example, as one focuses down on a spine, the first impression as one encounters the spine is a bright point, but it becomes broader and darker as one focuses down. By contrast, a hole in the surface is first encountered as a dark point (see Fig. 5.10). I also teach beginning students to exploit "edge analysis". If LO analysis and general impression indicate small spines, always carefully focus up and down on the outer edge of the grain. If one's interpretation of spines is correct, no matter if they are only 0.5 µm long, they will show on the edge as this is silhouetted in mid-focus. Similarly, if one has interpreted pits or micropits, they will not project in mid-focus on the edge, and careful focus will usually show them as tiny channels through the ektexine. (Reticulate sculpture causes beginners special difficulty, as the muri of the reticulum often simulate spines or baculae in mid-focus of the edge; but of course, LO analysis of the surface will not show spines!) However, the misinterpretation of reticulate as echinate or baculate, and the misinterpretation ("reversal") of LO analytical observations (that is, negative for positive, and vice versa), are the most persistent problems I have had in teaching practical palynological microscopy to generations of Penn State students. Figure 5.11 shows diagrammatically what can be done with LO and edge analysis.

S.E.M study of exine surfaces is the most sophisticated manner for sculpture study. As shown in Figure 5.12, S.E.M micrographs even offer the possibility of separating species of genera, and genera of difficult families such as Gramineae or Chenopodiaceae. The only drawback is that the method requires

Table 5.3 Principal sculpturing types for spores and pollen.

A. Positive sculptural elements absent
  B. Surface smooth   *psilate*
  BB. Diameter of pits <1 μm   *micropitted*

  BBB. Surface pitted, diameter of pits ≥1 μm   *foveolate*
  BBBB. Surface with grooves   *fossulate*
    (includes *negatively reticulate*)

AA. With positive sculptural elements
  B. Sculptural elements approximately isodiametric along the surface of the
    palynomorph (but may extend upward)
    C. No dimensions ≥1 μm   *scabrate*
    CC. At least one dimension ≥1 μm
      D. Sculptural elements not pointed
        E. Lower part of element constricted
          F. Greatest diameter along surface of palynomorph equal to or greater than height
            of element: elements globular   *gemmate*

          FF. Height of element greater than greatest diameter of projection: elements
            club-shaped   *clavate*

        EE. Lower part of element not constricted
          F. Greatest diameter along the surface of palynomorph equal to or greater than
            height of element: elements wart–like   *verrucate*

FF. Height of element greater than greatest diameter of projection: elements rod-shaped .... *baculate*

DD. Sculpturing elements pointed .... *echinate*

BB. Sculptural elements elongated along the surface of the palynomorph (length at least twice the breadth)

  C. Elements irregularly distributed .... *rugulate*

  CC. Elements approximately parallel to each other .... *striate*

BBB. Sculpturing elements forming a reticular (net-like) pattern (elements <1 μm but resolvable as reticulum: *microreticulate*) .... *reticulate*

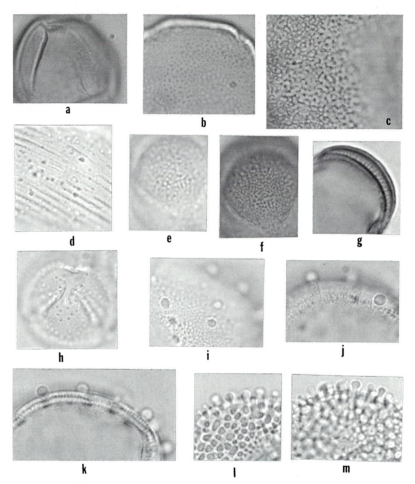

Figure 5.10 Photomicrographs (2,400 ×, oil, except as noted) of extant plant principal spores/pollen sculptural types. Many of these types are easier to illustrate with S.E.M. pictures, but as most users of this book will have access on a regular basis only to a light microscope, it is more useful to illustrate what they will see. Elements of structure are also illustrated. (a) Psilate (= laevigate): *Phoradendron serotinum* (Raf.) M. C. Johnston, Texas. High focus. Some texture is observable; a truly smooth exine surface does not exist. (b) Psilate(!), internally reticulate (= intrareticulate): *Pterocarya stenoptera* D.C., China. Low focus of surface. (c) Foveolate (and rugulate!): *Alangium platanifolium* (S. & Z.) Hanus, China. High focus. The "holes" are observable as black spots. (d) Fossulate (see striate!): *Schizaea digitata* S.W., Philippines. High focus. Fossulate sculpture refers to the grooves, i.e., negative sculpture. Obviously, striate sculpture, referring to the ridges, is more or less indistinguishable. (e) Scabrate: *Artemisia douglasiana* Bess., California. High focus. The bright spots are scabrae at high focus. (f) As (e). Low focus. Some of the dark spots are scabrae that are bright in high focus (LO analysis!). (g) As (e). Mid focus. Shows columellae in side view, under the

96

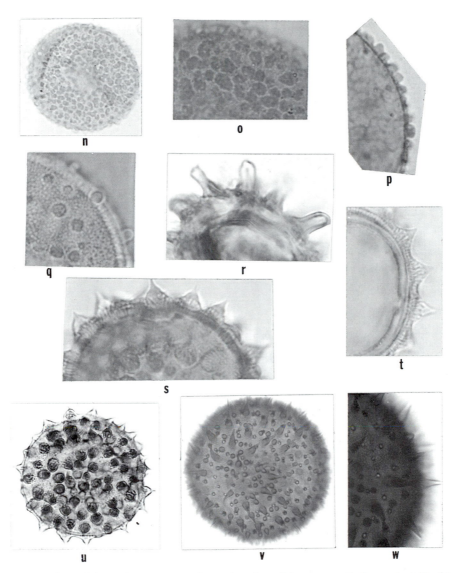

tectum. Hints of the scabrae are seen along the edge of the tectum ("edge analysis"). (h) Scabrate: *Cleome spinosa* Jacq., Venezuela. High focus. These scabrae are at the limit of resolution (1 µm). (i) gemmate (plus micropitted): *Hyphaene crinita* Gaertn. (cult.), South Africa. High focus. (j) As (i). Lower focus, near edge. (k) As (i). Edge focus, showing gemmae and columellae of ektexine. (l) Clavate: *Ilex cassine* L., Florida. High focus. (m) As (l). Lower focus, showing several clavae on the edge and the dark appearance of claval stalk in low focus. (n) Verrucate: *Sciadopitys verticillata* Sieb. & Zucc. (cult.), 1,000 ×. High focus. (o) As (n). Slightly lower focus at 2,400 ×. The verrucae are internally complex. (p) As (n). Low focus of edge (edge analysis!), confirming that major sculpture type is verrucate. (q) Gemmate–baculate (plus pitted): *Hyphaene thebiaca* (L.) Del. (cult.), Curaçao. Mid focus. Illustrates that distinction

97

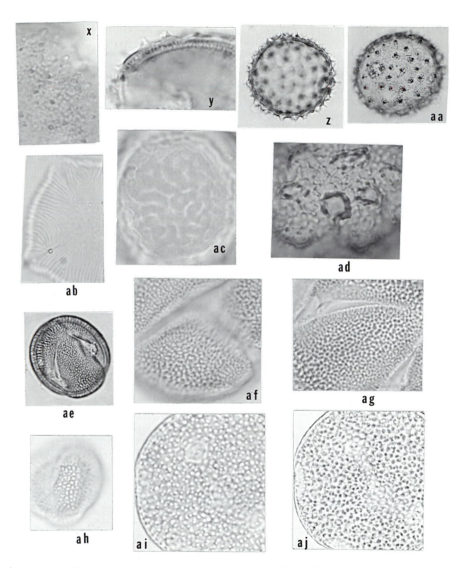

between sculpturing types may be blurred: some of the elements are gemmae, some bacula. (r) Baculate (perispore): *Cystopteris fragilis* (L.) Bernh., Mexico. Mid focus. This sculpturing is perisporal, but as sculptural terms are purely descriptive of external texture, the same terms are used as for exine. (s) Echinate: *Sphaeralcea lindheimeri* Gray, Texas. Mid focus. Note that echinae are biform (broad base narrowing abruptly to slender spine) and that the columellate underpinning of the spines is evident in edge analysis. (t) Echinate: *Senecio ampullacens* Hook., Texas. Mid focus. See (s), an unrelated taxon. Echinae not biform, but columellate ektexine under the spines is evident. (u) Echinate: *Abutilon incanum* (Link) Sweet, Texas. High focus, 1,000 ×. (v) Echinate (more or less biform and heteromorphic): *Althaea rosea* Cav. (cult.), high focus, 600 ×(!). As is common for malvaceous pollen, this is a huge form. See (w). (w) As (v), 1,000 ×. The exine is so thick that at higher magnifications it is hard to photograph.

considerable additional pretreatment of specimens and ready availability of a good scanning electron microscope. Relatively few palynologists really consider S.E.M as routine.

T.E.M. (transmission electron microscopy) offers additional possibilities for elucidating problems of sculpture and structure interpretation, because thin sections displaying internal features are studied, whereas S.E.M. can elucidate only superficial features, except on edges of broken surfaces. As an illustration of the nuances of sculpture amenable to study with electron microscopy, Bolick *et al.* (1984) and Salgado-Labouriau (1984) debate the interpretation of tiny holes in the *spines* of composite pollen!

Interference contrast (Nomarski) microscopy requires special condensers and objectives for the bright-field light microscope and provides an S.E.M-like sort of contrast that is very useful for thin-walled palynomorphs that appear featureless in ordinary bright-field microscopy.

Structure

Angiosperm pollen exines apparently developed their ektexine layer in the course of evolution by the terminal fusion of granular and rod-like elements placed upon the endexine. These elements have variously fused in evolution, so that the outer surface of most angiosperm pollen is really a secondary surface. The pioneers of palynological microscopy long ago recognized this and called the new surface, where it really does comprise a cover, the tectum.

The picture demonstrates the several different sizes of spines and that a fine microsculpture exists between spines. (x) Echinate: *Valeriana officinalis* L., Sweden. The echinae appear in mid-focus (of the surface) as dark blobs. The columellate structure of the ektexine shows through as a confused pattern. (y) As (x). Mid focus of grain, showing biform echinae and columellate structure. (z) Echinate: *Sphaeralcea angustifolia* (Cav.) D.Don., Texas. Mid focus, 1,000 ×. Note biform echinae and their appearance in mid-focus. See (aa). (aa) As (z), 1,000 ×. Higher focus to show the surface appearance in optical section of the columellae of the echinae bases (LO analysis!). (ab) Striate (see fossulate!): *Cuphaea cordata* R. & P., Peru. High focus. As explained for (d), striate forms and fossulate forms are hard to distinguish from each other and really depend on whether the grooves (fossulate) or the ridges (striate) are emphasized. Striate as a morphological type has broader bands (taeniae) and should perhaps be called taeniate to limit confusion with striate sculpturing. (ac) Rugulate: *Ulmus scabra* Mill. (cult.). High focus. (ad) Rugulate (perispore): *Dryopteris cristata* (L.) A. Gray, Germany. High focus. Note variation in size of rugulae. (ae) Reticulate: *Gentianella amarella* L., Finland. High focus, 1,000 ×. Reticulate sculpture is often easier to recognize at lower power because at high magnification the structure underneath may confuse. See (af) and (ag). (af) As (ae), 2,400 ×. High focus. Reticulum consists of ektexinous blocks fastened together. (ag) As (af). Lower focus of surface showing the reticulum and (small dark points) columellae under it. (ah) Reticulate: *Sinapis nigra* L., Pennsylvania. High focus. Large pattern reticulum on small pollen grain. (ai) More or less psilate(!): *Saponaria officinalis* L., New York. High focus. See (aj). The bodies that appear to be gemmae are instead columellae, of which the "heads" appear as bright spots in high focus. (aj) As (ai). Lower focus of surface. The dark spots are not gemmae but columellae. Illustrates the importance of "edge analysis": edge is entire or may have depressions.

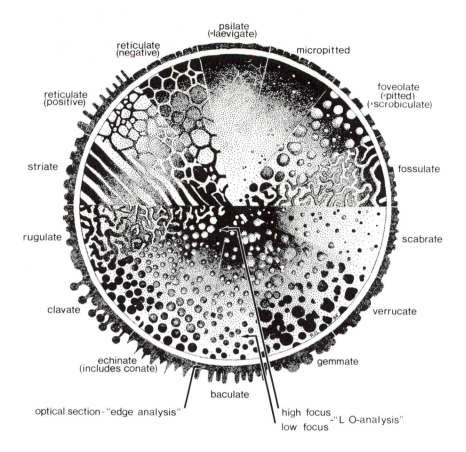

Figure 5.11 Sculpture types as seen at various levels of focus. Low focus is shown toward the outside of the grain, high focus toward the inside. Analysis of sculpture by comparing these levels is called "LO analysis", following Erdtman. For example, scabrae appear as bright spots at high level of focus and become dark as one focuses down through them. Pits appear dark in high focus and become brighter as one focuses down. Edge analysis (focusing on the surface of the exine at the outer edge of the grain, in mid-focus) provides a check on conclusions drawn from LO analysis. Clavae, for example, will show up clearly. On the other hand, beginners will often identify the muri of reticulate sculpture seen in such side view as bacula or echinae. Edge analysis must always be checked by LO analysis.

(Erdtman's use of tegillum for one kind of such surface (<80% coverage) and tectum for others (>80% coverage) is not a helpful distinction and has not been much used.) When a tectum exists, the sculpture, of course, sits atop it or comprises the surface features of the ends of its elements. These elements are usually basically rod-like (columellae), but they may be broadened below the top or arranged in various complex patterns internally. The interstices

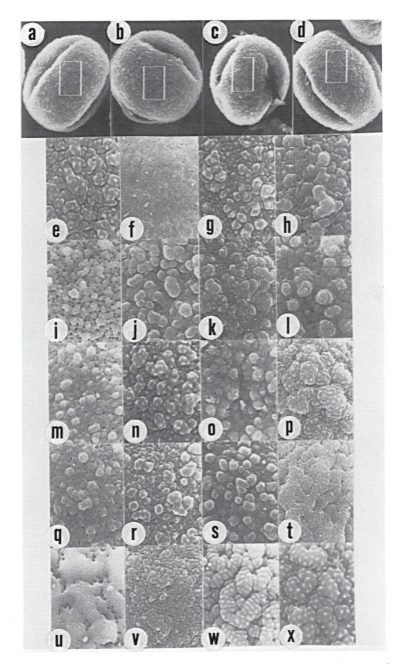

Figure 5.12 (a)–(d) Lower-magnification (ca. 500 ×) S.E.M. pictures of several species of oak pollen, to show the sizes of the blocks illustrated below. (a) *Quercus incana*; (b) *Q. pumila*; (c) *Q. imbricaria*; (d) *Q. laurifolia*. (e)–(x) Surface features of pollen

101

between columellae are filled with various substances, such as recognition-signal compounds and lipids in the living pollen. Muller (1984) has observed that pollenkitt may be present on the surface of pollen and make it sticky, but if deposited in tectum cavities it makes the pollen powdery. Thus, columellate reticulate pollen is usually but not always adapted to insect pollination.

These internal patterns of the ektexine comprise the structure (see Figs. 5.13–5.15). Often the terms for structure are quite similar to those for sculpture, and the prefix "infra-" is added to the sculptural terms to make it clear that the terms refer to internal matters, e.g., infrareticulate. Truth to tell, structure is very difficult to study with light microscopy and most often all one can ascertain is that there is structure, that columellae exist (often visible from edge analysis), and perhaps that the pattern is infrareticulate, infrastriate or whatever.

Additional notes on sculpture/structure

Reticulate vs. pitted (or microreticulate vs. micropitted!)   Students often have trouble differentiating these. In nature, there actually is a gradation. For convenience I consider porous sculpture with the area consisting mostly of holes (lumina) and less than half of solid walls (muri) as microreticulate or reticulate. Where there is more area of solid wall than of holes, we speak of micropitted or foveolate, depending on size of the openings. Some students have found it useful to compare the two situations with a sieve ("strainer"), which is reticulate (comprised mostly of holes), and a colander (comprised mostly of a solid surface, with more or less widely spaced holes), which is foveolate.

Mixed sculpture   This is very common and often confusing, requiring careful microscopy to understand. In the Devonian, spores with biform appendages, e.g., sharp spines on top of mamillate protuberances, are common. In angiosperm pollen, it is common in Compositae to have spines (echinae) on top of a reticulum. Clavae can also occur on top of a reticulum, e.g., in *Neobuchia*, in the Bombacaceae. It is common to have a mixture of different sorts of sculptural elements, e.g., gemmae and scabrae in *Grevillea banksii*, in the Proteaceae.

---

exines of a series of species of oak (*Quercus*, subgenus *Lepidobalanus*), as shown by S.E.M., magnification about 5,000 ×. Such sculptural differences of the exine in some instances permit identification of species or species groups in genera and of genera in families such as grasses, in which separation even of genera is difficult in light microscopy. In light microscopy the exines of oaks are usually too similar for routine separation. (e) *Quercus velutina*; (f) *Q. virginiana*; (g) *Q. velutina*; (h) *Q. palustris*; (i) *Q. georgiana*; (j) *Q. prinus*; (k) *Q. durandii*; (l) *Q. macrocarpa*; (m) *Q. lyrata*; (n) *Q. marilandica*; (o) *Q. bicolor*; (p) *Q. myrtifolia*; (q) *Q. ellipsoidalis*; (r) *Q. velutina*; (s) *Q. macrocarpa*; (t) *Q. virginiana* var. *minima*; (u) *Q. georgiana*; (v) *Q. virginiana* var. *minima*; (w) *Q. phellos*; (x) *Q. laurifolia*.

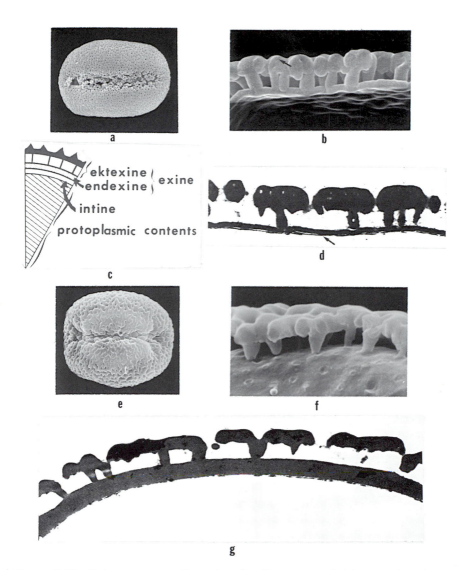

In the figure, the following labels appear:

a

b

ektexine }
endexine } exine

intine

protoplasmic contents

c

d

e

f

g

Figure 5.13 Exine structure of acetolyzed pollen, as revealed by scanning electron microscopy (S.E.M.) and transmission electron microscopy (T.E.M.) of two pollen grains of the Myristicaceae. (a) Distal view S.E.M. (2,500 ×) of pollen of *Virola elongata* (Benth.) Warb., showing sulcus and reticulate sculpture. (b) Same grain, S.E.M. view of "cut face" of exine (16,000 ×), seen from the side. Compare with (c). The barrel-like columellae extend from the nexine to the outside, where the heads ("capitals") form a "roof" or, in Latin, tectum (arrow). (c) Diagram of section of whole pollen grain. Only the exine survives acetolysis or fossilization. (d) T.E.M. of thin section of same grains as (a) and (b). Arrow points to nexine (17,000 ×). (e) Distal view S.E.M. (2,500 ×) of pollen of *Compsoneura capitellata* (A. Dc.) Warb., showing sulcus and rugulate–reticulate sculpture. (f) Same grain, "cut-face" S.E.M. view of exine showing "piano-leg" columellae with "capitals" fused to form the tectum. The narrowed bases of columellae pull away from the endexine leaving depressions (16,000 ×). (g) T.E.M. of thin section of same grain, showing same features as (f) (14,000 ×).

Spores/Pollen Wall Stratification per Erdtman:

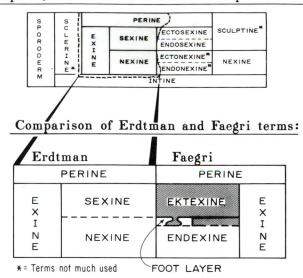

Comparison of Erdtman and Faegri terms:

* = Terms not much used   FOOT LAYER

Figure 5.14 The Erdtman wall stratification classification and comparison of Erdtman and Faegri terms for wall stratification. In practice, only the terms in the lower diagram are much used. Note that as Erdtman's terms are purely locational, "foot layer" is only required in the Faegri classification, where the ektexine–endexine distinction is drawn on observed, e.g., by staining, chemically based differences. On the other hand, sexine–nexine terms do not imply that such difference has been proven and are often therefore more in keeping with observation. The foot layer is a basal part of the ektexine, lying either on top of the endexine or interdigitating with it, as shown. If sexine–nexine terminology is used, the foot layer is best regarded as the top of the nexine. Many palynologists prefer to use "exospore" (= exosporium) and "perispore" (= perisporium) for the layers in spores similar to exine and perine for pollen. The term corresponding to intine would then be "endospore". (However, this term is also used in another sense: see Glossary.)

Additional notes on spores/pollen wall morphology

Differences between groups of embryophytes   Most of what has been presented so far about general wall morphology has to do with dicot angiosperm pollen. Monocot angiosperm pollen, though mostly based on a monosulcate pattern instead of a tricolpate pattern, does not differ enough in sculptural patterns to require separate treatment. Gymnosperm pollen is another story. Though some gymnosperm pollen approaches a columellate condition (e.g., the Mesozoic fossil, *Corollina*), the exine mostly lacks the angiosperm sculpture/structure difference that largely depends on columellate structure. Saccate gymnosperm pollen frequently has very different saccus sculpture from that of the corpus, and the corpus sculpture may differ markedly on proximal and distal surfaces. Pronounced positive sculpturing

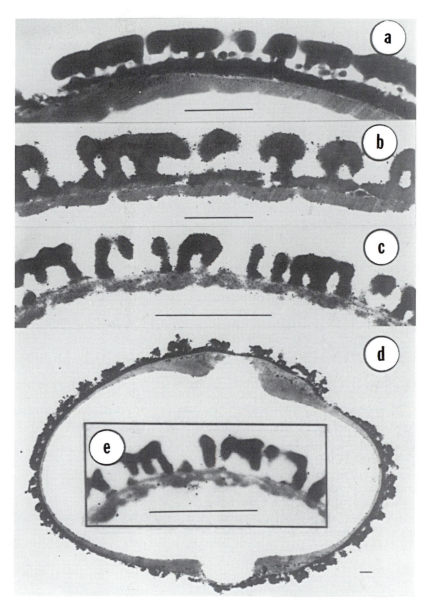

Figure 5.15 Transmission electron microscope (T.E.M.) demonstration of structure of exine of pollen of extant Primulaceae. (a) *Naumbergia thyrsiflora*: thick tectum supported by reduced columellae. Note well-developed foot layer and prominent endexine (see Fig. 5.14). (b) *Omphalogramma vincaeflora*: columellae massive. (c) *Primula officinalis*: no foot layer. (d) *Stimpsonia chamaedryoides*: oblique section of whole grain. Note endexine thickening adjoining apertural areas. Some angiosperms have endexine only in these areas. (e) *Primula veris*: no foot layer. Scale lines = 1 μm.

such as echinate or baculate is rare in gymnosperms but does occur. Sacci are not hollow sacs, but have a lining consisting of a meshwork of strands; according to the definition, this is structure. In some gymnosperm pollen, e.g., commonly in the Taxodiaceae, a perine (= perisporium) occurs external to the exine, sloughing off rather easily and therefore not usually found on fossils. The perine is sporopolleninous, however.

Ferns and other non–seed–bearing embryophytes have relatively homogeneous exines, not displaying the pronounced sculpture/structure distinction of angiosperms. However, the sculpture terms used for angiosperms can be applied without real difficulty. The most strikingly different feature of these spores is the prevalence of perines, the outer wall layer just mentioned as present in some gymnosperms. The perine often sloughs off partially or entirely in acetolysis preparations and can be presumed to do the same during fossilization. On the other hand, some perines are tough and adherent and survive acetolysis. The outer wall layer of some fossil spores may be perinous. The ektexine of angiosperm pollen could perhaps be regarded as a sort of perine, but this is very speculative.

Wall stratification   As explained earlier, I use routinely the terms of Faegri & Iversen (1975) for the wall layers. This scheme of terminology was developed mostly for angiosperm pollen. Faegri's endexine is described as differing chemically from the ektexine, the latter containing more dense sporopollenin and staining more deeply. Stained angiosperm exines thick enough to permit observation of this feature show the two layers only at ca.400× or more magnification. For that matter, the two layers can be seen also in unstained, acetolyzed preparations, because acetolysis affects the two layers differently. Erdtman used instead the terms sexine and nexine, purely "geographic" terms with no clearly defined fixed boundary between them: the sexine is above and has sculptural elements, the nexine is beneath and does not. In practical palynology, it might have been better to use Erdtman's sexine and nexine because by so doing one is admitting that he/she has not necessarily observed the chemical differences between ektexine and endexine. As used by most people, the two sets of terms are more or less synonymous. A complication is that T.E.M work has substantiated early light microscopy suggestions that the ektexine may have "roots" that extend down into the endexine, or this lower ektexine may be a distinct layer below the columellae called the foot layer, and the Erdtman classification has no provision for this. The sexine in Erdtman's terminology cannot by definition occur below the arbitrary sexine/nexine boundary, i.e., below the structural elements. It should be noted, however, that some palynologists use "foot layer" as if it were the bottom layer of the sexine. Erdtman's terminology for wall layers also is handicapped by excess baggage. The sexine is further subdivided into a theoretical ectosexine and endosexine, and the nexine into ectonexine and endonexine, which terms are really not useful or necessary in practical palynology. (As noted by Misset et al. (1982), the stratification in the area of apertures frequently is different from that away from them. The aperture membrane may have more of one layer, be lacking one layer entirely, or have one layer greatly modified.) Erdtman's term "sporoderm" for the whole pollen wall, including perine, exine, and intine, is, however, rather useful. Some authors have also found it advantageous to speak of the sclerine, when they are sure a fossil wall is either exine or perine but not sure whether both are present. Figure 5.14 shows the correspondence between the usable erdtmanian terms and those of Faegri (see also Fig. 5.13).

Some palynologists carefully avoid using the "-ine" terms for spores. Instead, for spores the terms equivalent to exine and perine are exospore (= exosporium) and perispore (= perisporium). This makes some sense because it is really difficult to be sure which layer of a pollen wall is homologous to a fern spore perispore, for example. It should also be emphasized that studies of Paleozoic and Mesozoic spores, pollen, and prepollen show wall layers that are not readily homologized with extant pollen wall layers. Abadie *et al.* (1977), for example, studied exines of the medullosan prepollen *Schopfipollenites* by S.E.M, T.E.M., and light microscopy and showed the presence of four layers, which are difficult to assign to ektexine and endexine. Among modern polypodiaceous ferns it has been shown that the perisporium is universally present, and this may even be true of ferns in general, including fossils, for which, however, the perisporium is unlikely to be preserved (Van Uffelen 1984, Tavera 1982, Moy 1986).

It has been suggested (Kress & Stone 1982) that monocotyledonous angiosperms have no acetolysis-resistant endexine at all, or have it only in the apertural region. That is to say, monocot acetolysis-resistant exine is practically all ektexine. However (see Guedes 1982), ektexine–endexine or sexine–nexine stratification seems general in the spores/pollen of vascular plants, based on ontogeny of the exine, and the absence of an ektexine–endexine separation where it occurs must be exceptional. The typical situation for angiosperms is for a columellate ektexine to support a superficial tectum, whereas in gymnosperms the outer exine is usually spongy (alveolar). The endexine of angiosperms is relatively homogeneous, whereas the inner exine of gymnosperms is typically laminate. A few gymnosperms and some angiosperms do not follow this pattern but have more or less granular exine structure. Taylor's (1982) study of Carboniferous medullosan seed fern prepollen (*Schopfipollenites* = the pollen of the pollen-bearing organs, *Potoniea*, and others), in comparison with extant cycad pollen, shows that the developmental stages by which the exine layers are formed are complex and can be very different. Extant cycad pollen apparently produces an outer sexine layer first, and a lamellate nexine last, whereas *Potoniea* pollen produced a lamellate nexine first, then a sculptured sexine. Zavada (1983) has shown that for *Zamia*, an extant cycad, sexine development begins in the tetrad phase and proceeds centripetally. This phase is followed by nexine development. All of this is quite in contrast to angiosperms, where a gametophytically controlled primexine is formed, perhaps providing a template for later sporopollenin deposition. Protosporopollenin is then deposited, a compound somewhat different from normal sporopollenin. Next, in angiosperms, the sporophytic tapetum of the anther is very active in altering the protosporopollenin of the exine, depositing sporopollenin and depositing tapetally derived compounds in the interstices of the exine. These compounds may be abundant and have to do with recognition and self-incompatibility. Thus, the sporophyte genome governs the structure and function of angiosperm pollen to a far greater extent than is true of gymnosperms, which at least typically lack recognition systems in the pollen exine. Mesozoic *Corollina* (= *Classopollis*) pollen was produced by coniferous plants, has complex columellate structure, may have had tapetal control as in angiosperms, and has recognition compounds in interstices, for reasons as yet not understood, as discussed by Taylor & Alvin (1984).

SPORES/POLLEN ORIENTATION AND SHAPE

Some special aspects of the shape and organization of spores/pollen of various groups will be dealt with later. The basic terms needed for written and oral description of spores/pollen shape and orientation are shown in Figures 5.6, 5.9, 5.16 & 5.17. They apply only to embryophyte spores/pollen.

Spore orientation and polarity

Embryophyte spores are normally produced in tetrads, the end products of meiotic division. The spores normally separate, and the individual fossil spores occur in sediments as monads. The orientation of such spores is given with reference to the center of the (usually not seen!) tetrad. The proximal surface of the spore is the surface that is or was toward the center of the tetrad, whereas the distal surface is away from the center. The haptotypic features, having to do with relics of the contact between the members of the tetrad, are found on the proximal face. The laesura is the main such feature. The basic laesura type is trilete, with three radii or arms. The monolete condition, which appeared much later in the fossil record and is still much less common, might be viewed as derived from the trilete condition by "loss" of radii, though such "loss" does not actually occur in ontogeny! The laesura may be complex, with thickened edges or other features such as raised lips alongside the main ridges (commissures) of the radii. The laesura, because it is a zone of weakness, also usually subsumes the function of dehiscence of the spore for germination, and the auxiliary features of the laesura probably are mostly related to this function. If a trilete spore is seen sideways, that is, from a lateral view, the area including the laesura will, if ideally preserved, appear more or less tent-like, and the distal surface thus seen will be more or less rounded. In practice, fossil spores are usually collapsed and folded and are only with difficulty "restored" in one's mind to the ideal shown in Figure 5.16. Monolete spores seen laterally are ideally the shape of a section of an orange. It is not best usage to refer to the proximal and distal surfaces of spores as "polar", because it is more precise to speak of them as proximal or distal. It is best to reserve the adjectives "polar" and "equatorial" for tricolpate pollen and derivations thereof, where we must use these terms because it is not possible to determine what is proximal or distal. However, the center of the proximal and distal surfaces can be and often are referred to as poles, and the term equatorial can in most cases be used without objection as synonymous with lateral.

108

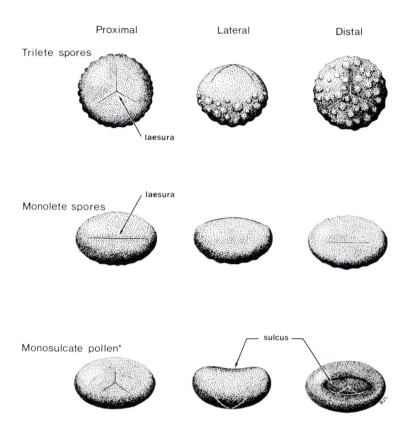

Trilete spores

laesura

laesura

Monolete spores

sulcus

Monosulcate pollen*

*laesura normally present only in prepollen

Figure 5.16 Orientation of spores and of monosulcate pollen (as well as forms obviously derived from monosulcates, such as monoporate and trichotomosulcate) is described in terms of the situation in the original tetrad. The side that was next to the center of the tetrad is the proximal side, the side away from it is the distal side. The scar on the proximal side of a spore, representing attachment to the other members of the tetrad, is the laesura, formed by intersection of the contact faces of the spore where it was against the other three members of the tetrad. While it is accurate to call the center of the proximal and distal sides of such grains the "poles", the use of "polar view" should be reserved for angiosperm pollen, where proximal and distal cannot be determined. One should be more precise, in calling polar views either proximal or distal where that is possible (trilete spores, monosulcate pollen, etc.). Monosulcate pollen has the sulcus on the distal side. A laesura is not normally present on the proximal side of pollen except in the case of extinct, primitive fossil pollen (= "prepollen"). "Monocolpate" is often used as synonymous with monosulcate, but monocolpate is better reserved for instances where the aperture is not obviously distal. As defined by Chaloner (1970b), prepollen has *only* the proximal laesura and no other aperture, but others have shown that a laesura and a sulcus can coexist, as would indeed be expected in the transition to distal germination. All of the forms illustrated here are models, not illustrations of actual specimens.

Pollen orientation

Primitive pollen from the Paleozoic ("prepollen") has haptotypic features like those of spores. For these prepollen grains the same observations as given under spore orientation apply. When the sulcus first appeared, it was always developed on the side away from the laesura, i.e., the distal surface, and had the function of permitting the development of a pollen tube whose function was presumably only haustorial. Germination to allow escape of the functional fecundating elements (spermatozoids) remained a function by dehiscence of the laesura on the proximal surface. With few exceptions, extant monosulcate pollen no longer bear a laesura, but the sulcus, from which the pollen tube emerges to serve both haustorial and fecundative functions, remains always on the distal surface. As is true for all spores, it is also not good usage, though not incorrect, to refer to the proximal and distal faces of monosulcate pollen as "polar" – proximal and distal are more precise. Monosulcate pollen can be spoken of as heteropolar, and tricolpate pollen as typically isopolar, whereas inaperturate or multiaperturate pollen grains are usually apolar.

With the advent of the dicotyledonous angiosperms, and tricolp(or)ate and triporate pollen, a more complex situation developed. Some families, such as Ericaceae, release their pollen in tetrahedral tetrads, the typical dicot tetrad. In these instances the orientation can be observed. One pole of each grain is on the outside, directly opposite the center of the tetrad, and the other is toward the center, 180° from the other pole. These are therefore the distal and proximal poles. However, once the individual pollen grains are released from the tetrad, the typical situation at anthesis, it is no longer usually possible to tell which of the poles is proximal, which distal. Therefore, the grain is described with reference to what can be observed: the poles and the colpi and/or pores. Pollen is seldom if ever perfectly spherical, but is described as if it were (see Fig. 5.17). Views from either pole are polar. Views from the side are not usually spoken of as lateral (though this is not wrong) but equatorial. Thus "polar" can be either proximal or distal. An equatorial view is, of course, a kind of lateral view. But it is best to keep the terminology clear by using polar and equatorial only for tricolpate pollen and its derivatives. The polar amb or limb (= outline) is what is seen in an equatorial view, and an equatorial amb or limb is what is seen in a polar view! The latter is what is meant by "amb" unless otherwise specified.

Orientation in illustrations of spores and pollen

Although Erdtman tried to establish standards in orientation of illustrations, he has been by no means universally followed. However, most of Erdtman's standards are sensible and might as well be adopted. Specific comments are as follows:

Trilete spores   In a proximal view, or a distal if the laesura shows through, one radial arm of the laesura should according to Erdtman point up as shown:

110

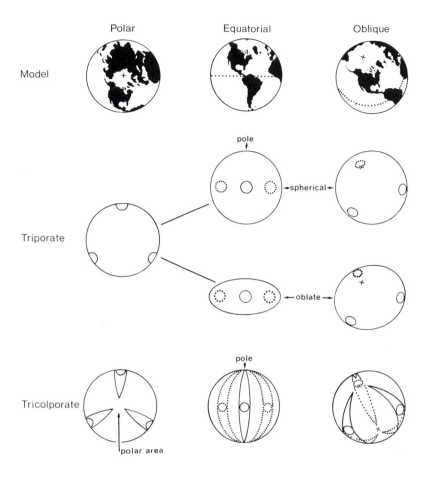

Erdtman Shape Classes–Equatorial View

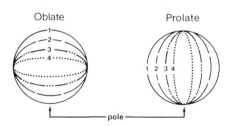

Figure 5.17 Description of the orientation of tricolporate and triporate pollen and multiaperturate forms of similar plan (4-stephanoporate, 4-stephanocolporate, for example) is a special problem because it is not usually possible to determine which part of the grain was proximal or distal with respect to the original tetrad. Instead, the proximal and distal sides are more or less identical. The centers of these two sides are termed "poles", and the other features are described in reference to them, e.g., an

111

In lateral views, the laesural prominence should logically (it is proximal) be down

but few authors follow Erdtman on this. Fortunately, this is not often a problem, as proximal–distal views are overwhelmingly more often illustrated.

Monolete spores   Erdtman illustrates the laesura as vertical in proximal–distal views

and follows logic in having the laesura down (proximal!) in lateral views

However, few authors follow Erdtman:

     and

are much more common in the literature.

---

imaginary line equivalent to the Earth's equator is called the equator. The pores of a triporate (P03) grain and a tricolporate grain (Pc3), for example, lie approximately on the equator by the definition used here. (If it can be shown that a three-pored grain has the pores elsewhere, it is 3-periporate.) In a view from the pole, as of the triporate grain on the left of the figure, the equatorial rim is seen as the "amb" or "limb" outline of the grain. It is best to reserve "polar" and "equatorial" for such multiaperturate pollen, where proximal and distal cannot be determined, and to use proximal and distal wherever it is possible to know. In a monosulcate grain such as *Magnolia*, for example, the sulcus is distal. Many palynologists, however, call the center of the proximal and distal sides of such a grain "poles". In any case, the line around the grain equidistant from these "poles" *is* the equator. Erdtman's shape classes shown in the lower part of the figure are handy for describing the shape of pollen as seen in outline, when the axis of the poles is perpendicular to the line of sight. However, fossil grains are flattened like little pancakes, and oblate grains, which normally are flattened pole-to-pole, can only be viewed from a pole, and strongly prolate grains can only be viewed in equatorial view, as of a tiny flattened American football. The shape classes therefore are best used only for relatively unflattened, non-fossil pollen.

Monosulcate pollen According to Erdtman, as the sulcus is distal, one would expect that it should be up in a lateral view

but he is very inconsistent in this. In distal views, the most common, Erdtman orients monosulcates with the sulcus vertical, as for laesurae of monolete spores

But most authors illustrate monosulcates with sulcus parallel to top and bottom of page, whether lateral

or proximal–distal

Bisaccates As the cappula is distal it should be expected to be up

in lateral views, and it *is* in Erdtman's illustrations and in those of a few other authors, but most authors, for some reason, always put the distal side (cappula) down

In proximal–distal views, most authors show the long axis parallel to the bottom of the page

Triporate, tricolpate, tricolporate, etc. In polar views, Erdtman usually illustrates one pore or colpus down

In equatorial views he shows the colpi vertical

Without realizing it, the author of this book has always disobeyed Erdtman on polar views, showing one pore or colpus up

I recall Erdtman berating me for publishing bisaccates "upside down", but apparently the colpus- or pore-up dictum for P03-Pc0 did not sink in.

It is certainly very disconcerting to illustrate spores and pollen randomly, and one should adopt a convention and stick to it. This includes drawings in laboratory notebooks of students!

Shape

The shape of angiosperm pollen is usually described with reference to the polar and equatorial dimensions. If the grain has a polar axis longer than the equatorial diameter, it is in the "prolate family"; if the equatorial diameter exceeds the polar axis, the grain is in the "oblate family". In my laboratory we only use this to describe quite pronounced variants from spherical, such as prolate and perprolate. Erdtman, however, characteristically sought elegance in the matter, and his numbered shape classes are shown in Figure 5.17.

MICROSCOPIC METHODS AND SPOROMORPH MORPHOLOGY

Nehemiah Grew's 17th-century observations of pollen morphology were made with a microscope that was hardly more than a hand lens. The advances of spores/pollen studies have followed closely the technological advances of microscopy. Transmitted light microscopy has not changed markedly for about a century, and the effective limit of magnification by this method, with a "medical style" binocular microscope, having an excellent oil immersion objective and good oculars, is about 1,500×. At least 95% of practical palynological study is still made with such equipment. However, for study of the fine details of sculpture of palynomorphs in general and for the study of very thick-walled or otherwise more or less opaque palynomorphs, the scanning electron microscope (S.E.M) is a must. For the detailed study of ultrathin sections showing exine structure, the transmission electron microscope (T.E.M.) is equally important. However, one should remember that day-to-day study is still done with the transmitted light microscope, and descriptions of palynomorphs only by S.E.M are not particularly helpful to the average palynologist who never, or only occasionally, uses the S.E.M. S.E.M.

114

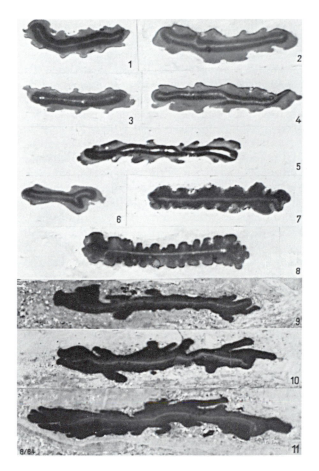

Figure 5.18 Spores in thin sections of bituminous coal, named as taxa by Stach (1964): (1)–(6) *Stratexinis ornatus*; (7), (8) *Thiessenexinis incisus*; (9)–(11) *Baculexinis raistrickii*. The two-layered nature of the exine is especially evident in (1)–(6). These taxa are undoubtedly synonyms of known dispersed spore forms.

pictures show only the external aspects of the palynomorph. T.E.M. pictures, plus S.E.M. photos, tell one the same things that one sees with the light microscope, but more highly magnified, and with greater resolution of detail. It is possible to make preparations of fossil or recent palynomorphs for study by both electron and light microscopy. Photos of a single palynomorph from both sides using both light and S.E.M. microscopy can also be made. It is also possible to study palynomorphs by light, S.E.M and T.E.M., though to do so for a single specimen is something of a tour de force (Ferguson 1977, Walker & Walker 1982)! Special light microscope attachments such as interference contrast ("Nomarski") allow detailed study, even of hyaline palynomorphs

115

with very low contrast. An "S.E.M.-like" effect is achieved. In some cases, fluorescence microscopy is also useful for palynomorphs of very low contrast and with very thin walls.

Although nearly all paleopalynology has been done with macerations of sediments, some studies have been done with rock thin sections, or even with polished surfaces studied by reflected light. The Green River Oil Shale (Eocene, western U.S.A.) defies ordinary maceration procedures. Thus, the famous studies by Wodehouse (1932, 1933), among the earliest paleopalynological investigations, were made from thin sections. So were many of the early studies of fossil dinoflagellates by Eisenack, Deflandre, and others. Dinoflagellate cysts were studied by them in thin sections of chert and flint. Many of the early investigations of spores in coal by Thiessen and others were made using coal thin sections (e.g., White & Thiessen 1913). Thiessen, Stach, and others have even classified the spores observed in thin section. It is obviously difficult to do this, as the sectioned spores tend to be more or less two-dimensional, and one works only with an outline. Stach, a coal petrologist, continued the study of coal spores in thin section and polished surfaces and even (Fig. 5.18) named species of spores based on such observations. As it is very difficult to compare these spore taxa with those based on whole, macerated spores, the use of such names is probably not desirable.

CHAPTER SIX

Stratigraphic palynology – Precambrian, Cambrian, and Ordovician

Other possible organizations of paleopalynological information might be as good, but my practice in teaching the subject will be followed here. The basic features of spores/pollen are included in Chapters 1–5. Beginning with this chapter we present the materials of paleopalynology systematically, beginning with the oldest palyniferous rocks. This will be followed by chapters on biostratigraphic methods and other matters. One problem this presents is that we must frequently digress into non–spores/pollen matters, but this seems to me more rational than avoiding the issue by relegating these subjects to appendices. The first such necessary digression relates to the acritarchs, because these are the first true palynomorphs (resistant-walled, right size range).

ACRITARCHS OF PRECAMBRIAN–ORDOVICIAN

The first robust-walled acritarchs appear in late Proterozoic rocks about 1.4 billion years old, although reports of forms more than 0.9 billion years old are rare (Vidal & Knoll 1983). The early forms seem to be linked with the appearance of eucaryotic organisms, including probable green algae, and oxygen levels in the atmosphere sufficient to terminate the dominance of reducing environments. Although microfossils are reported from rocks as old as 3.4 billion years (see Fig. 6.1), these are moneran remains, representative of the only kingdom of organisms which does not produce true palynomorphs. Some would call such microfossils, as well as sphaeromorph "robust-walled" (= sporopolleninous) acritarchs, "cryptarchs" (see Fig. 6.2). The non-sporo-polleninous spheres and filaments are not within the purview of paleopalynology, because the walls are not resistant enough to survive maceration.

Stromatolites also extend back in the fossil record about as far as the evidence for sedimentary rocks, but they were not abundant until about 2.5 billion years ago (Schopf 1977). These stromatolites were surely the product of mat-forming monerans, and they seem homologous with blue-green algal stromatolites of present-day Shark's Bay, Western Australia. Blue-green algal stromatolites thus range from the early Precambrian to the present, but their heyday was the late Precambrian and earliest Paleozoic, before O_2 levels in the atmosphere reached 10% of present levels ("*PAL*").

"Phytic" Eras	Conventional Eras	Periods	Years ago x 10⁹	Events revealed by fossils
"Cenophytic"	Cenozoic	Neogene	0.03	Herbaceous plants: e.g. grasses & composites
	– – – – – – –	Upper Cretaceous	0.08	Angiosperms dominate
	Mesozoic	Lower Cretaceous	0.13	First angiosperms
"Meso-phytic"		Triassic	0.22	Gymnosperms (Ginkgophytes, Cycadophytes, Coniferophytes) dominate
– – – – – – –	– – – – – –	Permian	0.26	
"Paleophytic"		Carboniferous	0.32	Lepidodendron, Calamites, etc., dominate
		Devonian	0.35	First seed plants (= gymnosperms)
		Upper Silurian	0.40	First vascular plants (= Tracheophytes)
	Paleozoic	Lower Silurian	0.42	First embryophytes (trilete spores; = land plants?) [O₂ = 10% PAL]
		Cambrian-Ordovician	0.5?	First protoembryophytes (= green algae?). Abundant & diverse acritarchs
"Proterophytic"	Late Proterozoic		0.8	[O₂ = ±1% P.A.L.]
	Middle Proterozoic		1.0	First eucaryotes. Nuclear division and tissues. Simple sporopolleninous acritarchs abundant
			1.4	First sporopolleninous acritarchs?
"Archeophytic"	Early Proterozoic		2.1	Photosynthesis; possible eucaryotes
	Late Archean		2.5	Stromatolites become ubiquitous
	Middle Archean		3.4	"Cryptarchs": procaryotes, cells, reproduction. Photosynthesis?
	Early Archean		±3.8	Oldest dated rocks on Earth
			4.5	Origin of Earth

(Phanerozoic spans Neogene through Cambrian-Ordovician; Cryptozoic = Precambrian spans Late Proterozoic through Early Archean)

Figure 6.1 High points in plant evolution, as related to paleopalynology. Compiled from various sources. Compare with time chart, Figure 2.1. Period and time information incomplete and not to scale here. The "phytic" eras, based on events in plant evolution, are discussed at various places in the text.

118

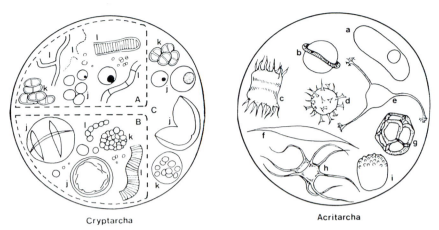

Cryptarcha   Acritarcha

Figure 6.2 Diver & Peat's (1979) proposal of the term "cryptarch" for Precambrian organic-walled microfossils (O.W.M.). The proposal would make all Precambrian sphaeromorphs ("j") "cryptarchs", whether sporopolleninous or not. Resistant-walled, non-sphaeromorph forms remain as acritarchs. *Cryptarcha*: (A) stromatolitic chert microbiotas; (B) and (C) shale microbiotas; (j) sphaeromorphs; (k) synaptomorphs; (l) nematomorphs. *Acritarcha*: (a) platymorph; (b) pteromorph; (c) diacromorph; (d) acanthomorph; (e) polygonomorph; (f) netromorph; (g) herkomorph; (h) prismatomorph; (i) oömorph.

It is probable that most sporopolleninous acritarchs, beginning with the oldest occurrences, represent reproductive cysts of green algae, and that they establish a probable phylogenetic link between the green algae and other green plants. However, the definition of acritarch is so general that inevitably the total is something of a "grab-bag", including, e.g., both marine and non-marine forms. Originally palynologists called all such things "hystrichosphaerids" (or even "hystrix"!). The name means "spiny spheres", but non-spiny bodies were always also included in the group. Evitt (1961, 1963) showed later from morphological studies, and Wall (1965) and Wall & Dale (1967) proved by culture of living forms, that many Mesozoic "hystrichosphaerids" were in fact dinoflagellate cysts. Evitt proposed "acritarch" (Greek *akritos* = doubtful; *arche* = origin) for all of the hystrichosphaerids which could not be shown to be dinoflagellates. This was taken up, and "hystrichosphaerid" dropped, by nearly all palynologists. As used by most palynologists, "acritarchs" are mostly marine green algal cysts, possessing extremely varied form. Some, e.g., the tasmanatids, have been referred with a high probability to particular groups of algae. *Tasmanites, Tytthodiscus*, various species of *Leiosphaeridia*, and other acritarchs, for example, have been referred to division Prasinophyta of the green algae (Tappan 1980). Others have suggested probable dinoflagellate relationships for other groups of acritarchs. Diver & Peat (1979) have proposed reserving the term acritarch only for those forms with a distinctive, algal-suggestive morphology. They would use the term "cryptarch" for other ancient organic-walled microfossils (O.W.M.). Thus, Precambrian sphaero-morphs would be cryptarchs, whether sporopolleninous or not. I do not favor this part of the Diver & Peat proposal (see Fig. 6.2).

119

Sporopolleninous ("robust") acritarchs first appear in rocks about 1.4 billion (10^9) years old, but did not become abundant until less than 1 billion years ago (Horodyski 1980). The oldest acritarchs are simple spheres, usually much folded and carbonized. The late Precambrian (= Proterozoic) acritarchs sometimes display a simple opening mechanism ("median split"), a linear rupture, and some of them are enclosed in sheaths. Low-diversity assemblages of these Proterozoic acritarchs seem to indicate coastal environments, whereas higher-diversity floras indicate more open-shelf situations (Vidal & Knoll 1983). A few of the late Precambrian forms achieve "megaspore" size; the form called *Chuaria circularis* Wolcott reaches 3,000 μm. In the Lower Vendian (latest Proterozoic), acritarchs with processes become common, polygono-morphic forms and double-walled forms appear, but mid-Vendian glaciations are correlated with a depauperization of this flora, and the terminal Precambrian flora is dominated again by simple sphaeromorphs, evidence of the earliest known episode of large-scale extinction (Vidal & Knoll 1983) (see Fig. 6.3).

Acritarchs were much diversified by earliest Paleozoic (post–Precambrian = Phanerozoic) time. A great variety of sculptural and structural complexity comes in during Cambrian time and reaches an especially great development in the early Ordovician. Figure 6.4 illustrates a variety of Precambrian and early Paleozoic forms.

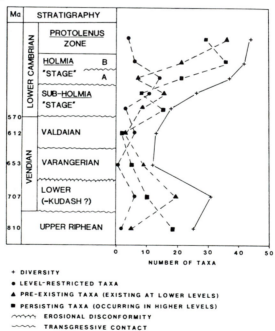

Figure 6.3 Diversity changes in late Precambrian to early Cambrian plankton (includes acritarchs but also non-sporopolleninous forms), showing the late Vendian extinction.

120

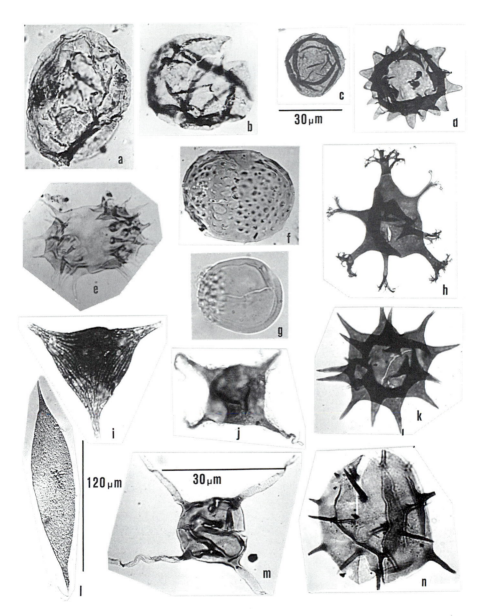

Figure 6.4 Precambrian and early Paleozoic acritarchs of various forms. Those in (a)–(c) and (g) are sphaeromorphs, (f) is a diacrodioid, and (g) is an oömorph. All others, in a broad sense, are acanthomorphs. Bars below (c) and next to (l) indicate approximate magnifications for those individual photomicrographs. Bar above (m) indicates approximate magnification for all other photos. (a) *Protoleiosphaeridium* sp.

The "purpose" of the sporopolleninous shell is intriguing. It may have been an ultraviolet light shield for the reproductive contents, which are less resistant to ultraviolet light than the rest of the life cycle. On the other hand, this would be a little strange, as sporopolleninous acritarchs became abundant just as the O_2 level of the atmosphere became high enough to screen out ultraviolet. More likely, these outer shells developed to protect the contents from oxidation, and from the ravages of oxygen-using bacteria and fungi, which became abundant about the same time. The "sporopollenin" of the wall appears to be the same substance as in spores, pollen, and dinoflagellate cysts, but this has not really been proven.

Morphology of acritarchs

Figures 6.4 and 6.6–6.9 illustrate various basic acritarch forms. Obviously, the acritarchs are artificially classified. The wall may be simple or complex. Excystment structures are often present and are quite diverse in form; when such an opening exists, it is called a pylome (see Cornell 1982). The most important time for acritarchs was late Precambrian to Silurian, though there are also abundant post-Silurian acritarchs, especially in marine rocks. The schemes for classifying acritarchs were mostly developed from studies of Paleozoic forms. The most important major categories are described below (Cramer & Diez 1979, Downie 1979); the pioneer investigators are pictured in Figure 6.5.

Sphaeromorphs (Sphaeromorphitae) (See Figs. 6.4 & 6.6.) As the name implies, these are mostly more or less spherical. Often, if the walls are thin, these are collapsed and folded, the folds simulating morphological features (which even have been interpreted by the unwary as laesurae). The walls may be psilate, scabrate, verrucate, rugulate, fossulate, but not truly spinous (echinate). Examples: *Tasmanites* and relatives (Precambrian to present), *Leiosphaeridia* and relatives (Precambrian to present).

Acanthomorphs (Acanthomorphitae) (See Fig. 6.7.) Main body is essentially spherical, with processes, from simple spines and baculae to complex branched processes. Some acritarch palynologists use the term "sculpture" for elements less than 5 µm long, "processes" for larger elements. Inner central body may be present. Possible germinal openings (not accidental slits) may be present. Sub-units (per Cramer & Diez 1979) are as follows:

(= *Leiosphaeridia* sp.), late Precambrian, U.S.S.R. (b) *Trachysphaeridium* sp., late Precambrian (0.8 billion years), Greenland. (c) *Kildinella* sp. (= *Leiosphaeridia* sp.), late Precambrian, U.S.S.R. (d) *Micrhystridium* sp., Lower Ordovician (Arenig), Morocco. (e) *Acanthodiacrodium* sp., lowest Ordovician (Tremadoc), U.S.S.R. (f) *Lophodiacrodium* sp., lowest Ordovician (Tremadoc), U.S.S.R. (g) *Ooidium* sp., lowest Ordovician (Tremadoc), U.S.S.R. (h) *Vogtlandia* sp., Lower Ordovician (Arenig), Morocco. (i) *Arkonia* sp., Lower Ordovician (Arenig), Morocco. (j) *Aureotesta* sp., Lower Ordovician (Arenig), Morocco. (k) *Polygonium* sp., Lower Ordovician (Arenig), Morocco. (l) *Cleithronetrum* sp., Bromide Fm., Middle Ordovician Oklahoma. (m) *Orthosphaeridium* sp., Upper Ordovician (Caradoc), England. (n) *Gorgonisphaeridium* sp., Maquoketa Fm., Upper Ordovician, Missouri.

a

b

Figure 6.5 The people pictured here have made landmark contributions to the practical use of (mostly Paleozoic) acritarchs for palynostratigraphy. All three are still active.

(a) Charles Downie, Sheffield, England. Downie's publications on Paleozoic sediments have demonstrated the utility of acritarch studies in unraveling the

123

Acanthomorphs proper Outline regular. Process distribution (symmetry) regular. Examples: *Baltisphaeridium, Micrhystridium, Multiplicisphaeridium.*

Netromorphs Outline fusiform. Example: *Leiofusa.*

Diacromorphs (Diacromorphitae) Outline bipolar. Processes may be distributed similarly or dissimilarly on the two poles. Example: *Acanthodiacrodium.*

Various other acritarchs (neither acanthomorphs nor sphaeromorphs) (See Figs. 6.8 & 6.9.) Wide variety of unusually constructed or ornamented forms, such as the "fenestrate" *Cymatiosphaera*, the double-sleeved *Riculasphaera*, Herkomorphs (Herkomorphitae) – more or less spherical bodies with surface divided into polygonal fields, Tasmanitomorphs (Tasmanititae) – *Tasmanites*, etc. (see Fig. 6.9). However (see above), tasmanitids can also be grouped with sphaeromorphs.

More complex classifications are available. There are more or less 300 described genera of acritarchs, but the number of genera used depends very much on the whim of the acritarch palynologist. Cramer & Diez (1979), for example, have stated that they could greatly reduce the number of generic units by severe application of the procedures for synonymy.

Downie (1973) has shown that acritarchs can be separated also on the basis of wall structure, as follows:

Tasmanitid Wall uniform, laminated, with narrow radial pores. Wall thickness often variable in a species, probably because of growth. Present in *Tasmanites, Baltisphaeridium*, and others.

Micrhystridian Wall homogeneous, simple, usually thin. Examples: *Veryhachium, Micrhystridium.*

Diacrodian Wall thin, simple, homogeneous. Tends to split into angular planes. Examples: all diacrodians (= diacromorphs).

Visbysphaerid Wall thin, homogeneous. Has capacity to develop an inner body closely attached to outer wall, forming a double wall. Example: *Visbysphaera.*

---

stratigraphy of otherwise difficult sequences (see Downie 1979). He also has contributed greatly to the systematics of these microfossils (see biographical notes in Sarjeant 1984).

(b) Fritz H. Cramer and Maria del Carmen R. ("Carmina") Diez. This husband (more or less Dutch) and wife (Spanish) paleopalynological team comprise one of the more fascinating stories in the field. It is difficult to be sure where legend ends and biography begins, and nobody as yet has attempted to do so in print. However, the Cramer–Diez team has, despite many vicissitudes, made great contributions to various aspects of paleopalynology, especially to acritarch studies. Their descriptive and cataloging work has been especially valuable (see Cramer & Diez 1979, Eisenack *et al.* 1973).

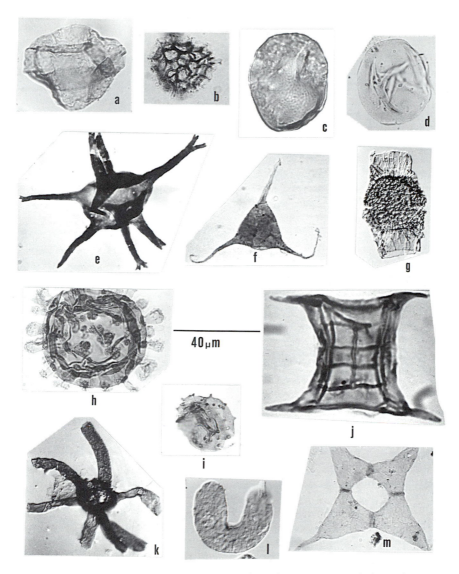

Figure 6.6 Silurian and Devonian acritarchs of various morphological groups. Approximate magnifications for most of the photos indicated by bars next to (h) and under (u). However, for (g) and (m) the bar should be read as 80 μm, and for (j) it should be read as 20 μm. (a) *Leiosphaeridia* sp., lowest Silurian, Pennsylvania. (b) *Cymatiosphaera densisepta* Miller & Eames, lowest Silurian, Pennsylvania. (c) *Leiosphaeridia* sp., lowest Silurian, Pennsylvania. (d) *Leiosphaeridia* sp., Lower Silurian, Missouri. (e) *Diexallophasis* sp., Silurian, Libya. (f) *Veryhachium*, Silurian, Libya. (g) *Carminella maplewoodensis* Cramer, Lower Silurian, New York. (h) *Visbysphaera* sp., Silurian, England. (i) *Buedingisphaeridium* sp., Silurian, Libya. (j) *Neoveryhachium* sp., Silurian, New York. (k) *Dilatisphaera laevigata* Lister, Silurian, England. (l) *Quisquilites buckhornensis* Wilson & Urban, Upper Devonian, Oklahoma (this form is referred, e.g.,

125

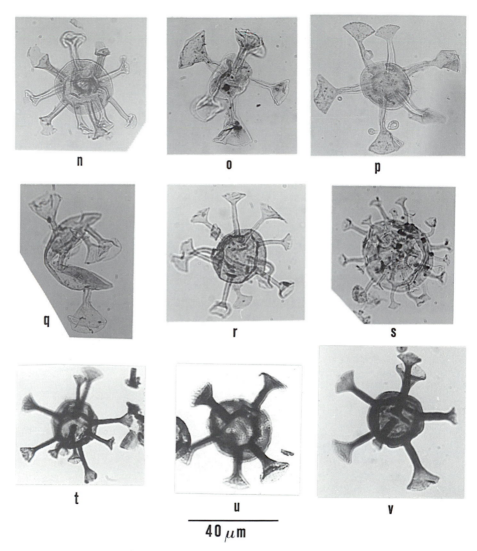

40 μm

by Tappan (1980), to the Tasmanitaceae – see Figure 6.9. (m) *Deflandrastrum* sp., Devonian, Libya (a coenobial form). (n)–(v) Various forms of the genus *Umbellasphaeridium*: (n)–(s) from the Bedford Shale, Upper Devonian, Ohio, (t)–(v) from the subsurface of Peru – (s) is a specimen of *U. deflandrei* (Moreau–Benoit) Jardiné *et al.*, and all others are representative of *U. saharicum* Jardiné *et al.* Part (q) shows equatorial (medial) excystment structure.

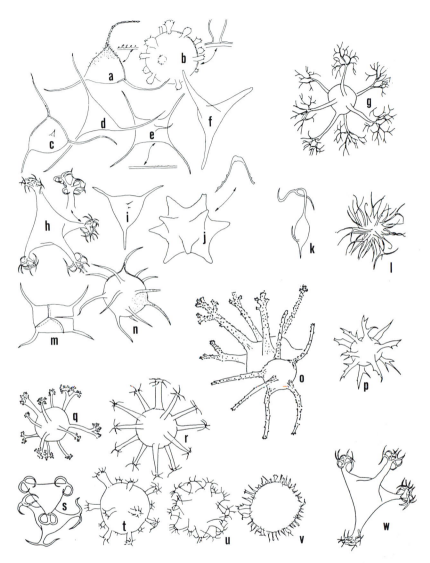

Figure 6.7 Diagrams of various acanthomorphic (in a broad sense) form-genera of Lower Paleozoic acritarchs. Magnifications vary greatly. Compare with Figures 6.4 and 6.6 for approximate scale. (a) *Villosacapsella*; (b) *Visbysphaera*; (c)–(f) *Veryhachium*; (g) *Multiplicisphaeridium* (see (o)–(w)); (h) *Vogtlandia*; (i) *Wilsonastrum*; (j) *Cordobesia*; (k) *Downiea*; (l) *Multiplicisphaeridium* (see (o)–(w)); (m) *Winwaloeusia*; (n) *Crassisphaeridium*; (o)–(w) *Multiplicisphaeridium* (see also (g) and (l)).

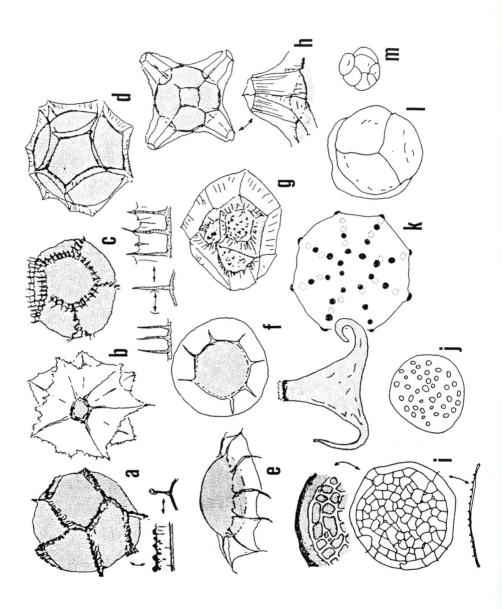

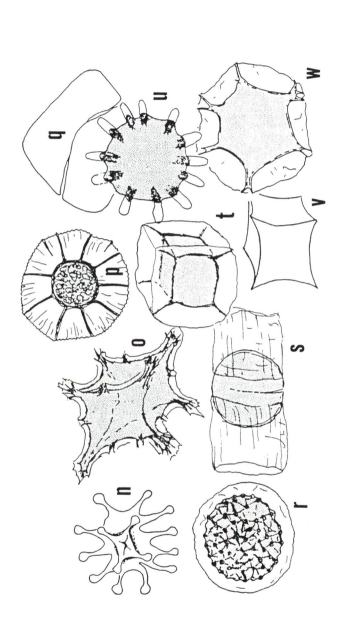

Figure 6.8 Diagrams of various form-genera of Lower Paleozoic acritarchs other than sphaeromorphs or acanthomorphs. Magnifications vary greatly – for approximate scale, compare with Figures 6.4 and 6.6. (a) *Cristallinium*; (b) *Conradidium*; (c) *Cymatiogalea*; (d) *Cymatiosphaera*; (e), (f) *Duvernaysphaera*; (g) *Muraticavea*; (h) *Daillydium*; (i) *Ovnia*; (j) *Perforella*; (k) *Pardaminella*; (l), (m) *Polyedrosphaeridium*; (n) *Polyplanifer*; (o) *Polyedryxium*; (p) *Pterospermella*; (q) *Pulvinomorpha*; (r) *Pterotosphaerula*; (s) *Riculasphaera*; (t) *Senzeillea*; (u) *Tornacia*; (v) *Staplinium*; (w) *Veliferites*.

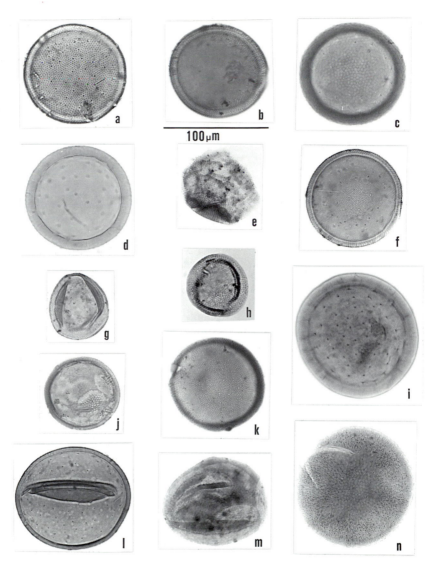

100μm

Figure 6.9 Tasmanitids: thick-walled "acritarchs" with walls perforated by canals, referable to the family Tasmanitaceae. Tasmanitids occur from early Paleozoic to present and have been shown to represent cyst-like parts of the life cycle of members of the green alga group, Prasinophyta, which has species resident in fresh to marine water. *Quisquilites* (Fig. 6.6l) is a Devonian "acritarch" usually put with the tasmanitids. (Because their biological relationship is known, the tasmanitids are not by definition acritarchs.) The forms illustrated here are from the Posidonienschiefer, Lower Jurassic oil shales of Germany. The tasmanitids have not been as abundant since the Jurassic as they were in nearshore marine environments of the Paleozoic and early Mesozoic. (a) *Pleurozonaria media* Mädler; (b) *P. suevica* (Eisenack) Mädler; (c), (d) *P. wetzelii* Mädler; (e) *P. media*, corroded, see (a); (f) *P. suevica*, see (b); (g) *P. suevica*, not fully developed,

130

Classification according to excystment style Shape of excystment opening can also be used to separate acritarchs, as follows:

Archeopyle Should be reserved for dinoflagellate cysts. A few acritarch excystment openings are similar.

Cyclopyle Opening circular in outline, sometimes very long. This is really synonymous with *pylome*, which is sometimes used in a general sense for acritarch excystment openings.

Epityche Excystment by a curving split allowing a flap to open (see *Veryhachium*).

Median split Splits into two more or less equal halves. Examples: many leiospheres, e.g., *Hemisphaeridium*.

Stratigraphic occurrence of early and mid-Paleozoic acritarchs

As already noted, sporopolleninous or "true", robust-walled acritarchs first occur in shales about 1.4 billion years old, and rather abundantly from a bit less than 1 billion years ago to the end of the Proterozoic. These are mostly simple, more or less psilate sphaeromorphs, mostly in the 20–40 μm size range. Despite the similarity of the forms, enough size and sculpturing difference exists to suggest the future possibility of some palynostratigraphic applicability of their study. For example, the range of size increases somewhat in the later Proterozoic (= late Precambrian). Earliest Phanerozoic acritarch floras continue the trend to more diversity in size (25–200 μm), but the most striking feature of early Cambrian acritarch floras is the advent of prominent and diverse sculpture: granulae, very short spines, and baculae. In the Ordovician the sculpture of many forms becomes much more pronounced (though low-sculptured and psilate sphaeromorphs continue to be present), with many sorts of spines and other processes, including some that are branched, and some that are nearly as long as the diameter of the body of the acritarch. In other words, the Ordovician is the heyday of the acanthomorphs of all sorts, such as diacromorphs and acanthomorphs proper.

The study of acritarchs has been applied to the solution of practical stratigraphic problems, and more progress can be expected in this direction. Cramer and Diez have shown that studies of acritarch distribution can also be used for study of problems such as probable paleoposition of continental plates in the early Paleozoic. Cramer & Diez (1974b) suggest two contrasting acritarch provinces in the early Ordovician. Vavrdova (1974) also described two similar acritarch provinces for the Ordovician, a Baltic province and a

---

see (b) and (f); (h) *P. wetzelii*, not fully developed, see (c) and (d); (i) *Tasmanites tardus* Eisenack; (j) *Pleurozonaria* sp.; (k) *P. suevica*, see (b) and (f); (l) *Tasmanites tardus*, not fully developed, see (i); (m) *Tytthodiscus* sp., corroded example; (n) *Tytthodiscus schandelahensis* (Thiergart) Mädler, peculiar preservation makes pore canals visible.

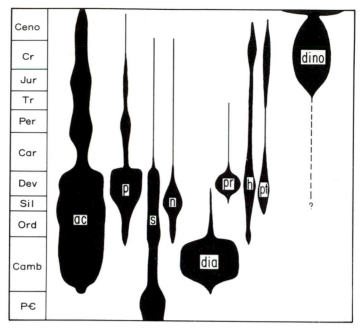

Figure 6.10   Geological range of principal acritarch groups and dinoflagellate cysts: ac, Acanthomorphitae; p, Polygonomorphitae; s, Sphaeromorphitae; n, Netromorphitae; dia, Diacromorphitae; pr, Prismatomorphitae; h, Herkomorphitae; pt, Ptermorphitae; dino, dinoflagellates.

Mediterranean province. Acritarchs have been especially useful also in Devonian studies. Playford & Dring (1981) have described a very diverse acritarch palynoflora from the marine Devonian of the Carnarvon Basin of Western Australia. Wicander (1983, 1984) has shown the potential usefulness of acritarch studies in the Devonian of North America. Figure 6.6 illustrates some of the Silurian and Devonian forms. Figure 6.10 shows the stratigraphic range of acritarchs in general, showing that they are largely but by no means exclusively a Paleozoic group.

CAMBRIAN–ORDOVICIAN CHITINOZOANS

Acritarchs were joined in Cambrian time by other organic resistant-walled palynomorphs, parts of, or elements of, various animals. The most important are chitinozoans (range: Cambrian to Carboniferous) and scoledoconts (range: Ordovician to present). Chitinozoans are composed of "pseudochitin", a C–H–O–N compound of uncertain structural formula which behaves much like chitin and sporopollenin (appearance and resistance to decay and to laboratory maceration of the enclosing rock). Always found in marine rock, chitinozoans occur much less abundantly per gram of sediment than do acritarchs and sporomorphs, and some special processing techniques should be applied. Paris (1984) reported a maximum of about 100/g being more common. Lithology is not as controlling a factor as it is for sporomorphs. Limestones, for example, are often productive, but so are marine shales. Unfortunately for us, they are thick-walled, and as a result are often opaque or nearly so and difficult to investigate with the light microscope. They mostly range in size from 50 to 250 μm, though there are some smaller than 50 μm, and a few "monsters" up to 600 μm or more. S.E.M. studies are best for elucidation of morphology but are not usually practicable for routine work such as counting. The biological affinities of chitinozoans are still not known, and various investigators have suggested as diverse provenance as graptolites (Jenkins 1970) and fungi (Loquin 1981). Fungal origin seems very unlikely if for no other reason than that resistant-walled fungal spores, and hyphae universally recognized as such, do not appear regularly until late Jurassic, although non-robust-walled fungi range to the Precambrian. An anomalous, saltating record would exist if chitinozoans were fungal. Also suggested have been Tintinnids and eggs of various unknown animals. An interesting circumstantial case (chemistry, common range, frequent association) can be made for derivation of chitinozoans from graptolites (Jenkins 1970), which would make chitinozoans probable chordate remains.

Morphology of chitinozoans

Figure 6.11 shows the basic morphology of typical chitinozoans (see also Fig. 1.1a, b). Chitinozoans are often referred to as "flask-shaped", but this and the term "mouth" (= oral) imply that we know more than we do about which end is "up". Specimens are sometimes encountered that consist of two or more units joined together, collar and mouth to base. Figure 6.11 shows several modes of joining chitinozoan units. Chitinozoans may be psilate or highly sculptured, with simple or complex processes, as is seen in Figure 6.11. Figure 6.12 presents photomicrographs and S.E.M. micrographs of various characteristic forms.

Stratigraphic occurrence of chitinozoans

Chitinozoans first occur in Cambrian rocks but are most abundant in Ordovician and Silurian. They continue to be modestly abundant in the Devonian but are rare by Carboniferous and gone by late Carboniferous time. Figure 6.13 shows the stratigraphic range of the more important chitinozoans.

CAMBRIAN–ORDOVICIAN SCOLECODONTS

Associated with acritarchs and chitinozoans in early Paleozoic marine sediments are often found the chitinous mouthparts (= "jaw apparatuses") of marine annelid worms (polychaetes). These mouthparts are called scolecodonts. Colbath & Larson (1980) showed that the chitinous layer of scolecodonts covers an inner $CaCO_3$ layer. Palynological maceration presumably destroys the $CaCO_3$, just as it does in foraminifera, though Colbath & Larson believe the organic layer may protect the carbonate from acid digestion. Identification

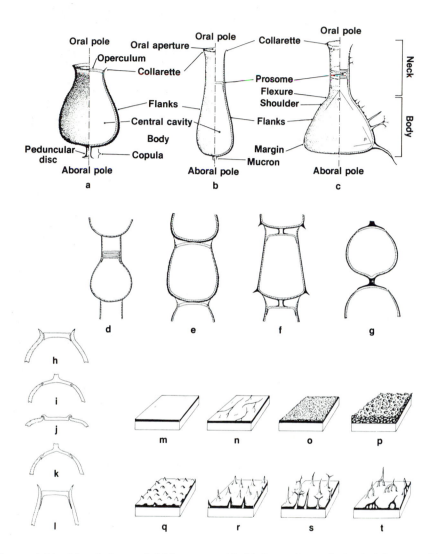

Figure 6.11 Morphology of chitinozoans. (a)–(c) General morphological features: (a) Desmochitinidae, (b) Conochitinidae, (c) Lagenochitinidae. (d)–(g) Different types of linear "colonies": (d) junction by simple juxtaposition; (e)–(f) junction by double adherence; (g) reinforced junction (black = periderm; hachured = ectoderm; stippled = endoderm). (h)–(l) Different types of fastening of operculum in the oral aperture: (h)–(k) mechanical assembly; (l) adherence (stippled = operculum; hachured = external and internal membranes). (m)–(t) Sculpturing types for chitinozoans: (m) psilate, (n) chagrenate, (o) tomentose ("felty"), (p) spongy, (q) conic verrucate, (r) echinate, (s) filiform, simple, and branched, (t) filiform, bi- or multipodal. The black layer represents the periderm and the white layer the ectoderm. Tests with (m)–(p) texture are regarded as glabrous, those with (r)–(t) texture as ornamented.

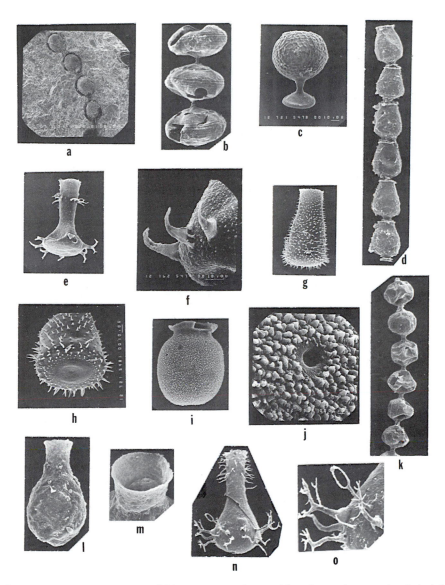

Figure 6.12 Representative chitinozoans as observed by photomicrography (ac)–(af) and scanning electron microscopy (all others). Compare with diagrams in Figure 6.11. Obviously, chitinozoans are an outstanding example of the advantages of S.E.M. for thick-walled palynomorphs, although some chitinozoans are more translucent than the specimens illustrated here. All specimens except (a) were obtained by rock maceration. (a) Chain of connected individuals of *Desmochitina* (*Desmochitina*) *bulla* Taugourdeau & Jekhowsky. S.E.M. of uncoated rock surface. Each unit about 230 μm long. Lower Ordovician, Czechoslovakia. (b) *Margachitina catenaria tenuipes* Paris, chain of three units. Note connecting "necks". Each unit about 75 μm long. Lower Devonian, France. (c) *Margachitina margaritana* (Eisenack), individual unit in orientation as in (b),

135

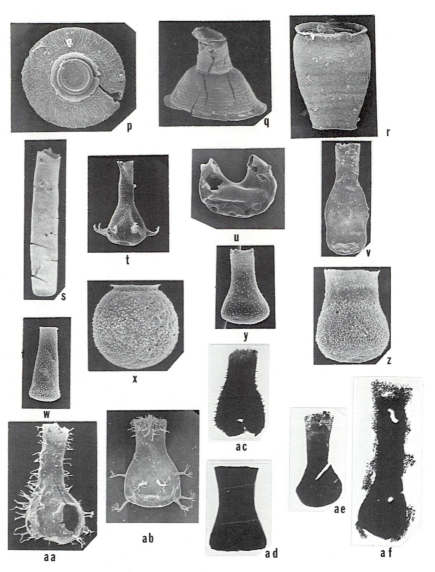

showing that the copula is linked to an operculum from both an aboral and an oral end of a chitinozoan unit. Fossil is about 95 μm long as shown. Silurian (Wenlockian), Sweden. (d) *Urnochitina urna* (Eisenack), chain of chitinozoan units linked by double adherence. Each unit about 100 μm long. Silurian (Pridolian), Czechoslovakia. (e) *Ancyrochitina gutnica* Laufeld, with filiform sculpture. Fossil is about 130 μm long. Silurian (Wenlockian), Sweden. (f) *Ancyrochitina* sp., two sizes of echinate sculpture. Large spines extend about 25 μm from surface. Silurian (Wenlockian), Sweden. (g) *Belonechitina wesenbergensis brevis* (Eisenack), with profuse echinate sculpture. Fossil about 150 μm long. Upper Ordovician glacial erratic, Baltic area. (h) Same species as (g), showing difference of sculpturing on aboral pole: echinate on flanks, verrucate

136

toward pole. Larger spines about 12 μm long. Upper Ordovician glacial erratic, Baltic area. (i) *Desmochitina* (*Pseudodesmochitina*) *minor* Eisenack, closely packed verrucate sculpture. Fossil about 90 μm long. Upper Ordovician glacial erratic, Baltic area. (j) Same specimen as (i), oral area, enlarged to show densely packed verrucate sculpture. Perforation shown about 5 μm in diameter. Upper Ordovician glacial erratic, Baltic area. (k) *Margachitina* sp., example of chain of units coupled by reinforced junction. Each unit about 70 μm long. Lowermost Devonian (Borshow), U.S.S.R. (l) *Lagenochitina deunffi* Paris, illustrating a form with large body, long neck, and tomentose sculpture. Specimen about 100 μm long. Ordovician (Caradocian), Portugal. (m) Same species and source as (n), specimen showing interior of neck with prosome at junction of neck and body. (n) *Gotlandochitina racheboeufi* Paris, a narrow flask-shaped form with branched filiform sculpture. Specimen about 165 μm long. Devonian (Emsian), France. (o) Detail of lower left side of (n), enlarged. (p) *Calpichitina lenticularis* (Bouché), oral view of specimen with highly developed collarette and operculum *in situ*. Specimen about 160 μm in diameter. Upper Ordovician, Libya. (q) *Cyathochitina* sp., specimen compressed dorsoventrally, presenting concentric structures on the body. Specimen about 200 μm long. Upper Ordovician, Portugal. (r) *Armoricochitina nigerica* Bouché, showing a carina and granulous sculpture. Specimen about 200 μm long. Upper Ordovician, Libya. (s) *Rhabdochitina magna* Eisenack, an elongated tubular form. Specimen about 600 μm long. Late Ordovician, Portugal. (t) *Ancyrochitina* sp., with narrow flask shape and two sorts of spines – detail of same shown in (f). Specimen about 140 μm long. Silurian (Wenlockian), Sweden. (u) *Parachitina curvata* Eisenack, possible chitinozoan (not yet so classified), with psilate sculpture. Specimen about 280 μm from left to right. Upper Ordovician glacial erratic, Baltic area. (v) Angochitininae group, specimen showing a constriction, possibly an abnormality (technical term: "teratoid"). Specimen about 220 μm long. Lowermost Devonian, France. (w) *Belonechitina robusta* (Eisenack), tubular form with short, scattered, complex, multipodal sculpture. Specimen about 275 μm long. Upper Ordovician glacial erratic, Baltic area. (x) *Desmochitina* (*Pseudodesmochitina*) *minor* Eisenack, unusually broad vase-formed. Ornate tomentose–spongy sculpture. Specimen about 80 μm long. Ordovician (Lower Caradocian), Portugal. (y) *Fungochitina fungiformis* (Eisenack), flask-shaped form with scattered, mostly echinate sculpture. Specimen about 140 μm long. Upper Ordovician glacial erratic, Baltic area. (z) *Eisenackitina rhenana* (Eisenack), broad form with tomentose sculpture. Specimen about 90 μm long. Upper Ordovician (Caradocian), Portugal. (aa) *Gotlandochitina marettensis* Paris, showing filiform sculpture. Specimen about 190 μm long. Devonian (Emsian), France. (ab) *Alpenachitina eisenacki* Dunn & Miller, specimen with long branching processes on the chamber and echinate sculpture on the upper neck. Specimen about 150 μm long. Middle Devonian, Libya. (ac) *Belonechitina* sp., abnormally large body (split and flattened). Specimen about 130 μm long, P.M. Lower Ordovician, France. (ad) *Cyathochitina varennensis* Paris. Specimen about 120 μm long, P.M. Lower Ordovician (Llanvirnian), France. (ae) *Sphaerochitina lycoperdoides* Laufeld, flattened specimen showing disorganized prosome. Specimen about 150 μm long, P.M. Silurian, Portugal. (af) *Muscochitina muscosa* Paris. The shaggy excrescences are indeed the loosely tomentose sculpture. Specimen about 260 μm long, P.M. Lowermost Devonian, France. Note that chitinozoans are governed by the zoological rules of nomenclature, which accounts for the differences from spores/pollen names. For example, *Eisenackitina rhenana* (Eisenack) means that Eisenack first named *rhenana*, in another genus, but the name of the transferring author does not appear after "(Eisenack)", as is required under botanical rules. Names in parentheses after the generic name are subgeneric names, as in (a) and (i). In trinominal names, as in (g), the second epithet (*brevis*) is the subspecific name (in botany, it would have to be labelled subspecies, variety, etc.).

137

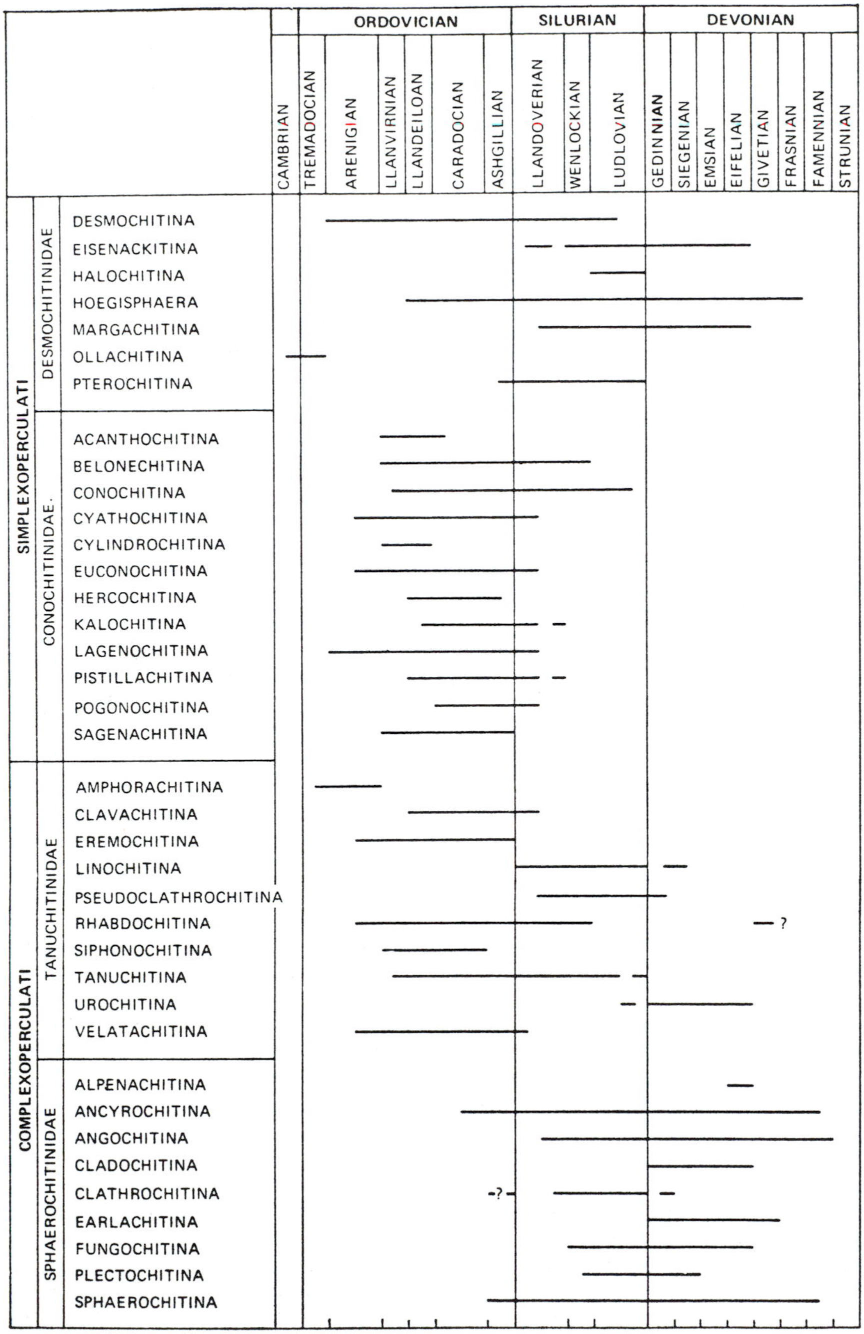

			ORDOVICIAN						SILURIAN			DEVONIAN							
		CAMBRIAN	TREMADOCIAN	ARENIGIAN	LLANVIRNIAN	LLANDEILOAN	CARADOCIAN	ASHGILLIAN	LLANDOVERIAN	WENLOCKIAN	LUDLOVIAN	GEDINNIAN	SIEGENIAN	EMSIAN	EIFELIAN	GIVETIAN	FRASNIAN	FAMENNIAN	STRUNIAN

SIMPLEXOPERCULATI

DESMOCHITINIDAE

- DESMOCHITINA
- EISENACKITINA
- HALOCHITINA
- HOEGISPHAERA
- MARGACHITINA
- OLLACHITINA
- PTEROCHITINA

CONOCHITINIDAE

- ACANTHOCHITINA
- BELONECHITINA
- CONOCHITINA
- CYATHOCHITINA
- CYLINDROCHITINA
- EUCONOCHITINA
- HERCOCHITINA
- KALOCHITINA
- LAGENOCHITINA
- PISTILLACHITINA
- POGONOCHITINA
- SAGENACHITINA

COMPLEXOPERCULATI

TANUCHITINIDAE

- AMPHORACHITINA
- CLAVACHITINA
- EREMOCHITINA
- LINOCHITINA
- PSEUDOCLATHROCHITINA
- RHABDOCHITINA
- SIPHONOCHITINA
- TANUCHITINA
- UROCHITINA
- VELATACHITINA

SPHAEROCHITINIDAE

- ALPENACHITINA
- ANCYROCHITINA
- ANGOCHITINA
- CLADOCHITINA
- CLATHROCHITINA
- EARLACHITINA
- FUNGOCHITINA
- PLECTOCHITINA
- SPHAEROCHITINA

Figure 6.13 Approximate ranges of the most important chitinozoan genera. The family groupings of the genera are quite different from those in Figure 6.11.

138

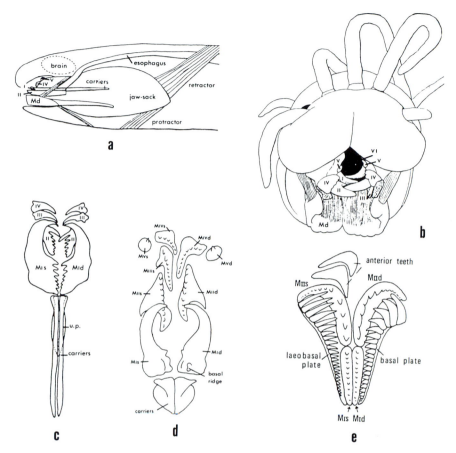

Figure 6.14 Scolecodont morphology. Scolecodonts are various parts of the chitinous "jaw" apparatus of the mouth–proboscis of a number of sorts of polychaetous annelid worms. These worms range from early Paleozoic to the present, but were most abundant in the Paleozoic. The various "teeth" and "jaw" parts are found as strengthened ridges and folds from a chitinous mouth–proboscis lining. Thus, the whole apparatus is hollow, filled in life with soft tissues. The various complex parts of an apparatus are connected in life by a thin lining of the proboscis, but the parts are easily disarticulated and thus quite different scolecodonts can and do represent the same species of animal. Under zoological rules, the correct name for a dozen different named scolecodonts is the first published of any of these, if they are shown to belong together. Form-taxa, for such items occurring dispersed, are not recognized in zoology. (a) *Eunice siciliensis* Grube, sagittal section of anterior portion, showing relationship of mandible (Md) and the maxillary apparatus (carriers plus maxillae I, II, III, and IV). The mandible is located ventrally and often has a calcareous cap. (b) Same species, frontal view, mandibles pulled down, showing maxillary apparatus behind. Parts of maxillae are numbered from posterior to anterior. (c) Diagram of maxillary arrangement of a prionognathid "jaw", viewed dorsally, showing long narrow carriers: u.p. = unpaired piece, MIs = first maxilla left (sinistral), MId = first maxilla right (dextral). (d) *Diopatria neapolitana* Ehlers, exploded diagram of labidognathid jaw

139

of the organic matter as chitin depends on appearance and behavior. Germeraad (1980) has noted the occurrence of scolecodont-like fossils in possibly non-marine sediment of Cenozoic age, but this is very exceptional.

Scolecodonts first occur in the Ordovician. They, like chitinozoans, vary a great deal; in size, they vary from around 100 to well over 200 μm. Also like chitinozoans, they are not particularly abundant in absolute terms in the marine shales in which they occur; they are found in the 100/g range, instead of 100–10,000/g as is typical of acritarchs or spores/pollen in a moderately productive sediment. Because of this, and the size of scolecodonts, special processing, including larger original sample size and less or no pulverizing, is recommended. Specimens may be hand-picked from the residues and mounted dry, as foraminifera, or mounted in a mountant on strew slides, as other palynomorphs. Scolecodonts are not quite as likely to be opaque as chitinozoans, but they frequently are thick-walled, and S.E.M. studies are a useful supplement to light microscopy. In Ordovician to Permian marine shales scolecodonts are especially prominent. These marine worm fossils persist in the modern oceans. They occur but are uncommon in Mesozoic and Cenozoic sediments (Jansonius & Craig 1971, Schäfer 1972, Germeraad 1980).

Complications of processing and microscopy are not the only reason that scolecodont studies are difficult. These "worm jaws" are not jaws but the chitinous mouth linings of the worms producing them. Unfortunately, one worm has more than one kind of mouth lining (see Fig. 6.14). Scolecodonts are also quite variable in morphology, making their systematics even more difficult. Edgar (1984) points out that the jaw apparatuses consist of three element groups: the anterior maxillae, the posterior maxillae, and carriers (see Fig. 6.14). The MI elements of the posterior maxillae are the most useful for diagnostic purposes. Note that scolecodonts are chitinous and produced by worms. They, or fragments of them, quite frequently are encountered in palynological macerations of Paleozoic marine sediments. Conodonts, which are sometimes confused with scolecodonts, are phosphatic, and never occur in palynological preparations, as they are dissolved by strong acids in our procedures; they are derived from some as yet not certainly identified organism. Although some conodonts and some scolecodonts resemble each other a bit, conodonts are considerably larger: 300–1,500 μm. Morphological terms for fossil scolecodonts are based on the presumed position in the annelid mouth, and on anatomical terms used by zoologists. See Figure 6.15 for illustrations of a variety of scolecodonts.

Scolecodont stratigraphy

First found in the Ordovician, scolecodonts range to present, but their greatest abundance is from Ordovician to Permian. More concerted work would

---

arrangement: MIs = left first maxilla, MId = right first maxilla, etc. Note that MIIIs is unpaired. (e) *Mochtyella polonica* K.-J. MIs = first left maxilla, etc., as explained for (c) and (d). Note the extra units: basal plate and its paired mate, the laeobasal plate, and anterior teeth (see Fig. 6.15).

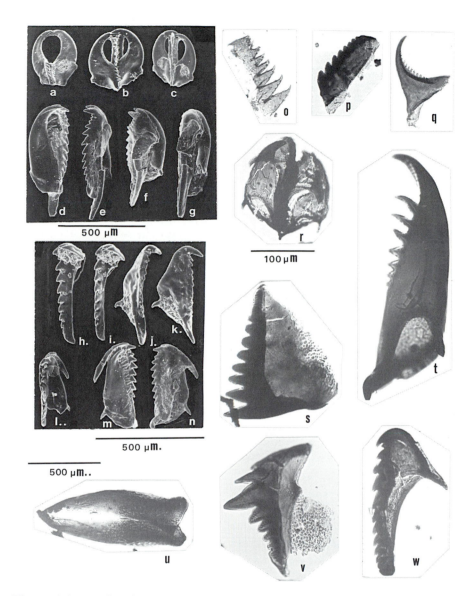

Figure 6.15 Scolecodonts, the chitinous "jaws" (= mouthpart elements) of marine polychaetous annelid worms. These fossils are usually found as isolated parts of the complex mouth lining of the worms, because they easily disarticulate; however, articulated or partially articulated apparatuses commonly occur. See Figure 6.14 for diagrams of the anatomical situation in the living worm. The maxillary jaws are usually numbered from the rear, the posteriormost maxillary elements being MI and MII. MIII is usually not paired. Other parts are named: basal tooth, mandibles, carriers, etc. (a)–(n) S.E.M. pictures of elements from Upper Devonian (Frasnian) of Northern Alberta; (r)–(w) photomicrographs of scolecodonts of same origin as (a)–(n); (o)–(q) scolecodonts from Ordovician, Anticosti Island, Quebec. Magnification for (a)–(g) and

141

doubtless improve their utility for practical palynostratigraphy; only a handful of specialists work on them. The fact that one species of worm may produce several genera of mouth linings is an annoyance but not a serious barrier to application, as the several taxa from one worm species will obviously have the same range!

(m), (n) indicated by bar under (d) and (e); magnification for (h)–(k) indicated by bar under (m) and (n) (bar and fossils set off by use of single dot with letters (h)–(k)); magnification for (l) (two dots) shown by bar under (l); magnification for (r)–(w) indicated by bar under (r); magnification for (o)–(q) is approximately the same as that for (a)–(g).

(a)–(c) *Xanthoprion albertensis* Jansonius & Craig, pair of MI, dorsal and ventral views. (d)–(g) *Elleriprion?* sp., MI and MII, left mandible: (d), (e) are dorsal views of assembled jaws with shaft of mandible below, (f) is oblique frontal view, (g) a full ventral view. (h)–(k) *Albertaprion* sp.: (h), (i) is a MIId, (j), (k) is an MId; (h)–(k) were articulated but separated in processing; (h), (i) would be put in genus *Leodicites*, (j), (k) in *Delosites* if found dispersed! (l) *Albertaprion?*, MId (posterior damaged) and MIID, dorsal view, (m), (n) *Albertaprion* sp. MIs and MIIs, oblique view of outer face, full lateral view of inner face. (o) *Mochtyella cristata* Kielan-Jaworowska. (p) *Marlenites millerae* Eller. (q) *Ungulites* sp. (r) *Albertaprion comis* (Eller), more or less complete maxillary apparatus: MId, MIId, MIs, MIIs. (s) *Paleoenonites triangularis*, showing squamate connecting tissue. (t) *Nereigenys* sp., MI. (u) A pair of carriers (see Fig. 6.14). (v) *Anisocerasites guttulatus* Taugourdeau. (w) *Leodicites divexus* Eller.

142

CHAPTER SEVEN

Silurian palynology

GENERAL DISCUSSION

In marine shales of earliest Silurian age, acritarchs, scolecodonts, and chitinozoans continue to provide the palynomorph assemblages encountered. (It is even possible that dinoflagellates were around.) However, in sediments from deltaic and other marginal environments of Llandovery age (Strother & Traverse 1979; see Figs. 7.1 & 7.2), such as the Tuscarora Fm. of Pennsylvania, there appear curious, apparently non–marine microfossils indicating the first experiments of plants with the sub-aerial milieu. Indeed, some researchers (Gray *et al.* 1982) have found these harbingers in rocks which they date as late Ordovician (Caradocian–Ashgillian). The suite of microfossils reported by Gray from late Ordovician of Libya seems essentially identical to that of the early Silurian of Pennsylvania and Virginia (Strother & Traverse 1979, Pratt *et al.* 1978). Either the Libyan material is really early Silurian or the material from Pennsylvania is really late Ordovician, or this type of palynoflora persisted for as much as 30 million years. As can be seen in Figure 7.2, the fossils consist of monads (these would be classified as acritarchs if found associated with more "normal" marine Silurian acritarchs; Richardson *et al.* (1984) have suggested the term "cryptospore" for the spore-like monads), dyads, triads, and tetrads, some with thicker exines than is ever normal for acritarchs. Forms such as *Tetrahedraletes* and *Nodospora* show peculiar thickenings along the contacts between members of the tetrad: a "contact ring". As pointed out by Johnson (1985), many of the dyads to tetrads have a membrane or extra wall enclosing the whole unit, perhaps suggesting algal origin. In rocks just a bit younger (Wenlock) monads with conspicuous, well-defined trilete laesurae are found: *Ambitisporites* sp., which was originally described by Hoffmeister (1959) from the Silurian of Libya, in rocks also somewhat younger than Llandovery. *Ambitisporites* is surely an embryophytic plant spore and indeed resembles closely some modern liverwort spores. In the Tuscarora Fm., along with abundant monads lacking laesurae, and the tetrads of *Tetrahedraletes*, etc., are also found puzzling pieces of cells and tissues: tubes of various kinds and sheets of cellular (pseudocellular?) material (Fig. 7.2). It is tempting to speculate that some of these represent other parts of the first sub-aerial plants. As is well-known, the first vascular plants do not occur until the very tiny upright plant, *Cooksonia*, of later Silurian time. *Ambitisporites* was soon joined by *Punctatisporites* and, soon after *Cooksonia*'s fairly common appearance in late Silurian time, about a dozen genera of land plant spores were occurring. Some of these, for example, the zonate *Ambitisporites*, do somewhat resemble certain liverwort spores. They might represent plants which, if they still occurred

Period	Approximate Duration in Years x 10^6	International Series Names	Paleobotanical/paleopalynological Developments
SILURIAN	6	Pridoli	Several taxa of simple vascular plants. About 20 genera of trilete spores, some >50 μm, some with well-expressed sculpture and other modifications.
	7	Ludlow	Simple vascular plants (Cooksonia) found in various localities. About 15 genera of small trilete spores.
	7	Wenlock	First, uncommon, small vascular plants. Common, small trilete spores: several genera, including Ambitisporites and Punctatisporites.
	10	Llandovery	Abundant tetrads and tissue fragments of probable land plants. Uncommon trilete spores; common spore-like tetrads, dyads, monads.
LATE ORDOVICIAN	10	Ashgill	Shales yield some spore-like tetrads and possible tissue fragments of land plants on maceration.
	10	Caradoc	

Figure 7.1   Land plants and palynomorphs derived from them, in late Ordovician and Silurian time. Source of chronological data: DNAG Geologic Time Scale, *Geology*, September, 1983.

144

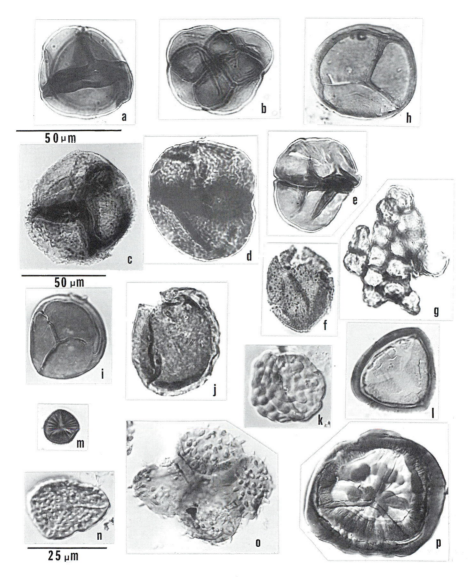

Figure 7.2 Silurian palynomorphs of non-marine origin, including obligate tetrads (a)–(c), (o), dyads and pseudodyads (d), (e), leiospheres (f), resistant-walled tissue fragments (g), double-walled monads (j), and trilete spores (h), (i), (k)–(n), (p). The 50 μm bar for (a) indicates magnification for all specimens except (c) and (n), which have their own bars. FM numbers in parentheses refer to specimen numbers of the British Museum (Natural History). (a) *Tetrahedraletes medinensis* Strother & Traverse, an obligate tetrad, Tuscarora Fm., Lower Silurian (Llandovery), Pennsylvania. (b) Same as (a), different view. Thickenings at contacts between individual spores or spore-like bodies produce the characteristic "pretzel" figures. (c) *Nodospora retimembrana* Miller & Eames, an obligate tetrad with conspicuously double-walled construction. Same source as (a). (d) *Pseudodyadospora rugosa* Johnson, an incompletely divided dyad-

145

today, would be classified as bryophytes. However, there is no evidence as yet that these spores are referrable to that group of extant plants, which includes mosses, liverworts, and hornworts. Figure 7.3 shows the distribution of kinds of sporomorphs found from the Llandovery stage through the Silurian and into the Lower Devonian. Richardson & Ioannides (1973) recorded a rich palynoflora of spore forms referred to 10 genera in Ludlow–Downton sediments of Libya, North Africa. Wood (1978) notes that the Ludlow of Indiana also has a considerable variety of spore types, reflecting the rapid evolution of the land flora. The relative abundance of acritarchs and other marine palynomorphs *vis à vis* spores is a useful indicator of marine influence, beginning with the Wenlock stage. For example, McGregor & Narbonne (1978) are able to hypothesize a sheltered nearshore marine environment for Ludlow beds of the Canadian Arctic, based on the mix of spores, acritarchs, chitinozoans, and scolecodonts. Richardson *et al.* (1981) emphasize that late Silurian spores are sensitive stratigraphic markers, as might be expected from the rapidly evolving plants which produced them. According to Richardson *et al.*, acritarchs of late Silurian time are more influenced by provinciality and local ecological factors. The early spores are more cosmopolitan and hence, when available, are better for stratigraphy. However, despite this fact, in the Silurian, the "norm" for palynology remains marine acritarchs (plus chitinozoans and scolecodonts). Figure 6.6 illustrates some of the characteristic Silurian acritarch forms. Sherwood-Pike & Gray (1985) have described probable ascomycete fungal remains from the Silurian, strongly suggesting the existence of land vegetation to provide sustenance for the ascomycetes.

"NON-SPORE" PALYNOLOGY

Many acritarch specialists have shown that acritarchs permit stratigraphic zonation of marine sediments throughout the Silurian. Dorning (1985), for example, has shown that acritarch assemblage zones can be established for the Ordovician and Silurian of eastern North America and northwest Europe with

---

like form. Same source as (a). (e) *Dyadospora murusdensa* (Strother & Traverse, a true dyad with complete cross walls. Same origin as (a). (f) *Leiospheridia acerscabrella* Johnson, a sac-like leiosphere. Same origin as (a). (g) Sheet of cells with resistant walls. Same origin as (a). (h) *Ambitisporites avitus* Hoffmeister, Lower Silurian (Llandovery), Libya, zonate trilete spore. (FM1). (i) *Retusotriletes warringtonii* Richardson & Lister, a non-zonate trilete spore with conspicuous curvaturae perfectae, but obviously not greatly different from *Ambitisporites*. Same origin as (h). (FM3). (j) Membrane-enclosed (= two-walled) monad. Same origin as (a). (k) Cf. *Synorisporites verrucatus* Richardson & Ioannides, trilete (not demonstrated in this view) spore with verrucae, Silurian (Wenlock–Ludlow), Libya. (FM12). (l) *Archaeozonotriletes chulus* (Cramer) Richardson & Lister, trilete zonate spore with widely gaping laesura. Same origin as (h). (FM5). (m) *Emphanisporites neglectus* Vigran, trilete spore with radiating exinal thickenings. Same origin as (k). (FM19). (n) Cf. *Brochotriletes* sp., trilete spore with foveolate sculpture. Same origin as (k). (FM22). (o) *Tetraletes variabilis* Cramer, obligate tetrad. Same origin as (k). (FM23). (p) *Emphanisporites splendens* Richardson & Ioannides, trilete spore with both radiating and circumpolar exinal thickenings. Same origin as (k). (FM30).

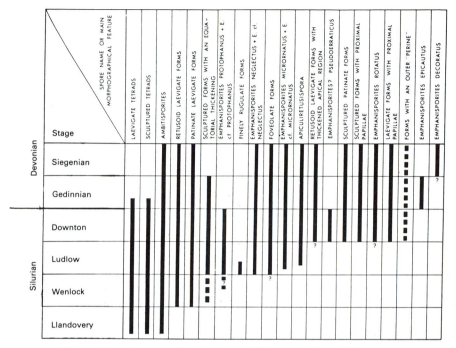

Figure 7.3 Ranges of spore types, Silurian to Lower Devonian.

an average duration of about a million years – about 55 zones for the two periods combined. On the other hand, as noted earlier, Silurian acritarchs were provincial and sensitive to local environments, so that Thusu (1973) could use *Deunffia* and *Domasia* complexes to suggest paleogeographical relationships of the source rocks, as Cramer (1969) had noted for Lower Silurian rocks of Pennsylvania. Acritarchs are so numerous and well-preserved in continuous sequences of marine sediments from Anticosti Island, Canada, that it has been proposed (Duffield & Legault 1982) to use stratigraphy based on them for standard general reference sections at the Ordovician/Silurian boundary and for the Lower Silurian.

Cramer (1970a) and Cramer & Diez (1972, 1974b) have described a series of acritarch-based belts for the Silurian of eastern North America, northwest Africa, and Europe. They suggested that these belts are parallel to paleolatitudes because they are determined by climates. Cramer (1970b) was even able to suggest a northwest movement of Pangea of 2–3 cm/yr.

In late Silurian time, more or less at the Ludlow/Downton (or Pridoli) boundary, about 405 million years ago (Gensel 1977), after the arrival of the little vascular plant, *Cooksonia*, vascular vegetation developed on land, and non-marine shales of that age and younger contain abundant land plant spores. The sporopolleninous exines "pioneered" by acritarchs and later by bryophyte-like green plants served presumably to protect the protoplasts of the haploid spores from desiccation, oxygen, ultraviolet light, and/or predation.

147

CHAPTER EIGHT

Devonian palynology

INTRODUCTION

The approximately 45 million years of the Devonian period was quite probably the finest hour of the embryophytic plants (see Figs. 7.3 & 8.1). Represented in the Gedinnian stage at the beginning of the period by only *Cooksonia*, the parade was soon joined in that stage by *Zosterophyllum*, a member of a primitive line probably related to the lycophytes, then in the Siegenian by the trimerophytes (*Psilophyton et al.*), many more lycophytes (*Baragwanathia*, *Drepanophycus*), and possible arthrophytes (= sphenopsids) such as *Protohyenia*. In the Emsian all of these lines continued, joined by cladoxylaleans (*Cladoxylon*) and progymnosperms (*Aneurophyton*). By the Givetian the protopterids were around (see Fig. 8.1). By latest Famennian time

(a)

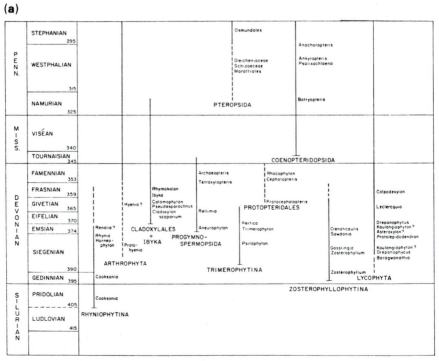

Figure 8.1 (a) Ranges of early vascular land plant groups, stressing the rapid evolutionary expansion in the Devonian, reflected also in the spore record, as shown in (b).

148

the progymnosperms had even produced true seeds. When this occurred, by definition, the microspore exine plus the microgametophyte thus developed inside it was a pollen grain. These primitive pollen grains, which morphologically were indistinguishable from microspores, are called prepollen.

All of the names mentioned in the preceding paragraph are for megafossil plants. The dispersed spores are separately named, and only relatively rarely do we find spores *in situ* in sporangia, giving us an idea of which *Sporae dispersae* go with which megafossil plants. A list of such correlations is provided in the next chapter. From what we know of modern plants, it is not surprising that one sort of megafossil plant may produce more than one sort of dispersed spore, and that one sort of spore may occur in several different megafossil taxa!

As can be seen in Figures 8.1 and 8.2 (more generic names could be added to Figure 8.2, and the ranges extended somewhat, but the overall picture is unchanged), the numbers of kinds of embryophytic spores increase steadily during the Devonian. Not surprisingly, this is in line with the increase in diversity of kinds of vascular plants, and with increasing variety and complexity of vascular plant organs. Figure 8.3 shows the concomitant diversification of spore features, both of general morphological characteristics and of exine construction. Chaloner (1970a) and others argue that the facts of steady increase in complexity and introduction of new forms of embryophytic plants from Lower Silurian through Devonian time, as reflected for example by spore morphology, bespeaks a strong likelihood of monophyletic origin.

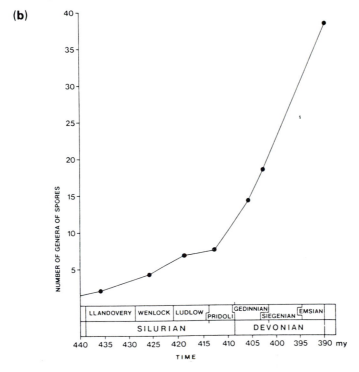

149

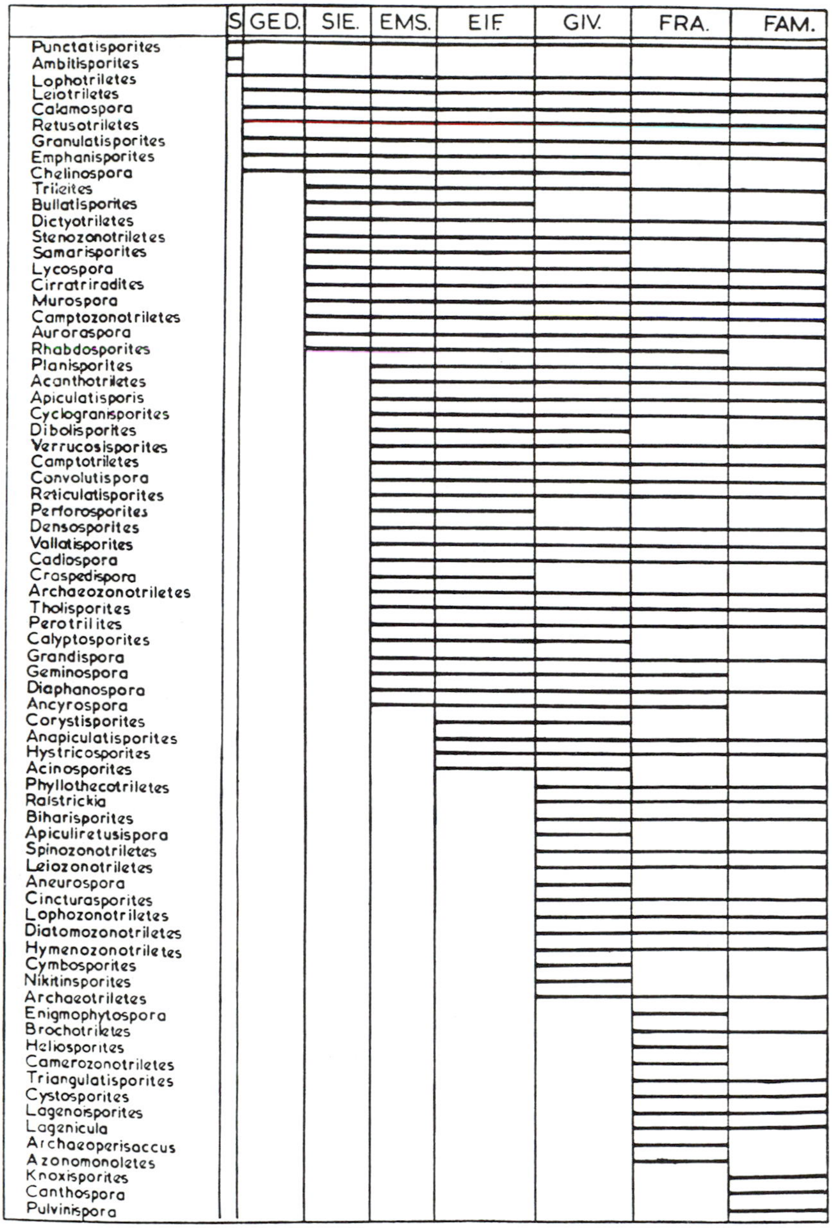

	S	GE.D.	SIE.	EMS.	EIF.	GIV.	FRA.	FAM.

Punctatisporites
Ambitisporites
Lophotriletes
Leiotriletes
Calamospora
Retusotriletes
Granulatisporites
Emphanisporites
Chelinospora
Triieites
Bullatisporites
Dictyotriletes
Stenozonotriletes
Samarisporites
Lycospora
Cirratriradites
Murospora
Camptozonotriletes
Auroraspora
Rhabdosporites
Planisporites
Acanthotriletes
Apiculatisporis
Cyclogranisporites
Dibolisporites
Verrucosisporites
Camptotriletes
Convolutispora
Reticulatisporites
Perforosporites
Densosporites
Vallatisporites
Cadiospora
Craspedispora
Archaeozonotriletes
Tholisporites
Perotrilites
Calyptosporites
Grandispora
Geminospora
Diaphanospora
Ancyrospora
Corystisporites
Anapiculatisporites
Hystricosporites
Acinosporites
Phyllothecotriletes
Raistrickia
Biharisporites
Apiculiretusispora
Spinozonotriletes
Leiozonotriletes
Aneurospora
Cincturasporites
Lophozonotriletes
Diatomozonotriletes
Hymenozonotriletes
Cymbosporites
Nikitinsporites
Archaeotriletes
Enigmophytospora
Brochotriletes
Heliosporites
Camerozonotriletes
Triangulatisporites
Cystosporites
Lagenoisporites
Lagenicula
Archaeoperisaccus
Azonomonoletes
Knoxisporites
Canthospora
Pulvinispora

Figure 8.2 Devonian spore ranges (S = Silurian; other abbreviations at top refer to Devonian stages (see Fig. 8.1)). Most forms occurring at the end of the Famennian also occur in the Carboniferous. Chaloner (1967) in publishing this chart noted that it is a fair indication of diversification of land plants in the Devonian, and this remains true despite the fact that some of the ranges have been extended by more recent information.

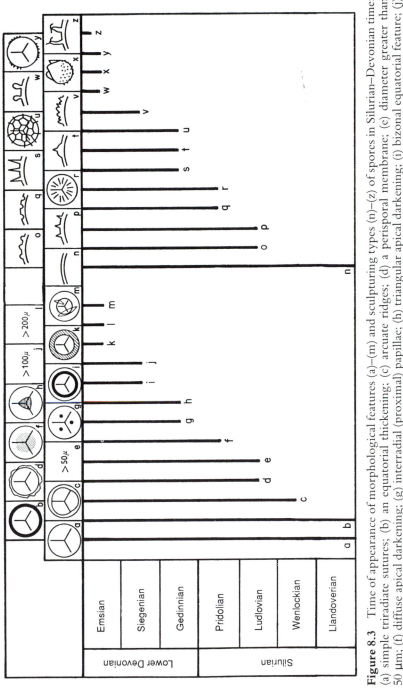

Figure 8.3 Time of appearance of morphological features (a)–(m) and sculpturing types (n)–(z) of spores in Silurian–Devonian time: (a) simple triradiate sutures; (b) an equatorial thickening; (c) arcuate ridges; (d) a perisporal membrane; (e) diameter greater than 50 μm; (f) diffuse apical darkening; (g) interradial (proximal) papillae; (h) triangular apical darkening; (i) bizonal equatorial feature; (j) diameter greater than 100 μm; (k) a cingulum; (l) diameter greater than 200 μm; (m) spore with saccus; (n) exine smooth; (o) exine papillate; (p) exine with cones; (q) exine reticulate; (r) radial ribs (muri) on the proximal face; (s) exine with spines; (t) exine with biform processes; (u) exine verrucose; (v) exine reticulate; (w) exine with clavate processes; (x) strong differentiation of prominent proximal unornamented face and distal, ornamented, hemisphere; (y) ornament concentrated interradially; (z) exine with grapnel-ended processes.

Regarding the developing provincialism of vegetation during the Devonian, studies of both spore and megafossil floras have demonstrated that detailed investigation reveals regional diversification, as would have been expected from great latitudinal spread of Devonian land masses (Raymond *et al.* 1985a). Use of multivariate techniques such as cluster analysis promises to elucidate much more about phytogeography from abundant Devonian spore data. In the later part of the early Devonian, evidence has already been summarized by Raymond *et al.* (1985a) for an equatorial–low latitude unit (North America–Eurasia), an Australian unit, and a south Gondwana unit. The reservation expressed by Raymond *et al.* (1985b) that palynofloral and megafossil plant data may not correspond phytogeographically does not agree with either modern or sufficiently studied fossil models. Indeed, fossil spores/pollen from sediments other than peat and/or coal generally do present, however, a more regionally representative picture than do megafossil floras.

In the late Devonian, despite the near-cosmopolitanism of some forms, particularly of the important latest Devonian marker, *Retispora lepidophyta* (Kedo) Playford, various taxa had restricted distribution, e.g., the early monolete form, *Archaeoperisaccus*, confined to a zone north of the presumed Devonian Equator, Hudson Bay to Scandinavia (McGregor 1979a).

PALEOZOIC SPORE MORPHOLOGY AND PERTINENCE TO THE DEVONIAN

In order to explain the course of Devonian spore developments, it is necessary to present some simple concepts about Paleozoic embryophytic spore morphology (see Fig. 8.4 for clarification). The fundamental type of embryophytic spore is trilete, as we have previously seen. When embryophyte spores first appear (excluding the obligate tetrad forms such as *Tetrahedraletes*, and monads which could be alete spores – "cryptospores") in the Llandovery–Wenlock, they are small (about 20–30 μm range), trilete, relatively psilate in sculpture, and simple. Some forms soon developed equatorial exinal thickenings (see Fig. 8.3). This makes the spores zonate. If the zona is thick, it is spoken of as a cingulum (Fig. 8.4). The contact areas between members of the original tetrad are the explanation for the trilete laesura's existence. These contact areas may be bounded on their outer sides by arcuate ridges (Figs. 5.6 & 8.4). This sort of feature is called a curvatura. If bounding the entire contact area, it is a curvatura perfecta; if fading out between ends of the radii, it is a curvatura imperfecta. Curvaturae are very important features of Devonian spore morphology. The zona and arcuate ridges were very early developments. Spores also quite soon began to include forms with an increase in size from the original 20–40 μm which still characterizes most homospores (= isospores). The 50 μm limit was reached in late Silurian time (Fig. 8.3). Increase in size of disseminules is a competitive advantage where environmental circumstances of dispersal favor greater investment in certain survival of a few disseminules, as opposed to random survival of a small percentage of many smaller ones, and may reflect dispersal other than by wind. At the end of the Silurian and early Devonian, exine modifications appear in greater diversity, presumably also

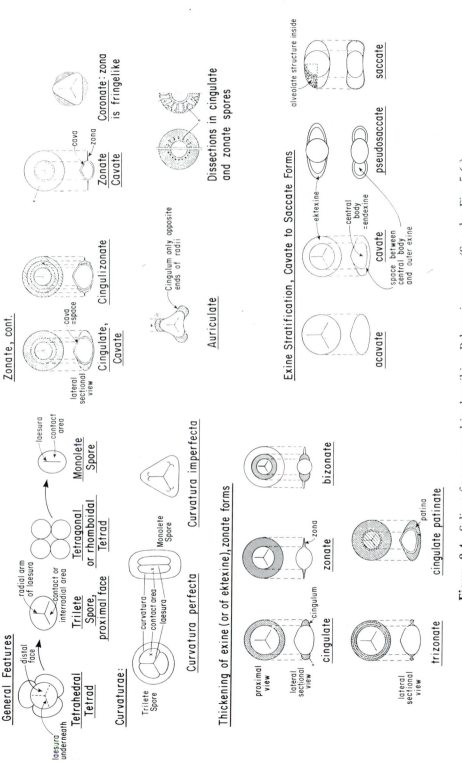

Figure 8.4 Salient features used in describing Paleozoic spores. (See also Fig. 5.6.)

reflecting dissemination requirements of some sort. Thickenings ("darkenings") of interradial areas of the proximal face of the spore (Fig. 8.3) came in, soon followed by interradial spot thickenings or papillae and triangular thickenings in the area of the intersection of the laesurae arms. In the Siegenian stage of the Devonian, there appears a further modification of the zonal equatorial feature that will continue to be important through the Paleophytic: bizonal, resulting from an equatorial extension that has two different thicknesses (see Figs. 8.3 & 8.4). In the Emsian, true cingula appear (like zonae, but thick), and spores with a saccus (or more exactly, pseudosaccus, because lacking internal structure of the "sacci"). Monolete spores are, from a phylogenetic point of view, derived from the basic trilete spore form, and they typically result from the spores being arranged in the original tetrad, not like a heaped-up tetrahedron, but with the centers of the spores more or less in one plane, like the sections of an orange (see Figs. 5.3 & 5.8). (Intermediate, dilete forms exist but are rare.) Monolete spores sometimes occur in modern ferns in the same sori as trilete spores. They have never been as important a spore type as trilete. They first occur in the Emsian.

Other important features of Paleozoic spore morphology not already mentioned in connection with the first appearance of morphological types are illustrated in Figure 8.4. The curvatura, already mentioned, is an important feature in Devonian spores, characterizing, for example, the genus *Retusotriletes*. Beginning already in the Devonian, Paleozoic spores show all sorts of changes on the equatorial extension or zonate theme, from simple zona or cingulum to bi- and trizonate and patinate (where a patina, or cingulum-like thickening, covers the distal surface of the spore). If the cingulum is confined to the ends of the laesura radii, the form is called valvate or auriculate; and if a fringe-like zona is seen (as in *Reinschospora*), the term coronate is employed to describe it (see Figs. 1.1 & 8.4).

It is common, beginning with late Devonian spores, for the exine to have layers which separate from one another. If the separation is relatively small, the condition is called camerate (cavate). If the cavea (separation) is relatively large and becomes vesicle- or blister-like, the spore or pollen grain is called pseudosaccate. If the vesicle or saccus is filled with alveolar or webby contents instead of being empty, the spore or pollen grain is saccate. (As far as is known, truly saccate forms are always pollen.) Intermediate forms occur which are hard to "shoehorn" into cavate (camerate), pseudosaccate, and saccate.

Along with the constructional changes noted during the Devonian, Chaloner (1970a) also pointed out (see Fig. 8.3) that exine sculpture changed in a regular progression. *Ambitisporites* of the Llandovery–Wenlock stages of the Silurian was smooth (psilate). Scabrate and slightly echinate (with coni) forms appeared soon after. Foveolate pitted exines appeared in latest Silurian time, as did radial ribbing (as in *Emphanisporites*) on the proximal surface. Truly spiny (echinate) sculpture began in the Gedinnian, along with biform appendages, e.g., spines on top of small mammae, and reticulate sculpture. Verrucose exines appeared in the Siegenian. Clavate sculpture is first encountered in the Emsian, along with spores with a bipolar difference in sculpture between proximal and distal faces. Sculpture confined to limited interradial areas also comes in during the Emsian, as well as exinal processes with curious "grapnel" hooks. Similar hooks are fairly common in both acritarchs and dinoflagellate cysts, but were never again to occur among spores, as far as I know. A rather complex terminology has grown up for description of the complexities of Paleophytic spore morphology. A compendium of such terminology, mostly from Smith & Butterworth (1967), is presented in Figure 8.4. However, if possible, students should also consult the very useful summary of such terminology in Grebe (1971). Figures 8.5 and 8.6 present illustrations of the more prominent genera of Lower, Middle, and Upper Devonian spores. Note that, in the next chapter, a listing of "Paleophytic" (see Fig. 2.1) spores per their "turmal" classification and as to their relationship to producing plants is given. This list also includes Devonian spores.

MEGASPORES, SEEDS, AND POLLEN

We have already noted that certain spores reached the 50 µm size range in late Silurian time (see Fig. 8.3). This already most likely represents a trend of some sort, as the average modern spore size is around 30–40 µm. A much clearer indication of trend is shown by the first 100 µm spores in Siegenian (Lower Devonian) time, and the first 200 µm spores in the Emsian. We have already noted that 200 µm is the arbitrary limit between "miospore" and "megaspore", but it works pretty well also as the size limit between the functional biological categories of megaspore and microspore (see Fig. 8.7). In other words, the 200 µm spores of late Devonian time were almost certainly true, functional megaspores. However, some genera such as *Hystricosporites* range well below and well above the 200 µm limit and are thus, in this sense, both megaspores and miospores! One Devonian form, *Cystosporites*, can be over a centimeter(!) in largest dimension, consisting of one very large functional spore and three aborted spores closely adhering to it. It is natural to speculate that, though this spore was produced by a lycopsid and was biologically a megaspore, it functioned from a practical point of view almost as a seed, hence "seed megaspore."

The trend toward existence of some much larger spores was a big feature of the Devonian. This increase in disseminule size certainly represented a stage in the move toward the seed habit. The large investment in a limited number of megaspores was perhaps an advantage for establishing a plantlet in a hostile environment. After late Devonian–Carboniferous time, functional, free megaspores were never again so important, though a few heterosporous non-seed plants, e.g., *Selaginella, Marsilea,* and *Azolla,* continue to produce them still. The study of megaspores is therefore especially rewarding in the Devonian and Carboniferous. Their potential stratigraphic and paleoecological usefulness has been amply demonstrated in such works as Chi & Hills (1976) in Arctic Canada, Higgs & Scott (1982) in Ireland, Scott & King (1981) in England, and Candilier *et al.* (1982) in North Africa. In our laboratory, Stolar (1978) has shown that the Devonian/Carboniferous boundary in central Pennsylvania can be demonstrated with megaspores as well as with miospores.

The techniques for processing and study of megaspores are different from

155

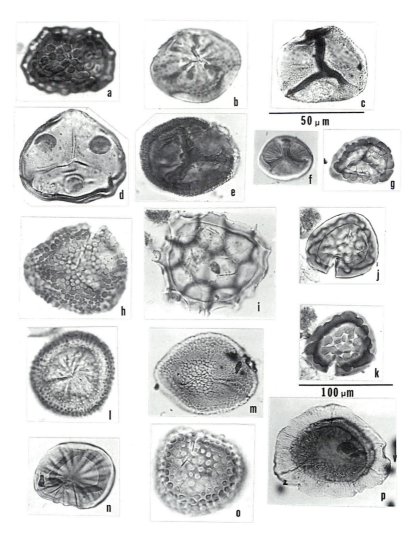

Figure 8.5 Photomicrographs of Lower and Middle Devonian spores. Magnification for (a)–(f), (h), (i), (l), and (n) indicated by bar under (c), except that for (b) the same line represents 25 μm. A separate line is provided under (k) for (g), (j), (k), (m), (o), and (p). Magnification for (r)–(y) indicated by bar under (v), except for (r) which has its own bar. (a) *Cymbosporites verrucosus* Richardson & Lister, Gedinnian, U.K. (FM29). (b) *Emphanisporites microornatus* Richardson & Lister, Siegenian, U.K. (FM35). (c) *Apiculiretusispora plicata* (Allen) Streel, Emsian, Quebec (GSC-15153). (d) *Retusotriletes maculatus* McGregor & Camfield, Siegenian, Ontario (GSC-41706). Note interradial thickenings. (e) *Perotrilites microbaculatus* Richardson & Lister, Gedinnian, U.K. (FM38). (f) *Emphanisporites epicautus* Richardson & Lister, Gedinnian, U.K. (FM37). (g) *Clivosispora verrucata* McGregor var. *verrucata*, Emsian, Quebec (GSC-15179). (h) *Verrucosisporites polygonalis* Lanninger, Emsian, Quebec (GSC-31989). (i) *Dictyotriletes emsiensis* (Allen) McGregor, Siegenian, Ontario (GSC-41734). (j), (k) Two focal levels of *Clivosispora verrucata* McGregor var. *convoluta* McGregor & Camfield, Siegenian,

156

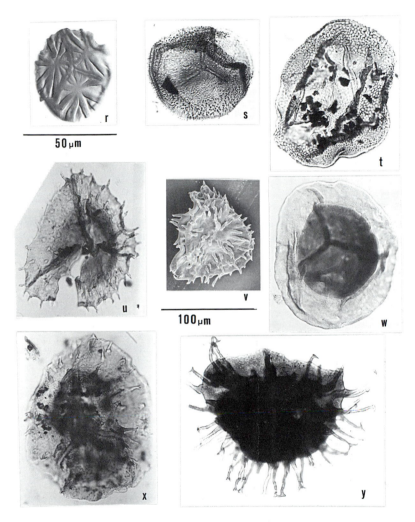

Ontario (GSC-27081). (l) *Emphanisporites decoratus* Allen, Siegenian, Ontario (GSC-27076). (m) *Dictyotriletes favosus* McGregor & Camfield, Emsian, Quebec (GSC-15181). (n) *Emphanisporites annulatus* McGregor, Emsian, Quebec (GSC-32002). (o) *Brochotriletes hudsonii* McGregor & Camfield, Siegenian, Ontario (GSC-41729). (p) *Camptozonotriletes caperatus* McGregor, Emsian, Quebec (GSC-15197). Note filmy zona, approaching saccus structure. (r) *Emphanisporites schultzii* McGregor, Emsian, Quebec (GSC-32008). (s) *Dibolisporites echinaceus* (Eisenack) Richardson, Eifelian, Quebec (GSC-31972). (t) *Dictyotriletes canadensis* McGregor, Emsian, Quebec (GSC-32001). (u) *Grandispora douglastownense* McGregor, Emsian, Quebec (GSC-32034). (v) *Hystricosporites micro-ancyreus* Riegel, Emsian, Germany. (w) *Rhabdosporites langii* (Eisenack) Richardson, Eifelian/Givetian boundary, Scotland (FM46). (x) *Ancyrospora loganii* McGregor, Emsian, Quebec (GSC-15282). (y) *Hystricosporites* cf. *corystus* Richardson, Emsian, Germany. Numbers in parentheses are Geological Survey of Canada (GSC) or British Museum (FM) specimen numbers.

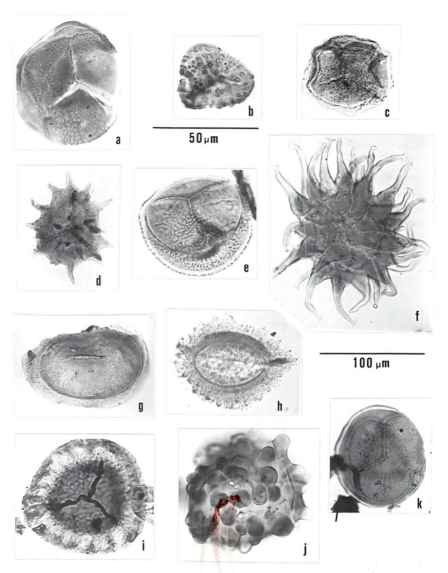

Figure 8.6 Middle and Upper Devonian spores. Magnification for (a), (b), (e), (g), (i), and (k) indicated by bar under (b). Magnification for (c), (d), (f), and (h) indicated by bar under (f). Devonian stage name for source immediately follows name of spore. (a) *Retusotriletes rugulatus* Riegel, Givetian, Melville Island, Arctic Canada (GSC-66379). (b) *Lophotriletes devonicus* (Naumova ex Chibrikova) McGregor & Camfield, upper Eifelian–Lower Givetian, Melville Island, Arctic Canada (GSC-66356). (c) *Perotrilites conatus* Richardson, Eifelian/Givetian boundary, Scotland (FM57). (d) *Ancyrospora melvillensis* Owens, Frasnian, Ellesmere Island, Arctic Canada (GSC-73295). (e) *Geminospora lemurata* Balme, Givetian, Melville Island, Arctic Canada (GSC-66296). (f) *Hystricosporites gravis* Owens, Givetian, Melville Island, Arctic Canada (GSC-15550).

158

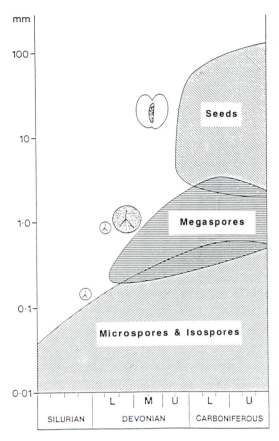

Figure 8.7 The transition from plants producing only isospores, to heterosporous plants with two kinds of spores, microspores and megaspores, came about in the Lower Devonian. It is probable that when the 100 μm (0.1 mm) size limit was crossed in the Lower Devonian, at least some of the 100 μm spores were functional megaspores. When the 200 μm boundary was crossed just a little later, most of the spores greater than 200 μm were surely functional megaspores. Free-sporing megaspores range up to about 5,000 μm (5 mm). Hence, in size, megaspores overlap with large isospores–microspores, and with small seeds. Seeds first appeared in uppermost Devonian and range in maximum size to about 100 mm (10 cm) by Upper Carboniferous.

(g) *Archaeoperisaccus ovalis* Naumova, Frasnian, Melville Island, Arctic Canada (GSC-73296). (h) *Archaeoperisaccus opiparus* Owens, Frasnian, Melville Island, Arctic Canada (GSC-73298). (g), (h) Very early monolete spores. (i) (?) "*Hymenozonotriletes denticulatus* Naumova" McGregor 1967 (GSC-49850). (j) *Lophozonotriletes lebedianensis* Naumova, Famennian, Bathurst Island, Arctic Canada (GSC-15422). (k) *Retusotriletes phillipsii* Clendening *et al.*, Famennian, New York (FM64). Numbers in parentheses are Geological Survey of Canada (GSC) type specimen numbers, except for (c) and (k), for which British Museum (FM) specimen numbers are given.

Figure 8.8 Latest Devonian and Carboniferous megaspores. Late Devonian to mid-Carboniferous was the heyday of sporopolleninous free-sporing megaspores. All of the pictures shown are S.E.M. micrographs, but megaspores can also be photographed with appropriate lenses by reflected light, as microphotographs. Transmitted light study is not so satisfactory, because of the thickness of the specimens, but can be useful for demonstration of a central body. All of these S.E.M. micrographs are at much smaller magnification than is true for photomicrographs of miospores elsewhere in this book. Magnification varies considerably and size is given with each caption. (a)–(l) are from latest Devonian, Ireland; (n)–(o) and (r) are from the Lower Carboniferous of

160

those for miospores, because of their size and abundance (see Appendix). Whereas paleophytic miospores commonly have a concentration of 5,000–10,000 per gram of sediment, megaspores are always much (several orders of magnitude) less abundant. Hills (1984) describes Devonian–Carboniferous megaspores with an abundance of 100/g as "extremely common"! Figure 8.8 shows some typical Paleozoic megaspores.

POLLEN VS. SPORE MORPHOLOGY, POLARITY, AND GERMINATION

It is necessary at this point to digress somewhat to review and expand what we have learned earlier of basic spores/pollen morphology, in order to have a clear

---

Scotland; and (m), (p), (q) and (s)–(u) are Carboniferous megaspores reworked into Tertiary, southern England. (a) *Sublagenicula* cf. *nuda* (Nowak & Zerndt) Dybova-Jackowicz *et al.* Lateral view, length 830 μm. The upper section is the contact area with a three-cornered "hat" appearance. This is termed the gula, but it is only modestly developed here. (b) *Auritolagenicula rugulata* Higgs & Scott. Lateral view, length 1,100 μm (photo is at low magnification!). The gula is here very fully developed, with auriculate extensions of the trilete rays. (c) *Triangulatisporites*(?) *leinsterensis* Higgs & Scott. Proximal view, diameter 1,120 μm. (d) Same species as (c). Specimen 1,000 μm in diameter but more magnified than (c). Proximal view. (e) Same species as (c), (d). Distal view, showing well developed zona, length 920 μm. (f) *Hystricosporites delectabilis* (McGregor) McGregor. Lateral view, length 600 μm. Note "grapnel hooks" on ends of processes, characteristic for the genus. Fragments of these processes are frequently found in miospore preparations. (g) *Hystricosporites* cf. *multifurcatus* (Winslow) Mortimer & Chaloner. Length 460 μm. (h) Same species as (g), detail. (i) *Ancyrospora furcula* Owens. A genus quite similar to *Hystricosporites*, with specimens sometimes well below the 200 μm usual cutoff for megaspores. Length of this specimen 380 μm. (j) Same species as (i). Proximal surface, demonstrating the zonate nature of *Ancyrospora*, a feature helping to distinguish it from *Hystricosporites*. (k) *Hystricosporites winslovae* Higgs & Scott. Proximal view, length 300 μm. A somewhat flattened specimen showing from the one complete process what has happened to the others. *Hystricosporites* process tips tend to be broken off. (l) *Hystricosporites winslovae* Higgs & Scott. Distal view, length 500 μm. (m) *Zonalessporites brassertii* (Stach & Zerndt) Potonié & Kremp. Proximal view, diameter 1,000 μm. (n) *Setispora subpaleocristatus* (Alvin) Spinner. Lateral view showing an erect laesura with anastomosed hairs along the radii, and scattered echinate sculpture on the distal side. Maximum dimension, 1,430 μm. (o) *Lagenicula crassiaculeata* Zerndt. Lateral view showing apical prominence or gula, being the "ruffled" contact area. Maximum dimension, 1,830 μm. (p) *Lagenicula horrida* Zerndt. Proximal view showing the contact area and modest gula. Diameter 800 μm (see (q)). (q) Same species as (p). Lateral view. Maximum dimension 960 μm. (r) *Lagenicula subpilosa* (Ibrahim) Potonié & Kremp forma *major* Chaloner ex Dijkstra. Lateral view showing the apical prominence or gula, and the pilose sculpture. Diameter 1,200 μm. (s) *Setosisporites* cf. *hirsutus* (Loose) Ibrahim. Obliquely proximal view showing the contact area, "ruffled" laesural radii, and strands of exine sculpture. Maximum dimension 580 μm. (t) *Setosisporites hirsutus* (Loose) Ibrahim. A different variety from that in (s). Proximal view showing contact area and small gula. Diameter 560 μm. (u) Same species as (t). Further illustrating sculptural variability of this species. Maximum dimension 600 μm.

grasp of what the morphology of Paleozoic spores/pollen means. Figures 8.9 and 8.10 help review the facts. Basic homosporous spores are what existed in Silurian and Lower Devonian time. The laesura is a scar representing the place of the spore in its original tetrad – in other words, it is a haptotypic feature. It also serves as a zone of weakness for the opening of the spore for the germination of the contents. As seen in Figure 8.10, the spore thus opens proximally for the gametophyte to begin to grow. Spermatozoids and eggs are produced by this bisexual gametophyte. Heterospory probably was under way when spores larger than 100 μm appeared in Siegenian time, but unquestionably was present when spores larger than 200 μm appeared in Emsian time. As can be seen from Figure 8.10, microspores germinate in the same manner as homospores, along the laesura on the proximal side of the spore. It is not shown in the diagram, but the megaspores also germinate in the same manner.

Seed plants appeared in the Famennian. The term "seed" means that a single megaspore germinates in the megasporangium to produce a megagametophyte, the egg of which is fertilized *in situ*. The associated microspores in the Famennian were by definition pollen grains (prepollen = pollen with morphological features of spores) which germinated on or very near the opening to the megasporangium, producing male gametophytes, the spermatozoids of

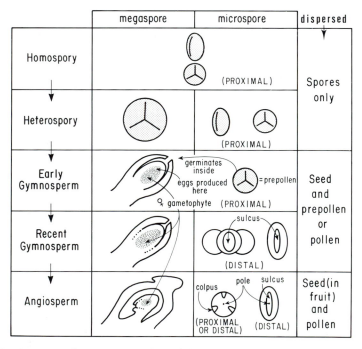

Figure 8.9 Spore evolution. Note that early gymnosperm prepollen germinated proximally, as do pteridophyte spores. Recent gymnosperm pollen germinates distally at the sulcus, and angiosperm pollen germinates either distally (sulcus) or laterally (colpus), or otherwise.

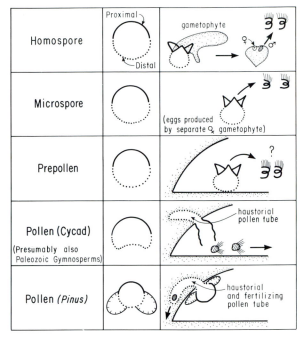

Figure 8.10 Diagram of evolutionary change in polarity of spore germination. Homospores germinate proximally by dehiscence along the laesura. The usually bisexual gametophyte produces eggs and sperm. Microspores germinate proximally just as homospores, but the male gametophyte is very reduced, producing practically nothing but sperm. Prepollen germinated in a pollen chamber of the female, seed-producing organ. Primitive gymnosperm pollen presumably germinated as modern cycad pollen – distally to produce a haustorial pollen tube but proximally to release sperm. In modern conifers, the pollen germinates distally to produce a pollen tube, but this acts both haustorially and to convey the gamete nuclei, the method also used by angiosperms.

which fertilized the egg of the female gametophyte, producing a seed. Note in Figures 8.9 and 8.10 that a prepollen grain behaved just as the microspore it is; it germinated proximally. A little later, presumably in the Pennsylvanian, gymnospermous pollen developed pollen tubes which emerged distally. Both fossil evidence and observations of some present–day cycads show that the pollen tube's function was originally haustorial, i.e., it acted as a "root" to absorb nutrients to support its life processes. The grain continued to open proximally for the generative purpose. Only later did more advanced gymnosperms abandon proximal germination of the pollen and used the pollen tube, produced distally from a germinal furrow (colpus), for both haustorial and generative purposes. Angiosperms, of course, do basically the same thing – the pollen tube of *Zea mays*, for example, comes out of the (distal) pore, and may traverse a half-meter of style ("silk"), "living off the land" haustorially and carrying its important message to the ovule.

NON-SPORE PALYNOMORPHS IN THE DEVONIAN

Acritarchs in marine Devonian rocks are important and useful palynomorphs. They have also been used in paleoecological reconstruction of paleoenvironments. Studies of them, however, are still scarce, and their undoubted stratigraphic paleoecological potential has not been realized as yet. Riegel (1974), for example, has shown that Devonian neritic facies of Europe have relatively depauperate acritarch assemblages compared with associated pelagic facies. Some marine Devonian acritarchs may well prove to be dinoflagellates eventually. Paleoecologic studies of Devonian acritarchs are still needed, although Nautiyal (1977) has identified acritarch facies areas that may be climatically controlled. Playford & Dring (1981) have demonstrated the usefulness of acritarch studies in Australia, and Wicander & Wood (1981) and Wicander (1983, 1984) have put Devonian acritarchs on the map in North America and have summarized the considerable potential stratigraphic use of acritarch studies in the North American Devonian. Chitinozoans and scolecodonts also continue to be significant palynomorphs in the marine Devonian.

DEVONIAN PALYNOSTRATIGRAPHY

Paleopalynology, especially spores/pollen-based, has been much used in connection with various Devonian biostratigraphic problems. McGregor in North America (Fig. 8.11), Streel, Richardson, and Clayton in northwest Europe, and Kedo in the U.S.S.R. have been particularly active in establishing palynostratigraphic frameworks for the Devonian. Figure 8.12 shows a summary of the application of palynostratigraphic methods to establishing the boundary between Middle and Upper Devonian. Figure 8.13 displays some of the prominent spore types employed in Upper Devonian to Lower Carboniferous palynostratigraphy. (See Figure 9.1 on p. 172: Mississippian in North America is equivalent to Lower Carboniferous in some classifications but includes some of the lower part of the Upper Carboniferous in others.)

One especially interesting palynostratigraphic problem has been that of the Devonian/Carboniferous (Mississippian) boundary. Particularly significant have been Streel's investigations on the stage type section in Belgium. The Carboniferous Congress in Heerlen in 1935 established the base occurrence of rock containing the ammonoid mollusc, *Gattendorfia subinvoluta*, as the boundary. Unfortunately, coupling this internationally accepted horizon with the information about type sections for the stages of the Upper Devonian and Lower Carboniferous has resulted in the boundary being set, not between the Tournaisian and the Famennian, as would be desirable, but well within the Lower Tournaisian, within its Tn1b subdivision. In other words, the lowest Tournaisian is Devonian. Streel showed that this boundary could be well-established palynologically in Belgium on the basis of the last appearance en masse of *Retispora lepidophyta* (alias *Hymenozonotriletes lepidophytus* or *Spelaeo-*

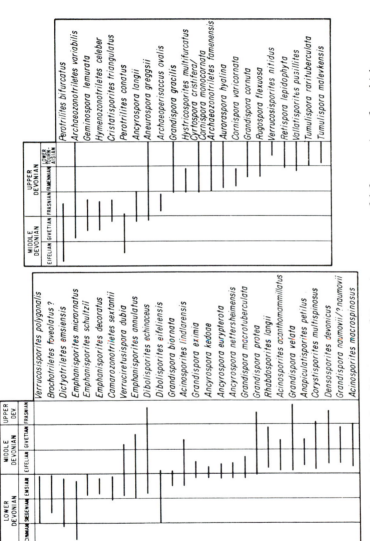

Figure 8.11 Stratigraphic ranges of Devonian spores useful for zonation, for which age ranges have marine faunal control: Lower and Middle Devonian to the left, late Middle and Upper Devonian to the right.

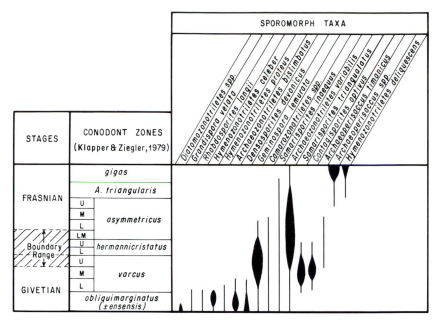

Figure 8.12 Graphical summary of palynological contributions to stratigraphy of the Middle vs. Upper Devonian. The internationally recognized standard is based on conodont zonation in Europe. The Givetian/Frasnian boundary is now established by international agreement at the base of the *asymmetricus* conodont zone. The spores/pollen ranges correlated with the conodont zones can be extended over wide areas and into non–marine sections such as the Catskill Magnafacies of New York–Pennsylvania, U.S.A.

triletes lepidophytus of various authors) in the uppermost Devonian, along with other forms such as *Cirratriradites hystricosus* and *Rugospora flexuosa* (= "*Hymenozonotriletes famennensis*"). Streel and I examined the classic American section at Horseshoe Curve, Pennsylvania, and other related sections in central Pennsylvania. Assuming that the European and American palynofloral zones are comparable, as would be expected from paleogeography, the Devonian/Mississippian boundary in central Pennsylvania (within the Tn1b) occurs well up in the Pocono Formation (Middle Sandstone unit), a rock unit traditionally regarded as Mississippian (see Fig. 8.14). The top of the *Retispora lepidophyta* horizon has been used for palynological location of the Devonian/Carboniferous system boundary in other places as well (see Fig. 8.15).

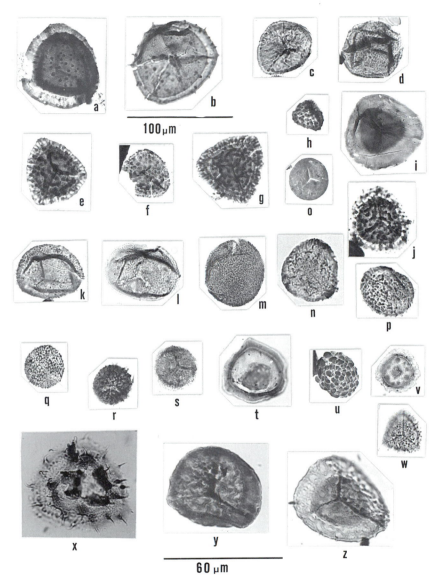

Figure 8.13 Characteristic spores of the Devonian/Carboniferous (Mississippian) boundary. Magnification for (a)–(w) indicated by bar under (b). Magnification for (x)–(z) indicated by bar under (y). Spores (a)–(w) are from the uppermost Devonian of the Schiefergebirge of the Rhine area, Germany: (a)–(j) are from LE Biozone in the Riescheid section; (k)–(w) are from LN Biozone in Seiler trench B. Spores (x)–(z) are from the uppermost Devonian of Centre County, Pennsylvania. (a) *Hymenozonotriletes explanatus* (Luber) Kedo. (b) *Grandispora echinata* Hacquebard. (c) *Rugospora flexuosa* (Jushko) Streel. (d) *Cyclogranisporites* sp. (e) *Vallatisporites verrucosus* Hacquebard. (f) Same as (e), to show variability. (g) *Vallatisporites pusillites* (Kedo) Dolby & Neves. (h)

167

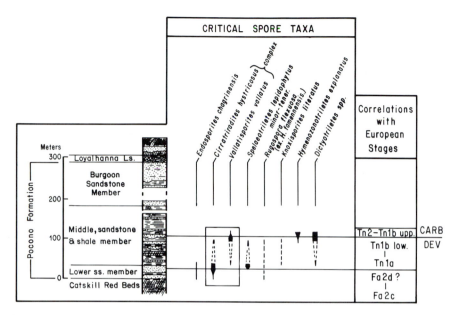

Figure 8.14 Horseshoe Curve, near Altoona, Pennsylvania: placement of Devonian/ Mississippian (= Lower Carboniferous) boundary in the Pocono Formation (in the Tn1b) by palynology. (N.B. *Spelaeotriletes lepidophytus* is the same taxon as *Retispora lepidophyta* and *Hymenozonotriletes lepidophytus*.)

Archaeozonotriletes minutus Kedo. (i) *Diducites mucronatus* (Kedo) Van Veen. (j) *Vallatisporites pusillites* (Kedo) Dolby & Neves; compare with (g). (k) *Apiculiretusispora verrucosa* (Caro-Moniez) Streel. (l) *Cyclogranisporites* sp.; compare with (d). (m) *Convolutispora ampla* Hoffmeister, Staplin & Malloy. (n) *Pulvinispora scolephora* Neves & Ioannides. (o) *Punctatisporites planus* Hacquebard. (p) *Pustulatisporites* sp. (q) *Camptotriletes paprothii* Higgs & Streel. (r) *Raistrickia* sp. (s) *Rugospora* sp. (t) *Tumulispora ordinaria* Staplin & Jansonius. (u) *Verrucosisporites nitidus* (Naumova) Playford. (v) *Lophozonotriletes triangulatus* (Ischenko) Hughes & Playford. (w) *Vallatisporites vallatus* Hacquebard. (x) *Cirratriradites hystricosus* Winslow – some regard this as part of a complex or "palynodeme" with (g), (j) and (w). (y) *Rugospora flexuosa* (Jushko) Streel; compare with (c). (z) *Retispora lepidophyta* (Kedo) Playford.

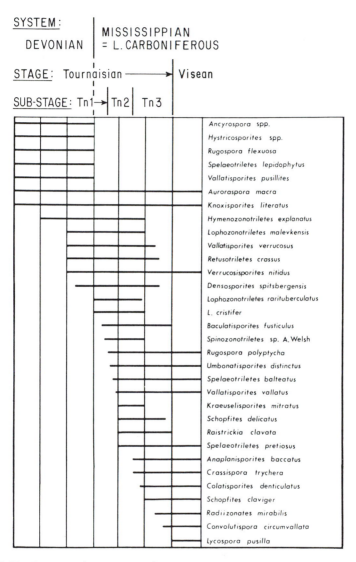

Figure 8.15 Stratigraphic ranges of important miospore taxa in Tournaisian and earliest Visean of central Ireland. The vertical lines represent palynologically based zones.

CHAPTER NINE

Carboniferous–Permian palynology to the end of the "Paleophytic"

INTRODUCTION

Carboniferous paleopalynology is, in one sense, where pre-Quaternary palynology all began. To be sure, Ehrenberg and Goeppert had both looked at palynomorphs of other pre-Pleistocene ages, but this is of little more than historical interest, as was Dawson's study of probably Carboniferous megaspores, which led ultimately to publication of the genus *Sporites* by H. Potonié in 1883. Reinsch in 1884 published the first photomicrograph of a fossil sporomorph, a Carboniferous form later named by Schopf *et al*. (1944) "*Reinschospora*" in his honor. Bennie & Kidston (1886) described Carboniferous megaspores.

Thiessen in the early 1900s described spores in Carboniferous coal thin sections and noted that they might be used stratigraphically. Fortunately, very few spores have been formally described and named from rock thin sections, as they are devilishly difficult to study in this "mountant". The best known examples of genera of fossil spores that were so described are various genera of Wodehouse (1932, 1933), such as *Momipites*, described from thin sections of Eocene Green River Oil Shale, and various genera such as *Baculexinis*, described by Stach (1957) from Carboniferous coal thin sections. (The latter are actually sections of the sporomorphs, whereas Wodehouse's specimens are whole spores/pollen in the rock thin sections.) The British "protopalynologist", Raistrick, made a concerted effort in the 1920s to employ fossil spores/pollen for the correlation of British Carboniferous Coal Measures. Two factors have conspired to keep from Raistrick as much credit as he clearly deserves for pioneer work. First, he did not name his groups of spores with formal, binomial Latin nomenclature, using instead code designations of letters and numbers. Many would feel he used uncommonly good sense, but the fact is that the people whose names pop up constantly in the paleopalynological literature, and are hence remembered, are mostly those who, rightly or wrongly, have named many taxa. Secondly, coal beds are notoriously difficult to correlate palynologically, mostly because coal beds are a very atypical biofacies, in which the palynoflora is nearly 100% autochthonously derived, not representing the regional flora. Nevertheless, in some sense Raistrick should be recognized as a founder, if not the father, of stratigraphic palynology.

When Robert Potonié began in the late 1920s and early 1930s to study fossil spores/pollen in detail, he studied both Cenozoic and Carboniferous coals, at first mostly the palynofloras of Paleogene brown coals (lignites) of central Europe, but a little later Carboniferous coals as well. At this point, World War II played an important role in developments. Potonié served in the German army on the Eastern Front and was a prisoner of war. While a prisoner, he developed his turmal system for artificial classification of sporomorphs. (Germs of the ideas were already present in his earlier publications and came in part from his students and associates, such as Ibrahim-Okay.) The system is especially useful for Paleozoic, notably Carboniferous, spores. Meanwhile, in North America, James M. Schopf and others had begun studying the spores/pollen of the American Pennsylvanian (see Fig. 9.1) coals in the 1930s. Schopf was very interested in taxonomy and nomenclature, and was always convinced that fossil sporomorphs should be described and named formally as paleobotanical entities representing fossil genera and species of plants that happened to be known provisionally from the spores/pollen. The first really comprehensive taxonomic treatment of Carboniferous spores was that of Schopf et al. (1944). This was also the first important comprehensive taxonomic treatment of fossil palynomorphs generally. Potonié's efforts in the 1930s with the Paleogene brown coal pollen/spores of Germany were somewhat confused by the problem that such pollen/spores represent largely plants with extant relatives. It is interesting that Potonié's turmal system is least useful for the sporomorphs he knew in most detail, those of the Cenozoic! In any event, we must now present the system because it is in well-nigh universal use for Carboniferous spores.

POTONIÉ'S TURMAL SYSTEM AND MODIFICATIONS OF IT

Perhaps because of the military setting in which the turmal system was developed, Potonié used terms from the Roman army for the categories in the system. The largest unit is the Anteturma, more or less like an army. In botanical terms, it is more or less equivalent to Class. Potonié set up an Anteturma Sporites for spores and an Anteturma Pollenites for pollen. Already he was in trouble, for the distinction between spores and pollen is functional, not morphological. Under the Anteturmas, the next smaller group is the Turma (= military division), dependent on major morphological features and more or less equivalent to Order in conventional taxonomy. Under Sporites are two large Turmas, Triletes and Monoletes, and two very small Turmas, Hilates (see Fig. 9.2) and Aletes. Under Anteturma Pollenites are Turmas Saccites (saccate) and Plicates (with colpi or colpi-like structures). The next smaller unit is the Subturma, taxonomically more or less equivalent to Family, based on smaller morphological differences, e.g., within Triletes whether member spores are zonate (Zonotriletes) or not (Azonotriletes). Note that some authors interpolate another category, Suprasubturma, between Turma and Subturma (see Fig. 9.2). Below the Subturma is the Infraturma, more or less equivalent to Subfamily, which in Sporites is based on major sculptural differences, e.g., Infraturma Apiculati (positive sculpturing consisting of projections) in the Subturma Azonotriletes. Under Infraturma is

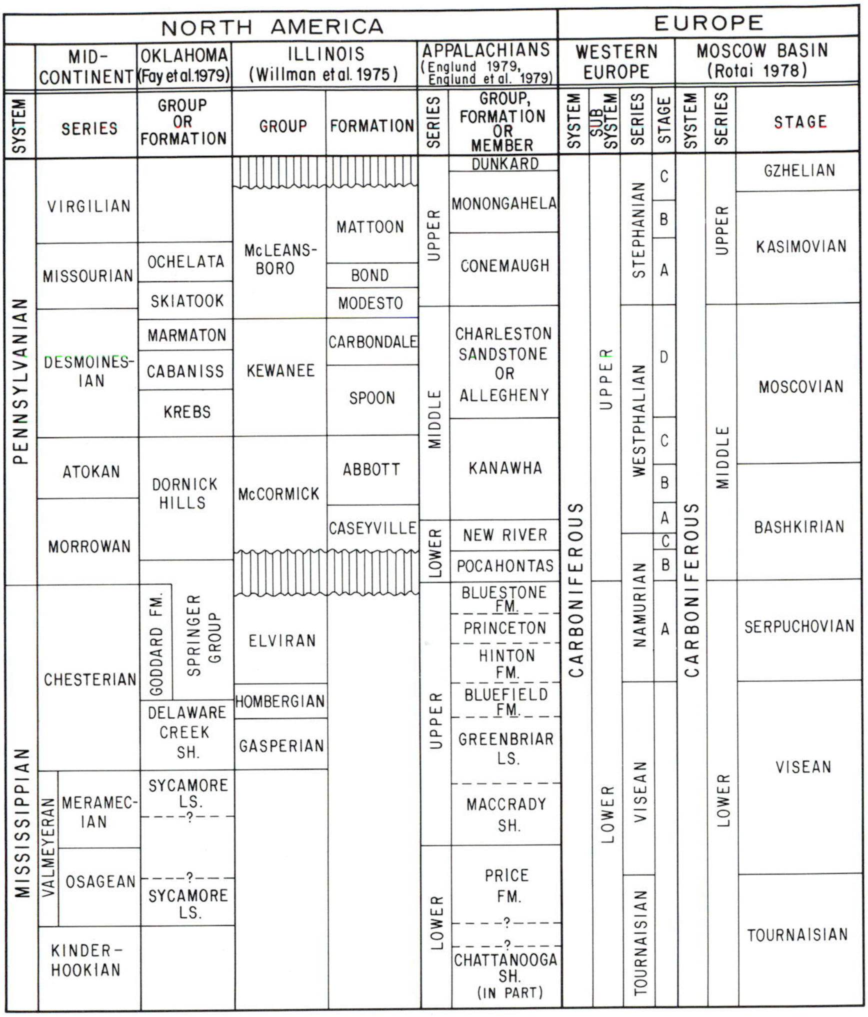

NORTH AMERICA							EUROPE						
	MID-CONTINENT	OKLAHOMA (Fay et al.1979)	ILLINOIS (Willman et al. 1975)		APPALACHIANS (Englund 1979, Englund et al. 1979)		WESTERN EUROPE				MOSCOW BASIN (Rotai 1978)		
SYSTEM	SERIES	GROUP OR FORMATION	GROUP	FORMATION	SERIES	GROUP, FORMATION OR MEMBER	SYSTEM	SUB SYSTEM	SERIES	STAGE	SYSTEM	SERIES	STAGE
PENNSYLVANIAN	VIRGILIAN		McLEANS-BORO	MATTOON	UPPER	DUNKARD	CARBONIFEROUS	UPPER	STEPHANIAN	C	CARBONIFEROUS	UPPER	GZHELIAN
						MONONGAHELA				B			KASIMOVIAN
	MISSOURIAN	OCHELATA		BOND		CONEMAUGH				A			
		SKIATOOK		MODESTO									
	DESMOINES-IAN	MARMATON	KEWANEE	CARBONDALE	MIDDLE	CHARLESTON SANDSTONE OR ALLEGHENY			WESTPHALIAN	D		MIDDLE	MOSCOVIAN
		CABANISS		SPOON						C			
		KREBS								B			
	ATOKAN	DORNICK HILLS	McCORMICK	ABBOTT		KANAWHA				A			BASHKIRIAN
	MORROWAN			CASEYVILLE	LOWER	NEW RIVER			NAMURIAN	C			
						POCAHONTAS				B			
	CHESTERIAN	GODDARD FM. / SPRINGER GROUP	ELVIRAN		UPPER	BLUESTONE FM.		LOWER		A		LOWER	SERPUCHOVIAN
						PRINCETON							
						HINTON FM.							
		DELAWARE CREEK SH.	HOMBERGIAN			BLUEFIELD FM.			VISEAN				VISEAN
MISSISSIPPIAN			GASPERIAN			GREENBRIAR LS.							
	MERAMEC-IAN (VALMEYERAN)	SYCAMORE LS. ---?---				MACCRADY SH.							
	OSAGEAN	---?--- SYCAMORE LS.			LOWER	PRICE FM.			TOURNAISIAN				TOURNAISIAN
	KINDER-HOOKIAN					---?--- ---?--- CHATTANOOGA SH. (IN PART)							

Figure 9.1 Comparison of Carboniferous stratigraphy of Europe and North America.

Subinfraturma, based on smaller sculptural differences, e.g., Subinfraturma Verrucati (verrucose) under Infraturma Apiculati. Under Anteturma Pollenites, however, Subturmas are based on how many sacci (in Saccites) or colpi (in Plicates) are present. Infraturmas under Saccites are based on haptotypic features – whether laesurae are present or not, etc. Infraturmas under Plicates are not much used.

First of all, it must be made clear that, while the turmal system is much used and is rather helpful for Carboniferous and Permian palynofloras, it is in my opinion much less useful for Mesophytic palynofloras, and is no use at all in the Cenophytic. Secondly, and very important to stress, is that the various

Figure 9.2 The Potonié Anteturma Sporites, as revised by Dettmann (1963) and Smith & Butterworth (1967). Note that Dettmann used the turmal classification for Cretaceous, and Smith & Butterworth for Carboniferous spores. Best usage is probably to restrict this classification to pre-Triassic material.

Diagnostic feature / Rank hierarchy:

Diagnostic feature → Rank	Classification
Anteturma	Categories — SPORITES
Turma	TRILETES · MONOLETES · HILATES · ALETES

Aperture (Turma): TRILETES

Stratification (Suprasubturma)	Aperture (Subturma)	Equatorial features (Infraturma)	Sculpture
ACAVATITRILETES	AZONOTRILETES	LAEVIGATI / APICULATI / MURORNATI	equatorial thickening and/or extension
ACAVATITRILETES	ZONOTRILETES	AURICULATI / TRICRASSATI / CINGULATI	
LAMINATITRILETES	AZONOLAMINATITRILETES	TUBERCULORNATI	
LAMINATITRILETES	ZONOLAMINATITRILETES	CRASSITI / CINGULICAVATI / PATINATI	
PSEUDOSACCITRILETES	MONOPSEUDOSACCITI		
PSEUDOSACCITRILETES	POLYPSEUDOSACCITII		
PERINOTRILETES			

Aperture (Turma): MONOLETES

Stratification (Suprasubturma)	Aperture (Subturma)	Infraturma
ACAVATO-MONOLETES	AZONOMONOLETES	LAEVIGATOMONOLETI / SCULPTATOMONOLETI
ACAVATO-MONOLETES	ZONOMONOLETES	
CAVATO-MONOLETES		

Aperture (Turma): HILATES

Stratification (Suprasubturma)	Aperture (Subturma)	Infraturma
ACAVATI-HILATES		
CAVATI-HILATES	AZONOCAVATIHILATES	EPITYGMATI / OPERCULATI
CAVATI-HILATES	ZONOCAVATIHILATES	

Aperture (Turma): ALETES

turmal categories are completely informal and do not enjoy the protection of the *International Code of Botanical Nomenclature* (Voss *et al.* 1983) or any other code. It is absolutely "dealer's choice" which version of Potonié's system one employs, or one can make up one's own; that is, indeed, one of the problems with use of the system. The individual units in the system are not subject to rules of priority, and one should not use author citations and dates for the units, which imply that they are validly published names subject to priority. Potonié in his *Synopsis* volumes, the main source for the system, does this. For example, he uses "Subinfraturma Nodati Dybova & Jachowicz 1957 . . .", and so forth. Authors also should not publish new turmal terms as if they were new taxa, e.g., "Turma Hilates turma nov." A curious side issue here is that some of Potonié's turmal terms also exist as validly published generic names, e.g., Sporites and Monoletes. Potonié incorrectly thought the use of these names as generic names could simply be suppressed. It is quite clear that Potonié and many who have followed him have at least apparently operated on the assumption that the turmal system is a formal classification with formally published names for the various categories.

Figure 9.2 shows in tabular form one noteworthy summary of the Potonié system for the Anteturma Sporites. All one need do is to reproduce such a table, or refer to it, or make up a modified one and present it, when explaining in a systematic presentation that one is using a turmal classification. Hardly any two palynologists use exactly the same version.

Here is my version of the turmal classification for late Paleophytic spores/pollen. Once again, note that the names in italics are validly published formal generic names. The turmal terms (Azonotriletes, etc.) are not validly published names and are absolutely informal! Figures of representative species of many of the genera appear in Figures 8.5, 8.6, 8.13, 9.3, 9.4 and 9.6.

"TURMAL" CLASSIFICATION OF "PALEOPHYTIC" (SILURIAN TO ABOUT MID-PERMIAN) SPORES AND POLLEN

This classification has been simplified from Smith & Butterworth (1967) and the first five volumes of *Synopsis der Gattungen der Sporae dispersae* by R. Potonié (1956–1970). Synopsis V (1970) – Potonié's last revision – was given special weight. The symbols are as follows: ★ megaspore genus (at least in part); † also appears in the other Anteturma; ‡ a one-letter difference occurs between these names and those of certain other genera.

I. Anteturma SPORITES

  A. Turma TRILETES
    1. Subturma AZONOTRILETES (no zone, cingulum, or auricle)
      a. Infraturma LAEVIGATI (more or less psilate)
        Calamospora
        Enigmophytospora★

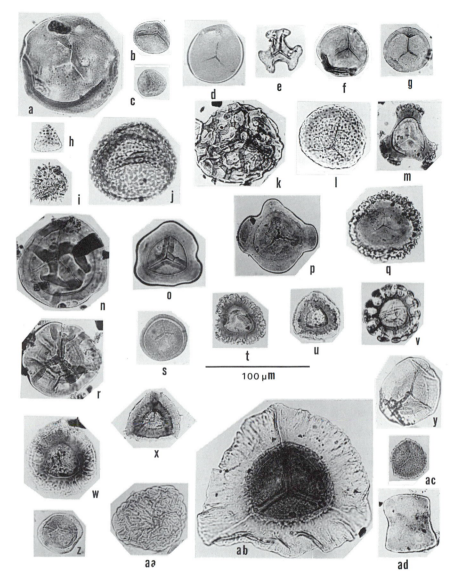

100 μm

Figure 9.3 Mississippian (Visean stage) sporomorphs, Maritime Provinces (MP) and Northwest Territories (NT), Canada, arranged according to the Potonié turmal system. Note that, according to some usage (Geological Society of America), Mississippian is essentially equivalent to Lower Carboniferous, and Pennsylvanian to Upper Carboniferous. Others regard part of the Upper Mississippian as Namurian and place that in the Upper Carboniferous (see Fig. 9.1). Magnification shown by bar below (t)–(v). (a) *Calamospora microrugosa* (Ibrahim) S.W. & B. (MP). (b) *Leiotriletes ornatus* Ishchenko (NT). (c) *Leiotriletes inflatus* (Schemel) P. & K. (MP). (d) *Punctatisporites* cf. *platirugosus* (Valts) Sullivan (MP). (e) *Waltzispora albertensis* Staplin (NT). (f) *Punctatisporites glaber* (Naumova) Playford (NT). (g) *Retusotriletes incohatus* Sullivan

175

Laevigatisporites★‡
Leiotriletes
Phyllothecotriletes
Punctatisporites‡
Retusotriletes
Trileites★
Triletes★

b. Infraturma APICULATI
 i. Subinfraturma GRANULATI (scabrate)
   Cyclogranisporites
   Geminospora
   Granisporites
   Granulatisporites
 ii. Subinfraturma VERRUCATI (verrucate)
   Cyclobaculisporites
   Kewaneesporites
   Schopfites
   Verrucosisporites
 iii. Subinfraturma NODATI (more or less echinate)
   Acanthotriletes
   Anapiculatisporites
   Anaplanisporites
   Aneurospora
   Apiculatisporis
   Apiculiretusispora
   Biharisporites★
   Grandispora
   Lophotriletes
   Planisporites
   Procoronaspora
   Spinosisporites
   Trimontisporites
   Tuberculatisporites★
 iv. Subinfraturma BACULATI (baculate)
   Ancyrospora

(MP). (h) *Anapiculatisporites minor* Butterworth & Williams (NT). (i) *Acanthotriletes castanea* Butterworth & Williams (NT). (j) *Convolutispora ampla* Hoffmeister, Staplin & Malloy (NT). (k) *Reticulatisporites cancellatus* (Valts) Playford (NT). (l) *Foveosporites insculptus* Playford (NT). (m) *Tripartites incisotrilobus* (Naumova) P. & K. (NT). (n) *Knoxisporites hederatus* (Ishchenko) Playford (NT). (o) *Murospora aurita* (Valts) Playford (NT). (p) *Murospora friendii* Playford (NT). (q) *Monilospora moniliformis* Hacquebard & Barss (NT). (r) *Camptozonotriletes velatus* (Valts) Playford (NT). (s) *Stenozonotriletes facilis* Ishchenko (MP). (t) *Densosporites subserratus* Hacquebard & Barss (NT). (u) *Densosporites bialatus* (Valts) P. & K. (NT). (v) *Densosporites duplicatus* (Naumova) P. & K. (NT). (w) *Densosporites* sp. (NT). (x) *Densosporites* cf. *landesii* Staplin (NT). (y) *Endosporites micromanifestus* Hacquebard (NT). (z) *Endosporites minutus* Hoffmeister, Staplin & Malloy (MP). (aa) *Rugospora* sp. (MP). (ab) *Cirratriradites solaris* Hacquebard & Barss (NT). (ac) *Perotrilites* sp. (MP). (ad) *Tetraporina horologia* (Staplin) Playford.

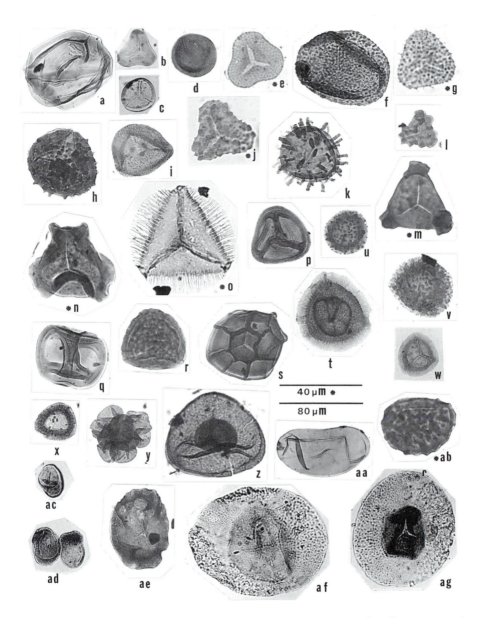

Figure 9.4 Upper Carboniferous (= Pennsylvanian) spores and pollen, arranged according to the Potonié turmal system. Specimens (b), (c), (ac), (ad), (af) and (ag) are referable to the Westphalian stage, Maritime Provinces, Canada. All the other photos are of specimens from Illinois and also are Westphalian. The magnification for items with an asterisk is indicated by the bar under (t) marked with an asterisk. Magnification for all other specimens is indicated by the other bar. (a) *Calamospora hartungiana* Schopf in S.W. & B. (b) *Leiotriletes adnatus* (Kos.) P. & K. (c) *Punctatisporites* sp. (d)

177

Dibolisporites
Hystricosporites★ (in part)
Nikitinisporites★
Raistrickia

c. Infraturma MURORNATI (more or less reticulate)

Camptotriletes
Convolutispora
Dictyotriletes
Emphanisporites
Microreticulatisporites

2. Subturma ZONOTRILETES (zone or cingulum, etc., present)
   a. Infraturma AURICULATI (auriculate)

   Ahrensisporites
   Mooreisporites
   Trilobozonotriletes
   Tripartites
   Triquitrites
   Valvisporites★

   b. Infraturma TRICRASSATI

   Diatomozonotriletes
   Reinschospora
   Triangulatisporites★
   Zonalessporites★

   c. Infraturma CINGULATI (cingulate)

   Ambitisporites
   Archaeozonotriletes
   Cadiospora
   Contagisporites★
   Knoxisporites
   Leiozonotriletes

Cyclogranisporites orbicularis (Kos.) P. & K. (e) *Granulatisporites granularis* Kosanke. (f) *Verrucosisporites sifati* (Ibr.) Smith & Butterworth. (g) *Acanthotriletes aculeolatus* (Kos.) P. & K. (h) *Apiculatisporis abditus* (Loose) P. & K. (i) *Crassispora kosankei* (P. & K.) Smith & Butterworth. (j) *Lophotriletes mosaicus* P. & K. (k) *Raistrickia crocea* Kosanke. (l) *Triquitrites sculptilis* Balme. (m) *Triquitrites* sp. (note that magnification is 1000 ×; this specimen is really about the same size as (l). (n) *Ahrensisporites querickei* (Horst) P. & K. (o) *Reinschospora triangularis* Kosanke. (p) *Cadiospora* sp. (q) *Knoxisporites triradiatus* Hoffmeister, Staplin & Malloy. (r) *Savitrisporites nux* (Butt. & Will.) Smith & Butterworth. (s) *Reticulatisporites reticulatus* (Ibr.) Ibrahim. (t) *Cirratriradites annulatus* Kosanke. (u) *Densosporites lobatus* Kosanke. (v) *Cristatisporites indignabundus* (Loose) P. & K. (w) *Lycospora pellucida* (Wicher) S.W. & B. (x) *Radiizonates striatus* (Knox) Staplin & Jansonius. (y) *Alatisporites hexalatus* Kosanke. (z) *Endosporites globiformis* (Ibr.) S.W. & B. (aa) *Laevigatosporites vulgaris* (Ibr.) Ibrahim. (ab) *Thymospora pseudothiessenii* (Kos.) Wilson & Venkatachala. (ac) *Torispora laevigata* Bhardwaj. (ad) *Torispora securis* (Balme) Alpern *et al.* (note that whether all or even any species of *Torispora* are monolete spores or some other sort of sporopolleninous-walled cell is still debated). (ae) *Schulzospora rara* Kosanke. (af) *Florinites similis* Kosanke. (ag) *Guthoerlisporites magnificus* Bhardwaj.

      Lophozonotriletes
      Reticulatisporites
      Rotaspora
      Savitrisporites
      Stenozonotriletes
      Vallatisporites
    d. Infraturma APPENDICIFERI (with appendages)
      Appendicisporites
      Elaterites
  3. Subturma ZONOLAMINATITRILETES (cavate, zonate)
    a. Infraturmae CRASSITI and CINGULICAVATI
      Cingulizonates
      Cirratriradites
      Crassispora
      Cristatisporites
      Densosporites
      Gondisporites
      Hymenozonotriletes
      Lycospora
      Radiizonates
      Samarisporites
      Simozonotriletes
    b. Infraturma PATINATI (patinate)
      Camarozonosporites
      Camarozonotriletes
      Cappasporites
      Tholisporites
  4. Suprasubturma PSEUDOSACCITRILETES (pseudosaccate, trilete)
    a. Infraturma MONOPSEUDOSACCITI
      Endosporites†
      Remysporites
      Schulzospora†
      Spencerisporites†
    b. Infraturma POLYPSEUDOSACCITI
      Alatisporites†
  5. Subturma PERINOTRILETES
      Perotrilites
      Vestispora
  6. Subturma LAGENOTRILETES
      Lagenicula★
      Setosisporites★

B. Turma MONOLETES
  1. Subturma AZONOMONOLETES
    a. Infraturma LAEVIGATOMONOLETI
      Laevigatosporites‡
      Latosporites

b. Infraturma SCULPTATOMONOLETI
   Columinisporites
   Punctatosporites‡
   Spinosporites
   Thymospora
   Torispora
   Verrucososporites
2. Subturma ZONOMONOLETES (zonate monoletes are very uncommon)
   Speciososporites

C. Turma HILATES (with hilum – a very small group)

D. Turma ALETES
   1. Subturma AZONALETES
      Fabasporites

E. Turma CYSTITES
   Cystosporites★

II. Anteturma POLLENITES

A. Turma SACCITES (one or more vesicles = sacci)
   1. Subturma MONOSACCITES
      a. Infraturma TRILETESACCITI
         Endosporites†
         Felixipollenites
         Guthoerlisporites
         Nuskoisporites
         Rhabdosporites
         Schulzospora†
         Spencerisporites†
         Sullisaccites
         Wilsonites
      b. Infraturma ALETESACCITI
         Archaeoperisaccus (see VESICULOMONORADITI)
         Cladaitina
         Florinites (sometimes has a vestigial trilete laesura)
         Perisaccus
      c. Infraturma VESICULOMONORADITI (saccate, monolete)
         Archaeoperisaccus
         Potonieisporites
      d. Infraturma SACCIZONATI
         Zonalasporites
   2. Subturma DISACCITES
      a. Infraturma DISACCITRILETI (bisaccates with trilete laesurae)
         Illinites

b. Infraturma DISACCIATRILETI (bisaccates without trilete laesurae)
   Parasporites
   Pityosporites
   Vesicaspora
3. Subturma STRIATITES (striate saccates)
   Lueckisporites
   Striatites
4. Subturma POLYSACCITES
   Alatisporites†

B. Turma PLICATES (pollen with one or more colpi, with or without pores)
   1. Subturma PRAECOLPATES (with several parallel folds, one being a true colpus)
      Marsupipollenites
      Schopfipollenites (= *Monoletes*)
   2. Subturma POLYPLICATES
      Vittatina
   3. Subturma MONOCOLPATES
      Entylissa

PALEOBOTANICAL MATTERS REGARDING THE LATE "PALEOPHYTIC"

During the Carboniferous, trends in plant evolution already established in the Devonian were expanded, with the establishment of widespread forests consisting of lycopsids such as *Lepidodendron* and *Sigillaria*, seed fern trees and shrubs such as *Medullosa*, sphenopsid trees and shrubs such as *Calamites*, ferns including both herbaceous forms and tree ferns such as *Psaronius*, tree and shrub cordaitaleans such as *Cordaites*, and primitive conifers. The seed plus pollen habit was by now dominant, and free-sporing megaspores began to assume a lesser importance. About Carboniferous and early Permian plants we know a great deal, partly from very widespread occurrence of compressed plant parts, but more especially from the fact that, as the name implies, the Carboniferous was a time of vast extent of coal-forming swamps. From studies of petrified peat (coal balls) containing exquisitely preserved fossil plants found in many Carboniferous coals, we know more about both anatomy-morphology and ecology of some Carboniferous plants than we do about many of their extant descendants. See, for example, current texts of paleobotany such as Stewart (1983) and Taylor (1981) for more information about coal-ball studies of Carboniferous plants. Various papers by Phillips *et al.* have contributed greatly to understanding of Carboniferous plant ecology (Phillips & DiMichele 1981, Phillips *et al.* 1974, 1985). Scheihing & Pfefferkorn (1984) have shown that study of modern plant taphonomy in a tropical delta can provide a model for the fossil plant association found in Carboniferous rocks.

One fallout effect of the relatively great amount of effort spent studying Devonian and Carboniferous petrified and compressed plant remains is that we know a considerable amount about which plants produced many of the taxa of dispersed spores ("Sporae dispersae"). Indeed, it is curious that we have far more such information than we do for Cenozoic dispersed spores/pollen! At least in part this is a product of the relatively poorer chances of fossilization of comparatively delicate angiosperm inflorescences and flowers than of gymnosperm and "pteridophytic" fructifications.

"PALEOPHYTIC" SPORES/POLLEN:
THE PLANTS THAT PRODUCED THEM

The following data are, with few exceptions, for spores/pollen that have been removed directly from fructifications – that is, the information is from *in situ* spores. Relationships of dispersed sporomorphs to megafossil plant taxa are ideally based on study of such *in situ* spores, but close and regular association of dispersed sporomorphs with particular megafossils is sometimes persuasive. The botanical relationships of the dispersed spores/pollen are organized below according to the "turmal" classification presented earlier in this chapter, to which the reader is referred for more information on the various categories and for the meaning of symbols. (Remy & Remy (1955) make the important point that, because of different stages of development and different states of preservation, the same taxon of fructification may contain quite different sorts of spores.)

I. Anteturma SPORITES

  A. Turma TRILETES

    1. Subturma AZONOTRILETES

      a. Infraturma LAEVIGATI

        Calamospora

Calamospora is the spore of *Calamites* and other sphenopsids such as *Sphenophyllum* (Good 1978), and of the equisetalean, *Paleostachya* (Gastaldo 1981a). It may be the spore of ferns (Pfefferkorn *et al.* 1971) such as *Scolecopteris* (Millay 1979). Also it comes from *Sawdonia*, a mid-Devonian zosterophylloid (Gensel *et al.* 1975), but the same plant also apparently produced spores referable to *Retusotriletes*. It is produced also by the Devonian trimerophytes, *Dawsonites* and *Psilophyton* (Allen 1980, Gensel 1980b; see also *Retusotriletes* and *Apiculiretusispora*). The heterosporous Devonian plant, *Enigmophyton* (referred by Gensel & Andrews (1984) to the "Barinophytaceae"), produced some microspores referable here (others are *Retusotriletes*; Allen 1980). Spores comparable to *Calamospora* were also produced by *Orcilla*, a Devonian zosterophyll (Gensel 1982b; cf. *Retusotriletes*), and by *Hostinella*, a Devonian rhyniophyte (Allen 1980). *Protobarinophyton* and *Barinophyton*, Devonian plants of uncertain affinity ("Barinophytaceae", according to Gensel & Andrews (1984)), were said to produce *C.* spores by Gensel (1980b) and by Taylor & Brauer (1983).

Enigmophytospora★

This megaspore was obtained from the Devonian plant, *Enigmophyton* (Allen 1980; see also *Phyllothecotriletes*), referred to "Barinophytaceae" by Gensel & Andrews (1984).

Laevigatisporites★‡

Laevigatisporites megaspores have been found in the sigillarian cones, *Mazocarpon* and *Sigillariostrobus* (see Scott & King 1981); but see also *Tuberculatisporites*.

Leiotriletes

Leiotriletes is, in part, a bryophytic spore (Potonié & Kremp 1956a). This spore can also belong to the ferns, *Sermaya* (Eggert & Delevoryas 1967), *Pecopteris* sp. (Laveine 1969), *Botryopteris* (Good 1979), *Donneggia* (Permian; Rothwell 1978), and other ferns (Bharadwaj & Venkatachala 1968), and probably to the lycopsids. Allen (1980) suggests that some late Silurian *Cooksonia* made spores referable here, as well as others identified as *Ambitisporites*.

Phyllothecotriletes

Phyllothecotriletes is produced by calamarians. Banks (1968; information from Vigran) says that the Devonian plant, *Enigmophyton*, made *P.* spores (as well as *Enigmophytospora*).

Punctatisporites‡

Punctatisporites is the spore of *Pecopteris* sp. (Laveine 1969), *Scolecopteris* (Millay 1979, 1982a), *Botryopteris* (Millay & Taylor 1982; but compare *Verrucosisporites*), the zygopteridalean fern, *Biscalitheca* (Millay & Rothwell 1983), the Pennsylvanian marattialean, *Araiangium* (Millay 1982b), and of other Filicineae (Pfefferkorn *et al.* 1971). It also occurs in certain Cycadofilicales (in which case, it would by definition be a pollen grain – or "prepollen", pollen that has laesurae and other evidence of spore-like habit), for example, *Telangium*, according to Eggert & Taylor (1968), and *Potoniea illinoensis* (*Punctatisporites kankakeensis* Peppers) according to Stidd (1978). It is also produced by certain of the Psilopsida, according to Gothan & Weyland (1964), and by Devonian *Zosterophyllum* (cf. Gensel 1982a). (Note that this is not the same as *Punctatosporites*. There is unfortunately a whole series of such one-letter differences between spore generic names. The "o" ones are monolete, the "i" ones trilete, an idea, I believe, of Ibrahim-Okay. Even palynologists can get confused by these one-letter orthographic variants, but they are perfectly legal.)

Retusotriletes

Retusotriletes spores were produced by a number of plants of various geological ages, for example, by *Renalia*, a Devonian plant close to rhyniophytes and zosterophylls (Gensel 1976). Gensel (1980a) says that *Psilophyton* produced these spores (*inter alia*). Banks (1968; data from Hueber) notes that the Devonian trimerophyte, *Dawsonites*, made *R.* spores and it was also produced (along with *Apiculiretusispora*) by *Trimerophyton* (Allen 1980). This spore type was also produced by *Zosterophyllum* (Gensel 1980b, Allen 1980), and by the Devonian zosterophyll, *Orcilla* (cf. *Calamospora*). Another zosterophyll, *Sawdonia*, was reported by Gensel *et al.* (1975) to produce both *Retusotriletes* and *Calamospora*. Some of the microspores of *Enigmophyton*, a Devonian plant of doubtful relationship, are referable here (Allen 1980; but see *Calamospora*, *Enigmophytospora*, and *Phyllothecotriletes*. This sort of spore was also produced by Lower Carboniferous sphenopsids, according to Scott *et al.* (1985b). Allen (1980) even refers the ca. 200 µm spores of the Upper Devonian non–vascular plant, *Foerstia*, to *Retusotriletes*.

Triletes★

The megaspore *Triletes* (not to be confused with Turma Triletes!) was produced by heterosporous lepidodendrids, per Brack-Hanes (1978), and by herbaceous lycopods as well (Chaloner 1954, Schlanker & Leisman 1969). Fortunately, the name is not much used now as a megaspore generic name, the genus having been subdivided. As a genus, *Triletes* can be confused not only with the Turma Triletes but also with the genera *Trilites* (echinate miospore, Triassic) and *Trileites* (psilate, retusoid megaspore, Devonian–Carboniferous).

b. Infraturma APICULATI

i. Subinfraturma GRANULATI

Cyclogranisporites

Cyclogranisporites is the isospore of *Pecopteris* spp. (Laveine 1969, Millay 1979) and of other Filicineae (Potonié & Kremp 1956a, Bharadwaj & Venkatachala 1968, Pfefferkorn *et al.* 1971), the marattialeans, *Acitheca* (*Pecopteris* foliage; Mapes & Schabilion 1979a) and *Scolecopteris* (Millay & Taylor 1984), and of *Noeggerathiostrobus*, a problematic pteridophyte perhaps related to *Archaeopteris* (Beck 1981). Millay & Taylor (1977) say that *C.* is the prepollen of the lyginopterid, *Feraxotheca*, Millay *et al.* (1978) say the same of the lyginopterid, *Crossotheca*, and Stidd *et al.* (1985) of the lyginopterid, *Schopfiangium* (but see *Verrucosisporites*). Various authors (Banks 1968, Gensel 1980b) mention that *C.* was produced as a microspore by *Archaeopteris*. Allen (1980) refers some *Rhynia* (early Devonian) spores here (but see also *Granulatisporites* and *Apiculiretusispora*).

Geminospora

Geminospora may be the spore of *Rhynia* (see Gensel 1980b). *G.* spores have been mentioned as a microspore type of various species of the Devonian progymnosperms, *Archaeopteris* (Gensel 1980b, Allen 1980) and *Svalbardia* (Allen 1980) (cf. *Aneurospora* under Subinfraturma Nodati, below).

Granulatisporites

Granulatisporites is the spore of certain Filicineae, such as *Botryopteris* (Good 1979) and *Renaultia* (Scott 1978), but it also can be the pollen (= prepollen) of some Cycadofilicales (Potonié & Kremp 1956a) such as the lyginopterid, *Feraxotheca* (Millay & Taylor 1977), and *Crossotheca* (Millay *et al.* 1978). Also it may be the spore of certain sphenopsids (Schopf *et al.* 1944), and of the early Devonian rhyniophyte, *Rhynia* (Allen 1980; spores of *Rhynia* spp. are also referred to *Apiculiretusispora, Granulatisporites*, and *Cyclogranisporites*).

ii. Subinfraturma VERRUCATI

Cyclobaculisporites

Cyclobaculisporites was produced by ferns such as Lower Gondwana *Dichotomopteris* (Lele *et al.* 1981).

Kewaneesporites

Kewaneesporites, a spore with odd, hollow verrucae, occurs in the sporangia of the fern, *Cyathotheca* (Taylor 1972, Mickle and Rothwell 1986), a Carboniferous plant of uncertain affinity.

Verrucosisporites

Verrucosisporites is a fern spore, for example, of *Biscalitheca* (Courvoisier & Phillips 1975, Bharadwaj & Venkatachala 1968), of the marattialean ferns, *Scolecopteris* (Jennings & Millay 1978), *Eoangiopteris* (Millay 1978), and *Millaya* (Mapes & Schabilion 1979b), of botryopterids (Millay & Taylor 1980, 1982; but compare also *Punctatisporites*), and can also be a seed fern prepollen (Bharadwaj & Venkatachala 1968), e.g., of the lyginopterid, *Schopfiangium* (Stidd *et al.* 1985).

iii. Subinfraturma NODATI

Acanthotriletes

Acanthotriletes spores (among others) have been found on *Botryopteris* fern foliage (Good 1979, Millay & Taylor 1982). Also it is said (see Gensel 1980b) to be made by the Devonian plant, *Eviostachya*. Species of the Lower Gondwana ferns, *Dichotomopteris* and *Neomariopteris*, produced *A.* spores, as well as other forms, according to Lele *et al.* (1981). Scott *et al.* (1985b) reported spores referable here or to *Apiculatisporis* from Carboniferous zygopterid ferns.

Anapiculatisporites

According to Ravn (1983, personal communication), spores illustrated by Baxter (1971) from the lycopsid fructification, *Carinostrobus foresmanii*, are referable to *A. spinosus* (Kosanke) Potonié & Kremp.

Aneurospora

Aneurospora is said to have been produced by the Devonian progymnosperms, *Aneurophyton* (Gensel 1980b) and *Archaeopteris* (Allen 1980; see *Geminospora*), and has been mentioned also as produced by the Devonian lycophyte, *Leclercqia* (Streel 1972). (*Streelispora* is a similar spore taxon, which Edwards *et al.* 1986 say was found in sporangia of the rhyniophyte, *Cooksonia*.)

Apiculatisporis

Apiculatisporis (= *Apiculatisporites*) spores were produced by the fern, *Corynepteris*, according to Galtier & Scott (1979), and by other Filicineae (Pfefferkorn *et al.* 1971); also as one kind of "large" spore by *Chaleuria*, a Devonian possible progymnosperm. (Another kind of *Chaleuria* large spore is referable to *Apiculiretusispora*. The small spores are *Camarozonotriletes* (Gensel 1980b).) Lele *et al.* (1981) note that a Lower Gondwana fern, *Dizeugotheca*, produced spores similar to *Apiculatisporis*, but also made spores referable to *Punctatosporites* (monolete!).)

Apiculiretusispora

This spore type was produced by *Pertica*, a Devonian progymnosperm ancestor (Granoff *et al.* 1976, Doran *et al.* 1978). *Psilophyton* produced *A.* spores (and other kinds) according to Gensel (1980a). The spores of the Devonian rhyniophyte, *Horneophyton lignieri*, have also been referred here, as have those of *Cooksonia*, the earliest vascular plant (Gensel 1980b, Allen 1980;

but see *Leiotriletes* and *Ambitisporites*). *A*. has also been mentioned by Gensel (1980b) as produced by the Devonian plants, *Krithodeophyton* and *Chaleuria* (as one kind of "large" spore; see *Apiculatisporis*). Spores referable to *A*. were produced by *Renalia* (but see *Retusotriletes*) and *Rhynia*, early Devonian rhyniophytes (Allen 1980; but see *Granulatisporites* and *Cyclogranisporites*). *A*. spores were produced by the zosterophyll, *Nothia* (Allen 1980), and by the Devonian *Trimerophyton robustius* (Allen 1980; see also *Retusotriletes*).

Biharisporites★
This megaspore was produced by various species of the Devonian progymnosperm, *Archaeopteris* (Gensel 1980b, Allen 1980; some of these plants also made megaspores referable to *Contagisporites*, or only made these, according to Allen).

Grandispora
Grandispora (and *Samarisporites* was produced by *Oocampsa*, a trimerophyte–progymnosperm, according to Andrews *et al*. (1975).

Lophotriletes
Lophotriletes (and others) have been found on *Botryopteris* fern foliage (Good 1979), and were produced by species of the Lower Gondwana ferns, *Dichotomopteris* and *Neomariopteris*, according to Lele *et al*. (1981).

Planisporites
Planisporites has been described from sigillarian (lycopsid) cones such as *Sigillariostrobus* by Chaloner (1953b), but is also said to be cycadofilicalean prepollen (Bharadwaj & Venkatachala 1968).

Trimontisporites
Trimontisporites is a spore of the fern, *Scolecopteris* (Millay 1979).

Tuberculatisporites★
This megaspore is produced by the lycopsid cones, *Sigillariostrobus* and *Mazocarpon* (see Scott & King 1981), but see also *Laevigatisporites*.

iv. Subinfraturma BACULATI

Dibolisporites
Dibolisporites was produced by the Devonian plant, *Calamophyton*, according to Bonamo & Banks (1966a).

Nikitinisporites★
This megaspore was probably produced by the late Devonian lycopod, *Kryshtofovichia* (Allen 1980; see *Archaeoperisaccus*).

Raistrickia
Raistrickia is the spore of *Pecopteris* sp. (Laveine 1969), of *Ankyropteris* sp. (Mickle 1980), and of other Filicineae (Schopf *et al*. 1944, Bharadwaj & Venkatachala 1968). Remy & Remy (1957) illustrate *R*. from *Senftenbergia*, a schizaeaceous Carboniferous fern, though they do not specifically make this identification (see *Convolutispora* below).

c. Infraturma MURORNATI

Camptotriletes
Camptotriletes is the spore of *Pecopteris* sp. (Laveine 1969) and of other Filicineae (Schopf *et al.* 1944, Pfefferkorn *et al.* 1971).

Convolutispora
Convolutispora is the spore of *Pecopteris* sp. (Laveine 1969) and of the zygopterid fern, *Biscalitheca* (Cridland 1966; but see *Verrucosisporites*). It is clear from the illustrations that Remy & Remy (1955) also found *C.* in *Senftenbergia*, a Carboniferous schizaeaceous fern, though the identification is not specifically made (see *Raistrickia*).

Dictyotriletes
Hamer & Rothwell (1983) found spores of this genus in *Phillipopteris*, a Pennsylvanian fern–like fructification. Scott *et al.* (1985b) illustrate beautifully preserved examples from Carboniferous botryopterid ferns. Ravn (1983, personal communication) states that the type species, *D. reticulatus*, is probably a lycopsid spore, but that other species are from ferns.

Emphanisporites
Emphanisporites spores may be produced by *Horneophyton* (Gensel 1980b).

2. Subturma ZONOTRILETES

a. Infraturma AURICULATI

Tripartites and *Triquitrites*
Tripartites and *Triquitrites* are spores of *Phlebopteris*, Matoniaceae, according to Potonié (1962).

Valvisporites★
Valvisporites was produced by *Polysporia*, a herbaceous lycopod (Chaloner 1958a, DiMichele *et al.* 1979), and by the closely related lycophyte, *Chaloneria* (Pigg & Rothwell 1983). Gastaldo (1981b) noted that this spore type occurs in *Lepidocystis*, a lycopsid reproductive organ (see *Endosporites*).

b. Infraturma TRICRASSATI

Reinschospora
Reinschospora is probably derived from certain Filicineae, according to Schopf *et al.* (1944).

Triangulatisporites★
This megaspore is found in *Selaginellites* cones (see Scott & King 1981).

Zonalessporites★
Zonalessporites is found as a megaspore in the heterosporous lycopod cone, *Sporangiostrobus* (microspores are *Densosporites*; see Chaloner 1962, Leisman 1970, and summary in Scott & King 1981).

c. Infraturma CINGULATI

Ambitisporites

This spore, along with *Punctatisporites*, ranges to earliest Silurian or latest Ordovician and the two are thus the earliest trilete spores in the fossil record. Allen (1980) says that *in situ* spores of some *Cooksonia* sp., the earliest vascular plant, and possibly of some *Rhynia* spp. (early Devonian) are referable to this genus (see *Leiotriletes*, *Cyclogranisporites*, *Granulatisporites*, and *Apiculiretusispora*).

Contagisporites★

These spores, certainly megaspores, though often well below 200 µm in size, were produced by various species of the Devonian progymnosperm, *Archaeopteris*, some of which also produce megaspores referable to *Biharisporites* (Allen 1980).

Knoxisporites

Knoxisporites specimens were found by Scott *et al.* (1985b) in sporangia of uncertain relationship but possibly filicilean. It had been earlier suggested that they are possibly the spores of certain Selaginellales.

Reticulatisporites

Reticulatisporites is possibly the spore of certain *Sphenophyllum* species (Schopf *et al.* 1944). Other *Sphenophyllums* have *Calamospora* spores.

Vallatisporites

Vallatisporites is said by Bharadwaj & Venkatachala (1968) to be a spore of lepidophyte cones.

d. Infraturma APPENDICIFERI

Elaterites

Elaterites is the spore of certain calamarians – found in *Calamostachys* by Baxter & Leisman (1967) and by Good & Taylor (1975).

3. Subturma ZONOLAMINATITRILETES
a. Infraturmae CRASSITI and CINGULICAVATI

Cirratriradites

Cirratriradites was produced as a microspore by *Selaginella*, according to Schlanker & Leisman (1969), and by *Selaginellites*, per Chaloner (1954).

Crassispora

Crassispora is the microspore of *Mazocarpon* and other sigillarian cones (Courvoisier & Phillips 1975, Bharadwaj & Venkatachala 1968).

Cristatisporites

Cristatisporites (if separated from *Densosporites*) was found as a microspore in *Sporangiostrobus*, a lycopod cone, by Chaloner (1962) and Leisman (1970).

Densosporites

Densosporites is the microscope of certain lycopsids (Chaloner 1958b, 1962). Potonié (1967) and Chaloner (1962) associate this genus with the lycopsid genus, *Porostrobus*. Chaloner (1962) and Leisman (1970) found *Denosporites* (or *Cristatisporites*, if that taxon is maintained) as a microspore in *Sporangiostrobus*, a heterosporous lycopsid cone. The Devonian possible lycopod, *Barrandeina*, produced spores referable here (Allen 1980; see *Samarisporites*).

Gondisporites

Gondisporites is the spore of certain lycopod cones (Bharadwaj & Venkatachala 1968).

Lycospora

Lycospora was unquestionably produced by a variety of lycopsids (see Balbach 1967, Courvoisier & Phillips 1975). Heterosporous lepidodendrids made *L.* as microspores, according to Brack-Hanes (1978). On the other hand, some apparently homosporous (*Lepidostrobus*) cones also produced *Lycospora* (see Thomas 1970). The Devonian plant, *Svalbardia*, has been mentioned as a producer of *L.* spores (see Gensel 1980b).

Radiizonates

Radiizonates (Courvoisier & Phillips 1975) was produced by *Sporangiostrobus*, a lycopod.

Samarisporites

Samarisporites (and *Grandispora*) were produced by *Oocampsa*, a trimerophyte–progymnosperm, according to Andrews *et al.* (1975). *S.* (and *Densosporites*) were made by the Devonian possible lycopod, *Barrandeina*, according to Allen (1980).

b. Infraturma PATINATI

Camarozonotriletes

Camarozonotriletes is mentioned as a "small" spore of the Devonian plant, *Chaleuria* (Gensel 1980b; see also *Apiculatisporis*).

Cappasporites

According to Courvoisier & Phillips (1975), *Cappasporites* was produced as a microspore by *Achlamydocarpon*, an arborescent lycopod. Chadwick (1983) also shows that these spores, seldom showing a trilete laesura, were produced in arborescent lycopods.

4. Suprasubturma PSEUDOSACCITRILETES

a. Infraturma MONOPSEUDOSACCITI

Endosporites†

Potonié & Kremp (1956a) put this in the Subturma Monosaccites as a pollen grain. Clearly this is because originally the genus was very loosely used, including forms now referred to *Florinites*, and other monosaccate genera. However, Chaloner (1953a, 1958a) has found spores referable to this genus as microspores in lycopsid fructifications, and Chaloner (1953a), Brack-Hanes & Taylor (1972) and DiMichele *et al.* (1979) have shown that *Polysporia*, a herbaceous lycopod, produced *Endosporites* microspores. Pigg & Rothwell (1983) note that the lycophyte, *Chaloneria*, produces this type of microspore. See *Valvisporites*★.

Schulzospora†

Schulzospora is a seed fern prepollen (Potonié 1962).

Spencerisporites†

Spencerisporites (= *Microsporites*) is the spore of the arborescent lycopsid fructification, *Spencerites* (Potonié & Kremp 1956a, Potonié 1962). Leisman & Stidd (1967) found, in addition to "normal" trilete spores, some monolete spores.

b. Infraturma POLYPSEUDOSACCITI

Alatisporites†

Potonié & Kremp (1956a) put this in the Subturma Polysaccites as a pollen grain and suggest cordaitean affinity. Schopf *et al.* (1944) suggested sphenopsid relationship. Note, therefore, that this form appears also as a pollen grain (Infraturma Triletesacciti).

5. Subturma PERINOTRILETES

Perotrilites

Perotrilites was produced by species of the Devonian fern, *Rhacophyton* (Andrews & Phillips 1968) and possibly by *Rhynia* (Gensel 1980b).

Vestispora

Vestispora is a spore of sphenopsids, found as a developmental stage of the same cones that produced *Calamospora*, according to Good (1977). However, this idea is disputed by Ravn (1983), who says *V.* was produced as a mature spore by the Sphenophyllales and other sphenopsids. Taylor (1986) reports *V.* from the sphenophyllaleans, *Sphenostrobus* and *Koinostachys*.

6. Subturma LAGENOTRILETES

Lagenicula★

Lagenicula is the megaspore of heterosporous lycopods, such as the late Devonian *Cyclostigma* (Chaloner 1968c), and of the Carboniferous cone, *Lepidostrobus* (Brack-Hanes 1981, Scott & King 1981). (Note that Brack-Hanes & Thomas (1983) have shown that *Lepidostrobus* should be subdivided, and the bisporangiate cones, such as those producing *Lagenicula*, would go into *Flemingites*.)

Setosisporites★

Megaspore of lycopsid cone, *Porostrobus* (see Scott & King 1981).

B. Turma MONOLETES

1. Subturma AZONOMONOLETES

a. Infraturma LAEVIGATOMONOLETI

Laevigatosporites‡

Laevigatosporites is, surprisingly, in part derived from sphenopsids such as *Bowmanites* (Courvoisier & Phillips 1975, Potonié & Kremp 1956a) and from lycopsids (Potonié 1967). Good (1978) notes that *Columinisporites* spores of certain sphenophyllaleans "become" *Laevigatosporites* if the perine is lost! Others are the spores of *Pecopteris* sp. (Laveine 1969, Pfefferkorn *et al.* 1971) and *Scolecopteris* (Millay 1979, 1982c, Millay & Taylor 1984). The often very abundant *L. minimus* was shown to be produced by the fern, *Zeilleria* (Thomas & Crampton 1971).

b. Infraturma SCULPTATOMONOLETI

Columinisporites

Columinisporites is the spore of *Peltastrobus*, a sphenophyllalean cone (Taylor 1986), and of other sphenophyllaleans, per Good (1978), but the same things are said to "become" *Laevigatosporites* if the perispore is lost. Riggs & Rothwell (1985) found *C.* spores in the sphenophyllalean, *Sentistrobus*.

Punctatosporites‡

Punctatosporites is the spore of *Pecopteris* spp. (Laveine 1969) and *Scolecopteris* (Lesnikowska & Millay 1985), but this is a rather generalized monolete spore, being produced (Lele *et al.* 1981) by the Lower Gondwana fern, *Dizeugotheca*, along with trilete spores. (*Scolecopteris* also makes *Punctatisporites*, a trilete spore.)

Spinosporites

According to Ravn (1983, personal communication), *S. exiguus* Upshaw & Hedlund 1967 apparently corresponds to a monolete spore illustrated by Millay (1979) for *Scolecopteris*. Millay & Taylor (1984) show that *S.* is also produced by *Scolecopteris*, a marattialean fern also making other trilete and monolete spores.

Thymospora

Thymospora (a synonym for *Verrucososporites*) was produced by the ferns, *Scolecopteris* and *Asterotheca*, according to Wilson & Venkatachala (1963). Millay & Taylor (1984) found *T.* spores *in situ* in *Scolecopteris*. Doubinger & Grauvogel-Stamm (1971) reported *T.* from *Pecopteris* fronds.

Torispora

Torispora was produced by sporangia of *Pecopteris* spp. (Laveine 1969). But some doubt that all *T.* species are spores. (*Crassosporites* is synonymous with *Torispora* according to Potonié (1960, p. 145.)

2. Subturma ZONOMONOLETES

Speciososporites

Speciososporites is the spore of *Pecopteris* spp. (Laveine 1969). (*Archaeoperisaccus*, cf. Vesiculomonoraditi under Pollenites, probably belongs here.)

C. Turma HILATES

A very small group, with hilum.

D. Turma ALETES

1. Subturma AZONALETES

Fabasporites

Fabasporites was produced by the marattialean fern, *Acaulangium*, according to Millay (1977). Potonié includes *F.* under monocolpate pollen, and the original author intended to describe *F.* as a monocolpate pollen grain. Others have considered it mono*lete*, but the current consensus seems to be that it is *alete*.

(Millay, quoted above, thought his *F.* spores might be immature *trilete* forms!)

E. Turma CYSTITES

Cystosporites★

Cystosporites, a large "seed megaspore", was produced by *Achlamydocarpon*, an arborescent lycopod (Leisman & Phillips 1979). Balbach (1966) had noted that *C.* megaspores were produced in *Lepidostrobus* cones. (See summary in Scott & King (1981). Compare *Lagenicula*.)

II. Anteturma POLLENITES

A. Turma SACCITES

1. Subturma MONOSACCITES

a. Infraturma TRILETESACCITI

Endosporites†

This has been included above under Monopseudosacciti.

Felixipollenites

Felixipollenites is prepollen of the cordaitean cone, *Gothania* (Taylor & Daghlian 1980), that is, of the *Mesoxylon* type of cordaitean (Trivett & Rothwell 1985). (See *Florinites*.)

Guthoerlisporites

Guthoerlisporites is a cycadofilicalean pollen grain (Bharadwaj & Venkatachala 1968).

Nuskoisporites

Nuskoisporites has been presumed to be a seed fern pollen grain, but has been found in Permian conifer cones by Clement-Westerhof (1974).

Rhabdosporites

Rhabdosporites, a trilete monosaccate, is the microspore (prepollen) of such Devonian progymnosperms as *Tetraxylopteris* and *Rellimia* (Bonamo & Banks 1966a, 1967, Bonamo 1977) and *Cathaiopteridium* (Allen 1980). Millay & Taylor (1974) suggest that saccate cordaite pollen evolved from this pseudo-saccate type.

Schulzospora†

Schulzospora is said to be a seed fern prepollen. (See listing above under Monopseudosacciti.)

Spencerisporites†

This is included in Monopseudosacciti.

Sullisaccites

Trivett (1983) reports this sort of prepollen from the cordaitean pollen cone, *Gothania*, but see also *Felixipollenites*.

Wilsonites

Wilsonites was produced by the Cycadofilicales (Potonié & Kremp 1956a, Bharadwaj & Venkatachala 1968).

192

b. Infraturma ALETESACCITI

Cladaitina

Cladaitina is the pollen of *Cladostrobus*, a probable cordaitalean (Maheshwari & Meyen 1975).

Florinites

Florinites, in a restricted sense, is the pollen grain of primitive conifers such as *Ernestiodendron, Lebachia,* and *Walchiostrobus* (Potonié & Kremp 1956a, Bharadwaj & Venkatachala 1968, Rothwell 1982), and of cordaiteans with *Cordaianthus* cones and endarch xylem (Trivett & Rothwell 1985). *Florinites* in this sense is alete. The name *Florinites* has also been used in a broad sense for monosaccate pollen, even of seed ferns, with or without laesurae (Bharadwaj & Venkatachala 1968). When further "split", *Florinites*-type pollen of *Lebachia* is *Potoniesporites,* that of cordaiteans with mesarch xylem and *Gothania* cones is *Felixipollenites.*

Perisaccus

Perisaccus is the pollen grain of cordaitaleans, according to Naumova (1953).

c. Infraturma VESICULOMONORADITI

Archaeoperisaccus

Archaeoperisaccus is a cycadofilicalean pollen grain, according to Naumova (1953). However, the microspore of the late Devonian lycopod, *Kryshtofovichia,* belongs here according to Allen (1980; see *Nikitinisporites*). As emended by McGregor (1969), *A.* is a monolete, camerate miospore (presumably micro-spore). Many have noted its close resemblance to *Aratrisporites,* a lycopod microspore. Note that Potonié regarded this form as alete and therefore listed it under ALETESACCITI. As now understood, *A.* should probably be under ZONOMONOLETES (Sporites). (See Braman & Hills 1985.)

Potonieisporites

Potonieisporites is the pollen of the coniferalean, *Lebachia (Rothwell 1982, Mapes 1983), and also of Walchianthus* according to Bharadwaj (1964), who concluded that *Sahnites* and *Vestigisporites* are synonyms of *Potonieisporites.*

2. Subturma DISACCITES

a. Infraturma DISACCITRILETI

Illinites

Illinites (Potonié & Kremp 1956a) was produced by coniferalean species (see *Florinites,* above) according to Bharadwaj & Venkatachala (1968).

b. Infraturma DISACCIATRILETI

Parasporites

Parasporites is pollen of the medullosan, *Parasporotheca* (Dennis & Eggert 1978, Millay *et al.* 1978).

Pityosporites

Pityosporites is the pollen of conifers (see *Florinites, Illinites,* above) according to Potonié & Kremp (1956a).

Vesicaspora

Vesicaspora was thought to be the pollen of Caytoniales by Potonié & Kremp (1956a) but more recently it is referred to the pteridosperms, *Callistophyton* (Hall & Stidd 1971, Rothwell 1972 – who describes pollen tubes of *Vesicaspora*!) and *Idanothekion* (Millay & Taylor 1970).

3. Subturma STRIATITES

Lueckisporites

Lueckisporites is the pollen grain of certain early conifers (Clement-Westerhof 1974), Caytoniales(?), Gnetales(?) (the latter seems doubtful), and other gymnosperms, according to suggestions of various authors.

Striatites

Striatites is the pollen grain of glossopterid plants (Potonié 1967) in part, but is thought by many also to be associated with various members of the gnetalean alliance.

B. Turma PLICATES

1. Subturma PRAECOLPATES

Monoletes (= *Schopfipollenites*)

Monoletes (= *Schopfipollenites*) pollen has been found in a variety of medullosan seed fern "anthers" or synangia (Bharadwaj & Venkatachala 1968, Leisman & Peters 1970, Millay *et al*. 1978). Specific examples are *Dolorotheca* (Dennis & Eggert 1978), *Sullitheca* (Stidd *et al*. 1977), *Stewartiotheca* (Eggert & Rothwell 1979, Millay *et al*. 1980), *Aulacotheca* (Eggert & Kryder 1969), *Rhetinotheca* (Rothwell & Mickle 1982), and *Boulaya* (Kurmann 1983).

2. Subturma POLYPLICATES

Vittatina

According to Potonié (1967), this pollen grain has been referred to the Gnetales by some authors and to the conifers by others.

3. Subturma MONOCOLPATES

Entylissa

Entylissa is widely recognized as the monocolpate pollen grain of various members of the ginkgoaelean and cycad alliances (Potonié & Kremp 1956a).

PALEOECOLOGY OF LATE PALEOZOIC SPORES

Most consideration of this subject has grown out of palynological studies of Carboniferous coals and their associated sediments. In a pioneer work in England, Smith (1962) showed a strong relationship between Carboniferous coal lithotypes and spore composition, finding four miospore assemblages, each assemblage being more or less associated with a distinctive coal petrographic type. In the initial phase of formation of a peat, *Lycospora* and associated forms dominate. In the terminal phases, *Densosporites* and associated forms prevail. The intermediate levels are dominated by "transition" palynofloras

194

of *Laevigatosporites, Densosporites*, and others, and "incursion" floras dominated by *Crassispora* and *Punctatosporites*. Butterworth (1966) has summarized the distribution of "densospores" (*Densosporites, Cingulatizonates, Cristatisporites, et al.*), trilete spores with a thick cingulum which is typically differentiated into thicker and thinner parts, giving the spore in proximo-distal view by transmitted light a "tire-within-a-tire" appearance. Most if not all are apparently spores of lycopods. They frequently occur in large numbers in coal seams and range stratigraphically from Devonian to Permian. They first occur abundantly in more northerly regions, but their distribution is displaced southward during the Carboniferous. They are most abundant and diverse in areas of slow subsidence. Coals formed under drier conditions are usually devoid of densospores. They are, for example, missing from the Carbondale Formation, a Pennsylvanian coal-bearing formation in Illinois with coals to 15ft (4.5m) thick. A decrease in humidity is thought responsible. Densospores disappear in the Upper Westphalian C in parts of Europe, also probably because of drop in humidity. On the other hand, in cyclothem situations, densospores often increase toward the top of coal seams, where the coal swamp was in the process of being transgressed (Habib 1966). Densospore production by lycopods requiring much moisture is the obvious explanation. Scott & King (1981), studying lycopod megaspores in relation to coal lithotypes in the Upper Carboniferous of England, showed that these spores also have relationship to level in the coal beds, with *Zonalessporites*, for example, relatively abundant in the upper part of the seams.

Phillips and DiMichele (1981), and Phillips *et al.* (1974, 1985) have studied the palynomorph content of shales and coals of Illinois Basin coal-bearing cyclothems. *Florinites* (from *Cordaites* trees and shrubs and similar forms) occurs in the lowest and highest parts of some coal seams, apparently coming from areas marginal to the swamp. *Lycospora* (mostly from *Lepidophloios*) tends to dominate (up to 80%), but decreases upward, being replaced by *Thymospora* (*Psaronius* fern) and, at the top, by *Laevigatosporites* (ferns) (see Fig. 2.2). *Cappasporites* (aborescent lycopods, cf. *Lepidodendron*) is never dominant but increases upward. *Calamospora* (*Calamites*) and *Vesicaspora* (seed ferns) are always minor constituents, interpreted by various authors as meaning that the producing plants lived outside the swamp on levees or point bars.

, Obviously, much of the underlying information depends on the vast array of data on Carboniferous megafossil plants. However, a succession in time is also at work, so that lycopod-dominated floras gave way to floras dominated by tree ferns (especially, *Psaronius* spp.) at the Westphalian/Stephanian boundary. Then *Lycospora* and *Densosporites* yield to *Laevigatosporites, Punctatisporites*, and other fern spores. This trend is apparently widespread in the late Carboniferous of Euramerica. Chaloner (1961, 1968a, b) and others have noted that *Florinites*, a monosaccate pollen form, is more abundant in deeper-water sediments, indicating upland origin for this buoyant pollen. (*Pinus* pollen sedimentation in the modern Great Bahama Bank shows somewhat similar distributions.) Others (Scott 1979) have pointed out that some *Cordaites* were perhaps mangrove-like, so the situation is probably not ecologically simple. However, *Florinites* is never a dominant constituent in coal palynofloras, though *Cordaites* was a heavy pollen producer, indicating that *Cordaites* was not a dominant plant in the swamps. Chaloner's work with Jurassic sediments showed that there is a regular relationship between general spores/pollen type and sediment type, with conifer pollen more abundant in marine sediments. Peppers & Pfefferkorn (1970) showed that the sedimentation of spores/pollen in Carboniferous coals and other sediments is really a complex of plant autecology (lycopods preferred swampy environments) and such factors as sporomorph preservation (thin-walled spores such as *Calamospora* do not preserve as readily as *Densosporites*). Overrepresentation of plants that produced many spores, and even the differentiated destruction effects of different maceration techniques, must be considered.

Spores/pollen studies of Carboniferous sediments have contributed considerably to phytogeography and may be expected to make bigger contributions in the future because of the massive amount of spores/pollen data compared with those from megafossil studies. As noted earlier, palynological data tend to be more regionally representative than those from megafossil floras (Raymond *et al.* 1985b, Sullivan 1967, Van der Zwan 1981). Through the early Carboniferous (Mississippian), phytogeographic diversity decreased and this trend continued into the later Carboniferous, probably as a result of moves toward assembly of Pangea II and warmer and/or moister conditions that resulted over large areas (Raymond 1985). Van der Zwan *et al.* (1985) used multivariate statistical methods (principal components) to link palynological assemblages of the Lower Carboniferous of Euramerica with climatic indications such as prevalence of evaporites or coal in the source rocks. In this way they were able to show a shifting southward of climatic zones, related to northward drift of Euramerica in the Lower Carboniferous.

As noted above, very detailed studies of the American late Carboniferous (= Pennsylvanian) coal swamps have shown similarly detailed local plant successions and migrations (Phillips *et al.* 1985). In the early middle Pennsylvanian, lycopod (*Cappasporites, Lycospora*) decrease was accompanied by cordaite (*Florinites*) and tree fern (*Punctatisporites, Laevigatosporites, Punctatosporites*) increase. Between middle and late Pennsylvanian, extinctions of coal-swamp lycopods permitted tree fern dominance. However, Phillips *et al.* note that corrective factors need to be carefully considered because of over- and underrepresentation of certain spores/pollen taxa. (The same is undoubtedly true for megafossil floras, but spores/pollen are orders of magnitude more numerous, making the statistics more obvious!) Tree fern spores and *Lycospora*-producing lycopods tend to be overrepresented in the coal-swamp sediments, whereas seed plants, particularly medullosan seed ferns, are underrepresented. *Florinites*, pollen of cordaiteans, is underrepresented in coal-swamp flora, on a percentage basis, because of lycopsid dominance.

In Gondwanaland, various studies have shown that the late Carboniferous glaciation in that area marked the beginning of provincialism in the floras, although the palynological data are as yet confined to Australia and southern South America (Truswell 1981) because of non-deposition, probably related to the glaciation. Azcuy (1975a, b) in Argentina noted that, although the South American palynofloras were of cosmopolitan aspect in the mid to late Carboniferous, by early Permian they were distinctly related to other Gondwana palynofloras.

COMMENTS ON TRENDS IN THE "PALEOPHYTIC" AND THE "PALEOPHYTIC"/"MESOPHYTIC" BOUNDARY

During the Lower Permian, most of the same sorts of plants persisted as dominated the Carboniferous. (See Figure 9.5 for Permian and Triassic subdivisions.) However, provincialism, already marked in the Carboniferous, becomes more noticeable in the Permian, and the correspondence between spore assemblages and megafossil floras is clearer (Truswell 1981).

Especially obvious is the Gondwanaland–Laurasia difference. The Permo-Carboniferous of Gondwana countries is characterized by the *Glossopteris* flora. *Glossopteris* itself produced striate *Protohaploxypinus* pollen, a pollen type that is a signature of the times to come in the "Mesophytic". (See Figure 2.1 for definition of informal plant-based "eras" such as "Mesophytic".) In general, the Gondwana floras of latest "Paleophytic" time are more "modern" (= more "Mesophytic") in aspect than are Laurasian floras of the same age. Powis (1981) has shown that the immediately preglacial sequences of various parts of Gondwanaland (Australia and South America) have similar palynofloras. The glacial palynofloras are also broadly similar: monosaccates and simple trilete spores, and the first occurrence of taeniate bisaccates. However, there are now marked regional differences, with two major palynofloral assemblage types:

(a) Australian type (dominant trilete spores, up to 10% monosaccate pollen), and

(b) Indian type (dominant monosaccates, few spores).

Antarctica, Australia, and South America yield Australian-type Permian palynofloras, while India has the Indian type, and Africa has some Australian

SYSTEM	SERIES	STAGE	(FREQUENTLY USED GERMAN TERMS)
TRIASSIC	UPPER	RHAETIAN* NORIAN KARNIAN	KEUPER
TRIASSIC	MIDDLE	LADINIAN ANISIAN	MUSCHELKALK
TRIASSIC	LOWER	SCYTHIAN	BUNTSANDSTEIN
PERMIAN	UPPER	TATARIAN KAZANIAN	ZECHSTEIN
PERMIAN	MIDDLE	KUNGURIAN	ROTLIEGENDES
PERMIAN	LOWER	ARTINSKIAN SAKMARIAN	ROTLIEGENDES

Figure 9.5 International Permian and Triassic subdivisions. * The status of the Rhaetian is currently disputed, with majority opinion of stratigraphers tending toward eliminating it or regarding it as merely a terminal substage of the Norian. Palynostratigraphers, however, still find it definable (see Visscher 1980).

197

type and some Indian type. Late Permian Gondwanaland palynofloras show increasing importance of bisaccate taeniate pollen.

The justification for using "Paleophytic" and "Mesophytic" (see Figs. 2.1 & 6.1) is that, despite the tremendous significance of the terminal Permian event to animals, especially marine animals, vascular plants do not seem to have changed as much at this interval as they did toward the middle of the Permian (during the Kungurian; see Fig. 9.5), where the older plants such as *Cordaites, Calamites*, and *Lepidodendron* virtually disappeared, and conifers such as various members of the Voltziales began to dominate in Laurasia, and glossopterids in Gondwana. The result of this is that Zechstein (Kazanian, Tatarian) palynofloras resemble Scythian (Lower Triassic) floras much more than they do Lower Permian floras (which are more like Upper Carboniferous floras). A striking aspect of this similarity is the striate (taeniate) pollen forms, which, as we have seen, appeared as pollen of the glossopterids in Gondwanaland earlier than taeniate pollen became abundant in Laurasia. Striate pollen has some combination of grooves, stripes or taeniae on the main body, which may run either parallel or perpendicular to the line connecting the centers of proximal and distal surfaces. Striate pollen may be either saccate or non-saccate (see Fig. 9.6). Some Upper Carboniferous forms, such as *Vittatina*, are striate, but the features explosively evolved in the middle Permian, and even the beginning student with an unknown sample crowded with striate pollen will know immediately "late Permian–early Triassic". Other sorts of conifer pollen, such as *Lueckisporites, Nuskoisporites*, and a number of other genera, also increase in abundance at the expense of the older Carboniferous forms after early Permian. All in all, from a paleopalynological point of view, the "Paleophytic"/"Mesophytic" boundary is a useful concept.

Remy, for example, in Gothan & Remy (1957), defines Känophytikum, Mesophytikum, and Paläophytikum with boundaries similar to my "Ceno-phytic", "Mesophytic", and "Paleophytic." However, Remy called the period before the Paläophytikum the Eophytikum, whereas I call it "Proterophytic", to parallel Proterozoic. It begins with the arrival in abundance of robust-walled acritarchs about one billion years ago, which agrees approximately with the probable arrival of eucaryotes (higher algae). Before the Eophytikum, Remy's oldest era is the Archaikum, which coincides with my "Archeophytic", in which only moneran remains are found. "Paleophytic" has unfortunately been used in quite different senses, and the whole "phytic" idea has had detractors as well as boosters, such as Frederiksen (1972) and Ash (1986). It seems to me reasonable to have "Paleophytic" run from the arrival of land plant spores at the Ordovician/Silurian boundary to the rise to dominance of advanced gymnosperms and their pollen in the middle of the Permian.

MORPHOLOGICAL COMMENT REGARDING CARBONIFEROUS–PERMIAN PSEUDOSACCATE AND SACCATE SPORES/POLLEN

Carboniferous palynofloras include a number of examples of pseudosaccate spores, such as *Alatisporites* and *Endosporites*, in which the apparent sacci are not true sacci but represent extensions of the cavate (camerate) condition, in which

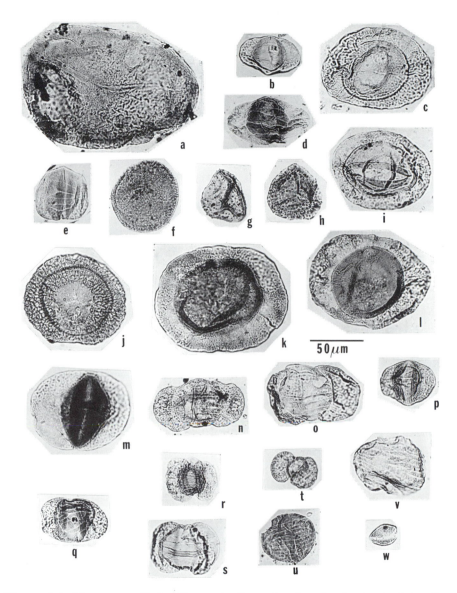

Figure 9.6 Uppermost Carboniferous (= Pennsylvanian, Stephanian Series) pollen (a)–(e) and lowermost Permian (Sakmarian Stage) spores (f)–(h) and pollen (i)–(w) of Canada. Items (g) and (w) are from Yukon Territory, all others from the Maritime Provinces. Nomenclature provided by H. Visscher and W. A. Brugman, Laboratory of Palaeobotany and Palynology, State University of Utrecht, The Netherlands. Magnification shown by bar under (l). (a) *Schopfipollenites ellipsoides* Potonié & Kremp. (b) *Illinites unicus* Kosanke; see (e), (p), and (q). This is a very variable taxon, of which *Complexisporites polymorphus* Jizba is a synonym. (c) *Potonieisporites novicus* Bhardwaj. (d) *Protohaploxypinus* sp. (e) *Illinites unicus* Kosanke; see (b). (f) *Cyclogranisporites vagus*

199

the space between two layers of exine balloons out to produce vesicles (blisters). A true saccus is internally "webby" – the ektexine has obvious structure. Truly bisaccate pollen also developed during the Carboniferous (see Fig. 9.7). *Vesicaspora*, for example, is bisaccate pollen produced by a seed fern, though at first glance it appears to be very like some extant conifer bisaccate pollen! Monosaccate pollen also developed during the Carboniferous, e.g., *Florinites*, a genus for pollen produced by cordaitalean and primitive conifers. Monosaccates become very important in the Lower Permian, when forms such as *Nuskoisporites* and *Lueckisporites*, and numerous other, often large, monosaccates are characteristic. Very revealing studies of Paleozoic saccates by Millay & Taylor (1970, 1974, 1976) have contributed greatly to our understanding of such pollen. As shown in Figure 9.7, there are transitional forms between monosaccate and bisaccate. Bisaccate forms are still important today in extant conifers such as *Pinus, Picea, Cedrus, Abies*, etc. Monosaccates also are found in extant conifers, such as *Tsuga*. (In my opinion, the layer of pollen wall of *Taxodium* and some other extant conifers that is usually called perine may really be the relict of a detachable saccus.)

LATE CARBONIFEROUS–PERMIAN MEGASPORES

Although free megaspores are never again as prominent as they were in late Devonian–early Carboniferous time, their study remains an important source of stratigraphic and especially of paleoecological information through the remainder of the Carboniferous, and the Permian, into the Triassic. S.E.M. pictures of some important forms of late Carboniferous megaspores are displayed in Figure 9.8. Dybova–Jachowicz *et al.* (1982) have demonstrated the importance of careful morphological analysis of these complex spores of the Carboniferous. Scott & King (1981) and others have shown the potential significance of megaspore studies in unraveling the environment of deposition of Carboniferous coals.

CARBONIFEROUS–PERMIAN ACRITARCHS

Aside from spores/pollen, the only palynomorphs of significance in late Paleophytic rocks are acritarchs, and they have been comparatively little studied, despite relative abundance in the marine sediments, and the great potential for palynostratigraphy that they represent. Outline drawings of some representative Permian and Triassic acritarchs are shown in Figure 10.5.

(Kosanke) Potonié & Kremp. (g) *Lophotriletes commissuralis* (Kosanke) Potonié & Kremp. (h) *Lundbladispora* sp. (i) *Potonieisporites grandis* Tschudy & Kosanke. (j) *Luberisaccites subrotatus* Dibner. (k) *Plicatipollenites indicus* Lele. (l) *Potonieisporites bhardwajii* Remy & Remy. (m) *Vestigisporites* sp. (= *Jugasporites omai* Helby). (n) *Protohaploxypinus* sp. (o) *Protohaploxypinus* sp. (p) *Illinites unicus* Kosanke; see (b). (q) *Illinites unicus* Kosanke; see (b). (r) *Striatopodocarpites* sp. (s) *Protohaploxypinus* sp. (t) *Platysaccus* sp. (u) *Vittatina vittifer* (Luber) Samoilovich. (v) *Vittatina costabilis* Wilson. (w) *Cycadopites* sp.

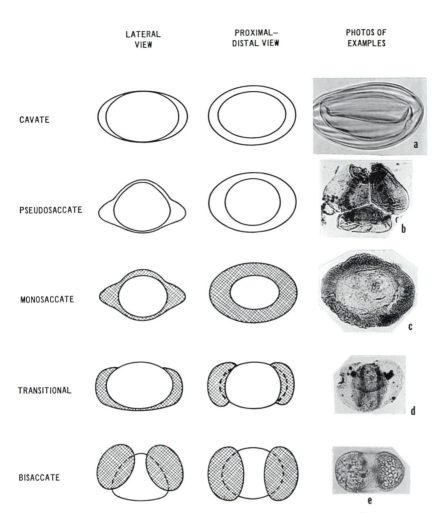

	LATERAL VIEW	PROXIMAL–DISTAL VIEW	PHOTOS OF EXAMPLES

Figure 9.7 Saccate vs. pseudosaccate and cavate spores/pollen morphology. Cavate spores/pollen have a central body of nexine separated by a space from the sexine. (a) Example shown in proximal–distal mid-focus is pollen of *Welwitschia mirabilis* Hook. (extant, Namibia), width 60 μm. Pseudosaccate spores have saccus-like structures which, however, are clearly merely a special case of cavate: the pseudosacci enclose hollow spaces between nexine and sexine. (b) Example shown is *Alatisporites trialatus* Kosanke (Westphalian D, Carboniferous, Canada), width 90 μm. Monosaccate pollen has an alveolate (spongy) envelope more or less surrounding the grain. The single saccus may enclose the whole grain as shown in the diagram, or may leave (usually distal) areas free of saccus, as in the illustrated grain (c) *Cordaitina* sp. (Westphalian C, Carboniferous, Canada), width 100 μm. Many, especially Mesozoic, forms trend toward bisaccate (= transitional) but the alveolate saccus is actually more or less continuous, especially on the proximal side of the grain. Illustrated form (d) is *Protohaploxypinus* sp. (Wolfcampian, Permian, Canada), width 70 μm. Bisaccate pollen has two well-demarcated, separate alveolate sacci. Illustrated form (e) is *Pityosporites* sp. (Albian, Lower Cretaceous, Canada), width 72 μm.

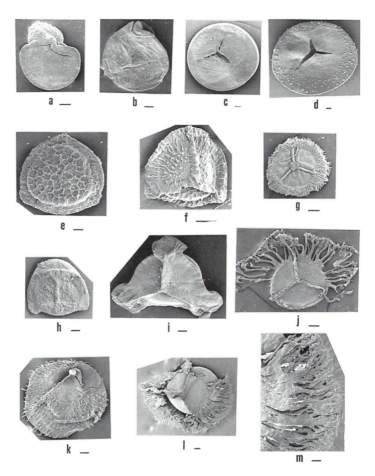

Figure 9.8 S.E.M. micrographs of late Carboniferous (Westphalian B) megaspores from Barnsley and Lidgett coal seams of the Yorkshire coal field. The small bar to the right of each letter indicates 100 μm. See also caption to Figure 8.8. (a) *Cystosporites varius* (Wicher) Dijkstra. Lateral view of an aborted form. (b) *Lagenoisporites rugosus* (Loose) Potonié & Kremp. Lateral view showing gula at top. (c) *Laevigatisporites primus* (Wicher) Potonié & Kremp. Proximal view. (d) *Tuberculatisporites* sp. Proximal view. (e) *Triangulatisporites regalis* (Ibrahim) Potonié & Kremp. Distal view. (f) *Triangulatisporites triangulatus* (Zerndt) Potonié & Kremp. Proximal view. Note zona here and in (e). (g) *Setosisporites hirsutus* (Loose) Ibrahim. Proximal view (see Fig. 8.8t). (h) *Valvisporites auritus* (Zerndt) Potonié & Kremp. Obliquely proximal view. (i) *Valvisporites appendiculatus* (Kowalewska-Maslankiewicz) Potonié & Kremp. Proximal view. Note auriculate structures reminiscent of those in miospore form, *Triquitrites*. (j) *Rotatisporites rotatus* (Bartlett) Potonié & Kremp. Proximal view of form with elaborate coronate zone. (k) *Lagenicula subpilosa* (Ibrahim) Potonié & Kremp. Obliquely proximal view showing rather small contact area and gula (see Fig. 8.8r). (l), (m) *Zonalessporites brassertii* (Stach & Zerndt) Potonié & Kremp. Proximal view, and detail of densely fimbriate zone, respectively (see Fig. 8.8m for a specimen of the species without the zone).

202

CHAPTER TEN
Permo-Triassic spores/pollen

INTRODUCTION

Late Permian, from the Kungurian stage on, is palynologically a time of saccates: monosaccates and bisaccates, and especially of striate bisaccates, surely bespeaking conifer or conifer-like gymnosperm dominance (plus glossopterids in Gondwanaland). This trend accelerated across the "Paleophytic"/ "Mesophytic" line. (See Figure 9.5 for Permian–Triassic time divisions.) However, Permian palynofloras follow those of the Carboniferous in showing great provincialization, and one really must look at the Permian floras both spatially and temporally to get an accurate picture. Hart (1974) and others (Truswell 1981) adopt provinces used by megafossil paleobotanists (Chaloner & Lacey 1973):

(a) Euramerican province (Chaloner & Lacey divide this into Atlantic, North American, and Cathaysian in Lower Permian),
(b) Angaran province (Central Asia, mostly),
(c) Cathaysian province (China, etc.), and
(d) Gondwana province (India, Africa, etc.).

To this should probably be added an Australian province, or Gondwanan subprovince, as Balme (1964) has ably demonstrated that that continent has had a distinct palynoflora since Carboniferous time, and Truswell (1980) shows that Gondwana can be further divided into subprovinces. Playford (1976) demonstrates that even late Devonian Australian palynofloras are to a high degree endemic. Some of Hart's (1971) basic palynofloral types for the Euramerican province are shown in Figure 10.1. The "Paleophytic"– "Mesophytic" change is very clear in the Euramerican and Cathaysian provinces, but is not so obvious in the Angaran province. It would appear to fall somewhere in the Talchis–Karbarbari transition in the Indian Permian. Hart (1974) later showed that the four major floras can be further subdivided. He also pointed out that the floras comprise palynofloristic zones based on paleolatitudes. The Cathaysian palynoflora, for example, would represent a tropical zone. Akyol (1975) and Horowitz (1973) have shown relationships between the Permian palynofloras of Turkey and Israel, respectively, and the Cathaysian palynoflora. Wilson (see Truswell 1981) and others have shown that the Permian palynofloras of Australia and New Zealand are very similar. But Gondwanaland Permian floras seem to have developed their own "intramural" palynofloral provincialism. Truswell, for example, has discussed the interesting case of *Dulhuntyspora* (see Fig. 10.3), a common and biostrati-

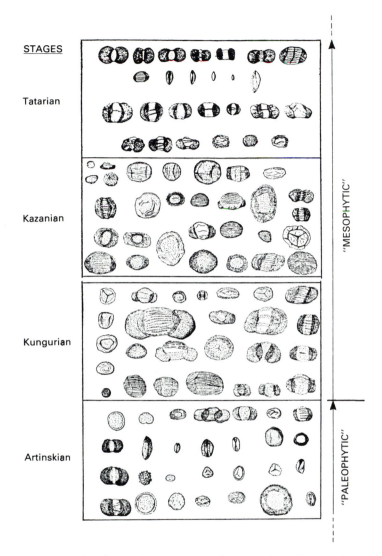

Figure 10.1 Generalized Permian sporomorph succession, Euramerican province, U.S.S.R. Note that an additional lowest Permian stage, the Sakmarian, exists. A "Mesophytic" era can be informally hypothesized to begin with the Kungurian stage and the coming to dominance of taeniate (striate) and bisaccate pollen, probably corresponding to a profound change in continental climate. The "Mesophytic" would continue until the beginning of the "Cenophytic", at the time of arrival of angiosperm pollen in the Neocomian, Lower Cretaceous (see Figs. 2.1 & 6.1). By the end of the Permian, conifer-like bisaccates with large sacci were beginning to predominate.

graphically important Australian late Permian trilete spore, which is essentially confined to Australia. Eshet & Cousminer (1986) report a succession in the Permo-Triassic of Israel. The Permian and Lower Triassic palynofloras are Gondwana-like, those of the Upper Triassic more like Laurasian floras.

STRIATES AND BISACCATES, PERMO-TRIASSIC HALLMARKS

When students in my introductory palynology course get a laboratory "unknown", they are always in (preliminary) luck if it is a late Permian or early Triassic sample, because the prevalence of striate bisaccate pollen is very characteristic practically worldwide of palynofloras of this time. The characteristic striate appearance of the corpus of such pollen is produced by deep parallel grooves (striae) and/or parallel straps (taeniae) in the exine. It would appear likely, though this is conjectural, that taeniae resulted from widening of striae, leaving the strap-like taeniae as exine islets. Both striate non-saccate or slightly saccate, e.g., *Vittatina*, and taeniate bisaccate, e.g., *Striatites*, pollen were well-established in the late Carboniferous and early Permian, but the "heyday" of bisaccate striate/taeniate pollen was in the late Permian and early Triassic. Glossopterid gymnosperms of Gondwanaland have taeniate pollen. The extant gnetaleans, *Ephedra* and *Welwitschia*, have striate pollen. The weirdly "spiral" taeniate form, *Equisetosporites*, of the late Triassic, has been found in conifer cones. Clearly, striate/taeniate pollen is a conifer–glossopterid–gnetalean "thing". Figure 10.2 presents an analysis of the nature of the taeniate/striate construction. Note that even certain extant dicot angiosperm pollen have adopted this strange pattern, which probably has a harmomegathic function for these angiosperms (family Acanthaceae) as an adaptation to the swelling and contracting caused by considerable losses and gains of moisture and likely had a similar function for Permo-Triassic gymnosperm pollen.

OTHER SPORE/POLLEN TYPES OF PERMO-TRIASSIC

Although bisaccate striate/taeniate pollen are the signature of this time, it is obvious that other spore/pollen forms are also important. Trilete spores, probably mostly of ferns, for example, continue to be important throughout the "Mesophytic" and some, e.g., *Dulhuntyspora* in Australia, are important stratigraphically. A group of sporomorphs that originated in the "Paleophytic" but became very significant in the Permo-Triassic and remain so through the "Mesophytic" are the monosulcates. *Entylissa* is one form, already present in the Carboniferous, but there are a number of others such as species of *Monosulcites* and *Cycadopites*. Figure 10.3 shows a range of typical Permo-Triassic sporomorphs from Australia. The stratigraphic range of some of the more important Permian–Triassic–Jurassic forms in Laurasia is given in Figure 10.4.

205

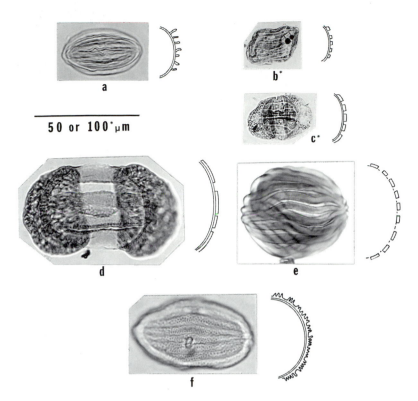

50 or 100*μm

Figure 10.2 Polyplicate and taeniate (= "striate") pollen morphology. One of the most interesting features of Upper Paleozoic and early Mesozoic pollen is the development in a wide range of gymnosperms of pollen, the corpus of which is covered with more or less numerous, more or less widely separated taeniae, or straps of exine. The feature is presumably harmomegathic in nature, allowing for expansion and contraction as required by moisture stress. Among modern plants, the feature is quite rare, being shown by the gnetaleans *Ephedra* (see (a)), *Welwitschia* (see Fig. 9.7a), and by a few angiosperms, such as some members of the family Acanthaceae (see (f)). In both instances, parallel evolution rather than homology to the Permo-Triassic forms is likely. (a) *Ephedra tweediana* C. A. Mey (extant plant, Uruguay). The plicae are rather thin and not much like fossil pollen taeniae. Note however (Fig. 9.7a) that *Welwitschia* has taeniae. (b) *Vittatina vittifer* (Luber) Samoilovitch (Permian of Yukon, Canada). A taeniate form which could accurately be described as striate, as the grooves are as conspicuous as the taeniae. *Vittatina* is one of the few non-saccate Permo-Triassic taeniate pollen forms. (c) *Protohaploxypinus* sp. (Permian of Yukon, Canada). This kind of pollen was made, e.g., by *Glossopteris*. The taeniae are perpendicular to the axes of the sacci in this form. In other genera they may be parallel to them, or helically arranged. (d) *Lunatisporites* sp. (= *Taeniaesporites*) (Permian, United Kingdom). The taeniae in this genus are characteristically four in number, comprising almost all of the corpus exine. (e) *Equisetosporites chinleanus* Daugherty (Triassic of Arizona). The exine seems to consist entirely of the taeniae, which interconnect and spiral around the grain. (f) *Nilgirianthus warrensis* (Dalz.) Bremak (extant plant of family Acanthaceae, India). The "taeniae" are strips of reticulate ektexine on a triporate grain (pores are subsurface in this specimen, but in related species examined, the pores break through to the surface and interrupt the taeniae).

206

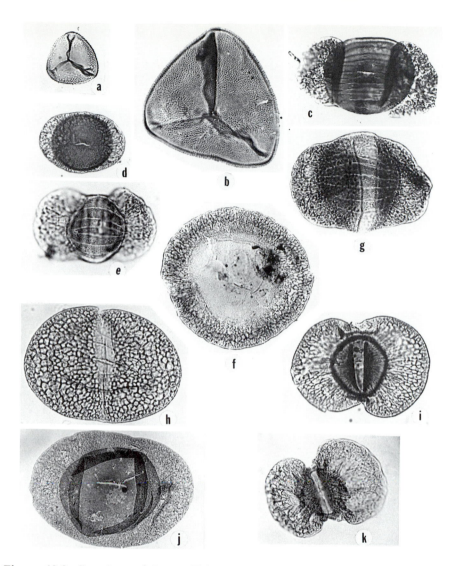

Figure 10.3 Permian and Lower Triassic spores/pollen from Australia. (a) *Microbaculispora tentula* Tiwari, 40 μm. (b) *Microbaculispora trisina* Balme & Hennelly, 85 μm. (c) *Striatoabieites multistriatus* Balme & Hennelly, 70 μm. (d) *Limitisporites* sp., 67 μm. (e) *Striatopodocarpites* sp., 60 μm. (f) *Cannaropollis janakii* Potonié & Sah, 115 μm. (g) *Protohaploxypinus amplus* Balme & Hennelly, 93 μm. (h) *Protohaploxypinus* sp., 102 μm. (i) *Platysaccus* sp., 89 μm. (j) *Potonieisporites* sp., 160 μm. (k) *Platysaccus* sp., 94 μm. (l) *Osmundacidites senectus* Balme, Lower Triassic, 78 μm. (m) *Lundbladispora willmottii* Balme, Lower Triassic, 80 μm. (n) *Marsupipollenites triradiatus* Balme & Hennelly, Upper Permian, 56 μm. (o) *Indospora clara* Bharadwaj, Upper Permian, 60 μm. (p) *Dulhuntyspora dulhuntyi* Potonié, Upper Permian, 88 μm. (q) *Dulhuntyspora* sp., Upper Permian, 80 μm. (r) *Taeniaesporites obex* Balme, Lower Triassic, 76 μm. (s) *Lunatisporites* sp., Lower Triassic, 80 μm. (t) *Lunatisporites pellucidus* Goubin, Lower Triassic 80 μm. (u) *Striatopodocarpites* sp., Upper Permian, 74 μm. (v) *Lunatisporites* sp., Lower Triassic, 83 μm. (w) *Lueckisporites virkkiae* P. & K., Upper Permian of Pakistan, 75 μm. (x) *Praecolpatites sinuosus* Balme & Hennelly, Upper Permian, 114 μm. (y) As (m), S.E.M., 100 μm. (z) *Weylandites lucifer* Bharadwaj & Salujha, Upper Permian, 38 μm. (aa)

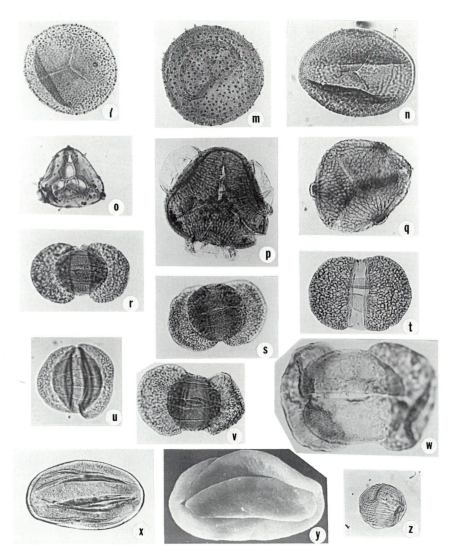

Dulhuntyspora sp., Upper Permian, 75 μm. (ab) *Dulhuntyspora dulhuntyi* Potonié, S.E.M., 85 μm. (ac) As (ab), photomicrograph, 84 μm. (ad) *Lundbladispora willmottii* Balme, Lower Triassic, 77 μm. (ae) *Weylandites lucifer* Bharadwaj & Salujha, Upper Permian, Pakistan, 54 μm. (af) *Lunatisporites pellucidus* Goubin, Lower Triassic, 72 μm. (ag) *Taeniaesporites obex* Balme, Lower Triassic, 72 μm. (ah) *Striatopodocarpites* sp., Upper Permian, 73 μm. (ai) *Marsupipollenites triradiatus* Balme & Hennelly, Permian, 60 μm. (aj) *Taeniaesporites* sp., Lower Triassic, 75 μm.

208

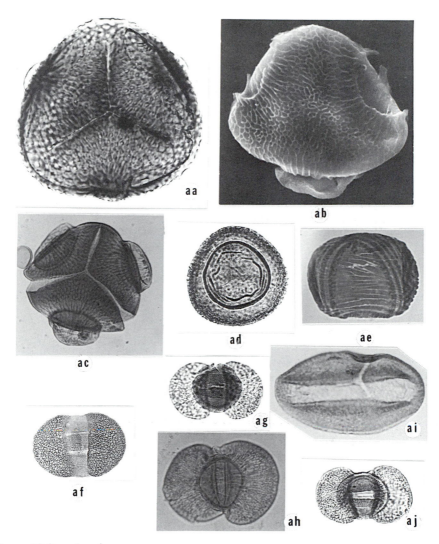

Figure 10.3 *continued*

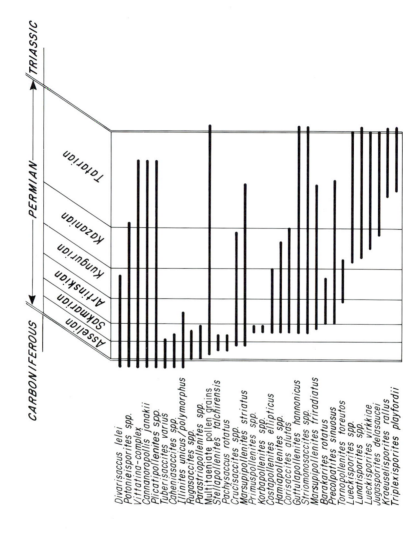

Figure 10.4 Tentative range chart of selected Permian sporomorphs. Some of the forms are restricted to various geographic provinces.

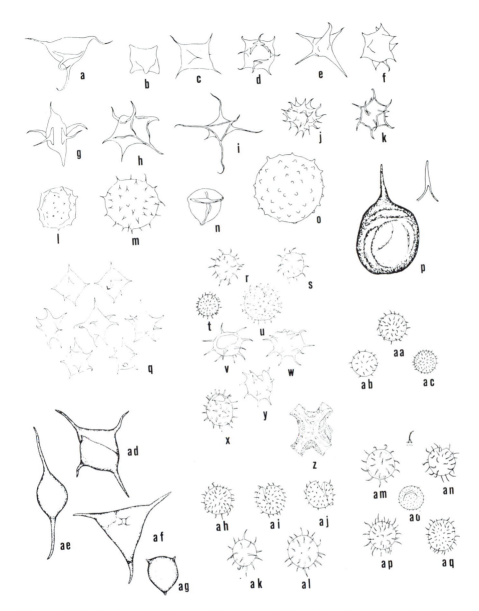

Figure 10.5 Outline drawings of Permian and Triassic acritarchs. (a)–(o) Upper Permian (middle and upper Zechstein of Germany (= Tatarian)); (q)–(z) Tatarian of Pakistan; (p), (aa)–(aq) early Triassic (Scythian and Anisian of Pakistan). (a) *Veryhachium hyalodermum* (Cookson) Schaarschmidt. (b) *V. quadratum* Schaarschmidt. (c) *V.* cf. *nasicum*. (d) *V. sedecimspinosum* Staplin. (e) *V. variabilis* Schaarschmidt. (f) *V. conispinosum* Schaarschmidt. (g) *V. cylindricum* Schaarschmidt. (h) *Polyedryxium kraeuselianum* Schaarschmidt. (i) *Polyedryxium* sp. (j) *Micrhystridium bistchoensis* Staplin. (k) *M.* cf. *albertensis* Staplin. (l) *M. microspinosum* Schaarschmidt. (m) *Baltisphaeridium brevispinosum* (Eisenack) Downie. (n) undetermined acritarch. (o) *Buedingiisphaeridium*

211

PERMO-TRIASSIC ACRITARCHS

Marine and near-marine Permian and Triassic sediments often contain abundant and diverse acritarchs. Jacobson *et al.* (1982) have noted that, for some as yet unexplained reason, though apparently sporopolleninous, they are more abundant in some phosphorites than in the associated mudstones (siltstones) where palynomorphs would be expected. The mudstones may even be altogether barren. Permo–Triassic acritarchs have considerable stratigraphic potential usefulness, as noted for example by Jacobson *et al.* (1982) and by Sarjeant (1970). Figure 10.5 shows some Permian and Triassic common forms.

---

permicum Schaarschmidt. (p) *Deunffia* sp. (q) Range of forms of *Veryhachium*(?) *riburgense* Brosius & Bitterli. (r), (s) *Micrhystridium breve* Jansonius. (t) *M. densispinum* Valensi. (u) *M. setasessitante* Jansonius. (v), (w) *M. circulum* Schön. (x) *M.* sp. (y) *M. karamurzae* Sarjeant. (z) *Polyedryxium* sp. (aa) *Micrhystridium teichertii* Sarjeant. (ab) *M. kummelii* Sarjeant. (ac) *M. densispinum* Valensi. (ad) *Veryhachium* cf. *bromidense* Loeb. (ae) *Leiofusa jurassica* Cookson & Eisenack. (af) *Veryhachium* sp. (ag) *Short-spined variant of V. valensii* Downie & Sarjeant. (ah)–(aj) *Michrystridium castanium* Valensi. (ak) *M.* sp. (al) *Solisphaeridium debilispinum* Wall. (am) *Michrystridium teichertii* Sarjeant; includes detail of spine above. (an) *M. jekhowskyi* Sarjeant. (ao) *M. balmei* Sarjeant. (ap) *M. jekhowskyi* Sarjeant. (aq) *M. castaninum* Valensi. Magnification various: (a) is about 40 μm wide, and this can be used to measure (a)–(o); (p) is 27 μm high; (ae) is 80 μm long, and this can be used for comparison with (ad)–(ag); (z) is 35 μm long; all the other figures are at about the same magnification, such that (aa) is about 20 μm in diameter. Note the details for spines of (p) and (am), a frequently used illustrative method for acritarchs. Because acritarchs often have delicate processes and hyaline walls and are consequently hard to photograph, line drawings are very commonly used to illustrate them.

CHAPTER ELEVEN

Triassic–Jurassic palynology

INTRODUCTION

Again, in this instance, the paleozoologically based "official" biostratigraphic divisions do not coincide with the palynological ones. The Buntsandstein or Scythian stage of the early Triassic "belongs" with the Zechstein in the late Permian, especially on the basis of common bisaccate striate pollen forms. Beginning in the Muschelkalk (Anisian–Ladinian), the once dominant striate bisaccates almost completely disappear. Indeed, very few striates/taeniates of any morphological sort survive after the mid-Triassic (exceptions: *Striatoabieites aytugii* Visscher, *Equisetosporites* (see Fig. 11.2), perhaps the striae of some Circumpolles forms?). Figure 11.1 illustrates a range of Triassic forms, including early *Triassic* taeniate forms such as *Lunatisporites* and *Protohaploxypinus*.

The late Triassic to Jurassic palynoflora is dominated by a bewildering array of non-striate bisaccate pollen (especially prominent in the Jurassic), monosaccate pollen, fern spores, monocolpate pollen, and various inaperturate pollen, as well as pollen variously provided with different sorts of apertures, etc., e.g., *Eucommiidites, Corollina* (= *Classopollis*). Figure 11.2 shows characteristic forms from the late Triassic, and Figure 11.3 gives ranges for some of the more important Triassic forms.

Triassic palynofloras display provincialism, the prominent boundaries seeming to be primarily latitudinal in origin (Truswell 1981). Australian palynologists have suggested that Triassic floras of the Southern Hemisphere can be plotted on a Triassic paleo-map to show such latitudinal zonation. Dolby & Balme (1976) show (see Fig. 11.4) a marked differentiation between the "Onslow microflora" of northwest Australia with western Europe, whereas the "Ipswich microflora" characterizes contemporary deposits of eastern Australia, New Zealand, and Antarctica. These probably were latitudinally controlled, the Onslow flora representing forest at 30–35°S, and the Ipswich flora a higher–latitude plant association.

Mid to late Triassic palynofloras of North Africa, North America, and Europe have much in common:

(a) The Keuper beds of central Europe: Karnian, Norian, and Rhaetian stages. Van der Eem (1983) has shown that careful analysis of the palynofloras reveals not only rather rapid floral evolution reflected in what he calls palynologic "phases" (comparable to but distinct from more rigidly defined "zones" for stratigraphic use), but also palynological differences that reflect fluctuating dry vs. wet local climatic conditions.

213

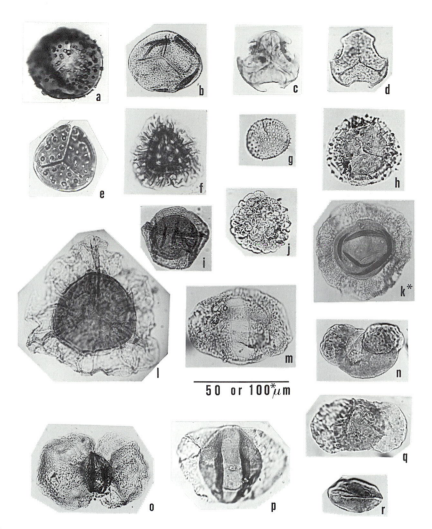

Figure 11.1 Representative Triassic sporomorphs from Poland. Magnification indicated by bar under (m). (a) *Cycloverrutriletes presselensis* Schulz, Scythian. (b) *Cyclotriletes microgranifer* Mädler, Scythian. (c) *Cornutisporites seebergensis* Schulz, Rhaetian. (d) *Triancoraesporites reticulatus* Schulz, Rhaetian. (e) *Paraklukisporites foraminis* Mädler, Rhaetian (ranges into Jurassic). (f) *Limbosporites lundbladii* Nilsson, Rhaetian. (g) *Anapiculatisporites telephorus* (Pautsch) Klaus, Ladinian (ranges Scythian–Lower Jurassic). (h) *Kraeuselisporites cuspidus* Balme, Anisian (ranges Scythian–Ladinian). (i) *Aratrisporites scabratus* Klaus, Ladinian (ranges at least Scythian–Karnian). (j) *Tsugaepollenites oriens* Klaus, Anisian. (k) *Heliosaccus dimorphus* Mädler, Ladinian. (l) *Semiretisporis gothae* Reinhardt, Rhaetian. (m) *Protohaploxypinus pellucidus* Goubin, Scythian. (n) *Microcachryidites fastidiosus* (Jansonius) Klaus, Anisian (ranges into Ladinian). (o) *Platysaccus papilionis* Potonié & Klaus, Scythian. (p) *Lunatisporites puntii* Visscher, Scythian. (q) *Triadispora plicata* Klaus, Anisian (ranges to Karnian). (r) *Cycadopites* cf. *folicularis* Wilson & Webster, Scythian. (More general ranges are given in parentheses.)

214

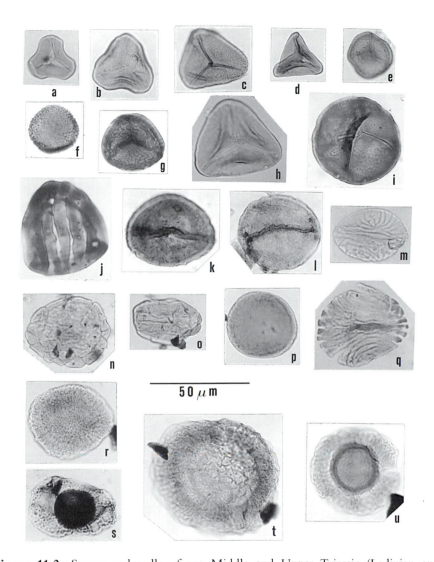

Figure 11.2 Spores and pollen from Middle and Upper Triassic (Ladinian and Karnian) rocks of North America. Magnification indicated by bars under (o) and (ae). (a)–(h), (m), (p), and (t)–(ak) are from the Petrified Forest Member, Chinle Formation, Petrified Forest National Park, Arizona. (i) and (j) are from Karnian portions of the subsurface Eagle Mills Formation, east–central Texas. (k) and (l) are from outcrop samples of the Coal Measures part (Karnian) of the Richmond Basin, Virginia. (n), (o), and (r) are from Ladinian to Karnian parts of the Fundy Basin, Martin Head, New Brunswick, Canada. (a) *Cyathidites minor* Couper. (b) *Gleicheniidites senonicus* Ross. (c) *Granulatisporites infermus* (Balme) Cornet & Traverse. (d) *Dictyophyllidites mortonii* (de Jersey) Playford & Dettmann. (e) *Stereisporites antiquasporites* (Wilson & Webster) Dettmann. (f) *Osmundacidites parvus* de Jersey. Distal view. (g) *Camarozonosporites rudis* (Leschik) Klaus. (h) *Dictyophyllidites harrisii* Couper. (i) *Todisporites* sp. (j) *Contignisporites cooksoniae* (Balme) Dettmann. (k) *Aratrisporites saturni* (Thiergart) Mädler. An odd monolete, zonate spore, showing prominent spines. (l) As (k). Specimen with spines

215

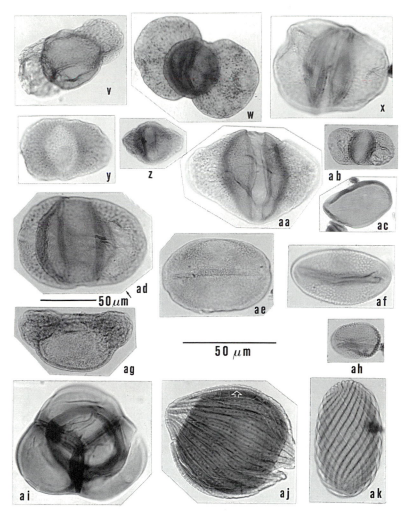

corroded. (m) *Brodispora striata* Clarke. See also (q). (n) *Camerosporites secatus* Leschik. (o) As (n). Different specimen. (p) *Pseudenzonalasporites summus* Scheuring. (q) *Brodispora striata* Clarke. See also (m). (r) *Patinasporites toralis* Leschik. (s) *Triadispora* sp. (t) *Patinasporites densus* Leschik. (u) *Daughertyspora chinleana* (Daugherty) Dunay & Fisher. (v) *Pityosporites oldhamensis* Dunay & Fisher. (w) *Platysaccus triassicus* (Malyavkina) Dunay & Fisher. (x) *Alisporites gottesfeldii*. See also (aa). (y) *Klausipollenites gouldii* Dunay & Fisher. (z) *Protodiploxypinus americus* Dunay & Fisher. (aa) *Alisporites gottesfeldii*. See also (x). (ab) *Vitreisporites pallidus* (Reissinger) Nilsson. (ac) *Lagenella martinii* (Leschik) Klaus. (ad) *Alisporites opii* Daugherty. Note different scale. (ae) *Ovalipollis pseudoalatus* (Thiergart) Schuurman. (af) *Granamonocolpites* cf. *luisae* Herbst. (ag) *Samaropollenites speciosus* Goubin. (ah) *Retisulcites* sp. (ai) *Pyramidosporites traversei* Dunay & Fisher. (aj) *Equisetosporites chinleanus* Daugherty. Note columellate structure (arrow). (ak) *Equisetosporites*? Specimens of this sort occur on the same slides as specimens such as (aj), and they often show similar tectate-like structure. They may represent a related but different polyplicate pollen grain. Intermediate forms occur.

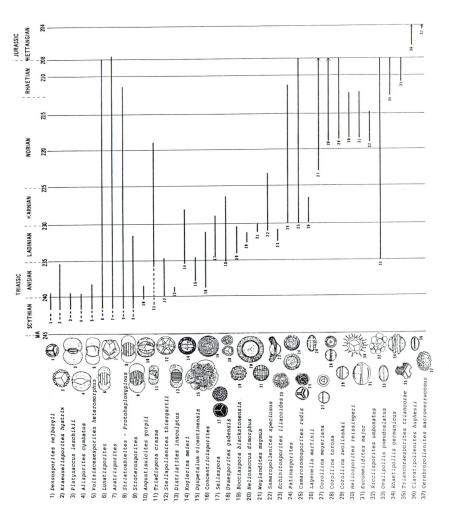

Figure 11.3 Triassic range chart for some selected sporomorphs.

1) *Densosporites nejburgii*
2) *Kraeuselisporites hystrix*
3) *Platysaccus leschikii*
4) *Alisporites cymbatus*
5) *Voltziaceaesporites heteromorphus*
6) *Lunatisporites*
7) *Aratrisporites*
8) *Striatoabieites - Protohaploxypinus*
9) *Stroterosporites*
10) *Angustisulcites gorpii*
11) *Triadispora crassa*
12) *Stellapollenites thiergartii*
13) *Distriatites insculptus*
14) *Kuglerina meieri*
15) *Dyupetalum vicentinensis*
16) *Concentricisporites*
17) *Sellaspora*
18) *Uvaesporites gadensis*
19) *Bocciaspora blackstonensis*
20) *Heliosaccus dimorphus*
21) *Weylandites magmus*
22) *Samaropollenites speciosus*
23) *Echinitosporites iliacoides*
24) *Patinasporites*
25) *Camarozonosporites rudis*
26) *Lagenella martinii*
27) *Corollina meyeriana*
28) *Corollina torosa*
29) *Corollina zwolinskai*
30) *Heliosporites reissingeri*
31) *Eucommidites major*
32) *Ricciisporites umbonatus*
33) *Ovalipollis pseudoalatus*
34) *Rhaetipollis germanicus*
35) *Triancoraesporites triancorae*
36) *Clavatipollenites hughesii*
37) *Cerebropollenites macroverrucosus*

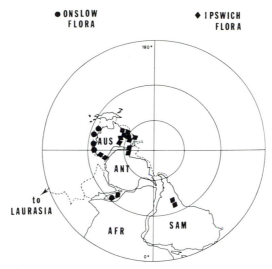

Figure 11.4 The Onslow and Ipswich mid to late Triassic floras, originally recognized in Australia (Gondwanaland), seen as if viewed from modern Australia. The Onslow flora, typical of western Australia, has affinities with Laurasia. The Ipswich flora of eastern Australia is Gondwanaland-distributed. Some prominent taxa in these floras are given in the table.

	Onslow flora (Euramerican affinities, many dry indicators)	Ipswich flora (Gondwanaland affinities, many moist indicators)
bisaccate pollen	*Falcisporites* (usually dominant)	*Falcisporites* *Alisporites*
	Infernopollenites *Staurosaccites* *Lunatisporites* *Ovalipollis* *Samaropollenites*	
monosaccate pollen	*Enzonalasporites* *Patinasporites*	
polyplicate pollen	*Ephedripites* *Decussatisporites*	
monosulcate pollen	*Cycadopites*	*Cycadopites*
trilete spores	*Camerosporites*	*Duplexisporites* (75% of some counts) *Osmundacidites* *Dictyophyllidites* *Uvaesporites*
monolete spores		*Aratrisporites, et al.*

218

(b) The Chinle Formation and Dockum Group of the American southwest: especially well-known, they have primarily Karnian, some Ladinian–Norian, floras. Some prominent representatives of the palynoflora are illustrated in Figure 11.2. Chinle and related palynofloras of the southwest are dominated by bisaccates such as *Alisporites, Falcisporites,* and *Klausipollenites,* and the odd spiral–striate form, *Equisetosporites,* as well as many taxa of monosaccates that are difficult to separate (*Patinasporites, Vallasporites,* etc.), trilete spore forms such as the odd "lumpy" *Camerosporites,* and other trilete forms such as *Deltoidospora* and *Dictyophyllidites* (Fig. 11.5), and monosulcate forms such as *Cycadopites* and *Lagenella.*

(c) The subsurface Eagle Mills Formation of northwest Texas, which includes Ladinian–Karnian sediments. In wells from the Eagle Mills, *Corollina* (= *Classopollis*) appears as a subdominant-to-dominant constituent toward the top of the section, but this probably means that drilling mud has caved from upwell of Jurassic–Cretaceous material, as there are also Cretaceous sporomorphs and dinoflagellate cysts. This is a common problem when using drilling cuttings rather than cores.

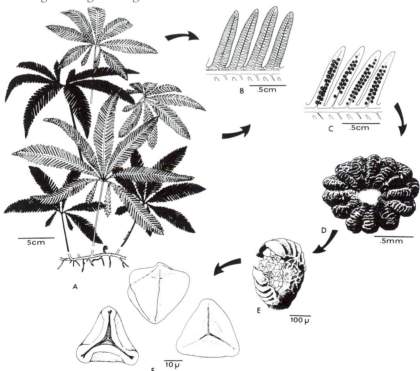

Figure 11.5 Diagrammatic reconstruction of the Chinle fern, *Phlebopteris smithii,* based on specimens collected in the Petrified Forest National Park, Arizona: (A) whole plant with suggested rhizome; (B) sterile pinnules; (C) fertile pinnules; (D) sorus; (E) sporangium; (F) details of spores, isolated by dissection of a mature sporangium. These *in situ* spores are referable to the dispersed spore species, *Dictyophyllidites harrisii* (see Fig. 11.2h). Studies of spores on pollen-bearing megafossil plants have made possible the assignment of many dispersed spore taxa to the producing plants.

219

(d) The Newark Supergroup of eastern North America (Fig. 11.6), which was deposited in a series of basins produced as a preliminary side effect of the separation of North America and Europe and Africa beginning about 190 million years ago. Mesozoic rocks of these basins outcrop from Nova Scotia to North Carolina, but related basins are found under the water of the Bay of Fundy and the North Atlantic, and in the subsurface of South Carolina, Georgia, and Florida. Closely related basins exist in Morocco (Cousminer & Manspeizer 1977).

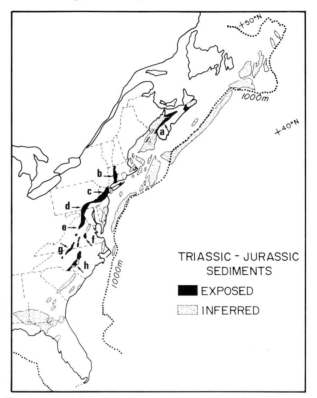

Figure 11.6   The "Newark" basins of eastern North America, which were formed as rift basins in the process of separation of Africa and North America. (Related basins are found in Morocco.) Location of 'inferred' basins is based on extrapolation from drilling information. Palynological and other evidence shows that basin sedimentation began along the whole front in Ladinian–Karnian (late Triassic) time. From central Virginia southward, sedimentation ended after a few million years, but from northern Virginia to Nova Scotia palynological evidence shows that it continued until well into the Jurassic, at least to Pliensbachian, a total of about 40 million years. The principal exposed basins are indicated by letters and arrows: (a) Fundy Basin (Nova Scotia and New Brunswick); (b) Springfield (Massachusetts)–Hartford (Connecticut) Basin; (c) Newark Basin (New Jersey); (d) Gettysburg Basin (Pennsylvania); (e) Culpeper Basin (Maryland–Virginia); (f) Richmond Basin (Virginia); (g) Danville (Virginia)–Dan River (North Carolina) Basin; (h) Deep River Basin (North Carolina). (Compare with Fig. 11.7.)

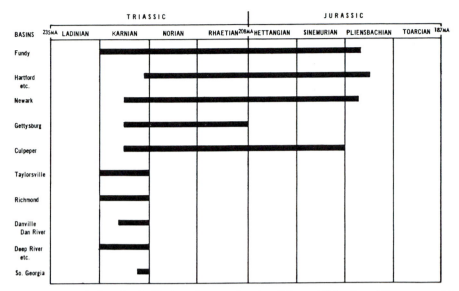

Figure 11.7 Palynologically based ages for the "Newark" basins. (Compare with Fig. 11.6.)

The sediments of the Newark Supergroup basins were presumed by most until fairly recently to be all Triassic in age. However, Cornet and associates (Cornet 1977a, b, Cornet & Olsen 1985, Cornet & Traverse 1975, Cornet *et al.* 1973, Ediger 1986b, Robbins 1982, Traverse 1986) have shown that the sediments range in fact from Karnian or even Ladinian (Deep River Basin, North Carolina; Richmond Basin, Virginia) to Liassic (= Lower Jurassic = Pliensbachian–Toarcian, see Fig. 11.7) in age (Hartford–Springfield Basin). This means that the Newark Supergroup spans more than 50 million years in time. The principal signatures of the younger age of the Rhaetian–Liassic parts of the sections are the complete dominance of various species of *Corollina* forms at many levels, along with other Liassic indicators such as species of *Callialasporites, Ischyosporites,* and others. Cornet has shown that species differences within the *Corollina* group can be used to typify various zones, which are then compared with European sections on the basis of other forms as well.

While the Newark Supergroup as a whole is very rich in palynomorph taxa (see Fig. 11.8), the Richmond Basin in particular is especially diverse, though the sediments are all Ladinian–Norian. In addition to the monosaccate, bisaccate, and trilete genera typical of late Triassic deposits of North America, some samples are rich in *Aratrisporites*, a wide-ranging genus of lycopod cavate monolete (micro-)spores. This genus, abundant in the Australian Triassic, is dominant at some levels in the Richmond Basin, indicating prevalence of marshy environments, a last gasp of dominance of the once mighty lycopsids (de Jersey 1982). Richmond Basin sediments also sometimes contain specimens of columellate (pseudocolumellate?) reticulate pollen with sulci, sulculi or even multiples of these in a very angiosperm-like manner (see Fig. 11.11). Odd,

221

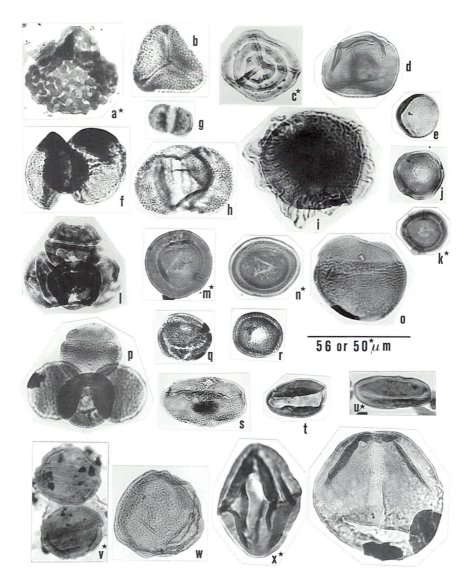

Figure 11.8 Latest Triassic–early Jurassic (Liassic) spores and pollen, North America. Magnification indicated by bar under (o). (a), (c), ((m) and (n) probably from Jurassic upwell caving), and (x) are from upper levels of the Eagle Mills Formation from a borehole in central Texas. Lower portions of this formation in the borehole yield palynomorphs of Ladinian–Karnian age. (k), (u), and (v) are from upper levels of a core (Chinampas Well) in the Bay of Fundy, New Brunswick. (l) is from the upper part of sediments of the Culpeper Basin, Virginia. All other photos are of sporomorphs from latest Triassic and early Jurassic formations in the Springfield–Hartford Basin, Connecticut and Massachusetts. (a) *Ischyosporites variegatus* (Couper) Schulz. (b) *Converrucosisporites cameronii* (de Jersey) Playford & Dettman. (c) *Polycingulatisporites*

222

very thick-walled, pollen occurring in tetrads and appearing trichotomosulcate ("*Placopollis*" – not yet formally named) also are found in the Richmond Basin sediments. In Europe (Fisher & Dunay 1981), latest Triassic indicators such as *Rhaetipollis* first show up in the Norian and persist into the Rhaetian and lowest Jurassic (see Figure 11.3 for ranges of some of the forms). The problem is somewhat bedevilled by arguments about the validity of the Rhaetian as a stage, some regarding it as a terminal substage of the Norian or not using it at all.

Orłowska-Zwolińska (1983) has traced the palynostratigraphic assemblages of the Upper Triassic of Poland, showing that the palynofloras of various levels can be assigned to "assemblages" characterized by important palynomorph types and correlating with the established ammonite zones for the Triassic. A *Heliosaccus dimorphus* assemblage (I) was followed upward by an *Ovalipollis–Triadispora* assemblage (IIb), succeeded by a *Toroisporis–Camarozonosporites–Aulisporites* assemblage (III), and that by a *Corollina* (= *Classopollis*) assemblage

---

mooniensis de Jersey & Paten. (d) *Todisporites rotundiformis* (Malyavkina) Pocock. (e) Circumpolloid pollen? Such apparent inaperturate forms have been given various generic names but may be endexines of *Corollina* (see (j)–(r)). (f) *Platysaccus* sp. (g) *Vitreisporites pallidus* (Reissinger) Nilsson. (h) *Alisporites thomasii* (Couper) Nilsson. (i) *Callialasporites* cf. *dampieri* (Balme) Sukh Dev. (j) *Corollina meyeriana* (Klaus) Venkatachala & Góczán. Distal view, upper focus. (Note: "*Classopollis*" is the most used generic name for this sort of pollen, but is a junior (= later) synonym of *Corollina*, which therefore has priority.) Note distal tenuitas. (k) *Corollina meyeriana*. Distal view in mid-focus, so that the distal, more or less circular tenuitas, the proximal triangular mark, and the pre-equatorial rimula all show. (l) *Corollina meyeriana*. Tetrad, with equatorial view of top pollen grain, showing especially the pre-equatorial rimula or groove, and the distal tenuitas in section at the very top. (m) *Corollina meyeriana*. Distal view, mid-focus, with the triangular proximal mark showing through, as well as the pre-equatorial rimula. (n) *Corollina torosa* (Reissinger) Klaus. View and focus as in (m), showing the prominent internal columellae in the equatorial area. (o) *Corollina itunensis* (Pocock) Cornet & Traverse. Lateral (= equatorial) view, high focus, emphasizing the large size and the prominent, numerous, and complex striations over close to half of the grain in the part below the rimula. (p) *Corollina torosa*. Tetrad showing the complex post-rimula, girdling striations, as in (o), on the top grain. (q) *Corollina murphyi* Cornet & Traverse. Oblique view showing the tenuitas at the top, the rimula as a crescent, and the thick and complexly structured exine. (r) *Corollina murphyi*. Distal view, mid-focus to show the widely spaced internal rods of the exine with the tenuitas also evident. (s) *Cycadopites reticulatus* (Nilsson) Cornet & Traverse. Obliquely distal view, high focus, showing the sulcus and reticulate sculpture. (t) *Cycadopites* cf. *jansonii* Pocock. Distal view, mid-focus, showing the widely gaping sulcus and psilate sculpture. (u) *Cycadopites* sp. with sulcus flared at ends. (v) *Corollina* sp. Lateral views of grains, upper one with tenuitas up, lower one with it down. This species has very thick, internally complex, exine and striate–reticulate sculpture. (w) *Araucariacites punctatus* (Nilsson) Cornet & Traverse. High focus of this inaperturate form. (x) *Pretricolpipollenites* cf. *ovalis* Danzé-Corsin & Laveine. Lateral distal view, high focus. These puzzling forms are, like *Eucommiidites* sp., probably best interpreted as monosulcate, with supplementary sulculi and folds. (y) *Araucariacites fissus* Reiser & Williams. An inaperturate form often, as here, with sulcus- or sulculus-like grooves and folds.

(IV), with a *Ricciisporites*-characterized assemblage (V) at the top. Orłowska-Zwolińska notes that assemblage I correlates with the Ladinian of western Europe, II–III with Karnian, and IV–V with Karnian–Norian (including Rhaetian). Orbell (1973) found the British "Rhaetic" to be divisible into a lower *Rhaetipollis* zone of latest Triassic age and an upper *Heliosporites* zone of early Jurassic (= Liassic) age. Both zones contained abundant *Corollina*. A general summary of all levels of Triassic (and Permian) palynostratigraphy in Europe is presented in Brugman (1983).

CIRCUMPOLLOID POLLEN

This sort of conifer pollen (turmal designation Circumpolles) first appears in the Ladinian stage of the Triassic and seems to characterize new gymnosperms of mid-Triassic to mid-Cretaceous time. As can be seen in Figures 11.9 and 11.10, this pollen has as its primary feature a girdling colpus-like thinning, which divides the pollen into two "hemispheres", one however usually smaller than the other. The more primitive representatives of the group (Fig. 11.9) have incomplete equatorial grooves. Some species have a trilete laesura or laesuroid marking on the proximal surface and/or a thin, often more or less triangular, distal colpus-like area, the tenuitas. The two hemispheres sometimes split apart and can then be found isolated (see Fig. 11.10). In some species, tetrads and dyads are very common; in others, only monads are normally found. The exine of the more advanced forms is probably unique among gymnosperms in having very well-developed nexinal columellae (see Fig. 11.10). However (Medus 1977), the structure of various circumpolloid forms differs from that of *Corollina torosa*, the "typical" form, and some forms seem to lack the multiple levels shown in Figure 11.10. The botanical relationships of circumpolloid pollen are very certainly known; they are the sort of pollen produced by a variety of Mesozoic conifers, especially by *Hirmeriella* (once known as *Cheirolepis*) and other cheirolepidiaceous conifers (whether produced by non-cheirolepidiaceous conifers is still debated). The ecological significance of the producing plants, often very widespread and dominant in Triassic–Cretaceous time, has been much discussed. At least some of them were apparently warmth-loving shrubs that grew in great thickets in the same sort of low-lying water-margin environment that produced extensive mudflats peppered with dinosaur tracks, as in the Triassic–Jurassic Hartford Basin. Francis (1983, 1984) shows that some *Corollina* producers were shrubs that tolerated semiarid conditions, and *Corollina* pollen distribution is correlated with evaporites. However, others may have been upland, xeric plants (Srivastava 1976a). That they were warmth-loving is indisputable, as they decline with increasing latitude. There seems also to be no doubt that they covered large areas, because the pollen often completely dominates (more than 90%) palynofloras. Some of such dominance characterizes acme zones which have stratigraphic significance. The first one in North America is Hettangian.

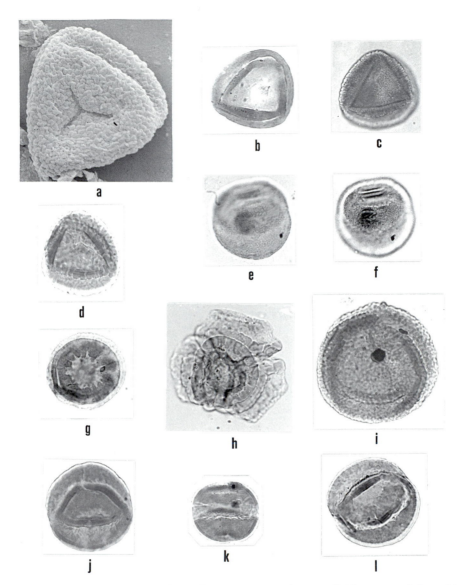

Figure 11.9 *Partitisporites* and *Duplicisporites*, two circumpolloid genera of the mid and late Triassic. *Praecirculina* has an incomplete semicircular equatorial furrow but is otherwise very similar to *Partitisporites novimundanus* Leschik ((e), (f)). *Camerosporites* (see Fig. 11.2 n, o) is another Triassic genus with probable circumpolloid relationship. *Corollina* (= *Classopollis*) is the circumpolloid pollen of the one coniferous taxon of this group that "made it big" (see Fig. 11.10). The magnification of these illustrations varies greatly, so the size is given for each specimen. (a) *Duplicisporites verrucosus* Leschik. S.E.M. micrograph, proximal view showing trilete mark. Range, Ladinian–Karnian. The circumpolloid equatorial furrow structure is tripartite (see (b)–(d)) in this genus. Size 35 μm. Karnian, Spain. (b) *Duplicisporites kedangensis* Schuurman. Distal view showing

225

Circumpolloid systematics

The basic genus is *Corollina* Malyavkina 1949, if one accepts that Malyavkina's (1949) simple line sketch (Fig. 11.10) is recognizable as this form. In my opinion, it is, and therefore the only correct name for the genus is *Corollina*, as this has (by far) priority. Unfortunately, Malyavkina's type specimen no longer exists. Pflug (1953) described and illustrated the form unequivocally, calling it *Classopollis*, and the fact that he thought it was angiospermoid is irrelevant to the nomenclatural problem. Inasmuch as botany does not have a commission that could make a binding decision on the matter of priority and adequate illustration, individual palynologists have to decide which name to use. (The decision should not, however, be made light-heartedly!) Perhaps the best solution would be to conserve the name *Classopollis* against *Corollina* legally, by means of the procedures outlined in the *International Code of Botanical Nomenclature* (Voss *et al.* 1983). In fact, most people use *Classopollis*. However, that is only one of the systematic problems. Another is how many genera to recognize. Some have elected to use just two genera, *Corollina* Mal. (= *Classopollis* Pfl.) for the forms with a striate belt, and *Circulina* Klaus for those lacking it. However, other genera have been described, such as *Gliscapollis* Venkatachala. As already mentioned, Cornet (1977a) used *Corollina* species and their relative dominance to typify some of the Triassic–Jurassic palynological zones in his Newark Supergroup work.

---

the tripartite equatorial furrow and the distal tenuitas. Range, Rhaetian–Cretaceous. Size 30 μm. Rhaetian, Italy. (c) *Duplicisporites granulatus* Leschik. Proximal view. Range, Ladinian–Karnian. Size 33 μm. Karnian, Spain. (d) *Duplicisporites verrucosus* Leschik. Proximal view. Data as for (a). (e), (f) *Partitisporites novimundanus* Leschik. Two levels of focus of obliquely lateral view, showing half of the equatorial furrow, which is bipartite in this genus, the semilunar segments nearly joining together (see (i)). Range, Ladinian–Karnian. Size 35 μm. Karnian, Spain. (g) *Partitisporites tenebrosus* (Scheuring) Van der Eem. Distal view showing complex structure of area around the tenuitas. Range, Ladinian–Karnian. Size 30 μm. Karnian, Spain. (h) *Partitisporites quadruplicis* (Scheuring) Van der Eem. Tetrad showing the flattened distal area in lateral view of top grain. Range, uppermost Ladinian–Karnian. Size (one grain) 35 μm. Karnian, Spain. (i) *Partitisporites novimundanus* Leschik. Mid-focus of distal view, showing bipartite equatorial furrow (see (e), (f)). Size 35 μm. Karnian, Spain. (j) *Partitisporites maljawkinae* (Klaus) Van der Eem. Distal view showing bipartite equatorial furrow. Range, Karnian. Size 35 μm. Karnian, Italy. (k) As (j). Lateral view showing somewhat complex equatorial furrow, and the distal tenuitas. (l) As (j). Distal view showing bipartite equatorial furrow and distal tenuitas. (Van der Eem (1983) grouped *Partitisporites novimundanus*, *Praecirculina granifer*, and "*Paracirculina*" *verrucosa* in the *P. novimundanus* morphon, and *Duplicisporites granulatus*, *D. verrucosus*, and *D. mancus* in the *D. granulatus* morphon.) Brugman and others in the Laboratory of Palaeobotany and Palynology at the University of Utrecht have been the first to make critical observations about the morphology, especially the structure of the equatorial furrows, in these forms.

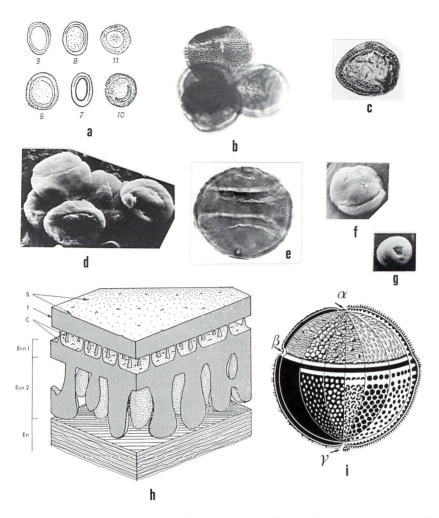

Figure 11.10 The remarkable pollen genus, *Corollina* Malyavkina (= *Classopollis* Pflug). (a) Malyavkina (1949) described from Lower Jurassic rocks of the Soviet Union the group of palynomorphs illustrated here collectively by her line drawings "9, 8, 11, 6, 7, 10". There can be little reasonable doubt that Malyavkina's drawings 6–11 represent what we now call in general "circumpolles" pollen, a group of distinctive, extinct conifer pollen with a pre-equatorial groove or rimula. However, she called numbers 10 and 11 *Corollina compacta* and numbers 6–9 two different species of *Circulina*. Pflug (1953) described from Lower Jurassic rocks of Germany the genus *Classopollis*, mistakenly describing the pollen as tri- and tetracolpate and believing them to be angiosperm-like. However, Pflug's photomicrographs are unequivocal. *Corollina* and *Classopollis* are almost certainly synonyms, and Malyavkina's names have priority. However, the fact that she used two generic names muddies the water, as differentiating *Circulina* from *Corollina* is a problem. Those who wish to accept Malyavkina's work lean toward suppressing the name *Circulina* and retaining only *Corollina*, but this is also controversial. Something like 80% of present palynologists now use the name *Classopollis*. Under the botanical rules, there is no way to declare that one or the other usage is "right" in a case like this. (b) Tetrad of *Corollina torosa* showing

227

COLPATE (SULCATE) FORMS IN THE TRIASSIC–JURASSIC

The encircling colpoid feature that characterizes circumpolloid pollen is very like such structures in the 1-syncolpate (or zonisulcate) pollen in the Angiospermae. Also found in Triassic–Cretaceous palynofloras are forms with other sorts of angiospermous-like ("angiospermid") sulci (colpi). For example, there are monosulcate forms such as *Retimonocolpites, Retisulcites,* and *Liliacidites,* with sculpturing suggestive of monocots (in contrast to typical *Cycadopites,* mostly scabrate to psilate, monosulcate pollen). Another important colpate form is *Eucommiidites,* first described by Erdtman from the early Jurassic of Sweden, and thought by him to be possibly angiospermous, because of the apparently tricolpate form (see Fig. 11.3). However, Couper (1956) and Hughes (1961) showed subsequently that the position and orientation of the colpi are wrong for a dicot tricolpate. Dicot PcOs have the three equal colpi on meridians connecting the proximal and distal poles and are 120° apart. *Eucommiidites* has a main colpus on one side and two subsidiary shorter colpi, which may be united to a single ring furrow, on the other side. Further, the three colpi are not equal in length, the central, or main, colpus being longer. *Eucommiidites* pollen has been found in the pollen chambers of gymnosperm seeds by Reymanówna (1968), and the exine has been shown to be of gymnosperm type. *Eucommiidites* is now known to range from Triassic to Cretaceous. *Pretricolpipollenites* pollen is another Triassic–Jurassic form best viewed as an angiosperm look-alike.

Cornet (1977a) has described, however, odd colpate forms from the late Triassic, especially from the Karnian of the Richmond Basin, which also have a columellate exine and in all are very angiosperm-like (Fig. 11.11). Of these, Cornet writes (personal communication): "One of the most interesting

---

the complex striate exine in a band encircling the top grain. The encircling groove just above the striate band is the rimula (arrow). (c) *Corollina torosa.* Mid-focus from distal side of single grain, showing the columellate structure of the exine. Rimula (arrow) also visible. Note similarity to Malyavkina's drawing 10 (see (a)). (d) A tetrad (left) and an extra grain of *Corollina meyeriana* in S.E.M. The rimula is clearly visible on all grains. (e) *Corollina zwolinskae* Lund. Lateral view of a peculiar *Corollina* with two rimulae or furrows, one pre-equatorial, one sub-equatorial (each shows through from the back and thus creates the appearance of four furrows!). Rhaetian, England. This form is diagnostic for the Rhaetian. (f) *Corollina meyeriana.* Obliquely proximal view in S.E.M., showing the faintly expressed triangular mark. Rimula visible below. (g) *Corollina meyeriana.* Distal view, S.E.M., showing the tenuitas or distal aperture. (h) Drawing illustrating the complex structure of *Corollina torosa* exine: S, sculpture; T, tectum; C, columellae; Ecn1–2 and En, layers of the outer exine. (i) Drawing showing the sculpture found on various circumpolles exines, displayed on the distal hemisphere (left to right): rugose, verrucose, mixed, 2-formed, echinate–baculate, echinate. The various sorts of internal structure found are shown in the lower hemisphere (left to right): massive, alveolate, reticulate, vermiculate, intrareticulate, punctate. Arrow with α, tenuitas; arrow with β, rimula; arrow with γ, proximal mark (in lateral section). The magnification is different for each illustration, but each palynomorph is about 30 μm in maximum dimension.

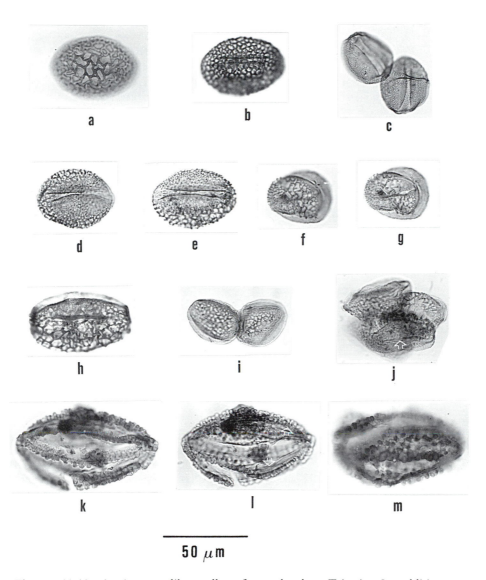

50 μm

Figure 11.11 Angiosperm-like pollen from the late Triassic. In addition to *Eucommiidites*, *Pretricolpipollenites*, and other late Triassic pollen with colpi-like furrows suggesting angiosperms, forms are found such as these from the late Triassic Richmond Basin, Virginia, which have not only such furrows but also angiosperm-like sculpture, and even tectate–columellate wall structure. As has been proven for *Eucommiidites*, all these forms are probably best viewed as monosulcates. Thus, these photos are oriented where possible as if monosulcate, according to Erdtmanian convention (presumed primary sulcus parallel to bottom of page). Whether this trend represents a line leading to angiosperms or is the result of convergent evolution is unknown. (a), (b) Two monosulcate forms. Proximal focus, emphasizing the monocot-like coarse reticulum. (c) Pair of monosulcate forms in distal focus with fine-pattern reticulum. (d), (e) Two

229

discoveries from the study of late Triassic palynofloras . . . has been the rare (usually less than 1%) but persistent presence . . . of angiosperm-like monosulcate and zonasulculate pollen grains . . .". A complex of about eight species of angiosperm-like pollen of late middle Karnian age from the Richmond Basin of Virginia includes highly reticulate–columellate monosulcates, zona-sulculates (trisulcates), and a reticulate–clavate pentasulcate (see Fig. 11.11).

Perhaps all of these Triassic–Jurassic forms should be viewed as an indication of a trend of various gymnosperms toward angiosperm pollen, and for all we know one or more of them may be somehow directly linked to flowering plant ancestry.

FURTHER NOTES ON TRIASSIC–JURASSIC SACCATES

Late Triassic and Jurassic sporomorph assemblages are dominated by gymnosperm pollen, Pv1, Pv2, P00–Pa0, and trilete fern spores. The monosaccates, *Patinasporites, Vallasporites, Enzonalasporites,* and others, are an interesting illustration of a morphon, or intergrading complex of forms. The bisaccates are extremely difficult to cope with systematically. Some of the more common sorts of monosaccate and bisaccate organization are shown in Figure 11.12. The bisaccates of the Triassic–Jurassic, because so numerous and of so many forms, are especially challenging. S. A. J. Pocock in a short course on "Mesophytic palynology" referred to their often large size and nondescript nature by calling them "big floppies".

JURASSIC PALYNOMORPH PALEOGEOGRAPHY

Paleopalynology shows that the Jurassic was not as homogeneous as once thought, though more cosmopolitan than most parts of the Phanerozoic. For example, Filatoff (1975) showed that Jurassic Pv3 (trisaccate) pollen is restricted to India, Australia, and Argentina. On the other hand, some conifers are indeed cosmopolitan, e.g., *Corollina* (= *Classopollis*), although there are latitudinal variations in its abundance (Truswell 1981). Dinoflagellate cyst studies have demonstrated marked provinciality in the Jurassic (Williams 1975).

---

monosulcates. Distal view and focus. Style of sulcus and relation of sculpture to it like monocots. (Pollen such as in (a)–(e) is usually put in the form-genus *Retisulcites*.) (f), (g) Two levels of focus of a distal view of one specimen that is zonasulculate, showing additional furrows, as well as the distal sulcus. (h) "Trisulcate" form in lateral view. The reticulate sculpture is modified near the sulcus. (i) Pair of "trisulcates" with reticulate sculpture. (j) Tetrad of "trisulcates", with columellate structure clearly showing at arrow. (k), (l) Different levels of focus of a "pentasulcate" pollen grain in which columellate structure is apparently present (arrows). (m) Another specimen of the "pentasulcate" form. Magnification shown by bar under (l).

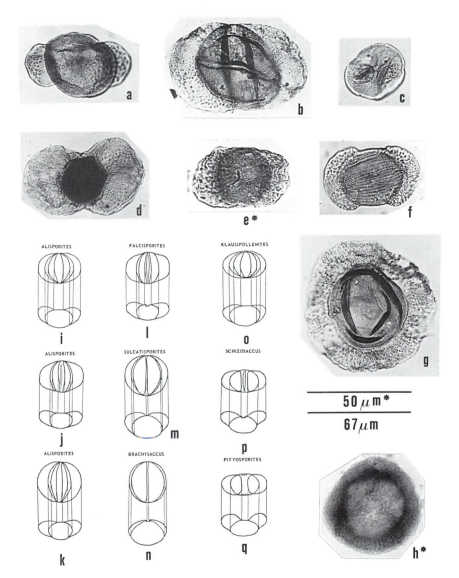

Figure 11.12 Triassic bisaccate and monosaccate diversity. The systematic problems encountered in the classification of saccate pollen are perhaps the most difficult in all of paleopalynology. Those of Triassic–Jurassic palynofloras are, as a group, among the most difficult. (a) *Microcachryidites fastidiosus* (Jansonius) Klaus. Anisian, Poland. Proximal view, high focus. (b) *Succinctisporites grandior* Leschik. Scythian–Anisian, Poland. Proximal view, mid-focus, showing characteristic monolete-like structure and saccus folds. This form is transitional between monosaccate and bisaccate. (c) *Minutosaccus gracilis* (Scheuring) Orlowska–Zwolińska. Ladinian, Poland. Proximal view, high focus. Forms with such tiny sacci are often classified as *Protodiploxypinus* spp. (d) *Platysaccus* cf. *papilionis* Potonié & Klaus. Scythian–Anisian, Poland. Mid-focus, illustrating very dense central body and large sacci. (e) *Triadispora crassa* Klaus.

231

MAJOR KNOWN* BOTANICAL RELATIONSHIPS OF "MESOPHYTIC" (LATE PERMIAN–EARLY CRETACEOUS) DISPERSED SPORES/POLLEN GENERA

The "turmal" classification is not as useful in the "Mesophytic" as in the "Paleophytic" because the system was designed by Potonié primarily for Paleozoic spores/pollen, and has not really been as well-developed for many of the "Mesophytic" sporomorphs, e.g., taeniate–striates, bisaccates, and mono-colpates. Furthermore, at least the broad botanical relationship is known for practically all "Mesophytic" sporomorphs and shoehorning into "turmal" categories is often time-consuming and frustrating. Therefore, I present the data for spores found *in situ* here in broad morphological categories, which follow roughly the turmal classification. With a few exceptions, I do not repeat the genera already treated under Paleozoic spores. However, such genera as *Apiculatisporites*, produced by ferns, and *Calamospora*, produced by a variety of sphenopsids, continue through the "Mesophytic".

I. SPORES (megaspores marked with asterisk)

A. TRILETE

(Fern fossils from "Mesophytic" rocks are sometimes referred to extant genera! Thus, Jurassic–Cretaceous spores obtained from osmundaceous fern megafossils are referred to as *Osmunda* spores by Krassilov (1978).)

---

* *in situ*, or very strong presumption from association.

Scythian–Anisian, Poland. Proximo–distal view, focus on the trilete laesura of proximal side. Lateral views show that sacci almost surround central body. (f) *Striatoabieites balmei* Klaus. Scythian–Anisian, Poland. Proximal view, mid-focus, taeniate (= polyplicate or "striate") form that survived into mid-Triassic after most of them had become extinct. (g) *Heliosaccus dimorphus* Mädler. Ladinian, Poland. Proximo–distal view, mid-focus, showing alete central body (otherwise resembles the earlier monosaccate form, *Nuskoisporites*). (h) *Patinasporites densus* Leschik. Karnian–Norian of Eagle Mills Formation, Texas. Proximo–distal view, mid-focus, showing the "bubbly" saccus that surrounds the grain. This sort of monosaccate is often very abundant in the late Triassic. The faint pseudo Y-mark visible would cause some to place this in *Vallasporites*. There are other forms of probably the same single genus, such as *Daughertyspora*. (i)–(q) Drawings by Bruce Cornet showing some of the variants on the Triassic–Jurassic bisaccate theme. Each pair of drawings presents a distal view above and the corresponding lateral view below: (i)–(k) variations within the *Alisporites* complex; (l) *Falcisporites*, which differs very slightly from *Alisporites* – the transverse sulcus displayed is not always demonstrable; (m) *Sulcatisporites*, despite the name, there is no visible sulcus; (n) *Brachysaccus*, can be difficult to distinguish from *Sulcatisporites*, but B. does have a sulcus; (o) *Klausipollenites* – in polar views the small, crescent-shaped sacci are distinctive; (p) *Schizosaccus*, which has sacci like *Pityosporites*, but S. is sulcate; (q) *Pityosporites*, grains resemble modern *Pinus* or *Cedrus* more than do any of the others illustrated. Scale under (g) applies only to photomicrographs.

Banksisporites★

Banksisporites was produced as a trilete megaspore by the Triassic lycopsids, *Cylostrobus* and *Selaginellites*, according to Helby & Martin (1965). (Microspores of *Cylostrobus* were monolete *Aratrisporites* and trilete *Lundbladispora*, respectively. See also entries for these two genera.)

Calamospora

Calamospora spores were so identified from Triassic fossils described by Halle as *Equisetites* in 1908 (Couper 1958).

Carnisporites

Carnisporites spores were produced by the fern *Cynepteris*, judging from illustrations in Ash (1969).

Cicatricosisporites

Cicatricosisporites spores were studied by Couper (1958) in the type slide of the Cretaceous schizaeaceous fern, *Ruffordia*. Couper also studied *C.* spores in preparations of other schizaeaceous ferns from Cretaceous to Eocene in age. Skog (1980, 1982) found *C.* spores in the Lower Cretaceous schizaeaceous ferns, *Ruffordia* and *Pelletixia*.

Clathropterisospora

Clathropterisospora was described by Zhang (1980) for dispersed spores identical with those obtained from Upper Triassic *Clathropteris* (Dipteridaceae) sporangia.

Cyathidites

Cyathidites spores were identified by Couper (1958) in preparations of Jurassic ferns probably referable to the Dicksoniaceae, and by Douglas (1973) from *Coniopteris*, in the Lower Cretaceous.

Sukh-Dev (1980) notes that *Cyathidites*-like spores were removed from Indian Lower Cretaceous *Onychiopsis* ferns (see also *Deltoidospora*).

Deltoidospora

Deltoidospora-like spores were obtained from Lower Cretaceous *Onychiopsis* ferns by Sukh-Dev (1980) (but see also *Cyathidites*).

Dictyophyllidites

Dictyophyllidites was obtained from Jurassic *Dictyophyllum* (cheiropleuriaceous? fern) by Couper (1958). Spores of this type were also obtained from Jurassic *Phlebopteris* and from Lower Cretaceous *Weichselia* ferns (Sukh-Dev 1980) (see also *Lametatriletes*). Litwin (1985) isolated spores referable here from Triassic *Phlebopteris* (Matoniaceae).

Dijkstraisporites★

Megaspores referred to *Dijkstraisporites* have been described from *Annalepis*, a lycopod cone also producing *Aratrisporites* microspores (see also *Tenellisporites*).

Gleicheniidites

Gleicheniidites spores were identified by Couper (1958) from figures of Jurassic gleicheniaceous ferns published by Harris.

233

Granulatisporites

Granulatisporites-type spores were identified by Litwin (1985) in Triassic *Clathropteris* (Dipteridaceae) material.

Horstisporites★

Horstisporites was found in the Triassic lycopsid cone *Skilliostrobus*, according to Ash (1979).

Klukisporites

Klukisporites spores were studied in preparations from Jurassic schizaeaceous ferns (*Klukia* and *Stachypteris*) by Couper (1958).

Lametatriletes

Spores similar to *Lametatriletes* (and to *Cyathidites* and *Dictyophyllidites*) have been removed from Lower Cretaceous *Weichselia* ferns (see Sukh-Dev 1980).

Lundbladispora

Lundbladispora was produced as a microspore by the lycopod *Selaginellites* (Helby & Martin 1965).

Marattisporites

Marattisporites spores were obtained from Triassic–Jurassic marattialean ferns by Couper (1958).

Matonisporites

Matonisporites spores were identified by Couper (1958) from figures of Triassic–Jurassic matoniaceous ferns (*Phlebopteris, Selenocarpus, Matonidium*) published by various paleobotanists. *M.* spores were also isolated from sporangia of *Phlebopteris* ferns (Ash *et al.* 1982) from the late Triassic of Arizona, and in Indian Jurassic *Phlebopteris* (*Dictyophillidites* also identified), according to Sukh-Dev (1980).

Minerisporites★

Minerisporites megaspores were obtained from *Isöetites* megafossils (Lower Cretaceous) (Sukh-Dev 1980).

Osmundacidites

Osmundacidites spores were identified by Couper (1958) from Jurassic osmundaceous ferns published earlier by Harris. Litwin (1985) found spores referable to this genus in *Todites* sporangia of Triassic age. Van Konijnenburg-Van Cittert (1978) referred *in situ* spores from *Osmundopsis* (osmundaceous megafossil) to *Osmundacidites*.

Paxillitriletes

Paxillitriletes is an isöetalean megaspore (Kovach & Dilcher 1985).

Punctatisporites★

Punctatisporites (or *Cyclogranisporites*) spores were described for the Triassic fern, *Anomopteris*, by Grauvogel-Stamm and Grauvogel (1980).

Tenellisporites★

Tenellisporites marcinkiewiczae Reinhardt & Fricke megaspores have been described by Grauvogel-Stamm & Duringer (1983) from the lycopod

234

fructification *Annalepis*, which produces also *Aratrisporites* microspores (these megaspores also have been referred to *Dijkstraisporites*).

Todisporites

Todisporites spores were obtained by Couper (1958) from Jurassic osmundaceous ferns, and from the Triassic fern *Wingatea* by Litwin (1985). Van Konijnenburg-Van Cittert (1978) said that *in situ* spores from *Todites* spp. (osmundaceous megafossil) are referable to *Todisporites*.

Triletes★

Triletes megaspores were removed from the Jurassic–Cretaceous probable lycopod, *Synlycostrobus*, by Krassilov (1978).

B. MONOLETE

Aratrisporites

Aratrisporites was produced as a microspore by the Triassic lycopsid *Cylostrobus*, according to Helby & Martin (1965). The same authors note the occurrence of *A.* also in *Lycostrobus*. Ash (1979) found *A.* as a microspore in the heterosporous lycopsid cone, *Skilliostrobus*. Grauvogel-Stamm & Duringer (1983) found spores close to *Aratrisporites minimus* Schulz in the lycopsid fructification *Annalepis zeilleri* Fliche (see also megaspore *Tenellisporites*). (Scott & Playford (1985) report *Aratrisporites* microspores attached to *Banksisporites* and *Nathorstisporites* megaspores.)

II. POLLEN

A. MONOSACCATE

Callialasporites

Callialasporites (also called *Applanopsis*) was produced in pollen cones of the Lower Cretaceous conifer, *Apterocladus* (Archangelsky & Gamerro 1967, Gamerro 1968).

Nuskoisporites

Nuskoisporites, a trilete monosaccate pollen form, was isolated from a late Permian conifer cone, *Ortiseia* (Clement-Westerhof 1974).

Patinasporites

Patinasporites, a *Tsuga*-like form, was obtained from *Pagiophyllum*-like conifer cones of the late Triassic of Pennsylvania by Cornet (1977a).

B. BISACCATE (including multisaccates, pseudosaccates, and striate–taeniate bisaccates)

Alisporites

Alisporites pollen of several species was found, and a new species of this dispersed pollen type described, by Grauvogel-Stamm (1978) from *Willsiostrobus* cones, a conifer from the early Triassic of France. (It does not affect the information here, but it would be better not to propose new *Sporae dispersae* taxa from megafossil plant specimens. The name of the megafossil taxon should simply be used for the palynomorphs.) *A.* pollen (but see *Favisporites* and *Lunatisporites*) was also found in Triassic pollen organs of *Pteruchus* (a pteridosperm) and *Masculostrobus* (a conifer) according to Townrow (1962). *A.* pollen (but see also *Pteruchipollenites*) was described in ultrastructural detail from *Pteruchus* pollen organs from the Triassic of Argentina by Taylor *et al.* (1984). Pollen of this sort also was described by Delevoryas & Hope (1973) from abundant late Triassic male cones associated with the ovulate conifer cone, *Compsostrobus. A.* (but see also *Platysaccus* and *Sulcosaccispora*) has been found on *Dicroidium*, a probable member of the Corystospermaceae. *A.* pollen was described from the conifer fructification *Lelestrobus* by Srivastava (1984).

Exiguisporites

Exiguisporites (as well as *Vitreisporites* and *Falcisporites*) pollen was reported from pollen sacs of *Caytonanthus* (Jurassic caytonialean) and *Harrisiothecium* (Triassic pteridosperm) by Townrow (1962).

Falcisporites

Falcisporites (as well as *Vitreisporites* and *Exiguisporites*) pollen was reported from pollen sacs of *Caytonanthus* (Jurassic caytonialean) and *Harrisiothecium* (Triassic pteridosperm) by Townrow (1962). Townrow (1965) also illustrates pollen of a corystospermaceous pteridosperm, which Balme (1970) recognizes as *Falcisporites*.

Favisporites

Favisporites (as well as *Alisporites* and *Lunatisporites*) pollen was found in Triassic cones of *Pteruchus* (a pteridosperm) and *Masculostrobus* (a conifer) by Townrow (1962).

Gigantosporites

Gigantosporites, a large, non–striate bisaccate, was found in a probable conifer cone of the late Permian by Clement-Westerhof (1974).

Illinites

Illinites (see also information under Paleophytic spores/pollen) was reported from cones of the Lower Triassic conifers *Aethophyllum* and *Willsiostrobus* by Grauvogel-Stamm (1978). She notes that others have called these palynomorphs *Chordosporites*, *Colpectopollis*, and *Sahnisporites*. (Grauvogel-Stamm & Grauvogel (1973) referred to such pollen from *Masculostrobus acuminatus*, a conifer, as *Parallinites*.)

Jugasporites

Jugasporites, a bisaccate having an odd corpus structure with a rent-like opening, was found by Clement-Westerhof (1974) in late Permian coniferous cones.

Kosankeisporites

Kosankeisporites pollen was described from Triassic sporangia of *Pteruchus*, a pteridosperm (but see also *Sulcatisporites* and *Pteruchipollenites*) by Townrow (1962).

Lueckisporites

Lueckisporites, a monolete(!) taeniate bisaccate in which the taeniae form most of the corpus, was isolated from late Permian coniferous cones of Italy by Clement-Westerhof (1974).

Lunatisporites

Lunatisporites (= *Taeniaesporites*), a taeniate bisaccate, was found in a conifer cone from the late Permian by Clement-Westerhof (1974). *L.* pollen (as well as *Alisporites* and *Favisporites*) was found in Triassic *Pteruchus* (a pteridosperm) and *Masculostrobus* (a conifer) cones by Townrow (1962).

Platysaccus

Platysaccus has been found in cones associated with *Dicroidium*, a probable member of the Corystospermaceae (Anderson & Anderson 1983) (but see also *Alisporites* and *Sulcosaccispora*).

Podocarpidites

Podocarpidites pollen (as well as *Vesicaspora*) was found in Triassic *Ruhleostachys* (a conifer or cordaite) cones by Townrow (1962).

Protohaploxypinus

Protohaploxypinus, a striate bisaccate, is known to have been produced by Permian glossopterid gymnosperms. As noted by Retallack (1980), Pant & Nautiyal (1960) illustrated a number of such palynomorphs, and some other sorts, from glossopterid seed pollen chambers. Gould (1981) noted the occurrence of *P.* pollen in *Arberiella*, the pollen sac organ of *Glossopteris*. However, this sort of pollen also has been found in the cones of a podocarp, *Rissikia* (Anderson & Anderson 1983).

Pteruchipollenites

Pteruchipollenites pollen was reported from the Triassic, in pollen organs of *Pteruchus*, a pteridosperm (but see *Sulcatisporites, Alisporites*) by Townrow (1962) and by Taylor *et al.* (1984). Possible *Pteruchipollenites* (see *Lunatisporites, Alisporites, Favisporites*) was described from preparations of Triassic *Pteruchus* by Couper (1958).

Rimaesporites

Rimaesporites Permian pollen was obtained from *Ullmannia* (a conifer) cones by Townrow (1962).

Satsangisaccites

Satsangisaccites pollen was described from Lower Triassic *Nidistrobus* pollen-bearing organs, an apparent pteridosperm, by Bose & Srivastava (1973).

Striatites

Striatites pollen is illustrated from the Triassic enigmatic (gymnospermous?) fossil, *Nidpuria*, by Pant & Basu (1979). Probable *S.* pollen was obtained from Permian *Arberiella* cones (Townrow 1962), a glossopterid gymnosperm.

Sulcatisporites

Pollen referable to *Sulcatisporites* (= *Lorisporites*) from Triassic *Pamelreuthia*, a pteridosperm, possibly a caytoniad, was studied by Townrow (1962). However, Townrow also noted that *S.*-like pollen was obtained from sporangia of *Pteruchus africanus*, a pteridosperm.

Sulcosaccispora

Sulcosaccispora (see *Alisporites* and *Platysaccus*) has been found in cones of *Dicroidium*, a probable member of the Corystospermaceae.

237

Triadispora

Triadispora, an unusual bisaccate with a trilete laesura, was found as mature pollen in the early Triassic conifer, *Darneya*, and in *Sertostrobus* conifer cones (but see *Inaperturopollenites*), by Grauvogel-Stamm (1978).

Vitreisporites

Vitreisporites (= *Caytonipollenites, Pityosporites*) *pallidus* is well-known to be the dispersed pollen of Jurassic *Caytonanthus* (Chaloner 1968b). The bisaccate nature of *V.* has always been a major stumbling block in efforts to connect the Caytoniales with angiosperm ancestry. *V.* pollen was studied by Couper (1958) in preparations of *Caytonanthus*. *V.* (and *Falcisporites* and *Exiguisporites*) pollen was reported from strobili referable to *Caytonanthus* and *Harrisiothecium* (a Triassic pteridosperm) by Townrow (1962).

Voltziaceaesporites

Voltziaceaesporites pollen was reported from early Triassic conifer cones, *Willsiostrobus* and *Yuccites*, by Grauvogel-Stamm (1978).

C. POLYSACCATE

Podosporites

Podosporites (*Microcachryidites*) trisaccate pollen was described by Vishnu-Mittre (1956) from Jurassic *Masculostrobus* male cones of apparent podocarpacean affinity, according to Balme (1964).

Trisaccites

Trisaccites pollen was found in cones of the Lower Cretaceous podocarpaceous conifer, *Trisacocladus*, by Archangelsky & Gamerro (1967) and by Baldoni & Taylor (1982).

D. INAPERTURATE

Araucariacites

Araucariacites pollen was identified by Couper (1958) in a preparation of the Jurassic conifer, *Brachyphyllum* (but see below under *Corollina*).

Exesipollenites

Pollen of *Exesipollenites* morphology was described from the type specimen of the Jurassic cyadeoid flower, *Williamsoniella lignieri*, by Harris (1974a).

Inaperturopollenites

Inaperturopollenites limbatus Balme pollen was shown to be produced by Lower Cretaceous *Brachyphyllum* cones, by Archangelsky & Gamerro (1967) and Gamerro (1968) (but see under *Corollina* below, and *Araucariacites* above). Pollen attributed to the genus *I.* was also found by Grauvogel-Stamm (1978) as immature grains in Lower Triassic *Darneya* conifer cones, of which the mature pollen was *Triadospora*. (Pollen in some of the illustrations appear circumpolloid. This pollen was later (Archangelsky 1977) transferred to a new genus, *Balmeopsis*.)

Perinopollenites

Perinopollenites pollen was identified by Couper (1958) in cone preparations of the Jurassic taxodiaceous conifer, *Elatides*. Harris (1973) confirmed this, noting that the pollen is quite variable from one cone to another.

E. CIRCUMPOLLOID (*Corollina, Circulina, Gliscapollis,* etc.)

Corollina

Corollina (= *Classopollis*) pollen was produced by the Lower Cretaceous conifer cone, *Tomaxiella* (Archangelsky & Gamerro 1967, Gamerro 1968). Couper (1958) identified it in preparations of Jurassic coniferous male cones of *Pagiophyllum*. He also identified *C.* from illustrations of pollen cones of *Hirmeriella* (= "*Cheirolepis*"). However, it should be emphasized (see Barnard 1968, Medus 1970) that circumpolloid pollen of various sorts have been obtained from male cones of a variety of Mesozoic conifers: *Brachyphyllum, Hirmeriella, Pagiophyllum,* and *Masculostrobus.* Some of these genera also have produced non-circumpolloid pollen. The primary association seems to be of *Hirmeriella* (Cheirolepidaceae) cones and *Corollina* pollen, however (see Francis 1983, 1984).

F. STRIATE (including taeniate, but not saccate)

Equisetosporites

Equisetosporites pollen was obtained from Triassic *Masculostrobus* (conifer) cones by Ash (1972). (However, see also under the trisaccate pollen of *Podosporites*.)

G. MONOSULCATE

Cycadopites

Cycadopites-like pollen was obtained from Jurassic *Sahnia* (Pentoxylaceae), according to Sukh-Dev (1980), and from *Lepidopteris* (Peltaspermales), per Anderson & Anderson (1983).

Monosulcites

Monosulcites pollen was identified from preparations of a Jurassic cycadalean fructification, *Androstrobus*, by Couper (1958) and, per the same publication, from Rhaetian–Lower Cretaceous ginkgoalean material, and from the Jurassic cycadeoids, *Williamsonia, Williamsoniella,* and *Wonnacottia*.

CHAPTER TWELVE

Late "Mesophytic" "non-pollen" palynomorphs

MEGASPORES

After the Carboniferous, free-sporing megaspores retreat ever more into the background, and in the modern flora they are made by only a few pteridophytic genera such as *Selaginella, Isöetes*, and *Azolla*. As in the modern flora, Mesozoic megaspores represent heterosporous lycopods and ferns. Nevertheless, Permian, Triassic, and later sediments do contain free megaspores, and they can be removed using the techniques described in the Appendix. Figure 12.1 includes illustrations of some Triassic megaspore forms and Figure 13.17 some Cretaceous forms. About 350 species of megaspores, referable to about 75 genera, have been described from Triassic to Cretaceous sediments (Sweet 1979). (Figure 12.2 summarizes internationally accepted Jurassic–Cretaceous stages.) These palynomorphs are clearly of potential biostratigraphic importance, but as yet too few assemblages have been studied to provide adequate biostratigraphic control over most of the world. Important studies have been made, for example, of Triassic megaspores of Australia (Dettmann 1961, Scott & Playford 1985), Europe (Orłowska-Zwolińska 1979), and the other continents, including Antarctica. Undoubtedly megaspore study is a field with a future. (See also discussion of megaspores in Chapters 8 and 9.)

ACRITARCHS

Marine and semi-marine sediments of Permian to Jurassic age often contain abundant acritarchs, but their study is not nearly as far advanced as is true of the "Paleophytic", especially of Cambro-Silurian time. The reason is simply that palynologists who have worked in post-Silurian, pre-Jurassic rocks have mostly specialized in spores/pollen. However, a fair amount is known about "Mesophytic" acritarchs, and it is clear that their study is very promising for the future. Figure 12.1 presents figures of some important forms.

DINOFLAGELLATES

It may well turn out that some of the acritarchs known from the "Paleophytic" are really dinoflagellate cysts, as a number have rather suggestively dinoflagellate-

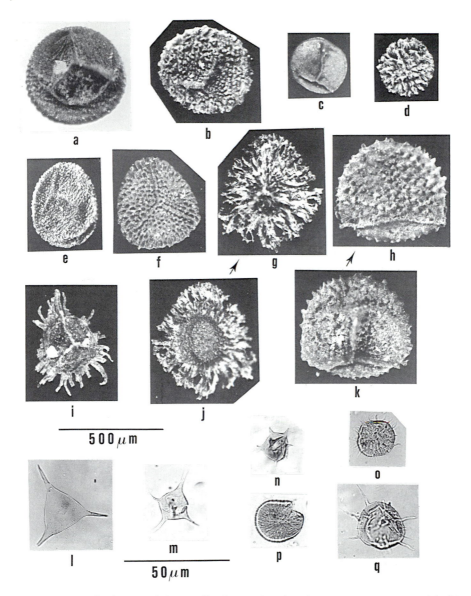

500 μm

50 μm

Figure 12.1 The large and the small of Triassic palynology. Free megaspores (a)–(k) reached their heyday in the late Devonian to early Carboniferous. As the seed habit took over in vascular plant evolution, free megaspores declined in importance, and today they are only produced by a handful of heterosporous ferns and lycopods. However, Mesozoic sediments still contain abundant megaspore remains. Methods for processing and photographing megaspores differ somewhat from methods normally employed for miospores, e.g., these photos were made using reflected light. S.E.M. in conjunction with transmitted light microscopy of cleaned (by HNO_3 or Schulze's reagent) specimens is also frequently used for megaspores (see Scott & Playford 1985).

241

like features. Indeed, one Silurian dinoflagellate has been described and is fairly widely accepted. However, the first unquestioned dinoflagellate cysts with all the required characteristics and clearly tied in with subsequent evolutionary developments are late Triassic (see Fig. 12.9).

The dinoflagellates are in today's environments an extremely diverse group of protists, including even forms that are parasitic in vertebrate guts. They are also very important constituents of the marine food chain and are the causative organisms of the infamous "red tides" that can make shellfish toxic to humans. However, the only dinoflagellates of importance to paleopalynologists are those which have a complex life cycle with a thin-walled, motile, characteristically flagellate stage usually called the theca or thecate stage, and a non-motile, thick-walled encysted stage, the cyst (see Figs. 12.4–12.7). The cysts, or at least the cysts we find as fossils, have their walls made of a substance apparently very similar to or identical to the sporopollenin of spores, pollen, and acritarchs: more or less the same color, the same response to carbonization, and the same staining reactions. This accounts for their preservation as palynomorphs. Although the dinoflagellates are currently classed in modern biology as protists (= kingdom Protista), it would seem probable that they are somehow related to the green algae and thus to other green plants. However (Spector 1984), the nuclei of dinoflagellates are so different from those of eucaryotes that some dinoflagellate specialists suggest calling them "meso-caryotes" to distinguish them from both procaryotes and eucaryotes, and to suggest that they are not closely related to any other group. When I first began my work as a palynologist, we did not know what the dinoflagellate cysts we found in marine sediments were, and lumped them with what we now call acritarchs as "hystrichosphaerids" (spiny spheres). It was one of the great achievements of palynology when Evitt (1961) (see Fig. 12.3) demonstrated on morphological grounds that many hystrichosphaerids are dinoflagellate cysts, because on first examination they do not look at all like thecae or

---

Acritarchs (n)–(q) had their apex of diversity in the early Paleozoic but continue to be abundant and important, especially in marine sediments, to present. These photographs are of typical Triassic forms (see line drawings in Fig. 10.5). Magnification for (a)–(k) shown by bar under (i), and for (l)–(o) and (q) by bar under (m); (p) is slightly more magnified – the specimen shown is 23 μm in maximum dimension. All specimens are from the Triassic of Poland. (a) *Verrutriletes utilis* (Marcinkiewicz) Marcinkiewicz, Rhaetian. Proximal view showing prominent contact scars. (b) *Verrutriletes litchi* (Harris) Potonié, Rhaetian. (c) *Trileites pinguis* (Harris) Potonié, Rhaetian. (d) *Echitriletes frickei* Kannegieser & Kozur, Karnian. (e) *Verrutriletes ornatus* Reinhardt & Fricke, Karnian. (f) *Horstisporites cavernatus* Marcinkiewicz, Rhaetian. (g) *Dijkstraisporites beutleri* Reinhardt, Ladinian. Proximal view; see also (j). (h) *Narkisporites harrisii* (Reinhardt & Fricke) Kozur, Karnian. Distal view, see also (k). (i) *Tenellisporites marcinkiewiczae* Reinhardt & Fricke, Ladinian. (j) *Dijkstraisporites beutleri* Reinhardt. Distal view of same specimen as (g). (k) *Narkisporites harrisii* (Reinhardt & Fricke) Kozur. Proximal view of same specimen as (h). (l) *Veryhachium reductum* (Deunff) Jekhowsky, Ladinian. (m) ?*Veryhachium irregulare* Jekhowsky, Scythian. (n) *Veryhachium dualispinum* Wall, Ladinian. (o) *Baltisphaeridium debilispinum* Wall & Downie, Ladinian. (p) *Baltisphaeridium aciculatum* Orlowska-Zwolińska, Ladinian. (q) *Baltisphaeridium longispinosum* (Eisenack) Eisenack, Ladinian.

SYSTEM	SERIES	STAGE	EUROPEAN "SERIES"
PALEOGENE	PALEOCENE	Danian	
CRETACEOUS	UPPER	Maestrichtian	
		Campanian	SENONIAN
		Santonian	SENONIAN
		Coniacian	SENONIAN
		Turonian	
		Cenomanian	
	LOWER	Albian	
		Aptian	
		Barremian	NEOCOMIAN
		Hauterivian	NEOCOMIAN
		Valanginian	NEOCOMIAN
		Berriasian	NEOCOMIAN
JURASSIC	UPPER	Volgian (±=Portlandian or Tithonian)	* MALM
		Kimmeridgian	MALM
		Oxfordian	MALM
		Callovian	DOGGER
	MIDDLE	Bathonian	DOGGER
		Bajocian	DOGGER
	LOWER	Toarcian	LIAS
		Pliensbachian	LIAS
		Sinemurian	LIAS
		Hettangian	LIAS
TRIASSIC	UPPER	Rhaetian **	

* An upper part of the Volgian, the "Purbeckian", usually regarded as uppermost Jurassic, often turns out to be Cretaceous, but if so is nevertheless classified as "MALM".

** If the Rhaetian is not recognized as a stage, this is late Norian.

Figure 12.2 Internationally recognized subdivisions of Jurassic–Cretaceous time/rock.

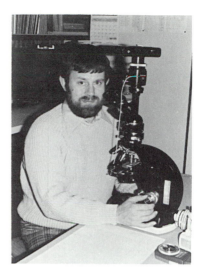

a

b

c

Figure 12.3 Several important persons in the modern study of fossil dinoflagellates.

(a) William R. Evitt, born 1923, fishing for freshwater dinoflagellates in a lake near his Stanford, California, laboratory, July, 1982. It was Evitt more than any other person who was responsible for recognizing from their morphology that many of the former "hystrichosphaerids" were in fact dinoflagellate cysts. It was Evitt who proposed dividing the "hystrichosphaerids" between those which are dinoflagellate cysts, and all others – the acritarchs.

(b) David Wall, born 1937, at the microscope. Dinoflagellate cysts are often so different from the thecal state of the organism that, despite Evitt's insightful proposals, someone had to grow thecal dinoflagellates from the cysts, in culture, and vice versa, to establish the life cycle and prove that Evitt's proposals were correct. Wall did this many

244

"normal" dinoflagellates. Indeed, at the time Evitt solved this riddle, I was working with hystrichosphaerids in marine sediments of offshore Florida and sent specimens to several of the then better known dinoflagellate experts, who would not accept them as dinoflagellates. They were accustomed to studying only the very different thecal forms. I said at the time that I would not accept "my" hystrichosphaerids as dinoflagellates unless somebody "hatched" one, to produce a thecal, motile dinoflagellate. Wall (1965) (see Fig. 12.3) promptly did just that, many times, and it is now known which "hystrichosphaerid" (= cyst) goes with which theca for many species pairs. Both cysts and thecae in many cases had names, and the nomenclatural problems created are somewhat troublesome. The situation is now that we no longer use the term hystrichosphaerid. Those former hystrichosphaerids which are known to be dinoflagellate cysts are called just that, and the others are now known as acritarchs (from Greek for "unknown origin"), a term introduced by Evitt (1963) for the non-dinoflagellate hystrichosphaerids.

The theca, or motile "normal" dinoflagellate, of the kind that makes palynomorph cysts, has a wall of cellulose, organized into a series of plates (see Figs. 12.4 & 12.5). The theca normally has a ventral furrow or sulcus with a flagellum in it. A more or less equatorial encircling transverse furrow (= cingulum) contains another flagellum. All of this is more or less irrelevant to paleopalynology because thecae apparently do not occur at all, or hardly at all, as fossils. Fossil dinoflagellates are cysts or "dinocysts" (a nickname not popular with some dinoflagellate experts). Some investigators of dinoflagellate cysts call the parts of cysts by terms similar to those used for thecae, but append the prefix "para . . .", yielding terms such as paraplate, parasulcus, etc., unless the term itself could only refer to a cyst (such as "archeopyle"). This seems to me an unnecessary complication to what is already complicated enough, but many specialists find it helpful and informative (see Fig. 12.4). Some palynologists believe that the cyst is not merely formed by dinoflagellates to survive bad times (though this is the majority view), but is formed by sexual conjugation. If this is true, cysts may be very roughly compared with algal zygospores and are hypnozygotes. Most encysting dinoflagellates today are marine, but some are brackish–water forms and a few are freshwater forms.

The cyst can be, for a one-celled shell, a very complex structure, reflecting the plate arrangement, etc., of the "parent" theca. On the other hand, some cysts, especially of freshwater and brackish-water forms, are more or less unadorned and uncomplicated "bags", affording only occasional glimpses of the presumed precursor thecal structures, such as plates. To prove that a fossil cyst was produced by a dinoflagellate, it is necessary to show that it has a true archeopyle, the place at which excystment occurs. To prove this, there must

---

times and for different dinoflagellates. From his work it was obvious that fossil dinoflagellates are always cysts.

(c) Isabel C. Cookson, 1893–1973, pioneer Australian paleopalynologist. Already well-known for her work with Cenozoic fossil pollen, Dr. Cookson turned her attention to dinocysts about 1953. She recognized the rich opportunities for Australian fossil dinoflagellate studies, and encouraged the interests of many others in this work.

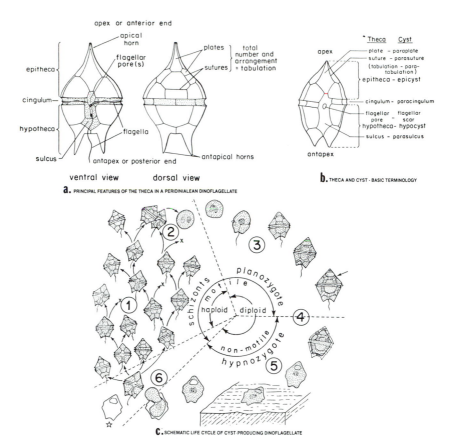

a. PRINCIPAL FEATURES OF THE THECA IN A PERIDINIALEAN DINOFLAGELLATE

b. THECA AND CYST - BASIC TERMINOLOGY

c. SCHEMATIC LIFE CYCLE OF CYST-PRODUCING DINOFLAGELLATE

Figure 12.4 Basic aspects of dinoflagellate cell construction and life cycle. The "vegetative" form with a cellulosic theca (wall) is the form encountered in many free-living dinoflagellates. Almost all fossil dinoflagellates represent the encysted phase, which may be quite different in appearance from the thecal form from which it derives, although at least part of the basic construction of the cyst, e.g., the archeopyle, can be shown to derive from specific plates of the theca. (a) Features of a dinoflagellate theca: the principal terms used are the same for theca and cyst, but for cysts, as seen in (b), some linguistic alterations are necessary. The prefix "para-" before sulcus and other terms is not used by all palynologists who study dinoflagellate cysts. The life cycle of the dinoflagellates that produce resistant-walled (sporopolleninous) cysts is variable, and that displayed in (c) is an average picture, not invariably followed even by one species. The x's in step 2 represent extra products of binary fission. Schizonts can act as gametes, and fuse to a zygote as shown in step 3. The thecae may be lost in this process, but a new theca is constructed for the diploid planozygote. As cell size increases, bands of new thecal material may appear along sutures (arrows). At step 4, the transition to a non-motile hypnozygote is shown. A cyst wall develops inside the hypnozygote thecal wall, and this wall then disintegrates. The cyst (hypnozygote) then settles as a sedimentary particle. Following a period of obligate dormancy, the protoplast excysts, as shown at the 5–6 interface, and meiosis yields new haploid cells which produce thecae and complete the cycle. The discarded, resistant cyst wall (with star) is now available for fossilization. Students should consult Evitt (1985) for more information about dinoflagellate morphology and biology.

246

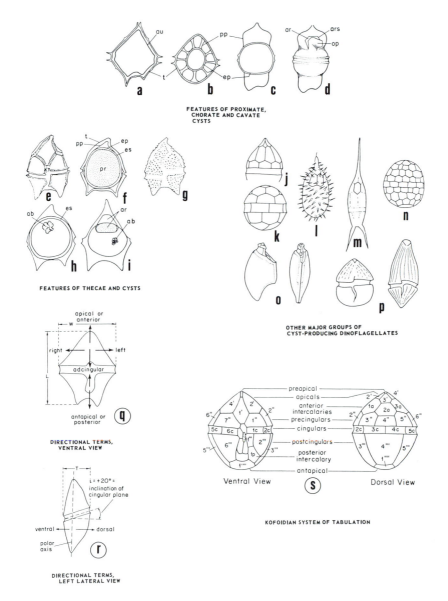

FEATURES OF PROXIMATE,
CHORATE AND CAVATE
CYSTS

FEATURES OF THECAE AND CYSTS

OTHER MAJOR GROUPS OF
CYST-PRODUCING DINOFLAGELLATES

DIRECTIONAL TERMS,
VENTRAL VIEW

DIRECTIONAL TERMS,
LEFT LATERAL VIEW

KOFOIDIAN SYSTEM OF TABULATION

Figure 12.5 More about basic features of dinoflagellate morphology, with special reference to cysts. (a) Longitudinal section of a proximate cyst (resistant wall = au, autocyst), which fills the theca (t), inside of which it was formed. (b) Section, as in (a), of chorate cyst, the contracted body of which is separated from the theca (t) by hollow processes (outer wall = pp, periphragm; inner wall = ep, endophragm). (c), (d) Cavate cyst with theca shed, in longitudinal section (c) and dorsal exterior view (d). Periphragm (pp) and endophragm are separated at each end but are appressed in the mid-section. A polygonal archeopyle (ar) opens through the periphragm. The operculum (op) is flap-like and adnate because the suture (ars) is incomplete (see Fig. 12.6). Evidence of the previous thecal tabulation is seen only in the archeopyle and

247

be a clear indication of the original plate boundaries of the piece of cyst that drops out for excystment, a trapdoor called the operculum. Without such plate boundaries, the archeopyle could be the exit aperture ("pylome") of an acritarch. Note that detached operculae are also sporopolleninous and can and do occur frequently as palynomorphs. The part of the theca in which the sulcus occurs is the ventral part (see Fig. 12.5), and the section opposite it is the dorsal part (the "front" and "back" of it!). The transverse furrow or cingulum separates two halves, the anterior and posterior (= episome and hyposome), which culminate in the apex and antapex, respectively. The plates are numbered for easy description in a scheme introduced by Kofoid in 1907 (see Fig. 12.5s). Note that not all the possible plate types exist in all dinocysts. (Figure 12.6 gives more information about tabulation–notation.)

Three principal kinds of dinocysts are recognized (see Fig. 12.5):

(a) Chorate cysts, in which the cyst is beset with processes. The cyst develops within the theca and only the placement of processes, the ends of which formed in contact with the thecal wall, and the eventual operculum–archeopyle production recall the original plate positions.

(b) Proximate cysts, in which the cyst forms directly within the thecal wall and has practically the same conformation as the theca. Special horns, when present, may show that there are, however, two cyst layers, the endophragm (from *phragma*, wall) and the periphragm. The plate boundaries of the theca, plus archeopyle and other features, are usually clearly observable, at least in some specimens.

(c) Cavate cysts, in which there is a very clear separation between inner wall (endophragm) and periphragm, resulting in a thick-walled inner body or

the paracingulum. (e)–(h) *Peridinium limbatum*, an extant freshwater species whose theca (e) shows plates separated by striate intercalary bands, whereas the cyst (g) shows the plates marked by sculptural fields, and the intercalary bands are represented by smooth strips. (f) An optical section of an encysted specimen, with theca (t) and two resistant layers of cyst wall (pp and ep), plus non-resistant contents, endospore (es) and protoplast (pr); other symbols as for (a)–(d). (h) Section of cyst (see (f)) showing accumulation body (ab) – these are also known as "eye-spots" and by other terms – probably metabolic wastes, found in both fossil and living cysts. (i) Fossil cyst of *Deflandrea* sp., with endophragm (stippled) showing through the archeopyle (ar). (j)–(p) Various fossil dinoflagellates not referable to the peridinioid or gonyaulacoid main lines to which most fossil cysts belong. (j)–(m) Dorsal views, showing large anterior intercalaries with characteristic knee-like points toward the cingulum: (j) *Rhaetogonyaulax* (late Triassic), (k) *Dapcodinium* (late Triassic–early Jurassic), (l) *Gochteodinia* (late Jurassic–early Cretaceous), (m) *Broomea* (late Jurassic); (n) *Suessia* (late Triassic), schematic dorsal view showing the numerous paraplates; (o) *Nannoceratopsis* (Jurassic) in right lateral and ventral views – tabulation is peridinialean only in the epicyst, and the hypocyst consists of just two large and two small paraplates; (p) *Dinogymnium* (late Cretaceous), ventral views of two examples, showing lack of paraplates in the resistant-walled test – this could represent a cell covering of the motile stage. (q) Directional terms, ventral view (L = length, W = width); adcingular = toward the cingulum. (r) Directional terms, left lateral view (T = thickness). (s) Tabulation (arrangement of plates) per the numbering scheme of Kofoid (see Fig. 12.6). Students should consult Evitt (1985) for more information.

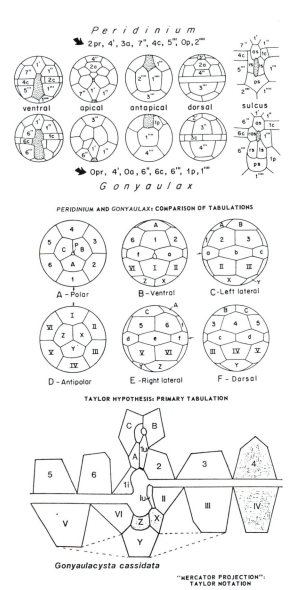

Peridinium

⬇ 2pr, 4', 3a, 7", 4c, 5"', 0p, 2""

ventral   apical   antapical   dorsal   sulcus

⬇ 0pr, 4', 0a, 6", 6c, 6"', 1p, 1""

Gonyaulax

PERIDINIUM AND GONYAULAX: COMPARISON OF TABULATIONS

A - Polar   B - Ventral   C - Left lateral

D - Antipolar   E - Right lateral   F - Dorsal

TAYLOR HYPOTHESIS: PRIMARY TABULATION

Gonyaulacysta cassidata

"MERCATOR PROJECTION":
TAYLOR NOTATION

Figure 12.6 More about tabulation–notation (see Fig. 12.5s for basic Kofoid system). At the top, the two common types within the Peridiniales: *Peridinium* and *Gonyaulax*. The arrows indicate the tabulation (= "paratabulation" for cysts) formulas, showing the number of plates in each series according to the Kofoidian scheme: 2 pr = two preapicals, 0p = no posterior intercalaries. In the sulcal area, as = anterior sulcal, rs = right sulcal, ls = left sulcal, ps = posterior sulcal, ras = right anterior sulcal.

In the middle of the figure is the hypothesis of Taylor as modified by Evitt, which assumes an ancestral gonyaulacoid model with a determinate number of symmetrically arranged plates. From such a model it is comparatively easy to derive a *Gonyaulax*-type cyst, and then to use the Taylor numbers and letters for it instead of the Kofoid

249

endocyst. Where there are multiple layers such as this, terms are often used to designate the layers (see Fig. 12.5).

To these I would add in very informal fashion:

(d) "Baggy cysts" ("saccocysts"?), such as those shown in Figure 12.7, in which only occasional or rare specimens show any trace of plate boundaries. Occasionally specimens are found with an archeopyle to prove dinoflagellate relationship, but except for the plates associated with the archeopyle, even these specimens demonstrate no plates. In many species the average specimen is thin-walled, and the collapsed cyst is folded or even extremely folded. In Neogene sediments of the Black Sea, I have found such cysts to be extremely abundant.

The nature of the archeopyle is very important to dinocyst classification. The principal features of archeopyles as described by Evitt are shown in Figure 12.8. Evitt (1985) also presents a rather complex but useful classification of archeopyles, an archeopyle formula, based on numbered positions of plates involved in archeopyle formation, such as type I for intercalary archeopyle, to which a prefix number showing the number of plates and subscript numbers indicating the specific plates involved may be added, such as $3P_{2-4''}$. The basic archeopyle letters used are:

A	= apical	HP	= postcingular
I	= anterior archeopyle	HI	= posterior intercalary
P	= precingular	HA	= antapical
C	= cingular	S	= sulcal

Triassic dinoflagellate cysts shown in Figure 12.9 are considered primitive by most investigators. Some forms, such as *Suessia* (Fig. 12.9g, h), have very numerous small plates, so many plates that it is difficult to assign code numbers to them according to the Kofoid scheme.

Figure 12.9 also illustrates a variety of Jurassic dinoflagellates. The Jurassic represents one of the dinoflagellates' "finest hours" from a palynological point of view, as the group was rapidly evolving and therefore very useful stratigraphically, at a time when spores/pollen and other palynomorphs are at some levels difficult to use for fine stratigraphy.

---

numbering system. The considerable modifications from the presumed ancestral model in the sulcal area are noted by showing derivation from the "ancestral" plate.

This can be seen in the "mercator" or "exploded" Taylor system (as modified by Evitt) projection below of a fossil cyst, in which 1u and 1i are derived from plate 1, and Iu from plate I. The Taylor notation has the advantage of keeping the plates in the same sequence, even if a plate has divided in evolution. The Kofoidian notation arbitrarily numbers plates in sequence as they are found, and a homologous plate may end up with quite different numbers in different species, if plates have been added by division. Students may consult Evitt (1985) for more information. The stippling in the bottom part was used by Lucas-Clark (1984) to emphasize particular plates in her discussion.

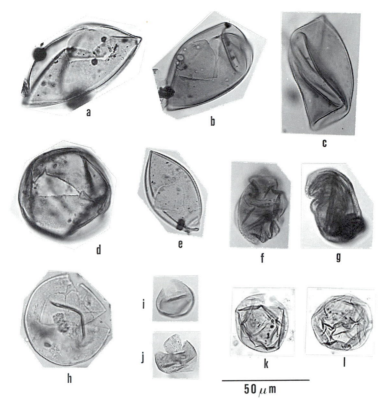

Figure 12.7 Students should not get the impression from neatly labelled diagrams and published dinoflagellate cyst photos that cysts are always easy to recognize as such! "Bag" dinoflagellate cysts of sorts encountered in great numbers at Neogene levels in cores of Black Sea sediment are shown here ((a)–(g) represent the most common kind, but are not necessarily one taxon). Thousands of specimens may be examined without finding even one with any evidence at all of plate boundaries. One might well identify the forms as P00 pollen grains. Then a level will be reached in which some specimens display obvious dinoflagellate archeopyles, mostly of the monoplacoid type ((h), however, is polyplacoid), also showing principal and sometimes accessory ((d), (h)) sutures (see Fig. 12.8). An adherent operculum is sometimes present ((a), (b), (e), (h), (j)). Sometimes from association and shapes a good identification can be made as to which plate(s) the archeopyle represents, as 2″, 3″, 4″ precingular plates for (h). However, systematic study, description, and later reidentification are very difficult with all of these forms. The student should look for a typical specimen with good archeopyle features. In many samples in the Black Sea sediment most of the baggy cysts are folded ((a)–(c), (e)), and some are crumpled as well as folded together ((f), (g)). These latter may have been "processed" in animal guts. Without having seen a few specimens with archeopyles, these would be completely unrecognizable as dinoflagellates. Some very small "baggy" types are also commonly found ((i), (j)). (Dinoflagellate experts who have examined these specimens agree that the archeopyle is almost certain proof that these are small dinoflagellate cysts.) Other forms may be leiosphaerid acritarchs ((k), (l)), but even these occasionally give tantalizing hints that they too are really "baggy" dinoflagellate cysts in very effective disguise.

251

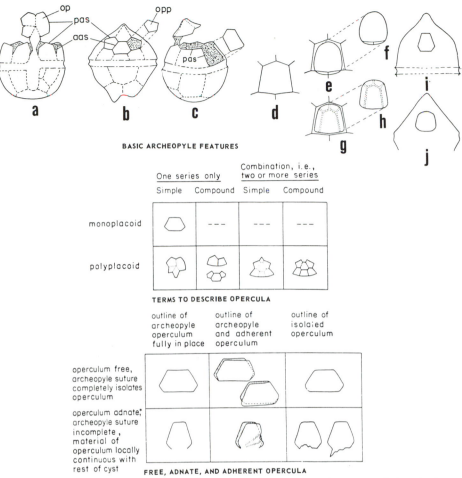

BASIC ARCHEOPYLE FEATURES

	One series only		Combination, i.e., two or more series	
	Simple	Compound	Simple	Compound
monoplacoid		- - -	- - -	- - -
polyplacoid				

TERMS TO DESCRIBE OPERCULA

	outline of archeopyle operculum fully in place	outline of archeopyle and adherent operculum	outline of isolaied operculum
operculum free, archeopyle suture completely isolates operculum			
operculum adnate, archeopyle suture incomplete , material of operculum locally continuous with rest of cyst			

FREE, ADNATE, AND ADHERENT OPERCULA

Figure 12.8 Dinoflagellate cyst archeopyle and operculum features and description. Archeopyles are the openings through which excystment occurs. In general, dinoflagellate cyst archeopyles are the "holes" left where opercula have been completely or partly removed. (As they are sporopolleninous, free opercula are frequently found as fossils in dinoflagellate cyst-rich sediments.) In (a)–(j), aas = accessory archeopyle suture; pas = principal archeopyle suture; op = operculum; opp = opercular piece of compound operculum. (a) Apical archeopyle, ventrally adnate operculum (see term explanations in tables at middle and bottom). (b) Intercalary archeopyle with compound operculum divided by accessory sutures into three opercular pieces. (c) Combination archeopyle with compound operculum, apical portion polyplacoid and adnate. (d) Archeopyle and paraplate congruent. (e)–(f) Archeopyle not congruent with paraplate. (g), (h) Reduced archeopyle involving paraplate, the surface of which bears accessory ridges related to overlap and growth of thecal plates ((h) is operculum). (i), (j) Angularity of opening. In (i), shape and position of opening are sufficient to indicate equivalence of the archeopyle to paraplate 2a of a peridinioid cyst, even in the absence of other tabulation. In (j), the rounded opening does not indicate paraplate equivalence. Evitt (1985) presents also a classification of archeopyles (= "archeopyle formula") based on the plates involved, e.g., 3P for a precingular archeopyle involving three plates. Students may consult Evitt (1985) for more details.

252

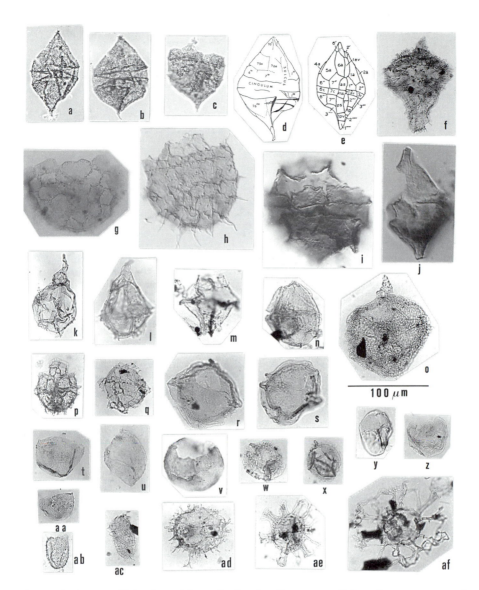

Figure 12.9 Late Triassic and Jurassic dinoflagellate cysts. Paleozoic occurrences of dinoflagellate cysts have been reported, and some Paleozoic acritarch forms may well prove eventually to be dinoflagellates. Nevertheless, the effective range of dinoflagellate cysts is late Triassic to present. The Triassic forms are somewhat unusual, in comparison to modern dinoflagellate cysts, but most of the Jurassic forms represent trends not greatly different from those of extant taxa. Magnifications for (k)–(af) shown by bar under (o); size for all others is given with their captions. (a)–(e) *Rhaetogonyaulax rhaetica* (Sarjeant) Loeblich & Loeblich, was, when described, the oldest known dinoflagellate cyst for which paratabulation could be demonstrated. (a)–(c) Specimens from the Rhaetian of England, length about 70 μm: (a) ventral view, showing the paracingulum and parasulcus; (b) obliquely ventral view; (c) cyst with the operculum of

253

apical plates missing. (d), (e) Drawings of ventral views: (d) shows the paratabulation as directly discernible in a whole specimen; (e) is an interpretative reconstruction of the paratabulation from a similar view. (f)–(j) Dinoflagellate cysts from marine shales of late Triassic age, North Slope, Alaska: (f) *Sverdrupiella spinosa* Bujak & Fisher. 86 × 78 μm. An oddly shaped spiny cavate cyst. (g) *Suessia* sp. 48 μm. The multiplate apically located archeopyle and the very numerous, somewhat irregular, plates are evident. (h) *Suessia* sp. 44 × 41 μm (+ 9 μm spines). Note the numerous plates as in (g), and the apically located archeopyle representing a number of epicystal paraplates. (i) *Dapcodinium* sp. Dorsal view showing the intercalary archeopyle. (j) *Shublikodinium arcticum* Wiggins, lateral view. The archeopyle in this cyst involves many of the apical, precingular, and intercalary paraplates. In the picture, only some of these plates have been shed, others still adhere. (k)–(af) Cysts from Pliensbachian–Toarcian ("PT"), Bajocian–Callovian ("BC"), and Oxfordian–Kimmeridgian ("OK") levels in the marine Jurassic of East Greenland. (k) *Gonyaulacysta* sp. (OK). (l) *Gonyaulacysta jurassica* var. *longicornis* (OK). Dorsal view, note precingular archeopyle. (m) *Atopodinium prostratum* Drugg (OK). (n) *Leptodinium* sp. (OK). Proximate cyst with clear indication of paratabulation, lateral view, archeopyle to the left. (o) *Acanthaulax scarburghensis* (Sarjeant) Lentin & Williams (OK). (p) *Meiourogonyaulax* sp. (OK). Ventral view, apical archeopyle. (q) *Scriniodinium* cf. *irregularis* (Cookson & Eisenack) Stover & Evitt (OK). (r) *Scriniodinium* cf. *playfordii* Cookson & Eisenack (OK). Cavate cyst, dorsal view showing precingular archeopyle. (s) *Scriniodinium crystallinum* (Deflandre) Klement (OK). Cavate cyst showing pericyst and endocyst walls very well. (t) *Mancodinium* sp. (PT). A cyst with an archeopyle involving the entire epicyst. (u) *Nannoceratopsis senex* van Helden (PT). (v) *Cassiculosphaeridia* sp. (OK). A cyst without much indication of paratabulation except in the area of the apical archeopyle where the paraplates are clearly demarcated. (w) *Stephanelytron redcliffense* Sarjeant (OK). The intricately interwoven processes reveal the tabulation. (x) *Kalyptea monoceras* Cookson & Eisenack (OK). (y) *Chytroeisphaeridia cerastes* Davey (BC). A relatively featureless cyst, but the precingular archeopyle leaves no doubt of its dinoflagellate affinity. (z) *Sentusidinium pelionense* Fensome (BC). A baggy cyst with clearly demarcated apical archeopyle. (aa) *Mancodinium semitabulatum* Morgenroth (PT). See (t). (ab) *Sentusidinium* sp. (OK). Apical archeopyle. (ac) *Prolixosphaeridium* sp. (OK). (ad) cf. *Systematophora* sp. (OK). Chorate cyst. (ae) *Compositosphaeridium costatum* (Davey & Williams) Dodekova (OK). Chorate cyst with double walls of processes clearly shown. (af) *Rigaudella aemula* (Deflandre) Below (OK). Chorate cyst with interconnecting processes, the outer limit of which indicates position of original thecal walls.

254

CHAPTER THIRTEEN

Jurassic–Cretaceous palynology: end of the "Mesophytic". Advent and diversification of angiosperms

---

INTRODUCTION

As has been noted already, one of the main palynological stories for the post–Liassic part of the Jurassic has to do with bisaccate, mostly conifer, pollen, which never, before or since, were so common. Many of them are large, 100 µm or more in length. As is true of bisaccates in general, these Jurassic Pv2s are very difficult taxonomically, and their classification can best be described as *ad hoc*. The classification of bisaccate pollen, later Triassic to Cretaceous, is badly in need of research. Spores, mostly of ferns, are also a very important part of non-marine Jurassic–Cretaceous palynofloras. A few of these forms from earliest Cretaceous are shown in Figure 13.1.

A number of pollen types found in Jurassic palynofloras are interesting as possible harbingers of the angiosperm condition, the fully developed arrival of which marks the end of the "Mesophytic" and the beginning of the "Cenophytic". One such form already mentioned is the circumpolloid pollen *Corollina* (= *Classopollis*, etc.). If it were not known that *Corollina* was produced by several conifers, at least partly very widespread, dominant shrubs, one would suspect angiosperm ancestry from the pollen, which has columellate exines and a circumpolloid colpal-like feature (see Fig. 11.10), which one could perhaps imagine evolving into a primitive angiosperm colpus.

Another such form already mentioned is *Eucommiidites*, which has two subsidiary shorter colpi in addition to the sulcus proper. As has been pointed out by Couper and others, the subsidiary colpi are not in the right position for the dicot tricolpate condition. Further, Doyle *et al.* (1975) showed that *Eucommiidites* has a thick, laminated endexine, an obligate gymnosperm character, whereas angiosperms lack laminated endexine, except sometimes in the apertural region. It is nevertheless interesting that such forms appear not too long before undoubted angiospermy arrives.

Cornet (1977a) has found a number of uncommon Triassic–Jurassic forms with multiple colpi and more or less syncolpate forms ("zonasulculate"; see Fig. 11.11) that sometimes have angiosperm-like, more or less columellate

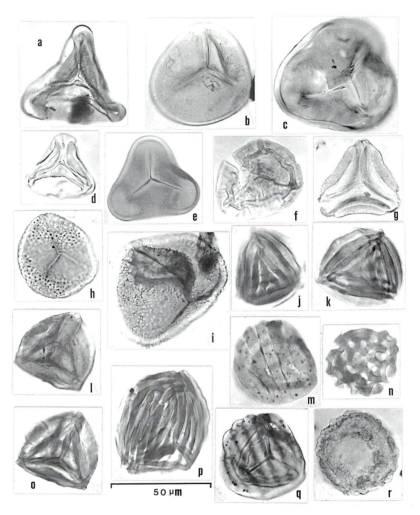

Figure 13.1 Typical earliest Cretaceous sporomorphs from the Hauterivian of Portugal: (a) *Auritulinisporites deltaformis* Burger; (b) *Dictyophyllidites equiexinus* (Couper) Dettmann; (c) *Deltoidospora germanica* Dörhöfer; (d) *Dictyophyllidites harrisii* Couper; (e) *Cardioangulina crassiparietalis* Döring; (f) *Coronatispora valdensis* (Couper) Dettmann; (g) *Gleicheniidites circiniidites* (Cookson) Brenner; (h) *Foveosporites subtriangularis* (Brenner) Döring; (i) *Pilosisporites* cf. *crassiangulatus* (Ivanova) Dörhöfer; (j) *Cicatricosisporites* cf. *hannoverana* Dörhöfer, distal view (see also (k)); (k) *Cicatricosisporites* cf. *hannoverana* Dörhöfer, proximal view (see also (j)); (l) *Plicatella parviangulata* (Döring) Dörhöfer, distal view (see also (o)); (m) *Contignisporites cooksoniae* (Balme) Dettmann, distal view (see also (q)); (n) *Ischyosporites pseudoreticulatus* (Couper) Döring; (o) *Plicatella parviangulata* (Döring) Dörhöfer, proximal view (see also (l)); (p) *Cicatricosisporites annulatus* Arch. & Gamerro; (q) *Contignisporites cooksoniae* (Balme) Dettmann, proximal view (see also (m)); (r) *Cerebropollenites mesozoicus* (Couper) Nilsson. Scale shown by bar under (p).

256

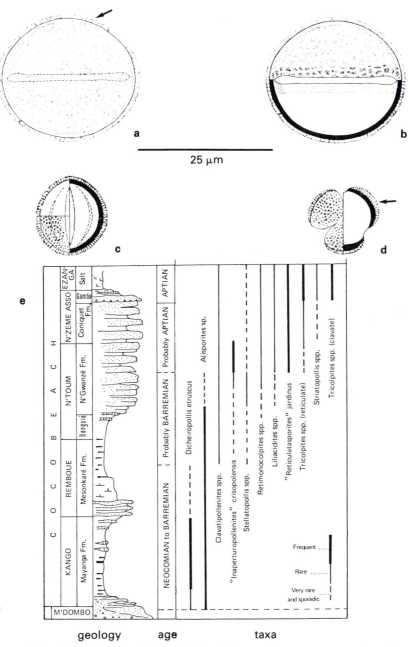

Figure 13.2 The "*Clavatipollenites*-link" and the earliest appearance of angiosperm pollen. Kemp (1968) observed, in Barremian–Albian rocks of England, monosulcate pollen of the type pictured in (a) (proximal view) and (b) (distal view). This pollen type clearly has columellate structure (see arrow). The species illustrated is *Clavatipollenites rotundus* Kemp. In Albian samples from the same area, the tricolpate form, *Tricolpites albiensis* Kemp, shown in (c) (equatorial view) and (d) (polar view) occurs commonly.

257

exines. One should also mention in this connection the Triassic–Jurassic form, *Pretricolpopollinites* (see Fig. 11.8), the multiple sulci (colpi?) of which are certainly "angiospermid". Of course, it should be noted that primitive angiosperm pollen probably was not multiaperturate but monosulcate, or perhaps 1-syncolpate ("circumpolloid" or zonasulculate, to use other terms descriptive of this condition), like some extant nymphaeaceous pollen. Another puzzle is that the Jurassic plants perhaps most like angiosperms are pteridosperms(?), the Caytoniales, the "fruits" of which are in fact "angio-spermous", that is, containing the seeds in a closed carpel. The pollen, however, is Pv2 (see Fig. 11.8). Could sacci somehow become colpi? Crane (1985), however, used cladistic analysis to demonstrate that the angiosperms' closest relatives seem to be the Gnetales, which have almost no fossil record. Two of the three extant genera make polyplicate pollen resembling forms found in the Jurassic and Triassic, however.

Toward the end of the Jurassic, pollen forms appear that may very well be truly angiospermous, although the megafossil supporting evidence is still lacking. As has been fully documented by Kemp (1968), Laing (1976), Hughes *et al.* (1979), and Hughes (1984), a number of forms with exines suggesting the characteristic columellate condition of angiosperms appear in the earliest Cretaceous of England. Similar observations have been made in other parts of the world. *Clavatipollenites*, a monosulcate to trichotomosulcate pollen grain, sometimes has more or less free clavae in the sexine, suggesting that lateral fusion of the clavae would produce a truly tectate columellate exine (see Fig. 13.2). Furthermore, it had long been hypothesized that a trichotomosulcate form could become tricolpate by simple projection of the "corners" of the three-cornered sulcus and "healing" of the distal side of the grain. Indeed, a tricolpate grain with exine structure–sculpture like *Clavatipollenites* does appear in the Aptian of England in sediments just above sediments containing *Clavatipollenites*. Such forms are *Tricolpites albiensis* Kemp (see Fig. 13.2). The *Clavatipollenites*-like tricolpate forms are joined worldwide in the Neocomian (see Fig. 13.2) by other primitive angiosperm monocolpate and tricolpate forms with just barely columellate exines. Figure 13.2e shows from Doyle *et al.* (1977) that in equatorial areas of Africa the advent of protoangiosperm pollen is even earlier than at high latitudes, and involves more forms. The oldest such things as yet reported are tiny inaperturate and monosulcate forms from the Hauterivian of Israel (Brenner 1984). Figure 13.3 shows some of the

---

This tricolpate, obviously dicot, grain has the same sort of structure (arrow) as *C. rotundus*, and it is likely that it evolved from such a monosulcate – one possibility would be through an intermediate trichotomosulcate form, though this is very controversial. (e) The very early appearance of primitive angiosperm pollen in equatorial Africa is shown here, from the work of Doyle *et al.* (1977). *Dicheiropollis* (circumpolloid) and *Alisporites* (bisaccate) are conifer pollen. Tricolpate pollen appeared much earlier in equatorial Africa than in Laurasia. After the appearance of *Clavatipollenites* (in a broad sense) in late Neocomian time, a number of other pollen with columellate-type structure appear, so that before the Albian there was already a large range of angiosperm pollen present. (Very recent work has turned up primitive angiosperm fossil pollen in the Hauterivian.) Representation of types of sedimentary rock follows conventional usage: dotted patterns for sandstone, black bands for coal, block vertical line patterns for limestone, horizontal lines for shale, and small circles for conglomerate.

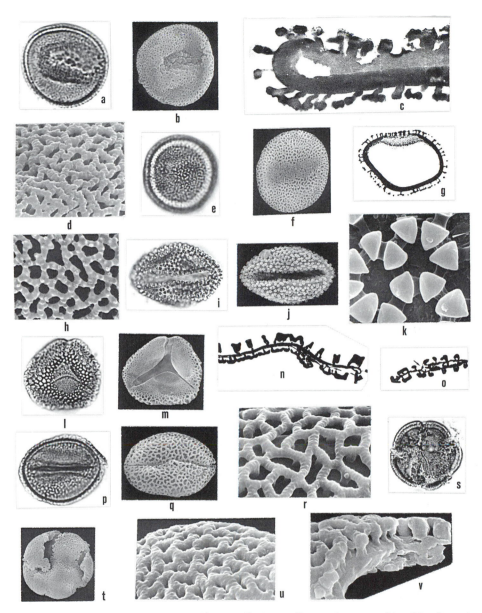

Figure 13.3 Early Cretaceous *Clavatipollenites* pollen and some of its friends and relatives. From the pioneering work of James W. and Audrey G. Walker we have learned many details of morphology of this ancestral angiosperm pollen form. Walker & Walker (1984) show that *Clavatipollenites* includes forms with two quite different pollen structures. *Clavatipollenites hughesii* Couper, illustrated here ((a)–(d)), has well-developed endexine – but only in the apertural regions – and is tectate–columellate. Pollen of the living taxon *Ascarina diffusa* A. C. Smith (Chloranthaceae) ((e)–(h)) is practically identical to *C. hughesii*, and there are other reasons for pinpointing this primitive angiosperm family as a living relic of the basal angiosperm at or near the

early protoangiosperm forms. Walker, Doyle, and others have shown that the evolution of the columellate angiosperm ektexine is the culmination of a long evolutionary process. Indeed, as Le Thomas (1980–81) has emphasized, some of the *Clavatipollinites*-type pollen of the Lower Cretaceous lacked ektexinous columellae, having instead a granular exine with no endexine, like modern Annonaceae. The arrival of the tricolpate condition is linked with the advent of a clearcut endexine–ektexine distinction in dicots, with a non-lamellar endexine typically present. (Gymnosperms typically have lamellar (= laminated) endexine.) Both the peculiar colpal arrangement of most dicots and the

---

ancestry of both monocots and dicots. Also shown are other key pollen types among the early Cretaceous angiosperms. *Clavatipollenites*, *Retimonocolpites*, and *Liliacidites* are very difficult to limit in light microscopy, but do show differences in electron microscopy. (a)–(d) *Clavatipollenites hughesii* Couper, Barremian–Lower Aptian, Potomac Group, Maryland (maximum dimension of pollen 23 μm): (a) photomicrograph, distal view; (b) S.E.M. micrograph, distal view, showing the sulcus and the reticulate exine organization; (c) T.E.M. micrograph of whole grain section, with sulcus on top, showing its (lighter) endexine, and individual columellae with laterally fused heads; (d) S.E.M. micrograph, much higher magnification than (b), with beaded muri. (e)–(h) *Ascarina diffusa* A. C. Smith, extant Chloranthaceae pollen for comparison with *C. hughesii* (maximum dimension 26 μm): (e) photomicrograph, distal view (see (a)); (f) S.E.M. micrograph, distal view, showing sulcus (see (b)); (g) T.E.M. micrograph, whole grain section, sulcus at top showing (lighter) endexine, and structure very like that of *C. hughesii* (see (c)); (h) S.E.M. micrograph, much magnified, beaded muri as in (d). (i)–(k) *Stellatopollis barghoornii* Doyle, Albian, Potomac Group, Delaware (maximum dimension 53 μm): (i) photomicrograph, distal view showing the triangular-appearing sculptural units in linked groups – such sculpture is seen in extant members of the quite advanced genus *Croton* (Euphorbiaceae), presumably because of convergent evolution, hence "crotonoid" sculpturing; (j) S.E.M. micrograph showing the same features as (i); (k) high-magnification S.E.M. micrograph showing that the triangular exine blocks seen in (i) and (j) are mounted on an underlying reticulum with circular muri. (l)–(n) *Liliacidites* sp., Albian, Potomac Group, Delaware (maximum dimension 33 μm): (l) photomicrograph, proximal view, medium-high focus; (m) S.E.M. micrograph, distal view, showing the trichotomosulcate morphology and reticulate sculpturing with larger pattern on proximal side; (n) T.E.M. micrograph of whole grain section, sulcus showing on underneath of right side – the additional layer interpolated here is endexine (arrow). (o)–(r) *Retimonocolpites dividuus* Pierce, Albian, Potomac Group, Delaware (maximum dimension 30 μm): (o) T.E.M. micrograph, one side of whole grain section, showing infolding (arrow) presumably the same as seen at edge of sulcus in (p) and (q); (p) photomicrograph, distal view – the apparent thick-bordered sulcus is due to infolding of the exine; (q) S.E.M. micrograph, distal view, showing sulcus with tucked-in-edge, and reticulate sculpture; (r) S.E.M. micrograph, high magnification, showing the corrugated band structure. (s)–(v) "*Stephanocolpites*" *frederickburgensis* Hedlund & Norris, Fredericksburgian (Albian) Oklahoma (maximum dimension 28 μm) (quotation marks refer to fact that the generic name commonly used for this form is illegitimate): (s) photomicrograph, polar view; (t) S.E.M. micrograph, polar view showing the irregular colpal margins; (u) S.E.M. micrograph, much higher magnification, showing that the densely organized reticulate-type sculpture is actually perforate (foveolate) because the perforations are far less than 50% of the surface – the exine is very thick, as demonstrated in S.E.M. micrograph of broken edge of grain in (v), with apertural region to the right.

complex exine structure of most angiosperms have undoubtedly to do with the reproductive processes that are critical to angiosperm success. The columellate exine provides a strong sporopollenin meshwork in the interstices of which a very complex set of organelles and compounds are dispersed (Rowley 1976, 1978, 1981). One function of some of these compounds is to provide sophisticated recognition mechanisms by which angiosperms are able to reject pollen-tube growth by unwanted pollen. The whole structural–chemical complex is clearly related to insect pollination, and the coevolution of insects and angiosperms is nowhere better illustrated than in palynology. Friis *et al.* (1986) have shown that floral structures (including *in situ* pollen) referable to the modern family Chloranthaceae (Magnoliales) were already present by Albian time. The pollen (see Fig. 13.3) is quite similar to pollen of the *Clavatipollenites* complex and seems clearly to have been insect-pollinated. However, Crane *et al.* (1986) have demonstrated that Albian sediments also contain flowers of non-magnoliid, higher angiosperms close to modern Platanaceae, probably bespeaking considerable pre-Albian evolution. Figure 13.4 demonstrates evolutionary trends in exine structures of gymnosperms and angiosperms.

With the arrival of undoubted tricolpate pollen, we can say with certainty that the "Cenophytic" has begun. Unfortunately for tidiness of the concept, some monosulcate Jurassic pollen might conceivably be angiospermous. Furthermore, a few late Jurassic to Neocomian clavate to columellate monosulcate forms could quite likely be pollen of angiosperms, though they could also represent gymnosperms whose pollen was "precociously" angiosperm-like ("mosaic evolution"). Even among extant dicot angiosperms, some presumed primitive forms have monosulcate pollen, e.g., *Magnolia*, and the monocots mostly have monosulcate pollen, or some sort of obvious derivative of Pa0 such as P01, Pv1 or P00. Indeed, the monocots, if they have a common origin with the dicots, as usually assumed, probably diverged very early (pre-Cenomanian) from them. Based mostly on analysis of characters of modern monocot groups, but partly also on diversity of fossil monocots, Walker (1986) suggests that this group had a temperate, Northern Hemisphere (Laurasia) origin, in contrast to the dicots, which quite clearly arose in the tropics of the Southern Hemisphere (Gondwana). At least one family (Nymphaeaceae) of modern dicots is difficult to assign to the dicots on the basis of pollen morphology, and another (Annonaceae) retains apparently primitive characters such as weakly granular exine structure and distally sulcate pollen, which this family of dicots shares with the monocots (Le Thomas 1980–81). Frederiksen (1980b) has stressed that monosulcate pollen has never been a dominant pollen type, though produced by six different plant orders, the vegetative parts of which are frequently common as fossils. He suggests overrepresentation of the leaves of these plants in the record.

Whatever it meant, the arrival of tricolpate (Pc0) pollen with columellate ektexine and non-lamellar endexine does provide a watershed. Brenner (1976) and others (Hickey & Doyle 1977; see Fig. 13.5) have shown that Pc0 pollen arrives first in equatorial, Southern Hemisphere areas in Barremian time; it reaches middle latitudes of the Northern Hemisphere in Aptian–Albian time and only penetrates to Arctic areas by Cenomanian time. Hickey & Doyle have shown from megafossil evidence that in the coastal plain of eastern North

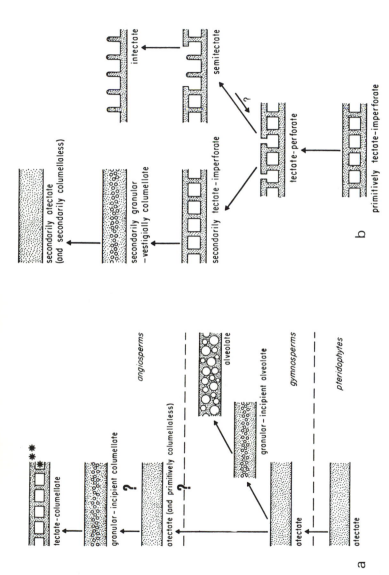

Figure 13.4 Evolutionary trends in outer exine (sexine or ektexine) structure. (a) Trends in vascular plants generally. Tectate sexine is always columellate because it is the columellae (single asterisk, upper left) which produce the extra "roof" or tectum (two asterisks). The "bubbly" (alveolate) outer exine of gymnosperms usually overlies a laminated nexine (not shown), whereas angiosperms have a non-laminated nexine or may have no nexine, or nexine only for part of the exine. (b) Trends in angiosperm pollen. "Intectate" refers to the presence of some indication of columellae but no tectum, whereas "atectate" refers to the total absence of columellae.

Columellate pollen is the characteristic angiosperm condition, and even forms lacking columellae are mostly derived from forms that had them. However, the ancestral angiosperms in earliest Cretaceous time (see (a)) were only incipiently columellate and probably derived from atectate gymnosperms. The question-marks in (a) indicate uncertainty as to whether the line between angiosperms and gymnosperms should be put where shown by the broken line or higher. It should be noted that Jurassic–Cretaceous or earlier pollen with columellae cannot be presumed to be either angiospermous or non-angiospermous on the basis of this feature alone.

America the *Clavatipollenites–Tricolpites* sort of flux was coterminous with a complex of primitive angiosperm leaves, suggesting a stream-margin pioneer flora, perhaps indicating that the angiosperms arrived at mid-latitude preadapted for marginal environments, perhaps where regular moisture supply is a problem, or in marginal marine environments. Doyle *et al.* (1977) showed, for example, that even *Tricolpites* was present in west Africa by the Aptian (Fig. 13.2), and a complex of possible monosulcate, columellate precursors by Barremian or earlier. Burger (1981) shows possible spreading pathways for *Clavatipollenites*, tricolpates and tricolporates, from usually equatorial or low- to middle-latitude origins. Others have other ideas: for example, the origin of angiosperms in coastal areas as mangrove-like plants (Retallack & Dilcher 1981a), or as quick colonizers of coastal river levees (Retallack & Dilcher 1981b); and we are far from an ultimate or unified solution to Darwin's "abominable mystery" of the origin of the angiosperms.

After the arrival of the angiosperms in the late Neocomian, their subsequent history, of great significance to paleopalynology, is reasonably clear. Doyle, Muller, and others have traced angiosperm (mostly, actually, dicot) pollen "evolution". It should be emphasized that it is the plants that evolve – pollen is only a single organ, representing one side of the haploid generation! However, Muller (1984) and others have called attention to the fact that mosaic evolution

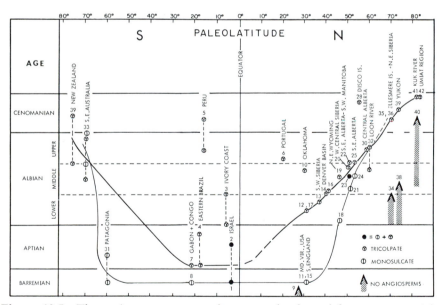

Figure 13.5 The angiosperms seem to have spread poleward from an equatorial or near-equatorial origin. This figure plots first occurrences of angiosperm pollen types. Note the poleward spread of tricolpates (all dicots), reaching high latitudes by Cenomanian time. Broken lines indicate a range where there is uncertainty about time of first appearances. Monosulcates (representing monocots and dicots) seem to have begun poleward migration before the tricolpates, which caught up later. Numbers associated with symbols refer to original publication sources, of which there is a list in Hickey & Doyle (1977).

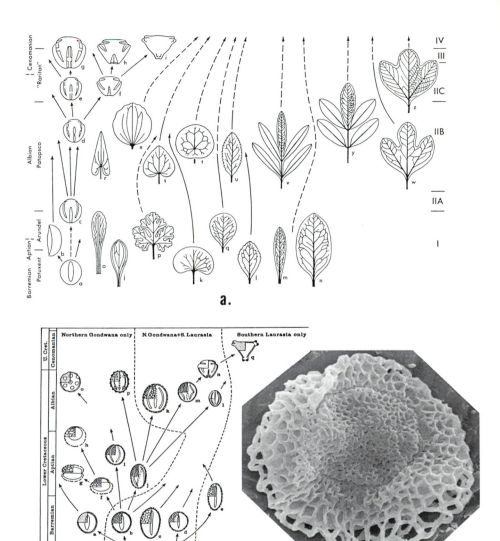

Figure 13.6 Adaptive radiation of angiosperms in about the middle of the Cretaceous (see Fig. 12.8: about 115 to 90 million years ago).

(a) The story from the Potomac Group of Maryland, eastern Coastal Plain, U.S.A., and vicinity: the pollen picture on the left and the coordinate leaf fossil picture on the right. "Patuxent", etc., at the left refer to the local lithologic units; I, IIA, etc., on the right refer to the pollen-based stratigraphic zonation (originally proposed by Brenner (1963)). Pollen types indicated: a, generalized tectate–columellar monosulcates (*Clavatipollenites*, *Retimonocolpites*, *Stellatopollis*); b, reticulate monocotyledonoid mono-sulcates (*Liliacidites*); c, reticulate to tectate tricolpates (*Tricolpites*); d, reticulate to tectate tricolporoidates (*Tricolpites*, *Tricolporoidites*); e, small, generally smooth-walled prolate tricolporoidates (*Tricolporoidites*); f, small, generally smooth-walled, oblate–triangular tricolporoidates (*Tricolporoidites*, *Perucipollis*); g, larger, smooth-walled to

264

does occur in Cretaceous angiosperms – that is, pollen evolving at different rates from other organs. Some peculiar combinations therefore occur, such as amentiferous inflorescences with monosulcate pollen (Dilcher 1979)! As can be seen in Figure 13.2, the first tricolpate pollen was small (less than 20 μm), and isodiametric. That is, the ratio of the pole-to-pole axis to the diameter at the equator is about 1. These are sometimes called "Longaxones". As can be seen in Figure 13.6, a trend to shortening of the polar axis was a very early development. Tricolporate (tricolporoidate) pollen appears in the late Albian. Triporate forms, which are usually bowl- to disk-shaped, were well-established at middle latitudes by Cenomanian time. In other words, "Brevaxones" forms, with a ratio of polar axis to diameter of 0.5 or so, had arrived. (Many extant angiosperms still have Longaxones pollen. Some are even perprolate, with a ratio of polar axis to diameter of 2 or more.) The patterns of evolution are displayed in Figure 13.7. Angiosperms were already the dominant land plants worldwide by late Cenomanian time (see Fig. 13.8), demonstrating their unique competitive edge. All of the major morphological variants of angiosperm pollen had appeared by late Cretaceous time. By that time the unique pollination relationships between various insect groups (especially Lepidoptera and Hymenoptera) and the angiosperms were well developed. Stanton *et al.* (1986) suggests indeed that selective pressures may

---

reticulate prolate tricolpor(oid)ates (*Tricolporopollenites*); h, larger, generally smooth-walled, oblate–triangular tricolpor(oid)ates (*Tricolporopollenites*); i, early members of the triangular triporate Normapolles complex (*Complexiopollis*, *Atlantopollis*). Leaf types indicated: j, acrodromous, narrowly obovate, monocotyledonoid (*Acaciaephyllum*); k, first rank, pinnately veined, reniform (*Proteaephyllum reniforme*); l, first rank, serrate (*Quercophyllum*); m, first rank, narrowly obovate (*Rogersia*); n, first rank, broadly elliptical (*Ficophyllum*); o, parallelodromous, elongate (*Plantaginopsis*); p, lobate reniform (*Vitiphyllum*); q, first rank, obovate (*Celastrophyllum*); r, campylodromous, sagittate (*Alismaphyllum*); s, actinodromous ovate–cordate–lobate ("*Populus*" *potomacensis*, *Populophyllum reniforme*); t, actinodromous, peltate (*Menispermites* "*tenuinervis*"); u, pinnately veined, serrate (*Celastrophyllum*); v, second rank, pinnatifid (*Sapindopsis magnifolia*); w, second rank, palinactinodromous, palmately lobed (*Araliaephyllum*); x, acrodromous, lobate elliptical (*Menispermites potomacensis*); y, third rank, pinnately compound, sometimes serrate (*Sapindopsis* spp.); z, third rank, palinactinodromous, palmately lobed (*Araliopsoides*, "*Sassafras*", etc.)

(b) Development of most major dicot pollen types during Barremian to Cenomanian time. Forms appearing later in the Cretaceous presumably had their origins in one of the lines shown. Non-Normapolles triporates, for example, probably also derived from "n". Pollen types indicated: a, tectate–granular relatives of *Clavatipollenites*; b, *Clavatipollenites* and *Retimonocolpites*; c, *Stellatopollis*; d and e, *Liliacidites* spp.; f, *Afropollis operculatus*; g, *A. zonatus*; h, *A. jardinus*; i, monosulcate with extended sulcus; j, early tricolpate; k, large, sculptured tricolpate; l, small, smooth tricolpate; m, tricolporoidate; n, oblate–triangular tricolporate; o, polyforate; p, tricolporoidate; q, early Normapolles; r, *Proteaephyllum reniforme*; s, *Ficophyllum*; t, *Rogersia*; u, serrate leaf; v, *Acaciaephyllum*; w, *Nelumbites*; x, pinnately dissected *Sapindopsis*; y, compound *Sapindopsis*; z, platanoid.

(c) *Afropollis operculatus* Doyle *et al.*, for comparison with pollen types f–h to the left in (b), distal view, diameter of grain 34 μm. Others of the ancestral types are illustrated in Figure 13.2.

well have been in the direction of "male fitness" – that is, in attractiveness of flowers to pollinators. The conversion of many angiosperms to wind pollination is a secondary phenomenon, especially of the Cenozoic. For example, the tricolporate form developed from the tricolpate, with intermediate tricolporoidate forms, apparently to provide a stronger "accordion pleat" for harmomegathic expansion and contraction, while retaining an efficiently thin area for germination. By the end of the Cretaceous, many forms are already referable with some confidence to modern families (see Muller 1981). Supporting this referability is work on *in situ* pollen from Cretaceous angiosperm flowers, such as that of Friis & Skarby (1982), Friis (1985b), and Basinger & Dilcher (1984).

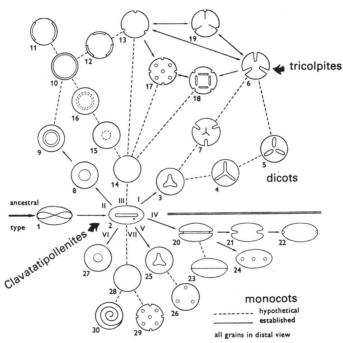

Figure 13.7 Probable paths for evolution of different aperture types in angiosperm pollen. The timeframe for these developments is seen in Figure 13.6. This scheme, proposed by Muller (1970), was mostly hypothetical, as shown by the dominance of broken lines, but later work has increased its plausibility. "Ancestral type l" represents a cycad or ginkgoid monosulcate with the sulcus flared at the ends. Type "2" refers to the sort of sulcus shown by *Clavatipollenites* (and also made by palms and canellaceous dicots). The roman numerals I–VII refer to distinct trend lines in evolution. The other, arabic, numerals refer to specific pollen morphological categories: (3, 4) trichotomosulcate; (5) distally tricolpate; (6) tricolpate; (7) trichotomosulcate and tricolpate; (8) distally monoporate; (9) distally operculate; (10) equatorially zonocolpate; (11) equatorially dicolpate; (12) equatorially tricolpate; (13) equatorially triporate; (14) inaperturate; (15) proximally monoporate; (16) proximally operculate; (17) periporate; (18) pericolpate; (19) tricolporate; (20) monosulcate; (21) dicolpate; (22) diporate; (23) meridionally zonocolpate; (24) distally meridionally triporate; (25) trichotomosulcate; (26) distally sub–equatorially triporate; (27) distally monoporate; (28) inaperturate; (29) periporate; (30) spiraperturate. See Muller (1970) for more details.

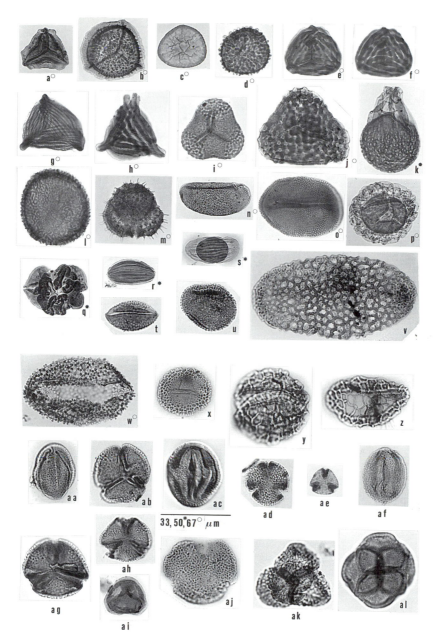

Figure 13.8 Early to middle Cenomanian (that is, about middle Cretaceous) spores/pollen from Peace River area, northwestern Alberta, Canada. Note that monosulcate forms (t)–(z) of rather primitive aspect and rather generalized tricolpate, tricolporoidate, and tricolporate forms (aa)–(aj), as well as obligate tetrads (ak), (al), have arrived in some abundance and diversity. Magnification is shown by line under (ac). (a) *Gleicheniidites bolchovitinae* Döring. Proximal view. (b) *Aequitriradites ornatus* Upshaw. Proximal view. (c) *Triporoletes cenomanianus* (Agasie) Srivastava. Distal view. (d) *Gemmatrileres clavatus* Brenner. Focused on distal surface. (e) *Cicatricosisporites*

267

NORMAPOLLES POLLEN

One of the early divergences from the basic tricolpate form of dicot pollen was the development in Cenomanian time of the triporate condition (P03), in which the germinal openings are more or less isodiametric pores (larger axis of opening less than two times the smaller axis), 120° apart, more or less on the equator of the grain. A variant on this theme was the development, also in the Cenomanian, of triporate (in a broad sense) pollen with internally complex pore structure (see Fig. 13.9). These are collectively called Normapolles. They diversified through the Cretaceous, remained, reduced in diversity in the early Paleogene, but were much reduced by early Eocene (Batten 1981b), and gone by early Oligocene (Hochuli 1984). Kedves (1983) notes that major trends in Normapolles evolution included decrease in number of exine layers and

---

crassiterminatus Hedlund. Focused on proximal surface. (f) Same specimen as (e). Distal view. (g) *Appendicisporites auritus* Agasie. Mid-focus, proximal–distal view. (h) *Appendicisporites insignis* (Markova) Khlonova. Distal view. (i) *Trilobosporites purverulentus* (Verbitskaya) Dettmann. Proximal view. (j) *Cibotiidites arlii* Srivastava. Distal view. (k) *Crybelosporites bellus* C. Singh. Lateral view. This spore has a spherical inner body and a loosely fitting, cavate outer coat, which is described by the non-committal term "sculptine". It is perhaps a perine. (l) *Rugulatisporites* sp. Distal view. (m) *Heliosporites kemensis* (Khlonova) Srivastava. A tetrad. (n) *Microfoveolatosporis pseudoreticulatus* (Hedlund) C. Singh. Proximo-lateral view. (o) *Reticulosporis foveolatus* (Pierce) Skarby. Proximal view. (p) *Crybelosporites pannuceus* (Brenner) Srivastava. Proximal view. See comments under (k). (q) *Rugubivesiculites multisaccus* C. Singh. Lateral view of unusual bisaccate grain. (r) *Equisetosporites fissuratus* Phillips & Felix. Mid-focus. (s) *Equisetosporites ambiguus* (Hedlund) C. Singh. Mid-focus. (t) *Liliacidites lenticularis* C. Singh. Distal view. (u) *Clavatipollenites tenellis* Phillips & Felix. Distal view. (v) *Liliacidites giganteus* C. Singh. Proximal view, with heads of columellae showing as dark dots. (w) *Stellatopollis largissimus* C. Singh. Distal view. (x) *Liliacidites tectatus* C. Singh. Distal view. (y) *Dichastopollenites dunveganensis* C. Singh. Lateral view of this strange, apparently zonisulculate grain, of which the two halves separated by the encircling sulculus frequently break apart and are found as separate fossils – see (z). (z) A single half ("hemisphere") of the same taxon as illustrated in (y). (aa) *Fraxinoipollenites constrictus* (Pierce) Khlonova. Equatorial view. (ab) *Foveotricolporites callosus* C. Singh. Polar view. (ac) Same taxon as (ab). Equatorial view. (ad) *Rousea doylei* C. Singh. Polar view. (ae) *Nyssapollenites nigricolpus* C. Singh. Polar view. (af) *Rousea candida* C. Singh. Equatorial view. (ag) *Foveotricolporites callosus* C. Singh. Polar view. (ah) *Retitricolporites pristinus* C. Singh. Polar view. (ai) *Phimopollenites tectatus* C. Singh. Polar view. (aj) *Phimopollenites megistus* C. Singh. Polar view. (ak) *Senectotetradites varireticulatus* Dettmann. Obligate tetrad, polar view of central grain. (al) *Foveotetradites fistulosus* (Dettmann) C. Singh. Obligate tetrad, showing "pretzel contact figure". All grains showing are in lateral views. For (aa)–(aj), note that with the advent of tricolpate pollen, it becomes important to guard against classifying polar and equatorial views as different taxa! They look very different (see (ab) and (ac)), and the sculpture is a good key to linking them. With modern pollen, it is desirable to examine the same specimen in different views by rotating it in a liquid mountant by tapping or pushing very gently on the coverslip, but fossil grains are almost always flattened, so tapping is seldom helpful. Tricolpate–tricolporate pollen is prevailingly flattened so as to produce polar views.

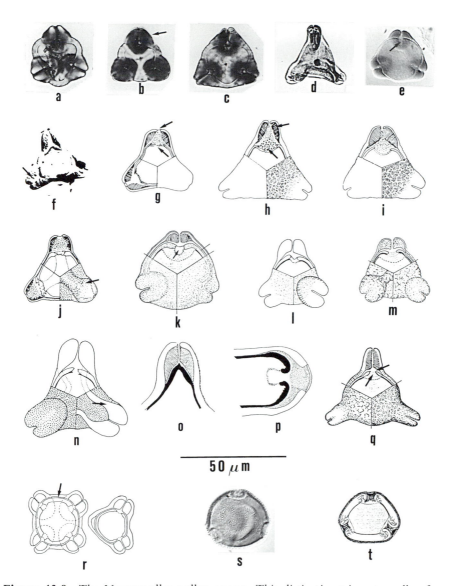

Figure 13.9 The Normapolles pollen group. This distinctive triporate pollen form characterizes the Normapolles palynofloral province of eastern North America and western Eurasia. The pore structure (germinal) is internally complex and is difficult to illustrate photographically. Hence the popularity of line drawings for illustration of such structure. Normapolles pollen first appears in about the middle of the Cretaceous and continues through the Paleocene into the Eocene. Magnification for the photomicrographs is indicated by bar under (o) and (p). The line drawings are at various magnifications, but the actual size of the specimens drawn is comparable to that of the photographed specimens. (a) *Oculopollis orbicularis* Góczán, Senonian, Hungary. Mid-focus, showing the pore canals (see also (b)). (b) *Oculopollis orbicularis* Góczán, Senonian, Hungary. High focus, showing the extraordinarily thick-walled oculus structures (arrow). (c) *Oculopollis parvooculus* Góczán, Senonian, Hungary. Mid-focus

as in (a). (d) *Complexiopollis vulgaris* (Groot & Groot) Groot & Krutzsch, Cenomanian, Spain. Mid-focus, showing complex internal pore structure. This is representative of the earliest Normapolles form. (e) *Trudopollis pertrudens* (Pflug) Pflug, Paleocene, Texas. Interference contrast photo, mid-focus, showing the oculus structures with prominent pore canal (arrow) (see also (k)). (f) *Basopollis basalis* (Thomson & Pflug) Pflug, Paleocene, Texas. Interference contrast photo, mid-focus showing the complex internal pore structure, a centripetally developed polyannulate annulus (arrow). (g) *Osculapollis aequalis* Tschudy, Campanian, Tennessee. Sections of drawing showing internal structure at different levels. The exine openings, which in Erdtman terms would be the pore (external) and os (internal), are shown by arrows. Normapolles experts usually call these openings germinals (exogerminal and endogerminal). (h) *Vacuopollis munitus* Tschudy, Maestrichtian, Missouri. Sections show germinal structure and sculpture pattern. The endexine is two-layered. There are internal verrucae (lower arrow) in the atrium between inner and outer germinals. The upper arrow points to the characteristic "baculate" structure of the wall of the vestibulum, a feature found in many Normapolles pollen. (i) *Extremipollis versatilis* Tschudy, Maestrichtian, Kentucky. Sectional view of germinal and surface view. Note baculate structure as in (h), but even more pronounced. (j) *Plicapollis retusus* Tschudy, Campanian, Tennessee. Sectional views and a surface view (lower right) showing the sculpture and a plica (arrow), a common Normapolles feature. (k) *Trudopollis pertrudens* (Pflug) Pflug, Paleocene, Germany. Optical section above showing prominent endogerminal (arrow) and surface features (left) and surface features reflecting internal structure (right) (see also (e)). (l) *Semioculopollis* sp., Maestrichtian, Tennessee. This has characteristic "oculi" (much thickened annuli) on one side of grain only. The slit-shaped exogerminals extend farther toward the poles on the side with the oculi. Lower right of drawing shows one side and lower left the obverse side. (m) *Pseudoculopollis principalis* (Weyland & Krieger) Krutzsch, late Cretaceous, central Europe. Lower left and right show appearance of opposite sides of grain. Note plica on one side (see (j)). (n) *Pseudoculopollis* sp., Maestrichtian, Tennessee. Oculus present on one side only. Arci (arrow) present only on side without oculi. Arci are thickenings similar to plicae, but arc-shaped. (o), (p) *Choanopollenites conspicuus* (Groot & Groot) Tschudy, Paleocene, Maryland: (o) the apertural region in polar view, (p) the same region in a sectional equatorial view. In (o) the densely packed baculae show very clearly. In (p) the exogerminal's opening appears funnel-shaped, though in polar view and surface expression it is a slit. The great difference of these two views of the same apertural area illustrates well the complexity of Normapolles morphology. (q) *Extratriporopollenites fractus* (Pflug) Pflug, Paleocene, Germany. This shows difference in sculpture on opposite sides of grain, atrium (lower arrow) and vestibulum (upper arrow). (r) As pointed out by Skarby (1968), many observed variations in basic Normapolles morphology are teratological ("freaks"): left, a 4-germinal form; right, a two-pored form. Extensive spaces between the outer and inner exine are a common feature of Normapolles, as shown here. The space (arrow) is called an interloculum. (s) *Thomsonipollis magnificus* (Thomson & Pflug) Krutzsch, Paleocene, Texas. Interference contrast photo. *Thomsonipollis* is a Cretaceous–Eocene form usually grouped with Normapolles forms because of the unusual, very complex, pore (germinal) construction. Whether *Thomsonipollis* really is related to Normapolles, however, is problematical. (t) Drawing of *Thomsonipollis* illustrating the multiple exine layers and the invagination of layers in connection with the germinal structure. (An important summary of Normapolles morphology is to be found in Batten & Christopher (1981).)

development of secondarily granular sexine from the columellate condition of the ancestral Brevaxones forms. Skarby (1968) has stressed the evident plasticity of this pollen form, showing that a number of different forms may occur in one anther of fossil flowers. She therefore doubts that it is very useful to proliferate generic names for Normapolles pollen, and would refer most to the one genus, *Extratriporopollenites*. Góczán *et al.* (1967), Tschudy (1975), and others have created many genera for these forms, and they have been much studied for palynostratigraphic and paleobiogeographic purposes. Batten & Christopher (1981) have published a useful summary of Normapolles morphology and a dichotomous key of 86 Normapolles and Normapolles-like genera. Recent studies of Senonian fossil flowers by Skarby (1981) and by Friis (1983) have shown that Normapolles pollen was produced by dicots close to the extant family, Juglandaceae. (One extant juglandalean genus, *Rhoiptelea*, family Rhoipteleaceae, makes pollen very similar to Normapolles.) Muller (1984) suggests that the extinct plants producing Normapolles pollen were a group at the family level in the order Juglandales. Friis notes that pollen from the Senonian flowers are referable to the dispersed form-genera *Plicapollis* and *Trudopollis*.

LATE CRETACEOUS ANGIOSPERM POLLEN AND *WODEHOUSEIA* AND *AQUILAPOLLENITES*

Late Cretaceous terrestrial palynofloras are rich in forms unlike known modern angiosperm pollen, yet possessing features seen in various modern families. Figure 13.10 illustrates some significant late Cretaceous pollen forms. Two outstanding examples of late Cretaceous pollen which are clearly dicotyledonous but not referable to a modern group are *Wodehouseia* and *Aquilapollenites* (see Fig. 13.11). *Wodehouseia* is a loaf-shaped pollen grain with a very thick exine and four pores, two on either side of the grain. *Wodehouseia* and related forms such as *Azonia* have been collectively referred to as the *Oculata* group (Wiggins 1976, Takahashi 1984). The group as a whole is mostly confined to the late Cretaceous, but *Wodehouseia* and *Aquilapollenites* species do get into the Paleocene. *Aquilapollenites* is a larger genus than *Wodehouseia*, with something like 80 species in the whole complex. Some palynologists, however, split the genus into a number of genera: *Triprojectus, Hemicorpus, Mancicorpus, Integricorpus, Bratzevaea*, and others (see Fig. 13.11) (Takahashi 1984). The group of genera, including *Aquilapollenites*, can be called "triprojectate" or the Triprojectacites group. It is mostly a late Cretaceous group, but a few Triprojectacites occur in the Paleocene and even Eocene (Choi 1983). Many variants on the basic theme exist, but that theme is a tricolporate grain in which the three colpi occupy the termini of branch-like extensions from the main body, which in turn has polar extensions. The three colpal and two polar extensions give the whole the appearance of a child's "jack". It has been suggested that *Aquilapollenites* was produced by the family Loranthaceae, which in the modern flora is mostly a parasitic family (mistletoe, for example).

271

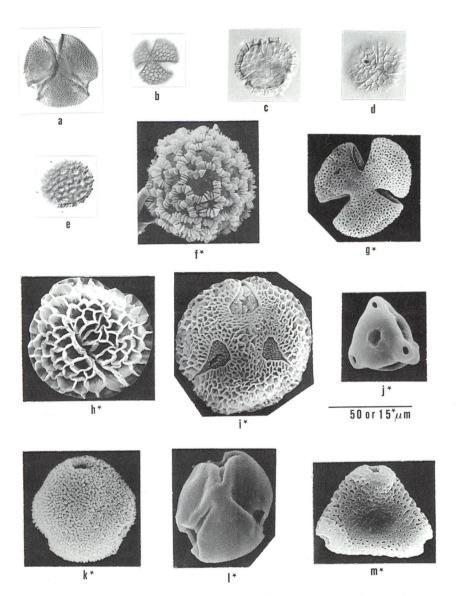

Figure 13.10 Illustrations of some significant late Cretaceous dicot pollen from Colorado, U.S.A. (a)–(e) Interference contrast photomicrographs; (f)–(m) S.E.M. pictures. Magnification indicated by bar under (j). Magnification for S.E.M. pictures is approximate. (a) *Tricolporopollenites* sp. Polar view. A very nearly syncolporate form with prominent colpal margos. (b) *Retitrescolpites* sp. Polar view. (c) *Retitrescolpites* sp. Polar view, mid–focus. (d) Same as (c). High focus. (e) *Erdtmanipollis pachysandroides* Krutzsch. (f) Same as (e). S.E.M. picture. (g) *Gunnera microreticulata* (Belsky *et al.*) Leffingwell. Polar view. (h) *Retitrescolpites* sp. Equatorial view. Compare with light pictures (c), (d) to see the advantages of S.E.M. for interpretation of sculpture. (i) *Libopollis jarzenii* Farabee *et al.* Polar view. (j) *Interpollis sapplingensis* (Pflug) Krutzsch. Polar view. (k) *Thomsonipollis magnificus* (Thomson & Pflug) Krutzsch. Polar view. (l) *Tricolporopollenites* sp. polar view. (m) *Proteacidites* sp. Polar view.

272

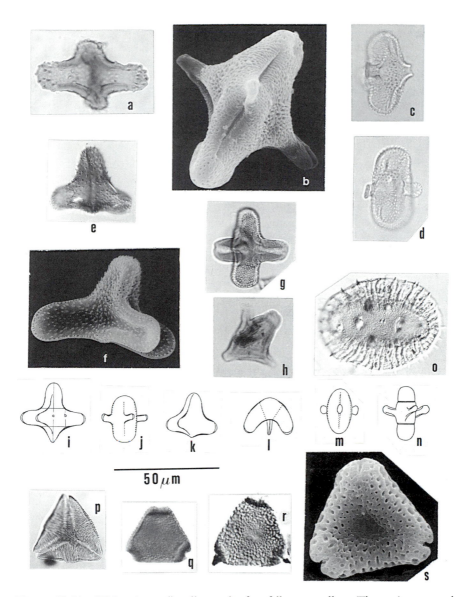

Figure 13.11 "Triprojectate" pollen and a few fellow travellers. The curious, mostly Cretaceous, characteristically multipronged pollen form, *Aquilapollenites*, is the form-genus which gives its name to the *Aquilapollenites* late Cretaceous palynofloral province. In a very broad sense, all of these pronged pollen could be shoe-horned into the single, original genus, *Aquilapollenites*. Various other genera are usually recognized, but palynologists have not yet agreed on how many, or which ones. The whole group comprises the "triprojectate complex". (i)–(n) Line drawings from Takahashi & Shimono (1982), illustrating some of the triprojectate genera they recognize. The colpal apparatuses are located on the ends of each of the (usually three) lateral arms. Associated with *Aquilapollenites* spp. and other triprojectates in the *Aquilapollenites* palynofloral province are a number of other genera. Examples of three of these genera, *Wodehouseia*,

CRETACEOUS PALYNOFLORAL PROVINCES

Palynological provincialism has been noted throughout the Cretaceous, and is especially marked in the late Cretaceous. Batten (1984) points out that Cretaceous floral provincialism is a continuation of Jurassic trends, in which the Northern Hemisphere was already broadly divided into two paleofloristic realms, the Siberian–Canadian and the Indo–European. The middle to late Cretaceous provinces are largely based on distinctive palynomorphs of restricted geographic and stratigraphic distribution, with broadly latitudinal boundaries.

Herngreen & Chlonova (1981) have published Neocomian through Senonian palynological provinces for the U.S.S.R. and vicinity; see Figure 13.12 for their provinces for early and late Cretaceous. One fact about provincialism long ago observed in the Northern Hemisphere is that palynofloras including Normapolles are found from mid-continent North America eastward to western Asia, whereas palynofloras including *Aquilapollenites* and *Wodehouseia* are found eastward from eastern Asia to central North America (see Fig. 13.12). However, isolated *Aquilapollenites* pollen grains are found in sediments from the Normapolles pollen province, suggesting that it may have been anemophilous. Toward the polar regions, the longitudinal precision is lost. Abundant *Aquilapollenites* occurs, for example, in the Eureka Sound

---

Cranwellia, and *Proteacidites*, are illustrated here. (a) *Aquilapollenites* sp. Rouse, late Cretaceous, Canadian Arctic. Equatorial view. (b) *Aquilapollenites trialatus* Rouse, Maestrichtian, Colorado. S.E.M., oblique equatorial view, 35 µm. (c) *Triprojectus echinatus* Mchedlishvili, late Cretaceous, Canadian Arctic. Equatorial view. (d) Same as (c). Different specimen showing one of colpal arms in optical section. (e) *Aquilapollenites quadrilobus* Rouse, Maestrichtian, Utah. Equatorial view, interference contrast. See also (f). (f) *Aquilapollenites quadrilobus* Rouse, Maestrichtian, Colorado. S.E.M., equatorial view, 44 µm. See also (e). (g) *Aquilapollenites rigidus* Tschudy & Leopold, Campanian, Montana. Equatorial view. Note the thickenings in the colpal arms as seen in optical section. (h) *Integricorpus* sp., late Cretaceous, Canadian Arctic. Oblique equatorial view. The poles are in the upper right and lower left. The colpal arms, of which one is shown in the center of the grain, have pronounced costae. (i) *Aquilapollenites*. Equatorial view. The small "*a*" refers to the length of the polar axis, and "*b*" refers to the width (thickness) of the equatorial (colpal) projection. Takahashi & Shimono use the ratio *a/b* as a character in describing the genera (see (a), (b)). (j) *Triprojectus*. Equatorial view (see (c), (d)). (k) *Hemicorpus,* Equatorial view (see (e)–(g)). (l) *Mancicorpus*. Obliquely polar view. (m) *Integricorpus*. Equatorial view (see (h)). (n) *Bratzevaea*. Equatorial view. (o) *Wodehouseia spinata* Stanley, Maestrichtian, Montana. Equatorial view. (p) *Cranwellia rumseyensis* Srivastava, Maestrichtian, Montana. Polar view. *Cranwellia* is believed by some to be related to the triprojectates. It is an important fellow traveller in any event. (q) *Proteacidites* sp., Maestrichtian, Colorado. Interference contrast, polar view. Note characteristic exinal thickenings associated with the pores (see (r)). (r) *Proteacidites* sp., Maestrichtian, Utah. Interference contrast, polar view (see (q)). (s) *Proteacidites* sp., Maestrichtian, Colorado. S.E.M., polar view, about 30 µm. Magnification for photomicrographs is shown by bar under (j) and (k). The line drawings are at various magnifications. For the S.E.M., the size for each grain is given with the listing.

CRETACEOUS PALYNOFLORAL PROVINCES

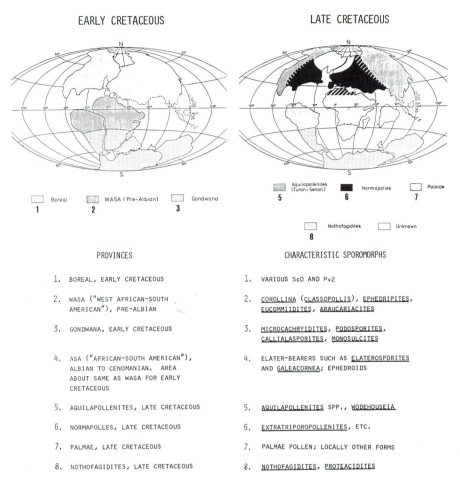

EARLY CRETACEOUS

N

S

1	2	3
Boreal	WASA (Pre-Albian)	Gondwana

LATE CRETACEOUS

N

S

5	6	7
Aquilapollenites (Turon.-Senon.)	Normapolles	Palmae

8	
Nothofagidites	Unknown

PROVINCES

1. BOREAL, EARLY CRETACEOUS

2. WASA ("WEST AFRICAN–SOUTH AMERICAN"), PRE-ALBIAN

3. GONDWANA, EARLY CRETACEOUS

4. ASA ("AFRICAN–SOUTH AMERICAN"), ALBIAN TO CENOMANIAN. AREA ABOUT SAME AS WASA FOR EARLY CRETACEOUS

5. AQUILAPOLLENITES, LATE CRETACEOUS

6. NORMAPOLLES, LATE CRETACEOUS

7. PALMAE, LATE CRETACEOUS

8. NOTHOFAGIDITES, LATE CRETACEOUS

CHARACTERISTIC SPOROMORPHS

1. VARIOUS ScO AND Pv2

2. COROLLINA (CLASSOPOLLIS), EPHEDRIPITES, EUCOMMIIDITES, ARAUCARIACITES

3. MICROCACHRYIDITES, PODOSPORITES, CALLIALASPORITES, MONOSULCITES

4. ELATER-BEARERS SUCH AS ELATEROSPORITES AND GALEACORNEA; EPHEDROIDS

5. AQUILAPOLLENITES SPP., WODEHOUSEIA

6. EXTRATRIPOROPOLLENITES, ETC.

7. PALMAE POLLEN; LOCALLY OTHER FORMS

8. NOTHOFAGIDITES, PROTEACIDITES

Figure 13.12 Cretaceous palynofloral provinces. See Figures 13.13–13.15 for photomicrographs of examples of many of the characteristic sporomorphs. (*Aquilapollenites*, Normapolles and *Nothofagidites* provinces are not much illustrated in Figures 13.13–13.15, because there are a number of illustrations of the pertinent forms in other figures.)

Formation of Ellesmere Island, which is on about the same meridian as Maryland, well within the Normapolles province if defined meridionally. However, as Batten (1982) has shown, the *Aquilapollenites*–Normapolles zone boundaries may be more or less latitudinal in the North Atlantic area. Srivastava (1981) has also pointed out that the late Cretaceous vegetational zones based on pollen data were at least semi-latitudinal rather than meridional. (*Wodehouseia*, though much less common than *Aquilapollenites*, is found in the *Aquilapollenites* province.)

Nichols (1984) has noted that, within the *Aquilapollenites* province in western North America of late Cretaceous time, subprovinces can be delineated on the basis of taxa occurring with *Aquilapollenites*:

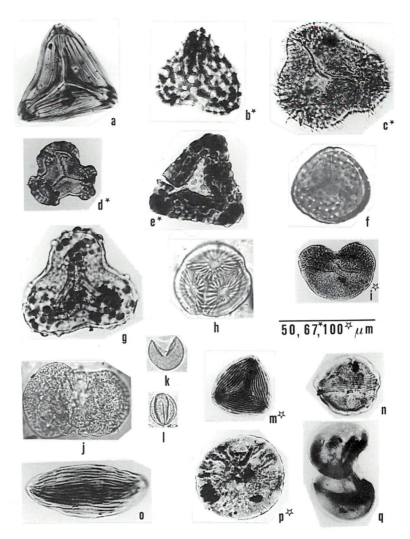

Figure 13.13 Pollen and spores characteristic of the early Cretaceous, Boreal palynofloral province ((a)–(l)), and of the pre-Albian early Cretaceous WASA (= West African–South American) province ((m)–(q)). See Figure 13.12 for more information about the Cretaceous palynofloral provinces. (a)–(e), and (i) are from the Upper Jurassic to Lower Cretaceous of the Netherlands; (g) is from the Aptian of Germany; (f), (h), and (j)–(l) are from the Soviet Union; (m)–(q) are from Brazilian boreholes. Magnification indicated by bar under (i). (a) *Cicatricosisporites abacus* Burger. Proximal view. (b) *Ischyosporites pseudoreticulatus* (Couper) Döring. Distal view. (c) *Pilosisporites trichopapillosus* (Thiergart) Delcourt & Sprumont. Proximal view. (d) *Trilobosporites hannonicus* (Delcourt & Sprumont) Potonié. Proximal view. (e) *Trilobosporites bernissartensis* (Delcourt & Sprumont) Potonié. Proximal view. (f) *Foveosporites cenomanicus* (Khlonova) Schvetzova in Bolkhovitina & Fokina. Proximal view. (g) *Impardecispora apiverrucata* (Couper) Venkatachala, Kar & Raza. Proximal view. (h) *Stenozonotriletes radiatus* Khlonova. Proximal view. (i) *Parvisaccites radiatus* Couper. Lateral view. (j) *Rugubivesiculites aralicus* (Bolkhovitina) Khlonova. Mid-focus, distal–

(1) *Expressipollis* subprovince (Arctic),
(2) *Callistopollenites* subprovince (Canadian plains),
(3) *Proteacidites retusus* subprovince (northern Rockies, U.S.A.),
(4) *Thomsonipollis magnificus* subprovince (southern Rockies, U.S.A.).

Wodehouseia spinata is associated also in subprovinces 1–3 but not in subprovince 4.

Brenner (1976) pointed out the existence of Cretaceous palynoflorally based provinces in the Southern Hemisphere, e.g., a northern Gondwana province and a southern Gondwana province, each with a distinctive palynoflora. Herngreen & Chlonova (1981) have summarized distribution data for a number of sporomorph taxa characterizing provinces from early to latest Cretaceous. Their summary diagrams of the provinces are shown in Figure 13.12, and representative pollen forms in Figures 13.13–13.15. The mid-Cretaceous ASA province (see Fig. 13.14) is characterized by *Galeacornea* and various elater-bearing pollen, and other unusual forms. In this province the earliest pollen unquestionably referable to dicot angiosperms appears (see Livingstone & Van der Hammen 1978). Srivastava (1983) has pointed out that, in latest Cretaceous time (Maestrichtian), there is mixing of elements from the previously distinct palynofloral provinces, caused by worldwide regression. Normapolles and *Aquilapollenites* producers, for example, were able to migrate to South America and Africa. Many of the significant early Cretaceous spore forms persisted for very long times, whereas the late Cretaceous pollen types were relatively short-lived, as is shown in Figure 13.16.

CRETACEOUS MEGASPORES

As noted earlier, free megaspores were never in center-stage in the post-"Paleophytic". However, heterosporous ferns and lycopods continued to produce megaspores that were preserved as fossils. Cretaceous megaspores are characteristic of specific environments of deposition and can be locally relatively abundant in fine sandy sediment. Hueber (1982) noted the relative abundance of such genera as *Arcellites* (Fig. 13.17m) and *Paxillitriletes* (Fig. 13.17e, f, j) in the lower Cretaceous Potomac Group of Virginia. Singh (1983) has described a rich megaspore palynoflorule from the mid-Cretaceous of Alberta, Canada (see Fig. 13.17 for representative forms), and Batten (1969, 1974) studied megaspores from the Wealden (Lower Cretaceous) of Great Britain, mostly representing lycopods growing on the Wealden delta, and showed their usefulness in reconstructing paleoenvironments.

---

proximal. (k) *Taxodiaceaepollenites hiatus* (Potonié) Kremp. Lateral view. (l) *Tricolpopollenites micromunus* Groot & Penny. Equatorial view. (m) *Cicatricosisporites australiensis* (Cookson) Potonié. Distal view. (n) *Corollina* (= *Classopollis*) sp. Lateral view. (o) *Ephedripites* sp. Lateral view. (p) *Araucariacites* sp. or *Inaperturopollenites* sp. (q) *Dicheiropollis etruscus* Trevisan. Lateral view. This form occurs normally as an apparent dyad.

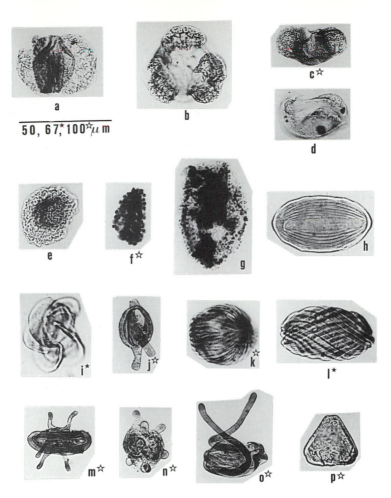

Figure 13.14 Characteristic pollen and spores of the early Cretaceous Gondwana palynofloral province ((a)–(d)), and of the later early to mid Cretaceous (Albian–Cenomanian) ASA (= African–South American) province ((e)–(p)). See Figure 13.12 for more information. (a)–(d) are from Albian levels in a borehole in Australia; (e)–(p) are from Albian–Cenomanian levels in Brazilian boreholes. Magnification indicated by bar under (a). (a) *Podocarpidites* cf. *ellipticus* Cookson. Proximal–distal view, mid-focus. (b) *Microcachryidites antarcticus* Cookson. Distal view. (c) *Alisporites grandis* (Cookson) Dettmann. Lateral view. (d) *Podosporites microsaccatus* (Couper) Dettmann. Oblique lateral view. (e) *Afropollis jardinus* (Brenner) Doyle *et al.* (f) *Reyrea polymorphus* Herngreen. (g) *Stellatopollis* sp. This genus is a representative early-appearing, monosulcate, angiosperm pollen grain. (h) *Ephedripites* sp. Distal view. (i) *Galeacornea causea* Stover. (j) *Sofrepites legouxae* Jardiné. (k) *Ephedripites elsikii* Herngreen. Obliquely lateral view. (l) *Ephedripites jansonii* (Pocock) Muller. (m) *Elaterosporites klaszi* (Jardiné & Magloire) Jardiné. (n) *Elaterocolpites castelaini* Jardiné & Magloire. (o) *Elateroplicites africaensis* Herngreen. (p) *Triorites africaensis* Jardiné & Magloire. Note that the odd forms (i), (j), and (m)–(o) are characteristic ASA forms and their peculiar elater or elater-like appendages probably were an adaptation to arid climate, a suggestion strengthened by association with ephedroid pollen ((h), (k), (l)). Indeed, *Elateroplicites* combines elaters with the polyplicate (taeniate) condition characteristic of *Ephedra* and relatives.

278

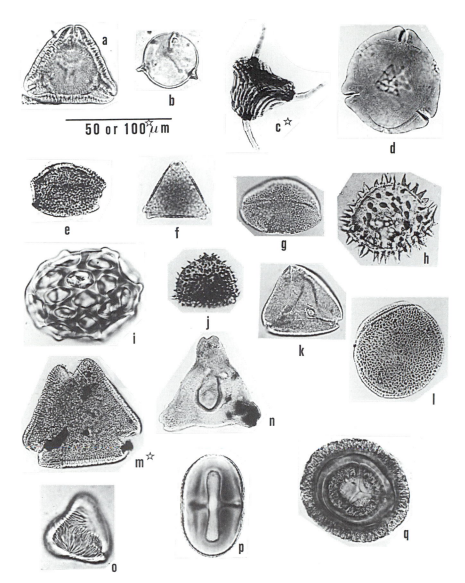

Figure 13.15 Spores and pollen of the late Cretaceous (Turonian–Senonian) *Aquilapollenites* palynofloral province ((a), (b)), the late Cretaceous Normapolles palynofloral province ((c), (d)), and the late Cretaceous (Senonian) Palmae palynofloral province ((e)–(q)). See Figure 13.12 for more information. For other representatives of the *Aquilapollenites* province, see Figure 13.11 and for other Normapolles province forms see Figures 13.9 and 13.10, (a) and (b) are from the Soviet Union; (c) and (d) are from the Senonian of Hungary; (e) and (h) are from the Senonian of Venezuela; (f), (g), (i)–(k) and (m)–(q) are from the Upper Senonian of Brazil; (l) is from the Senonian of Nigeria. Magnification indicated by bar under (a) and (b). (a) *Borealipollis bratzevae* Khlonova. Polar view, mid-focus. (b) *Orbiculapollis globosus* (Khlonova) Khlonova. Polar view. (c) *Appendicisporites tricuspidatus* Weyland & Greifeld. Distal view. (d) *Pseudopapillopollis praesubherzynicus* (Góczán) Góczán. Polar view, mid-focus. This is a

279

CRETACEOUS FERN SPORES

Fern spores continue to be important in Cretaceous non-marine sediments, as they are in the Jurassic. Markova (1966), for example, shows that in Siberian sediments fern spores referable to the family Schizaeaceae reached a peak of abundance in the Hauterivian–Barremian, just before the arrival of tricolpate angiosperm pollen on the scene. Although it is not conventional to use modern fern generic names for Cretaceous fern spores, most of the Cretaceous forms do seem to be similar in form to modern fern spores, though there are exceptions. This is in contrast to Cretaceous pollen, which is mostly different from the pollen of Cenozoic plants, even of those believed to be related. On the other hand, the demonstration by Hughes and coworkers (Hughes & Moody-Stuart 1966, 1969, Hughes & Croxton 1973) that fern spores have diversity and rapid evolution enough to be the basis for successful stratigraphic use in the British Cretaceous shows that Cretaceous ferns were still an important and dynamic floral element.

"TERMINAL CRETACEOUS EVENT"?

It is well-known that marine organisms suffered some sort of crisis at the end of the Cretaceous ("K–T boundary"). Indeed, the total extinction of ammonites is one *raison d'être* for the period (and era) boundary at this point. Marine vertebrates and invertebrates generally were decimated. Many kinds of land animals also perished: all dinosaurs and pterosaurs, indeed all land animals except the relatively small mammals, small reptiles and amphibians, and, of course, arthropods. Many authors have suggested a worldwide catastrophe such as, an epidemic of volcanoes, or the collision with the Earth of an asteroid–meteorite or comet. One might expect that vegetation must have been affected by such an event, if there were one. Paleopalynological evidence has been mostly hard put to demonstrate this sort of catastrophe. Perhaps the reason could be that most vascular plants responded with some resilience to the desperate but temporary conditions. Tschudy & Tschudy (1984) have shown that, at many locations in the northwest U.S.A. where sediments

---

Normapolles form in which some of the characteristic structures (see Fig. 13.9) are not as strongly expressed as in other taxa. (e) *Retidiporites magdalenensis* Van der Hammen & Garcia. Equatorial view. (f) *Proteacidites sigalii* Boltenhagen. Polar view. (g) *Retimonocolpites* sp. Distal view. (h) *Spinizonocolpites echinatus* Muller. Proximal–distal view, mid-focus. (i) *Buttinia andreeva* Boltenhagen. (j) *Echitriporites trianguliformis* Van Hoeken-Klinkenberg. Polar view. (k) *Cupanieidites* sp. Polar view showing the syncolpate morphology. (l) *Proxapertites operculatus* (Van der Hammen) Van der Hammen. Proximal–distal view. This genus is zonisulcuate (has a ring furrow: in a sense, monosulcate and syncolpate). (m) *Foveotricolpites irregularis* Herngreen. Polar view. (n) *Aquilapollenites sergipensis* Herngreen. Polar view. (o) *Scollardia srivastavae* Herngreen. Polar view. (p) *Crassitricolporites brasiliensis* Herngreen. Equatorial view. (q) *Gabonisporis vigourouxii* Boltenhagen. Proximal view.

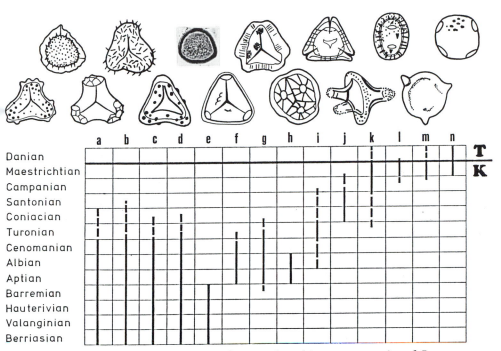

Figure 13.16 Stratigraphic ranges of some selected important species of Cretaceous spores and pollen, illustrating that early Cretaceous spore forms tended to persist for long periods, whereas later Cretaceous taxa tended to have shorter periods of stasis, facilitating both stratigraphic and palynofloral (provincial) subdivision. Broken lines mark sporadic occurrence. Except for (f), the taxa are here represented by line drawings. They are mostly illustrated by photographs elsewhere in the book. (a) *Impardecispora apiverrucata* (Couper) Venkatachala *et al.*; (b) *Aequitriradites spinulosus* (Cookson & Dettmann) Cookson & Dettmann; (c) *Impardecispora trioreticulosa* (Cookson & Dettmann) Venkatachala; (d) *Pilosisporites verus* Delcourt & Sprumont; (e) *Trilobosporites bernissartensis* (Delcourt & Sprumont) Potonié; (f) *Coptospora paradoxa* (Cookson & Dettmann) Dettmann; (g) *Triporoletes singularis* Mchedlishvili; (h) *Asbeckiasporites borysphenicus* (Voronova) Theodorova-Shakhmundes; (i) *Stenozonotriletes radiatus* Khlonova; (j) *Borealipollis bratzevae* Khlonova; (k) *Aquilapollenites unicus* (Khlonova) Khlonova; (l) *Wodehouseia spinata* Stanley; (m) *Orbiculapollis globosus* (Khlonova) Khlonova; (n) *Ulmoideipites krempii* Anderson.

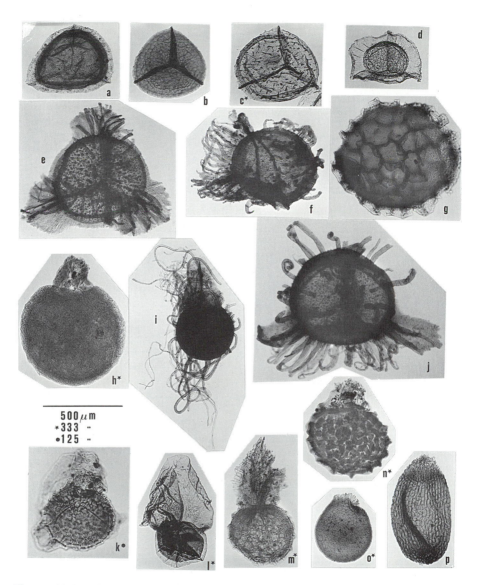

Figure 13.17 Cretaceous megaspores. These megaspores, from the Cenomanian of the Peace River area, northwestern Alberta, illustrate that free-megaspore-producing heterosporous ferns and lycopsids were still a factor in mid-Cretaceous time, though free-sporing megaspores were never as diverse or numerous again as in the late Devonian and Carboniferous. Magnification indicated by bar under (h). (a) *Minerisporites* sp. Proximal view. (b) *Minerisporites dissimilis* Tschudy. Proximal view. (c) *Henrisporites angustus* Tschudy. Proximal view. (d) *Minerisporites pterotus* C. Singh. Proximal view. This species has a membranous zona. (e) *Paxillitriletes dakotaensis* (Hall) Hall & Nicolson. Proximal view (see also (f) and (j)). (f) As (e) and (j). Lateral view. (g)

282

straddle the K–T boundary, there is indeed a temporary blotting out of many pollen types and a "fern spike" of great abundance of monolete and trilete fern spores, perhaps representing an expansion of ferns into decimated forests. After this brief interlude most but not all of the late Cretaceous forms come back, to fade out later in the Paleogene, perhaps because of evolutionary pressures triggered in part by the K–T event. Among the forms that do not come back after the K–T event are *Proteacidites* spp., and some kinds of *Aquilapollenites* (see Fig. 13.18). However, other kinds of triprojectates – *Aquilapollenites* in a very broad sense – persist well into the Paleocene, and Choi has recently described a new genus of them in the Eocene of the Canadian Arctic. Many Normapolles forms drop out at the end of the Cretaceous, but others persisted until they terminate in the Eocene. Some forms, such as *Casuarinidites granilabrata*, actually become more abundant in the Tertiary (Fleming 1984). As was true at the end of the Permian, the animal extinction event seems to be not very profoundly correlated with broad-scale plant extinctions. However, Wolfe (1986) argues for extensive extinction, at least among evergreen angiosperms at the K–T boundary.

DINOFLAGELLATES AND ACRITARCHS OF JURASSIC–CRETACEOUS

Dinoflagellate cysts first appear abundantly and very recognizably in paleo-palynological preparations of marine sediments in late Triassic. In the Jurassic they are very abundant and very fast evolving, making them ideal subjects for palynostratigraphy. Williams and others (Williams 1977, Bujak & Williams 1979) have described the very rapid evolution of Jurassic dinoflagellates. This is very fortunate for palynologists, as non–marine Jurassic sediments, dominated by rather difficult–to–work–with Pv2s and ScOs, can be troublesome for stratigraphy. Charts in Wilson & Clowes (1980) show the range of principal general forms, late Triassic to Neogene, Figure 13.19 illustrates some Cretaceous dinoflagellate cysts. The Jurassic dinoflagellate palynofloras show considerable provinciality, and zonations based on them cannot be extended worldwide

---

Erlansonisporites erlansonii (Miner) Potonié. Distal view. (h) *Molaspora fibrosa* C. Singh. Lateral view. (i) *Ariadnaesporites cristatus* Tschudy. Lateral view. The tangling, tubular threads presumably served to attach the spore to substrates for germination or to trap microspores. (j) As (e) and (f). Oblique view showing the long, hooked processes of the distal surface. (k) *Balmeisporites glenelgensis* Cookson & Dettmann. Obliquely lateral view showing laesura on proximal surface. (l) *Ariadnaesporites antiquatus* C. Singh. Lateral view. (m) *Arcellites reticulatus* (Cookson & Dettmann) Potter. Lateral view. The apical leaf-like appendages on this megaspore (see (e), (l), (n) and (k)) are a characteristic megaspore feature, probably having to do with "capture" of microspores. (n) *Rugutriletes comptus* C. Singh. Lateral view. (o) *Ricinospora pileata* (Dijkstra) C. Singh. Lateral view. (p) *Spermatites ellipticus* Miner. Lateral view. Although in the same size range and found in the same preparations, *Spermatites* is presumably not a megaspore, but a seed of which the seed cuticle with cellular structure is preserved.

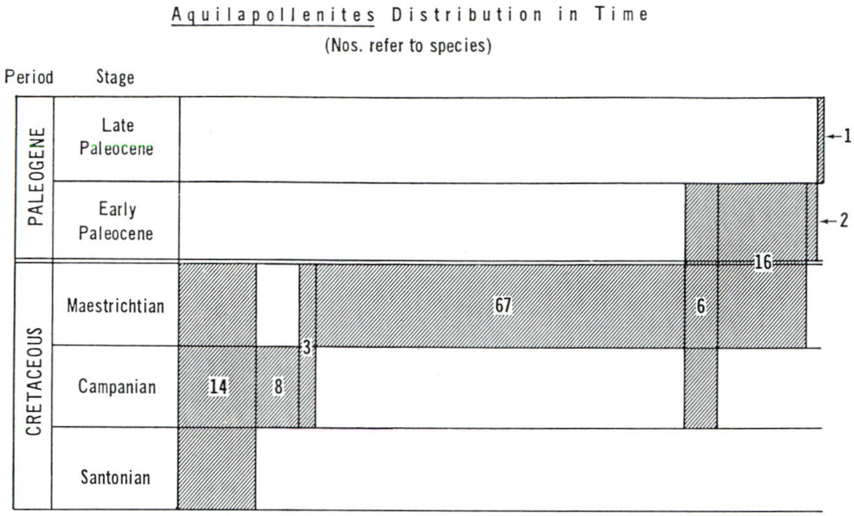

Aquilapollenites Distribution in Time

(Nos. refer to species)

Period   Stage

PALEOGENE	Late Paleocene		←1
	Early Paleocene		←2
		16	
CRETACEOUS	Maestrichtian	67	6
		3	
	Campanian	14 8	
	Santonian		

Summary

Occurrence	No. of Species
Cretaceous only	92
Both Cretaceous & Paleogene	22
Paleogene only	3

Figure 13.18 The behavior of *Aquilapollenites* (broad sense) pollen across the Cretaceous/Cenozoic ("K–T") boundary illustrates well what happened to palynofloras across this period and era boundary. *Aquilapollenites* was decimated at the end of the Cretaceous, but many species survived, after the setback, into the Paleocene. All but a tiny remnant became extinct several millions of years later, about in the middle of the Paleocene. Similar phenomena apply to other Maestrichtian pollen genera that survived into the Paleocene.

(Williams 1975). Dinoflagellate-based palynostratigraphy remains important in marine sediments of the Cretaceous and also ties in with the non–marine spores/pollen zonations, which from the end of the Neocomian on, because of rapid angiosperm evolution, are also quite well-controlled. Truswell (1981) has summarized this situation, and has noted that the Cretaceous dinoflagellate work has also been used to indicate factors in sedimentation such as sediment source and energy levels.

Acritarchs continue to be important in marine sediments of late Mesozoic age. Some characteristic forms from the Cretaceous and Paleogene of the Canadian Arctic are shown in Figure 13.20. Habib & Knapp (1982) have shown that some very small (less than 10 μm) acritarchs of the Cretaceous can be of considerable stratigraphic importance. Conventional palynological investigation often misses forms so small. As mentioned elsewhere, study of "Mesophytic"–"Cenophytic" acritarchs is a wide-open field.

284

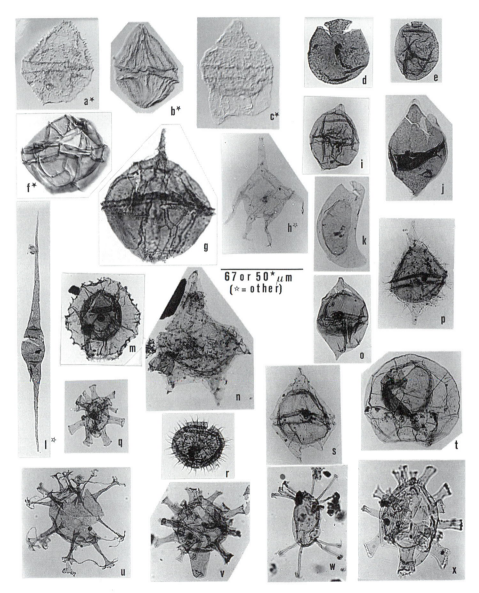

Figure 13.19 Selected dinoflagellate cysts of early and middle Cretaceous age from North America. (a)–(c) are interference contrast photomicrographs of specimens from the early Cretaceous of Wyoming. All others are bright-field (except as noted) photomicrographs of specimens from the Cenomanian of northwestern Alberta, Canada. Size of specimens indicated by bar under (h). Size of specimens marked "other" indicated in caption. (a) *Chichaouadinium vestitum* (Brideaux) Bujak & Davies.

285

NOTE ON CLASSIFICATION
OF JURASSIC–CRETACEOUS SPORES/POLLEN

The taxonomy of (that is, the classification of) Mesozoic, especially post-Triassic, spores/pollen is a very difficult matter. Although some, especially older Soviet palynological works, use names of extant plants for some mid-Mesozoic pollen, most of the forms are usually placed in form-genera: *Corollina*, *Ovalipollis*, *Triadispora*, *Aquilapollenites*, *Wodehouseia*, etc. It is possible to follow some rough grouping: all trilete spores together, all bisaccate pollen

---

Dorsal view of spiny proximate cavate cyst showing intercalary archeopyle. (b) *Dinogymnium* sp. Ventral view of this characteristically pleated taxon of proximate cysts. Many species of this genus were formerly known as *Gymnodinium* spp. (c) *Ascodinium verrucosum* Cookson & Hughes. Dorsal view of cavate cyst showing separation of the large apical operculum. (d) *Batiacasphaera macrogranulata* Morgan. Proximate cyst with apical archeopyle. (e) *Fromea granulosa* (Cookson & Eisenack) Stover & Evitt. Proximate cyst with apical archeopyle. (d) and (e) are "bag-like" cysts with very limited expression of paratabulation. (f) *Leptodinium modicum* Brideaux & McIntyre. Dorsal view of proximate cyst with clearly shown paratabulation and precingular archeopyle. (g) *Cribroperidinium intricatum* Davey. Ventral view, proximate cyst with clearly expressed paratabulation and apical horn. (h) *Muderongia pentaradiata* C. Singh. Proximate cyst with one apical, two cingular, and two antapical horns. Specimen 140 μm long. (i) *Subtilisphaera ?inaffecta* (Drugg) Bujak & Davies. A slightly cavate cyst with a small apical horn and a clearly demarcated paracingulum. (j) *Trithyrodinium rhomboideum* C. Singh. Dorsal view of proximate cavate cyst with intercalary archeopyle – the three intercalary paraplates are released separately. The picture shows one of these paraplates partially separated. (k) *Wallodinium anglicum* (Cookson & Hughes) Lentin & Williams. Lateral view of curved cavate cyst with a small endocyst and apical archeopyle. Operculum shown separating. (l) *Odontochitina singhii* Morgan. Unusual cavate proximate cyst. There is an apical horn and two antapical horns, one of which is vestigial. Length of cyst 400 μm. (m) *Catastomocystis spinosa* C. Singh. Dorsal view of proximate cavate cyst with precingular archeopyle. The endocyst is dark and smooth, and its wall (the endophragm) is closely appressed to the dorsal side of the pericyst wall (the periphragm). (n) *Endoceratium pentagonum* C. Singh. Proximate cavate cyst with apical archeopyle, dorsal view, with open archeopyle suture. (o) *Alterbidinium daveyi* (Stover & Evitt) Lentin & Williams. Proximate cavate cyst, mid-focus, lateral view. (p) *Palaeohystrichophora infusorioides* Deflandre. Proximochorate, cavate cyst, dorsal view, with prominent cingulum. (q) *Discorsia nanna* (Davey) Duxbury. Skolochorate cyst. (r) *Cleistosphaeridium* cf. *aciculare* Davey. With apical archeopyle, paratabulation not evident. (s) *Subtilisphaera hyalina* C. Singh. Cavate proximate cyst, dorsal view, with prominent cingulum, paratabulation not evident. (t) *Stephodinium australicum* Cookson & Eisenack. Cavate proximate cyst with small endocyst, lateral view, clearly delimited paraplate boundaries. (u) *Oligosphaeridium trabeculosum* C. Singh. Skolochorate cyst, apical archeopyle. The stringy connections between the processes represent the approximate level of the original thecal wall. (v) *Florentinia* cf. *deanei* (Davey & Williams) Davey & Verdier. Skolochorate cyst with precingular archeopyle, dorsal view. (w) *Bourkidinium psilatum* C. Singh. Skolochorate cyst with apical archeopyle, processes with filiform, recurved spines. (x) *Florentinia cooksoniae* (Singh) Duxbury. Skolochorate cyst with precingular archeopyle, dorsal view.

286

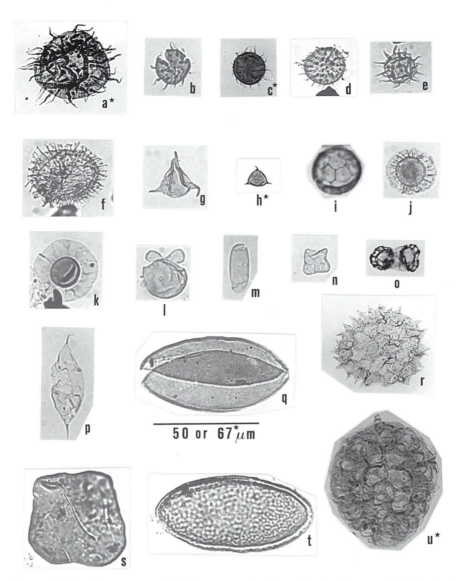

Figure 13.20 Cretaceous and Paleogene acritarchs and miscellaneous algal-derived palynomorphs. Marine shales containing palynofloras usually have among these floras various bodies that survive sedimentation, post-depositional processes, and laboratory maceration. Some of these are acritarchs – algal cysts and cyst-like bodies not referable to the dinoflagellates. Others are such colonial algae as *Pediastrum* (r), *Botryococcus* (illustrated elsewhere in the book), and *Palambages* (u). Although acritarchs are especially rich in numbers and diversity in early Paleozoic marine rocks, they have remained important to the present. Algal bodies, e.g., *Botryococcus* and *Pediastrum*, and acritarchs occur also in non-marine rocks but are not as abundant or diverse there as in sediment generated in marine environments. Magnification indicated by bar under (q). Most of the specimens illustrated are from northern Canada (AH = Axel Heiberg

287

together, etc. It would be possible to develop a "turmal" listing, and indeed many Mesozoic palynologists do use modifications of Potonié's turmal system. The problem ceases to be a problem by Eocene–Oligocene, when enough certainty can be attached to family assignments to use simply an Engler–Prantl, or similar, modern botanical classification scheme, plus broad morphological groupings for form-genera of unknown relationships.

Island; EI = Ellesmere Island). (a) *Baltisphaeridium* sp., Cenomanian, northern Alberta; (b) *Micrhystridium fragile* Deflandre, Paleocene–Eocene, AH; (c) *Micrhystridium recurvatum* Valensi, Cenomanian, northern Alberta; (d) *Baltisphaeridium* sp., Santonian, AH; (e) *Micrhystridium breve* Jansonius, Paleocene, AH; (f) *Baltisphaeridium fimbriatum* (White) Sarjeant, Paleocene–Eocene, AH; (g) *Veryhachium reductum* (Deunff) Jekhowsky, Santonian, AH; (h) *Veryhachium reductum* (Deunff) Jekhowsky forma *breve* Jekhowsky, Cenomanian, northern Alberta; (i) *Pterosphaeridia* sp., Campanian, AH; (j) *Pterospermella microptera* (Deflandre & Cookson) Eisenack & Cramer, Campanian, EI; (k) *Pterospermella australiensis* (Deflandre & Cookson) Eisenack & Cramer, Campanian, EI; (l) *Sigmopollis psilatus* Piel, Paleocene, EI; (m) *Navifusa* sp., Paleocene, EI; (n) *Tetraporina* sp., Campanian, EI; (o) *Bipatellifera clathrata* C. Singh, Cenomanian, northern Alberta; (p) *Leiofusa jurassica* Cookson & Eisenack, Paleocene, EI; (q) *Pilospora parva* (Cookson & Dettmann) Filatoff, Campanian, EI; (r) *Pediastrum* sp., Paleocene, Mississippi; (s) *Petalosporites quadrangulus* Agasie, Campanian, EI; (t) *Ovoidites ligneolus* (Potonié) Potonié, Eocene, EI; (u) *Palambages* sp., Campanian, EI.

CHAPTER FOURTEEN
Paleogene palynology

INTRODUCTION

Pollen and spore floras of the Cenozoic are systematically very well-known, for a number of reasons. One is that the German brown coals were among the first sediments to which paleopalynological methods were applied. Indeed, the first fossil pollen grains (*Alnus*) ever illustrated (by Goeppert in 1844) were from these beds. Pollen is comparatively easy to separate from lignitic coals. Potonié (1934) obtained his early Tertiary floras from the same sort of German brown-coal-bearing sediments as Goeppert investigated. In North America, other Paleogene sediments were targeted early: the Eocene Green River Oil Shale by Wodehouse (1932, 1933), even before the Potonié studies. The early to late Paleocene Fort Union Formation or Group (Wilson & Webster 1946), and the late Paleocene Wilcox Formation or Group (very little published until Elsik 1968a, b), both lignite-bearing, were inviting problems. My doctoral work (Traverse 1951, 1955) on the Brandon Lignite of Vermont came about fortuitously, because Professor E. S. Barghoorn was working on that lignite when I needed a doctoral problem. Another factor favoring Cenozoic studies is that sediments of this age are the oldest from which palynofloras can be obtained that offer strong possibilities for reliable paleoecological analysis, based on comparison with modern related taxa. During the Paleogene, modern plant distribution began to take shape, as was amply demonstrated by megafossil paleobotanists, such as Axelrod (1958) and others, before paleo-palynology played a major role in paleobotany. For example, in the Northern Hemisphere there were already in the Paleocene–Eocene "Arcto-Tertiary" elements in more northerly locations such as Ellesmere Island in the Canadian Arctic. From these elements came eventually the modern temperate forests. Although Wolfe (1977) and others have cast doubt on the validity of the "Arcto-Tertiary" flora concept because it is oversimplified, and some of the data on which it was originally based are flawed, the existence of temperate plant taxa of the families Betulaceae, Fagaceae, Ulmaceae, Juglandaceae, etc., in the far north during the Paleogene is amply illustrated by studies of Arctic Paleogene palynofloras. The far northern deciduous forests of Paleogene time were of course unlike any modern deciduous forests that contain some of the same elements (Wolfe 1980). On the other hand, middle- and low-latitude floras included elements of "paleotropical" vegetation which featured taxa now components of the subtropical and tropical forests. The Cenozoic mountain building resulted in perhaps the highest stand of the continents in geologic history, creating rain shadows from prevailing moisture sources. At the same time there was an increase of seasonality (hot–cold; wet–dry) and a very great increase in proportions of the Earth characterized by "abnormal geography": huge semiarid areas, as well as deserts and glaciers. Above all, there was a general decline in world temperatures (see Fig. 14.1) related in part to some of the above-mentioned factors, in part probably also to astronomical factors causing cyclic decline in effective insolation. Already in the Paleogene the development of precursors of semiarid and desert vegetation such as in the Madro-Tertiary flora (Axelrod 1958) and the Cordilleran flora (Axelrod & Raven 1985) indicate the response of angiosperms to Cenozoic climatic trends. As pointed out by Frederiksen (1985), some sorts of plant communities from the Paleogene have been much more thoroughly studied than others. For example, coastal brackish-water environments and coastal-plain peat-swamp communities are much better known than upland communities of the interior.

Figure 14.1 shows the overall chronology and major paleopalynological high points of the Cenozoic 66-million-year time segment. The traditional division of the Cenozoic into Tertiary (about 64 My) and Quaternary (about 2 My) periods is a relict of the old division of Earth history into Primary (Paleozoic), Secondary (Mesozoic), Tertiary, and Quaternary. The "Primary" and "Secondary" are now very little used, and Tertiary and Quaternary should join them in the dustbin. If that were done, the Neogene and its last epoch, the Pleistocene, should run to the present, and both Quaternary and Holocene would be unnecessary terms (the last 10,000 years would be "present interglacial" or some such term). However, this must be officially accomplished by stratigraphers, if they agree that it is a correct idea.

As at present recognized, the Paleogene period is about 32 million years long. Whether land vegetation experienced a terminal Cretaceous event of any significance is not yet certain. However, it is quite clear that there was a climatic crisis on land at the Cretaceous/Paleocene boundary, and the characteristic Cretaceous palynofloras were greatly altered at least locally and temporarily. For example, *Aquilapollenites* (see also Fig. 13.18) managed to make it across the boundary but was much diminished in the Paleocene. Normapolles survived the terminal Cretaceous event in diminished numbers and persisted through the whole 30 million years of the Paleocene and Eocene epochs as well, finally passing out of the picture in the Oligocene.

PALEOGENE CLIMATIC MATTERS

As can be seen in Figure 14.1, geomagnetic data and oxygen-isotope ($\delta^{18}O$) measurements provide, respectively, control on stratigraphic position and presumed world temperatures for our conclusions about the nature and timing of floral evolution in the Cenozoic. Radiometric and biostratigraphic methods are also helping to provide information on the timing of events. Another important matter has been the gradual acceptance that Cenozoic time–rock segments must depend ultimately on stratotypes. Before these developments, palynologists' conclusions about the meaning of Cenozoic floras were sometimes wrong simply because a correct stratigraphic position "call" was not possible.

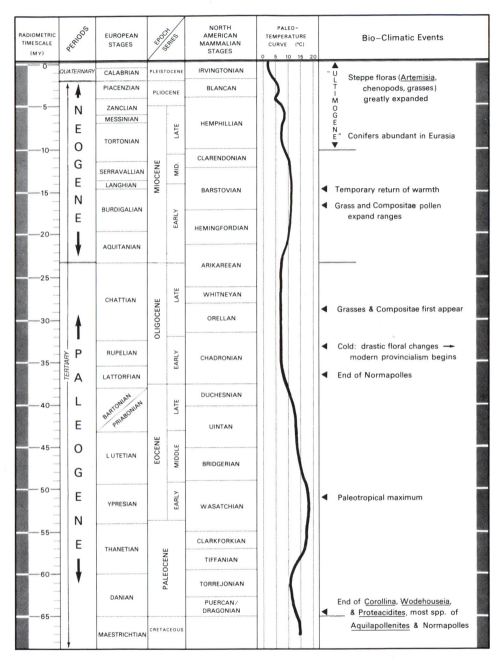

RADIOMETRIC TIMESCALE (MY)	PERIODS	EUROPEAN STAGES	EPOCH SERIES	NORTH AMERICAN MAMMALIAN STAGES	PALEO-TEMPERATURE CURVE (°C) 0 5 10 15 20	Bio–Climatic Events
0	QUATERNARY	CALABRIAN	PLEISTOCENE	IRVINGTONIAN		" U L T I M O G E N E " Steppe floras (Artemisia, chenopods, grasses) greatly expanded
		PIACENZIAN	PLIOCENE	BLANCAN		
5	N E O G E N E	ZANCLIAN				
		MESSINIAN	LATE	HEMPHILLIAN		Conifers abundant in Eurasia
		TORTONIAN				
10			MID.	CLARENDONIAN		
		SERRAVALLIAN	MIOCENE			
15		LANGHIAN		BARSTOVIAN		◄ Temporary return of warmth
		BURDIGALIAN	EARLY			◄ Grass and Compositae pollen expand ranges
20		AQUITANIAN		HEMINGFORDIAN		
				ARIKAREEAN		
25	P A L E O G E N E	CHATTIAN	LATE	WHITNEYAN		
30			OLIGOCENE	ORELLAN		◄ Grasses & Compositae first appear
35		RUPELIAN	EARLY	CHADRONIAN		◄ Cold: drastic floral changes → modern provincialism begins
		LATTORFIAN				◄ End of Normapolles
40		BARTONIAN PRIABONIAN	LATE	DUCHESNIAN		
				UINTAN		
45		LUTETIAN	MIDDLE			
			EOCENE	BRIDGERIAN		
50		YPRESIAN	EARLY	WASATCHIAN		◄ Paleotropical maximum
55		THANETIAN		CLARKFORKIAN		
			PALEOCENE	TIFFANIAN		
60				TORREJONIAN		
		DANIAN		PUERCAN/ DRAGONIAN		◄ End of Corollina, Wodehouseia, & Proteacidites, most spp. of Aquilapollenites & Normapolles
65		MAESTRICHTIAN	CRETACEOUS			

Figure 14.1 Cenozoic framework for palynostratigraphy. Neogene, as used in this book, includes the Quaternary, per usage, for example, of Berggren *et al.* (1985). According to most sources, however, the Quaternary is separate from the Neogene. The Neogene and Paleogene can be regarded as sub-units of the Tertiary or as periods

291

As Figure 14.1 shows, the late Paleocene and early Eocene were mostly very warm, characterized by truly tropical floras in England and warm temperate floras even in what is now the Canadian Arctic. Maximum tropicality is at the early/middle Eocene boundary. The early Eocene is sometimes called the paleotropical maximum, as this was the time of greatest expansion of the paleotropical flora in the Northern Hemisphere. The last stand of Normapolles occurred much later. Note that, whatever other disagreements exist, all paleobotanists and most other paleontologists are agreed on the Eocene tropical expansion.

Also recognized by paleontologists in different specialties, and shown by oxygen-isotope data, is pronounced cooling beginning in the earliest Oligocene, though there is some disagreement about its duration within the Oligocene. In any event, the Oligocene was a watershed time, as this cooling was both preceded by, and followed by, significantly warmer times. (Collinson *et al.* (1981) note that evidence from British fossil plants suggests that cooling began in about the middle of the Eocene, rather than being sudden, at the end of the Eocene, as others have suggested.) It was during the Oligocene that very marked drastic floral changes leading to the development of modern plant associations began. Temperate deciduous forests expanded greatly. Grasses (family Gramineae) and composites (family Compositae) first appear, though not as important floral elements. Conifer forests apparently developed in mountainous areas. It is evident that local variations are superimposed on rather general climatic alterations shown in Figure 14.1, such as Oligocene cooling. It is also evident that the overall Cenozoic story shows a fluctuating, but progressive, cooling trend to the Pleistocene.

"POSTNORMAPOLLES"

The decline of Normapolles was accompanied in the late Cretaceous and Paleocene–Eocene by the appearance of many forms of triporate pollen that seem likely to have been produced by plants that evolved from Normapolles producers. The probability that this is so is increased by Skarby's (1981) and Friis' (1983) studies indicating the probability of a Juglandales alliance for the plants that produced Normapolles pollen. These newer sorts of triporates are called "Postnormapolles" and included *Carya*-like pollen (*Caryapollenites*) and *Myrica*- and *Engelhardia*-like pollen (*Momipites*), among others. Most of the Postnormapolles pollen, as Normapolles, was probably produced by amentiferous trees and shrubs. (Pflug (1953) introduced Postnormapolles as a suprageneric unit or "Stemma" for fossil porate pollen lacking the Normapolles special structural features.)

---

of the Cenozoic era. The paleotemperature plot is much smoothed out and averaged for high-latitude ocean-surface-water temperature, based on oxygen-isotope data measured from foraminifera, from the work of Shackleton & Kennett (1975) and other sources. In the epoch/series column, 10,000 years B.P. to present is the Holocene, but the scale presents the word from appearing, as is true of the final North American mammalian stage, the Rancholabrean, which however represents a considerably longer period of time than does the Holocene.

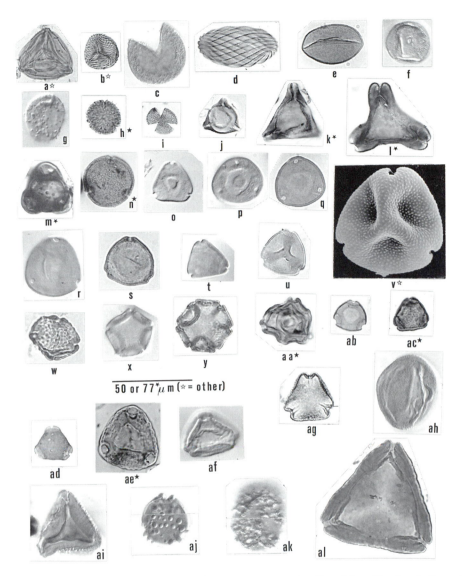

Figure 14.2 Some characteristic Paleocene spores and pollen. Magnification indicated by bar under (x) and (y), except for (a), (b), and (v), the sizes for which are given with the names of the taxa. (N.E.R. = Paleocene levels of D.S.D.P. cores at Ninetyeast Ridge, Indian Ocean; T.R.F.U. = Tongue River Member, Fort Union Formation, late Paleocene of Wyoming and Montana; E.F. = Evanston Formation, early Paleocene, Wyoming; W.G. = Wilcox Group, late Paleocene, Texas.) (a) *Appendicisporites crassicarnatus* Harris. Proximal view, spore 48 μm, N.E.R. (species typical of Paleocene levels, but this particular specimen is from an Eocene–Oligocene level in N.E.R.). (b) *Cicatricosisporites sp.* Proximal view, spore 70 μm (note small magnification), W.G. (c) *Taxodiaceaepollenites* sp. Lateral view showing characteristic splitting-open, T.R.F.U. (see also (f)). (d) *Equisetosporites jansonii* Pocock. A polyplicate form, W.G. (e) *Arecipites pseudotranquillus* Nichols *et al*. Distal view of monosulcate grain, W.G. (f) *Taxodiaceae-*

293

CHARACTERISTIC PALEOGENE SPORE/POLLEN FLORAS

Figure 14.2 illustrates some representative Paleocene spores/pollen from various localities, especially from the Fort Union Formation of Wyoming. A variety of Postnormapolles triporates, a few sorts of Normapolles, various Pc0 and Pc3 forms, and fern spores characterize the Paleocene palynoflora. The latest Paleocene Wilcox Group of Louisiana, Arkansas, Texas, and adjoining Gulf Coast areas of Mexico and the U.S.A. is represented by specimens from Texas. This is a considerably different association, with Postnormapolles, especially *Momipites*, very similar to modern genera, abundant palm pollen, and forms probably referable to mostly tropical families. Figure 14.3 shows

---

pollenites sp. Distal view of germinal papilla (not a pore), T.R.F.U. (see also (c)). (g) *Pandaniidites radicus* Leffingwell. Interference contrast photo, showing the baculate sculpture, T.R.F.U. (h) ?*Tubulifloridites* sp. Polar view, N.E.R. (i) *Tricolpites* sp. Polar view, E.F. (j) *Nudopollis terminalis* (Thomson & Pflug) Pflug. Polar view, W.G. ((j)–(m) and (al) are representative of continuing Normapolles influence in the Paleocene.) (k) *Nudopollis endangulatus* (Pflug) Pflug. Polar view, late Paleocene, South Carolina. (l) *Choanopollenites discipulus* Tschudy. Polar view, early Paleocene, Alabama. (m) *Interporopollenites turgidus* Tschudy. Polar view, early Paleocene, Missouri. (n) *Intratriporopollenites pseudinstructus* Mai. Polar view, late Paleocene, South Carolina. (o) *Momipites leffingwellii* Nichols & Ott. Polar view of juglandaceous form thought by Nichols to be ancestral to *Caryapollenites*, E.F. (p) *Caryapollenites veripites* (Wilson & Webster) Nichols & Ott. Polar view, interference contrast photo emphasizing the "moat-and-island" in the center of the polar area (see (q)), T.R.F.U. (q) Same species as (p), bright-field photo showing opercula on pore membranes, W.G. (r) *Caryapollenites imparalis* Nichols & Ott. Polar view, interference contrast photo, T.R.F.U. (s) *Triporopollenites* sp. Polar view, W.G. (t) *Momipites wyomingensis* Nichols & Ott. Polar view, T.R.F.U. (u) *Momipites wodehousei* Nichols. Polar view, W.G. (v) *Momipites triradiatus* Nichols. Polar view, S.E.M. micrograph, grain 24 μm, W.G. (w) *Ulmipollenites undulosus* Wolff. Polar view of 5-stephanoporate form, W.G. (x) *Alnipollenites verus* Potonié. Polar view, interference contrast photo, emphasizing the arci (thickened bands arching between pores), T.R.F.U. (y) *Alnipollenites* sp. Polar view, bright-field (see (x)), W.G. (aa) *Paraalnipollenites confusus* (Zaklinskaya) Hills & Wallace. Distorted polar view, late Paleocene, Alabama. (ab) *Momipites tenuipolus* Anderson. Polar view, showing thinned polar area and large vestibules associated with pores, W.G. (ac) *Pseudoplicapollis serena* Tschudy. Polar view, showing vestibular areas around pores, early Paleocene, South Carolina. (ad) *Momipites dilatus* (Fairchild) Nichols. Polar view, W.G. (ae) *Interpollis paleocenica* (Elsik) Frederiksen. Polar view, early Paleocene, South Carolina. (af) *Ulmipollenites tricostatus* (Anderson) Frederiksen. Polar view, interference contrast photo, T.R.F.U. (ag) *Nyssapollenites* sp. The tricolporate form is shown clearly even in polar view, W.G. (ah) *Nyssapollenites* sp. Equatorial view, interference contrast photo, T.R.F.U. Despite reference to same form-genus, this form clearly lacks the sort of clearcut pore (ag) would display in equatorial view. (ai) *Insulapollenites rugulatus* Leffingwell. Polar view, interference contrast photo showing the syncolpate morphology. T.R.F.U. (aj) *Pistillipollenites mcgregorii* Rouse. Polar view, T.R.F.U. Two-level mosaic interference contrast photo, the top part showing lower-level focus and the arrangement of gemmae around one of the three pores. (ak) *Erdtmanipollis pachysandroides* Krutzsch, T.R.F.U. Two-level mosaic interference contrast photo, showing the muri of the reticulum consisting of small exine blocks. (al) *Choanopollenites eximius* Stover. Polar view of very large Normapolles-type pollen (see (l)), W.G.

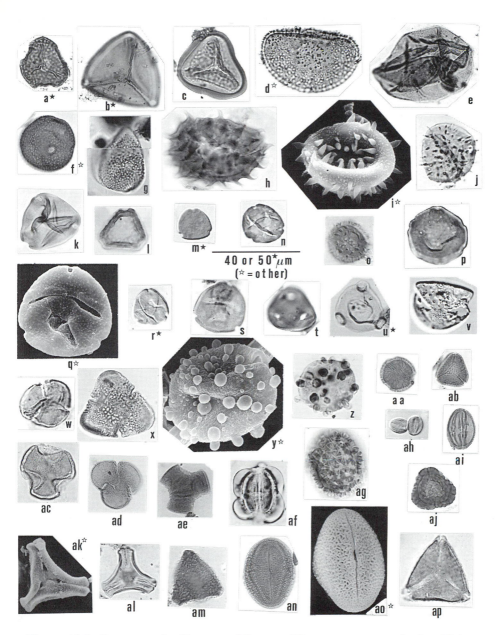

Figure 14.3 Representative Eocene and Eocene–Oligocene sporomorphs. In addition to the forms shown, bisaccate coniferous pollen resembling modern genera is common. Magnification shown by bar under (m) and (n). A few forms are at other magnifications, and their size is given in the caption. Some of the forms are from Eocene to Oligocene levels of D.S.D.P. cores at Ninetyeast Ridge, Indian Ocean (N.E.R.). Some others are from the Middle Eocene of San Diego, California (M.E.S.D.). The others are separately identified. (a) *Foveotriletes palaequetrus* Partridge. Proximal view, N.E.R. (b) *Deltoidospora* sp. Proximal view, N.E.R. (c) *Polypodiaceoisporites* cf. *tumulatus* Partridge. Proximal view, showing pronounced labra along the

laesural rays and heavily sculptured interradial contact areas, N.E.R. (d) *Punctatosporites varigranulatus* Kemp. Lateral view of monolete spore, length 42 μm, N.E.R. (Because such monolete spores are very generalized, this generic name is used for monolete spores from Carboniferous to Cenozoic! *Deltoidospora* for smooth trilete spores such as (b) is also a much used generic name.) (e) *Araucariacites australis* Cookson. A collapsed, thin-walled P00 form, such as several extant conifers make, N.E.R. (f) *Milfordia hungarica* (Kedves) Krutzsch. Distal view of this P01, diameter 42 μm, M.E.S.D. (Such Restionaceae-like forms presage modern grass pollen.) (g) *Longapertites* sp. Composite photomicrograph of two focal levels, length 35 μm, M.E.S.D. (a pear-shaped Pa0). (h) *Mauritiidites* sp. Distal view, length 42 μm, M.E.S.D. (Echinate Pa0s resembling this are found among modern palms.) (i) *Echiperiporites rotundus* Kemp. P03–echinate, pore showing at lower left, S.E.M. micrograph, diameter 20 μm, N.E.R. (see (o)). (j) *Pandaniidites* sp. Early Eocene, Wyoming. (k) *Plicatopollis plicata* (Potonié) Krutzsch. Polar view, showing vestibulate pore structure, early or mid-Eocene, Wyoming. (l) *Ulmipollenites tricostatus* (Anderson) Frederiksen. Polar view, early Eocene, Wyoming. (m) *Momipites coryloides* Wodehouse. Polar view, early Eocene, Wyoming. (n) *Platycarya platycaryoides* (Roche) Frederiksen & Christopher. Polar view, showing characteristic curving arcoid streaks (= pseudocolpi) (see (q)), early Eocene, Wyoming. (o) *Echiperiporites rotundus* Kemp. Photomicrograph for comparison with S.E.M. micrograph of (i), diameter 19 μm, N.E.R. (p) *Polyatriopollenites vermontensis* (Traverse) Frederiksen. Polar view of stephanoporate form, early or mid-Eocene, Wyoming. (q) *Platycarya platycaryoides* (Roche) Frederiksen & Christopher. Polar view, S.E.M., of P03 for comparison with (n) and (r), diameter 18 μm, early Eocene, Wyoming. (The pseudocolpi are thin-walled depressions.) (r) Same data as for (q), photomicrograph. (s) *Platycarya* sp. Mid-Eocene, Mississippi (see (n), (q), and (r)). (t) *Anacolosidites reklawensis* Elsik. Polar view of periporate form (six-pored, three pores on each hemisphere), early Eocene, Alabama. (u) *Corsinipollenites oculis-noctis* (Thiergart) Nakoman. Polar view of onagraceous-type pollen with remains of viscin threads near the pole, M.E.S.D. (v) *Dicolpopollis* sp. Polar view of odd dicolpate form, early Eocene, Alabama. (w) *Tricolpites* sp. Polar view, mid-Eocene, Wyoming. (x) *Bombacacidites nacimientoensis* (Anderson) Elsik. Polar view, M.E.S.D. (The colporate apertures are situated not at the angles of the amb – the more normal position – but midway between them; this is a normal apertural position for Bombacaceae.) (y) *Pistillipollenites mcgregorii* Rouse. S.E.M. micrograph (compare with photomicrograph in (z)), diameter 28 μm, early Eocene, Wyoming, characteristic gemmate sculpture. The grain is P03; Crepet & Daghlian (1981) showed this dispersed pollen form to have been produced by an extant plant belonging to the modern gentian family. (z) Same data as for (y). (aa) *Simpsonipollis mulleri* Kemp. Polar view, striate sculpture, N.E.R. (ab) *Myrtaceidites* sp. Polar view of Pcs form similar to extant Myrtaceae, N.E.R. (ac) *Cercidiphyllites* sp. Polar view, mid-Eocene, Wyoming. (ad) *Tricolpites reticulatus* Cookson. Polar view, N.E.R. (The generic name "*Tricolpites*" really is not very helpful, as a wide range of more or less Brevaxones tricolpate forms can be put in it – see (w) and (ae). Note also that, though this taxon is common in the Eocene, this specimen came from a Paleocene level at N.E.R.) (ae) *Tricolpites asperamarginis* McIntyre. Polar view of form with gaping colpi, N.E.R. (af) *Nuxpollenites crockettensis* Elsik. Equatorial view of form with extraordinarily thickened ektexine, mid-Eocene, Alabama. (ag) *Erdtmanipollis pachysandroides* Krutzsch. This displays the characteristic blocks of ektexine making up the surface reticulum, M.E.S.D. (ah) *Cupuliferoidaepollenites minutus* (Brenner) Singh. Equatorial views of small, more or less smooth tricolporoidate, probably fagalean forms that characterize many parts of the Cenozoic, early Eocene, Wyoming. (ai) *Rhoipites microluminus* Kemp. Equatorial view, N.E.R. (see also (an) and (ao)). (*Rhoipites* is a broad form-generic concept for such Pc3 forms, a very generalized dicot pollen type common throughout the Cenozoic.) (aj) *Myrtaceidites oceanicus* Kemp.

some characteristic Eocene sporomorphs of mid-latitude North America and elsewhere. At this level, tricolporates of more or less tropical affinity dominate, such as pollen referable to Nyssaceae, Cornaceae, Tiliaceae, and Bombacaceae, plus many sorts of Postnormapolles, palm pollen, bisaccate, and inaperturate conifer pollen. Megaspores continued to be produced by Paleogene lycopods and heterosporous ferns, as they are today, and they are found in Paleogene sediments. For example, *Minerisporites* was produced by isöetaleans, and megaspores of various species of the still extant heterosporous fern *Azolla* (Salviniaceae) also occur and have been described (see Collinson 1980, Collinson *et al.* 1985).

Nothofagidites

In the Southern Hemisphere a characteristic constituent of late Cretaceous and Cenozoic palynofloras is very distinctive stephanocolpate pollen related to the modern southern beech, *Nothofagus*. This fossil pollen, at least in the Paleogene, is usually referred to the form-genus, *Nothofagidites*. Figure 14.4 illustrates significant forms.

Oligocene palynofloras

Oligocene palynofloras include forms carried over from the Eocene, such as nyssoid pollen, many triporates, and many forms referable to modern families but living far out of the modern range of those families. The cooler circumstances of Oligocene time are reflected in the first abundant appearance of grass and grass-like pollen (the first reliable record of grass pollen is Paleocene) and of composite pollen (Muller 1981). Figure 14.5 illustrates some characteristic Oligocene spores/pollen.

Fungal spores in the Paleogene

The fungi have a long fossil record, probably to the Precambrian Bittersprings Limestone Chert, Australia, about one billion years old (Schopf 1968, Schopf & Blacic 1971). A maceration-resistant assemblage of probable ascomycete fungi, consisting of both spores and hyphae, has been documented from the Silurian (Ludlow) of Sweden (Sherwood-Pike & Gray 1985). Fungi were certainly well-established by Carboniferous time: many and well-preserved fungal remains have been reported from Pennsylvanian coal balls, e.g., by Stubblefield *et al.* (1983). *Microsporonites cacheutensis* Jain, from Permian and Triassic rocks (Jain 1968, Ecke 1984) may very well be a chitinous-walled

---

Polar view of Pcs pollen with heavy verrucate sculpture, N.E.R. (ak) *Gothanipollis* cf. *gothanii* Krutzsch. Polar view, an oddly syncolporate form. S.E.M. micrograph for comparison with photomicrograph in (al), diameter 24 μm, N.E.R. (al) Photomicrograph, same data as for (ak). (am) *Proteacidites* cf. *symphonemoides* Cookson. Polar view, N.E.R. (an) *Rhoipites grandis* Kemp. Equatorial view for comparison with (ao), N.E.R. (ao) Same data as (an), S.E.M. micrograph, length 45 μm; (ap) *Boehlensipollis* sp. Polar view of Pcs with very small polar area, middle Eocene, Wyoming.

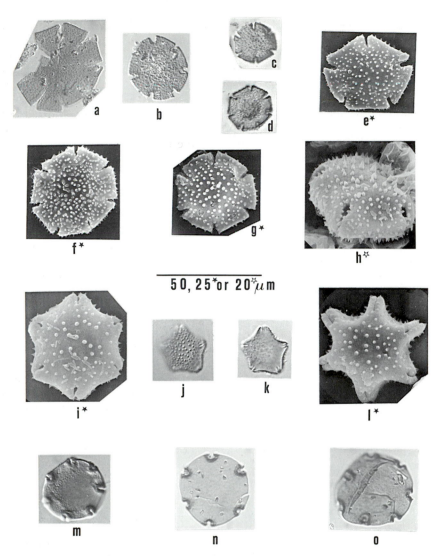

50, 25*or 20*μm

Figure 14.4 *Nothofagidites*, an important constituent of Southern Hemisphere latest Cretaceous and Cenozoic palynofloras. This pollen type is clearly represented in the extant flora by the southern beech, *Nothofagus*. The pollen of *Nothofagus* is very distinctive and diagnostic, but the fossils are usually referred to the form-genus, mostly on the general and debatable principle held by most that fossil material should never be placed in an extant genus. Nevertheless, the range of *Nothofagidites* is assumed to indicate that the genus *Nothofagus* extends back to the Santonian (late Cretaceous) (Muller 1981). *Nothofagus* (and *Nothofagidites*) pollen is referable to three groups or types (Cookson 1959): (1) *menziesii*-type (a), with unrimmed, usually gaping colpi; (2) *brassi*-type (b)–(l), relatively small, more or less angular amb, the colpus margins firm though unthickened (it is this type that extends back to Santonian; the other two types begin in the Maestrichtian); (3) *fusca*-type (m)–(o), rimmed apertures, actually colpate but resembling pores. Magnification shown by bar under (g). The specimens come from:

298

fungal spore. Nevertheless, robust chitinous-walled fungal disseminules, separable from the substrate rock by maceration, are not regularly encountered until late Jurassic (Elsik personal communication 1982). They become somewhat more abundant during early Cretaceous time and are first really abundant and diverse by late Cretaceous and Paleogene time. (The assertion by a few palynologists (Loquin 1981) that early Paleozoic chitinozoans are fungal remains is very unlikely to be true. The links of chitinozoans with the animal kingdom are more convincing.) This rise of chitinous-walled fungal parts is so coincident with the origin and rise to dominance of the angiosperms that it is logical to wonder if there might not be a connection, such as adaptation to requirements of parasitizing, digesting or otherwise making a living from the flowering plants.

Fungi can and do live in all sorts of substances, including, for example, palynological samples. When I first worked on a sample of Brandon Lignite in 1947, the sample was one that had long been stored in the Harvard museum. The palynological residue was rich in a variety of fungal spores. However, freshly collected samples obtained the next summer yielded relatively few fungal spores. However, Elsik (1976) and others have pointed out that this fact does not at all preclude the stratigraphic use of fungal spores. It is possible to prevent fungal growth in samples by storing them in alcohol or some other fungiostatic substance, or by storing them frozen. In my experience, hardly any palynologists routinely do this. More importantly, however, the fungi, especially ascomycetes, produce many kinds of very distinctive spores and spore-like bodies, and they have evolved rapidly during the Cenozoic, with many forms useful as marking boundaries such as the top of the Paleogene. They cannot be confused with extant fungi. The Cenozoic range of some species of these genera is potentially very useful stratigraphically. Modern

---

sub-basaltic sediments at Bungonia, New South Wales, Australia, early Eocene (N.S.W.); Ulgnamba Lignite, Hale River Basin, central Australia, middle to late Eocene (C.A.); Ross Sea, Antarctica, recycled specimens of uncertain age (R.S.). All except (h) are polar views. (a) *Nothofagidites asperus* (Cookson) Stover & Evans, R.S. (b) *Nothofagidites lachlanae* (Couper) Truswell. 7-Colpate form with bluntly echinate sculpture, interference contrast, R.S. (c), (d) *Nothofagidites* spp. Two gold-coated specimens prepared for S.E.M., interference contrast, showing that light microscopy is possible on such specimens without removal of gold, N.S.W.; (c) same specimen as (g); (d) same specimen as (f). (e) *Nothofagidites emarcidus* (Cookson) Harris. S.E.M. micrograph, a 5-colpate specimen showing the broad-based spines, N.S.W. (see also (g), (j) and (k)). (f) *Nothofagidites vansteenisi* (Cookson) Stover & Evans. S.E.M. micrograph, a 7-colpate specimen, N.S.W. (g) *Nothofagidites emarcidus* (Cookson) Harris. S.E.M. micrograph, a 6-colpate form. (h) *Nothofagidites* sp. Probably *N. emarcidus*. S.E.M. micrograph, obliquely equatorial view, showing at higher magnification the broad-based biform echinae and the relatively simple colpal structure, C.A. (i) *Nothofagidites falcatus* (Cookson) Stover & Evans. S.E.M. micrograph, C.A. (j), (k) *Nothofagidites emarcidus* (Cookson) Harris. Interference contrast, two levels of focus (see (e) and (g)), C.A. (l) *Nothofagidites falcatus* (Cookson) Stover & Evans. S.E.M. micrograph of form with protruding colpal structures, C.A. (m) *Nothofagidites* sp. *Fusca*-type with pore-like colpi, N.S.W. (n) *Nothofagidites* sp. Corroded specimen, 6-pored form, R.S. (o) cf. *Nothofagidites flemingii* (Couper) Potonié. Shows folds developed by thin-walled specimens, R.S.

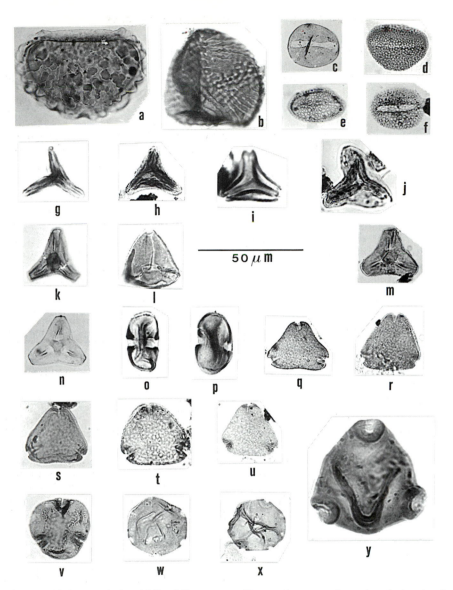

Figure 14.5 Typical middle Oligocene pollen and spores from boreholes in the Bristol Channel, England. By Oligocene time most sporomorphs are very similar to spores or pollen of extant genera and can with some degree of confidence be referred to extant families. Nevertheless, most paleopalynologists use form-generic names for Oligocene sporomorphs. Usually specific names are used with the form-generic names, but opinion is divided as to the usefulness of such specific names in the Oligocene. Most of these photos are from Boulter & Craig (1979). Boulter & Craig did not use specific names for these forms, only giving form-generic references. Magnification indicated by bar under (i). (a) *Polypodiidites* Ross. Monolete spore, lateral view. (b) *Cicatricosisporites* Potonié & Gelletich. Trilete spore, obliquely proximal view. (c) *Monocolpopollenites* Pflug & Thomson. Monosulcate pollen, distal view. (d) *Arecipites* Wodehouse. Monosulcate pollen, proximal view, mid-focus. (e) *Arecipites* (see (d)). Proximal view,

300

contaminating fungal spores in, say, an Eocene sample are only an annoyance. It is true, however, that fungal spores are a very abundant constituent of the palynomorph load in the atmosphere. *Alternaria*-type spores (Fungi Imperfecti) are an important pollinosis vector, and various fungal spores *in toto* may in some situations be as abundant as embryophytic spores/pollen encountered in spore/pollen traps used by aerobiologists (see Ogden *et al*. 1974). The presence of fungal spores in Pleistocene samples may be an indication that the sample is no longer good for radiocarbon dating of bulk sediment (Barnosky personal communication 1986). For this reason, such samples should be frozen after collection.

Figure 14.6 gives a classification of fungal palynomorphs, based on the principal morphologic features of fungal spores and hyphae, permitting their description and separation. The classification is a simplified version of that of Elsik (1979), the nestor of fungal palynology (see Fig. 14.7). It must be emphasized that "spore" is a very general term that palynologists have come to think of as representing only the meiotically derived spores of the embryophytic plants. Fungal "spores" are many sorts of propagules: unicellular, multicellular; sexual, asexual. The chitinous–walled fungal propagules we encounter as palynomorphs are prevailingly produced by ascomycetes and imperfect fungi, although basidiomycetes are also significant. Uncarbonized fungal spores in recent sediments are usually a darker color than sporopolleninous exines from the same sample. This color results from a melanin pigment (Elsik personal communication). However, fresh fungal spore walls can also be colorless (see Fig. 18.1).

Figure 14.8 illustrates some of the important Paleogene fungal forms. The multicellular fructifications (= "fruit bodies") and the germlings of the epiphyllous family Microthyriaceae (Ascomycetes) are especially characteristic

---

high focus. (f) *Arecipites* (see (d)). Distal view, high focus. (g) *Boehlensipollis* Krutzsch. Polar view of this oddly tricolpate, nearly syncolpate pollen grain. (h) *Boehlensipollis* (see (g)). A form that is apparently fully syncolpate. (i) *Boehlensipollis* (see (g)). A form with kyrtome-like structures. (j) *Boehlensipollis* (see (g)). A more robust form. (k) *Gothanipollis* Krutzsch. Polar view of tricolpate pollen with weakly developed apertures. The pollen is most characterized by the "trilete" ridge structure, and by the "polar pad" showing as a dark oval in center of photo. (l) *Boehlensipollis* (see (g)). A less angular form. (m) *Gothanipollis* (see (k)). (n) *Gothanipollis* (see (k)). A form in which the "polar pad" is three-pronged. (o) *Mediocolpopollis* Krutzsch. Tricolporate pollen in which the exopores and endopores are much more prominent than the colpi, prominent costal thickenings, equatorial view. (p) *Mediocolpopollis* (see (o)). Equatorial view showing the sinuous costal thickenings. (q) *Porocolpopollenites* Pflug. A tricolporate form with very short colpi and therefore large polar areas, polar view. Some extant members of the extant family Symplocaceae make very similar pollen. (r)–(u) *Porocolpopollenites* (see (q)). (v) *Tiliaepollenites* Potonié. A tricolporate form in which the costal thickenings give the grain a "padded triporate" appearance in polar views such as this. Virtually identical with extant *Tilia* pollen. (w) *Polyatriopollenites* Pflug. A 6-stephanoporate form in polar view, each pore provided with an annular thickening. Practically identical pollen is produced by the extant genus *Pterocarya* (Juglandaceae). (x) *Polyatriopollenites* (see (w)). A seven-pored specimen. (y) *Corsinipollenites* Nakoman. With "automobile tire"-like thickenings around the pore structures, polar view. Virtually identical pollen is made by members of the extant Onagraceae.

301

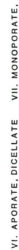

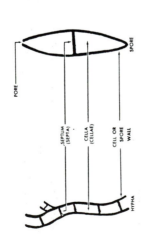

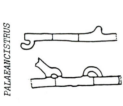
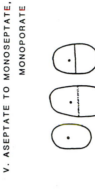

SPORES & FRUIT-BODIES: I–XVII

BASIC MORPHOLOGY

PORE
SEPTUM (SEPTA)
CELLA (CELLAE)
CELL OR SPORE WALL
SPORE
HYPHA

HYPHAE

PALAEANCISTRUS

I. APORATE, ASEPTATE

INAPERTISPORITES

II. MONOAPERTURATE, ASEPTATE

PORTALITES

III. DIAPERTURATE, ASEPTATE

FOVEODIPORITES

IV. ASEPTATE, APERTURE NUMBER VARIABLE

EXESISPORITES

V. ASEPTATE TO MONOSEPTATE, MONOPORATE

DICELLAESPORITES

VI. APORATE, DICELLATE

FUSIFORMISPORITES

VII. MONOPORATE, DICELLATE

DIDYMOPORISPORONITES

VIII. DIAPERTURATE, DICELLATE

DYADOSPORITES

IX. TRICELLATE

DIPOROPOLLIS

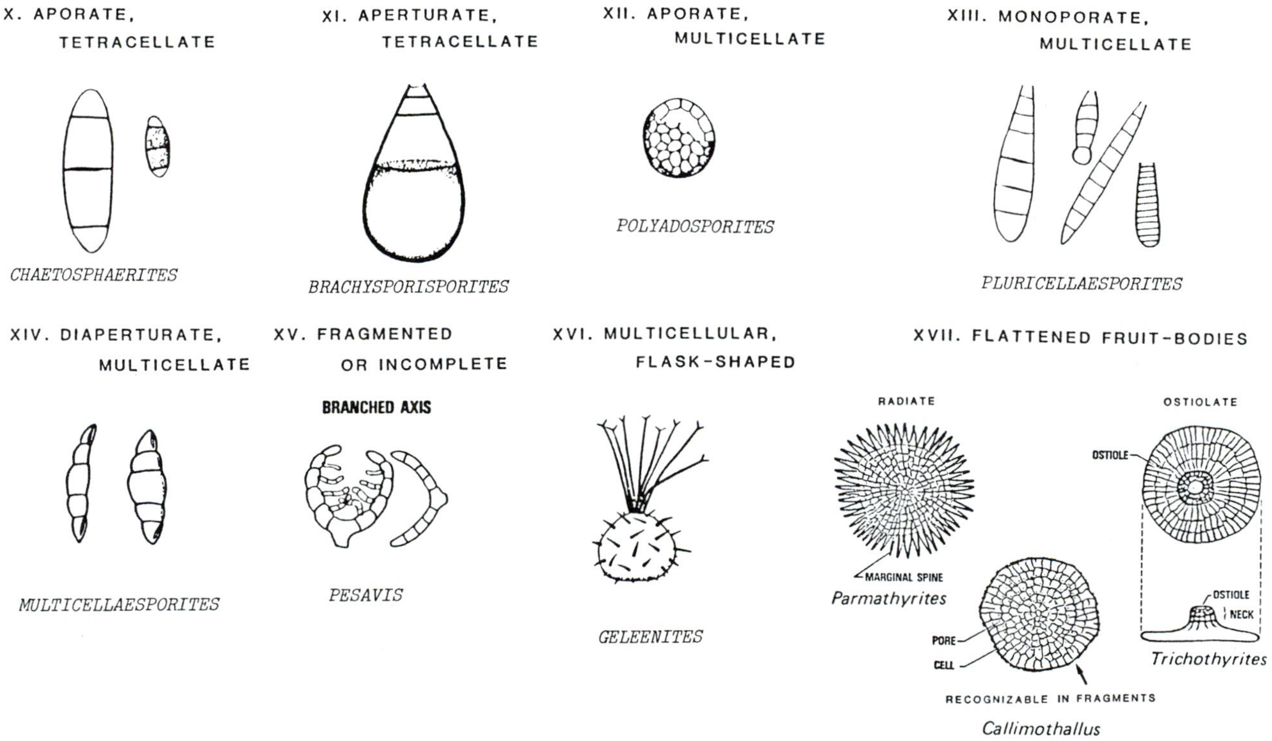

Figure 14.6 Simplified Elsik classification of fungal palynomorphs. Only one example is shown for each large class. In the expanded version of the classification, each class is subdivided into a number of sub-units, based on smaller morphological features, and on sculpture. Compare with Figure 14.8 for photomicrographs of fossil representatives of some of the classes.

Figure 14.7 William C. Elsik, born 1935, in his laboratory at Exxon Co., Exploration Department, Houston, Texas, 1985. Elsik more than any other person has been responsible for promoting the systematic study and stratigraphic use of fossil fungal spores, fruiting bodies, and hyphae.

of some Paleogene sediments. Fossil fungal spores are often referred to artificial form-genera, the names of which stress the number of pores and/or cells, or chambers (they usually communicate by holes in the septa between chambers): *Diporicellaesporites, Monoporisporites, Dicellaesporites*, etc. The pores (see Fig. 14.6) at the ends of fungal spores may be either true exit pores or traces of the attachment point of the spore to the producing organ. (Ascospores are formed free in an ascus and lack attachment scars; most other fungal spores have them.) "Germinals", more or less random openings on the sides of fungal spores, may be produced by biosolution from within the wall, and there may be one or several such "germinals", in addition to the fixed pore. Whereas pores are often accentuated by processing, septa are sometimes lost in fossilization or during laboratory processing (Elsik 1979). For paleopalynological purposes, an observable opening in a fungal spore is a "pore" regardless of original biological function. Most fossil fungal spores are inaperturate or monoporate. (Fungal hyphae and mycelia occur abundantly in palynological macerations of Cenozoic sediments, especially those of deltaic origin.) The maceration process for fungal spores is the same as for

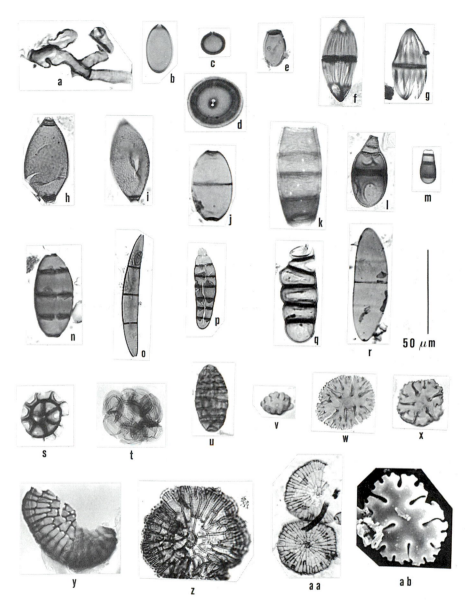

Figure 14.8 Fossil fungal spores, fruiting bodies, and hyphae. Although the kingdom Fungi has a fossil record back to the Proterozoic, chitinous-walled, "robust", fungal remains that survive maceration are not abundant until the Cretaceous. They especially characterize the Cenozoic, from which these specimens were obtained. Compare with Figure 14.6 for the abbreviated Elsik classification according to which these fossils are mostly arranged. All specimens except (b), (c), (f), (g), (k), (m), (p), (s), (t), (x), and (y) come from Eocene–Oligocene levels of Deep Sea Drilling Project Site 254, Ninetyeast Ridge, Indian Ocean. (b) (c), etc., are also Cenozoic, as separately noted. (a) Branching, septate hyphae. (b) *Polyporisporites* sp. Despite the name, a monoporate form, middle Eocene, Arkansas. (c) *Portalites* sp. Monoporate, middle Eocene, Arkansas. (d)

305

sporopolleninous sporomorphs. Fungal spores seem in my experience to be a little more resistant to geothermal alteration and to oxidation than are sporopolleninous fossils. They occur reasonably well-preserved in some Franciscan Melange samples from California, in which most of the embryophyte spores and dinocysts are badly carbonized. However, Elsik (personal communication) says that fungal spores may be either more or less resistant to both carbonization and oxidation, depending on as yet poorly understood factors. He also says that fungal spores do not change gradually in color as carbonization progresses but instead become suddenly black. They take biological stains only if they have been oxidized.

Muller (1959) noted some time ago the abundance of fungal spores in sediments of the Orinoco River Delta, and I have observed in my research on recent sediments a similar phenomenon in Gulf Coast U.S.A. deltas. Fungal spores are abundant in sediments in which organic matter (such as wood fragments, cuticles, and other tissue pieces) abound, presumably as a reflection of saprophytic fungi at work. Deltaic sediments provide such substrates. The study of Paleogene and Neogene fungal spores deserves much more attention, as studies by palynologists such as Smith (1978) in England, Ediger (1981) in Turkey, and Germeraad (1979) in Jamaica have shown. Smith comments on the obviously difficult nomenclatural problem with (usually multicellular) fungal spores ("fruit bodies") because of their great variability in form.

Exesisporites sp. Pore number variable (1–2), Paleogene, Arkansas. (e) *Monoporisporites abruptus* Sheffy & Dilcher. Monoporate. (f) *Fusiformisporites* sp. Aporate, dicellate, Eocene, Texas. (g) *Fusiformisporites crabbii* Rouse. Aporate, dicellate, Paleogene, Texas. (h) *Foveodiporites anklesvarensis* Varma & Rawat. Diporate. (i) Same as (h). Different specimen, interference contrast photo emphasizing dense internal granulation of wall. (j) *Dyadosporites* sp. Diporate, dicellate. (k) *Diporicellaesporites* sp. High focus showing transverse sculptural striae of the wall, mid–Eocene, Arkansas. (l) *Brachysporisporites pyriformis* Lange & Smith. Monoporate, tricellate. (m) *Brachysporisporites* sp. See (l), Eocene, Texas. (n) *Diporicellaesporites* sp. Shows flaps of ruptured septa. (o) *Diporicellaesporites* sp. (p) *Pluricellaesporites* sp. Monoporate, multicellate, Paleogene, Arkansas. (q) ?*Pluricellaesporites* sp. A phragmospore (one with two or more transverse septa) of uncertain relationship. (r) A tetracellate spore, non-porate, dicellate but with "shadow bands" suggestive of incipient or incomplete additional septa. (s) *Involutisporonites* sp. Spiral form, multicellate, mid–Eocene, Arkansas. (t) *Polyadosporites* sp. Aporate, multicellate, Paleogene, Arkansas. (u) *Dictyosporites* sp. Multiple cells in irregular series. (v) *Desmidiospora* type. Invaginations and central hyaline spot characteristic (see (w)–(ab)). (w) Same as (v). More developed stage (see (v), and (x)–(ab)). (x) *Desmidiospora* type. As (v), (w), late Eocene, Texas. (y) *Callimothallus pertusis* Dilcher. Microthyriaceous flattened fruit body, recognizable even from fragments. Note pores in the individual cells. (z) *Paramicrothallites* sp. Microthyriaceous, flattened fruit body. (aa) *Paramicrothallites* sp. (see (z). (ab) *Desmidiospora* type. S.E.M.

TAXONOMY AND NOMENCLATURE OF PALEOGENE SPORES/POLLEN

An outsider to our field might expect that in the "Cenophytic"–Cenozoic the difficult problems of classifying and naming fossil sporomorphs would become somewhat easier because the fossils belong prevailingly to the Angiospermae and to conifers such as *Pinus*. On the contrary, however, the paleopalynological taxonomic problems in the Cenozoic are the worst of all, and this has always been true. The most important reasons for this perplexing situation are as follows.

Identity problems per se   Even extant angiosperm pollen is for the most part only identifiable to the genus (sometimes only to the family). For all we know, this is also true of, say, Devonian spores, but palynologists feel less compunction about frequent genus transfers, such as *Spelaeotriletes lepidophytus* (Kedo) Streel = *Retispora lepidophyta* (Kedo) Playford = *Hymenozonotriletes lepidophytus* Kedo in the Devonian, than they do about referring an Oligocene pollen grain to *Nyssa* today and some other genus tomorrow. For most palynologists, use of form-generic names is somehow less controversial than risky use of a perhaps incorrect modern generic name.

Scarcity of fossil angiosperm flowers   Another sort of identity problem with angiosperms is that only a relatively few fossil angiosperm flowers have been found, and therefore identification of fossil angiosperm pollen has so far depended almost entirely on comparison with modern reference material.

Mosaic evolution and problems of range   It has long been noted that the various organs of angiosperms have evolved at different rates; this is called "mosaic" evolution. Pollen resembling the pollen of an extant genus might.come from a plant that was otherwise quite different, though in the Cenozoic fossil flowers have usually yielded pollen not too unlike the pollen of modern relatives of the flower-producing plants (see Crepet 1979). On the other hand, Wing (1981) and Hickey & Wing (1983) reported early Cenozoic *Platycarya* pollen and other organs coming from plants with leaves not referable to *Platycarya* or in some cases even to its family, Juglandaceae. Potonié (1951, 1956b, 1975) long ago warned about this problem as part of his bill of particulars against use of extant generic names for fossil pollen.

It should also be noted that during the Cenozoic angiosperm genera were rearranging their distributions more than they were becoming extinct or evolving new forms. (It is important to remember in this connection that palynology really can dependably recognize only generic differences.) For example, Germeraad *et al.* (1968) show the ranges of taxa of Cenozoic palynomorphs used in practical stratigraphy by Royal Dutch/Shell, and it is clear that many forms have quite different ranges when the forms are studied in the Caribbean, West Africa, and the Indonesia area.

Philosophic objections Some object that, as the pollen grain is a haploid organ representing only the brief gametophytic segment of the life cycle, it is inappropriate to refer fossil pollen to extant genera, or to create species of such genera for fossil pollen. Some say this represents an unfair extension of the circumscription of extant taxa. Of course, bryologists have always based taxa on gametophytic–haploid information, and megafossil paleobotanists have for a century without controversy referred fossil leaves to extant genera, often as "leaf-species" thereof.

Potonié was the primary writer on these matters, and his prolific productivity of publications on the subject has substantially carried the day. It is largely forgotten that his earlier writings on the subject of what to do with Cenozoic spores/pollen were badly confused about the purpose and methods of botanical nomenclature. He long advocated what amounted to three parallel systems of classification and nomenclature:

(a) *Natural*, where reference to extant taxa is certain. The modern generic name could be used, but naming of new species based on fossil pollen was to be avoided. Example: *Pinus silvestris*.
(b) *Half-natural*, where reference to an extant taxon is suspected but not proven. A generic name is coined that purports to show the alleged relationship. Example: *Betulaceaepollenites*.
(c) *Artificial*, where the relationship is not known at all, and a form-generic name based on morphological features is created. Example: *Tetracolporites*.

Erdtman, not a paleopalynologist, had ideas very similar to Potonié's on these matters (see Potonié *et al.* 1950). Erdtman objected to the use of extant generic names for fossil pollen – at least, pre-Pleistocene. He perhaps was even the source of Potonié's ideas on this matter (Potonié said not). He published very few names but managed to confuse nomenclature by proposing simultaneous publication of "natural" and "half-natural" generic names for pollen; he even published a few such non-binomial names. Unfortunately, coworkers of Erdtman (e.g., Cookson 1947) actually did this also for a time, with confusing results.

However, all names of plant taxa, fossil or extant, are governed by one code, the *International Code of Botanical Nomenclature* ("ICBN") (Voss *et al.* 1983), and it provides a means of referring to the population concerned. It is not illegal to formulate privately certain rules for making up generic or specific names, but the rules have no standing in the *Code*. Under the present *Code*, "half-natural" names are just form-generic names, the same as "artificial" names, providing they are validly published. The real question is whether it is allowable to use modern generic names for fossil spores/pollen, or whether one must always use form-generic names. The question of how to *coin* the form-generic names and how to list them subsequently is all smokescreen. At the moment it is clear that those who favor use of only form-generic names in the Cenozoic are by far in the majority. For my part, I still feel that, where the generic reference is absolutely clear, there is no reason at all to avoid the extant generic name. For example, association with other organs makes it very clear that the *Nyssa* pollen in the Brandon Lignite described by Traverse (1955) could only be *Nyssa*, and that the transfer of these fossils to *Nyssapollenites* is a ridiculous exercise. However, I emphasize that my view is a minority view. Further, I would also draw the line somewhere. Because of the known time range of genera, I would hesitate to call a Jurassic fern spore *Aneimia*, even though it seems inseparable morphologically from that genus. If the megafossil evidence thoroughly supported the reference, however, I see no defensible reason to avoid calling the spore *Aneimia*! In Pleistocene palynology, extant names are almost always used, even if only to family or other suprageneric units. Of course, this is mostly because the pollen identifications can be made, but it is also true that the purpose of Pleistocene palynology is mostly paleoclimatological and paleoecological interpretation, for which "*Tricolporites* sp." is not of much help! An ecological study in the Paleogene, such as Collinson's (1983), makes sense of the pollen record because the megafossils are coordinately studied and referred mostly to extant plant taxa. Without the megafossil information, "Tricolporopollenites spp.", even with suggestions of natural affinity, would be rather unhelpful. However, as Boulter & Wilkinson (1977) have pointed out, great numbers of late Cretaceous to present angiosperm pollen consist of tricolpate, tricolporate, and triporate forms that are very difficult to distinguish consistently, and neither the application of modern plant names nor the application of form taxa is very useful. Boulter & Wilkinson suggested a non-Linnaean, non-binomial approach by using a grid system based on basic morphological type, size, and sculpture. Though this suggestion has considerable merit for dealing with such forms, it has not been widely adopted. Hughes (1970, 1975) has suggested broad-scale substitution of such a non-binomial classification approach to palynological systematics. Hughes' proposal, unlike Boulter & Wilkinson's, calls for the virtual abandonment of formal description of palynomorph taxa by traditional methods. Both the Boulter & Wilkinson and the Hughes suggestions represent practical approaches to dealing with the huge bulk of paleopalynological data on difficult-to-separate forms, particularly in the "Cenophytic". Such approaches will be more widely used if their authors avoid conflict with standard paleobotanical nomenclature and emphasize instead the practical purposes of the methods.

"CENOPHYTIC" (MORE OR LESS CENOMANIAN–PLEISTOCENE) *SPORAE DISPERSAE* BOTANICAL RELATIONSHIPS

As noted earlier in this chapter, the botanical affinity of "Cenophytic" spores/pollen (mostly angiosperm pollen) is still known mostly from comparison with modern reference pollen slides, because so few fossil flowers have as yet been studied. Thus, very curiously we at present know far more about the relationship of Carboniferous dispersed spores/pollen to the producing fructifications than we do about similar Cretaceous–Paleogene dispersed spores/pollen. (In the Neogene there is not really much doubt about the generic, or at least familial, reference of all the abundant and important

309

forms encountered – they are extant.) The "Cenophytic" begins with the arrival of undoubted angiosperm fossils in the record, both reticulate-veined leaves and more or less columellate pollen of angiosperm character. This event is time-transgressive, occurring earlier toward the Equator (about Barremian) than at middle latitudes (about Aptian) or high latitudes (about Cenomanian). Despite the angiospermous nature of much Cretaceous and Paleogene pollen, however, it is not possible to refer with certainty any pollen to extant angiosperm taxa smaller than order until latest Cretaceous. Although some angiospermous families are recognized in the Cretaceous, most are first identified in the Paleogene. Not until about 20 million years ago (Muller 1981, personal communication) do all angiosperms encountered belong to extant families. Beginning about 10 million years ago almost all angiosperms were referable to extant genera. Many very important Paleogene and Cretaceous angiosperm pollen forms are still enigmatic as to relationship, awaiting the discovery of flowers bearing them, e.g., *Aquilapollenites* and *Wodehouseia*. In the "Cenophytic", fern spores are usually referable to extant genera, or at least to extant families. Conifer and other gymnosperm pollen is often quite like extant genera, but cone studies have shown that this resemblance often reflects parallel evolution.

The following very short list presents data on "Cenophytic" spores/pollen whose botanical relationship has been proven paleobotanically. In many hundreds of other cases, e.g., *Nyssapollenites* and *Alangiopollis*, the botanical relationship is nearly certain but has not been proven by finding fossil spores/pollen-bearing organs. In many instances, the presumed botanical relationship shows in the name, as discussed under taxonomic considerations above. However, only in the relatively few instances in which flowers have been studied is the evidence absolutely certain. The complexity of the situation is illustrated by the work of Crepet & Daghlian (1980), in which Eocene castaneoid inflorescences yielded beautifully preserved pollen appearing identical to modern *Castanea*, though the generic assignment of the inflorescence is by no means so certain; this is an illustration of mosaic evolution. Another problem is that megafossil paleobotanists working in the Cenozoic are not especially interested in the referability of *in situ* pollen/spores from their inflorescences to taxa of dispersed sporomorphs. The relationship to modern forms is a more pressing problem. For example, Manchester & Crane (1983) describe in great detail the pollen of a fagaceous plant of Oligocene age, and compare the pollen with pollen of the genera of the modern beech–oak alliance. They do not, however, even mention whether the pollen if dispersed would be *Quercoidites, Cupuliferoipollenites* or something else. The literature includes hundreds of generic names for dispersed Cretaceous–Cenozoic spores/pollen, replete with taxonomic reference to extant genera or families. Much of this information is probably all right, but it is humbling to see from this list how little of it is backed up by megafossil evidence. Paleobotanists are at present making progress in the search for fossil flowers, and there will undoubtedly be much more information before many more years. In the meantime, the following short list is representative of how little we know from *in situ* evidence of "Cenophytic" *Sporae dispersae*.

I. SPORES

A. TRILETE

Deltoidospora

Deltoidospora spores were described from Eocene–Oligocene polypodiaceous ferns close to *Acrostichum* by Collinson (1978).

II. POLLEN

A. TRICOLPATE–TRICOLPORATE

Pistillipollenites

Pistillipollenites (brevicolpate) pollen was obtained from flowers of Paleocene–Eocene age, apparently referable to the family Gentianaceae, by Crepet & Daghlian (1981). This very interesting and important pollen form has been described in detail by Rouse & Srivastava (1970).

Tricolpites

Pollen similar to *Tricolpites minutus* (Brenner) Dettmann was found in staminate flowers of *Platanus*-like plants of the Lower Cretaceous (Potomac Group) of Maryland by Crane *et al.* (1986).

Tricolporites

Pollen referable to *Tricolporites* was obtained from *Actinocalyx* (probably ericalean) flowers of the Upper Cretaceous by Friis (1985b). Friis (1985a) also reported obtaining pollen referable to *Tricolporites* in *Scandianthus* flowers (probably saxifragalean).

B. TRIPORATE–STEPHANOPORATE

Momipites

Momipites pollen has been obtained by Crepet *et al.* (1975) from *Eokachyra*, a form-genus for Eocene juglandaceous catkins. Crepet *et al.* (1980) reported it also from *Eoengelhardia*, another Eocene juglandaceous catkin. Juglandaceous-like triporate pollen of this sort is also found in the literature under *Engelhardia* or *Engelhardtioipollenites*.

Extratriporopollenites

Extratriporopollenites pollen in the broad sense (which includes most Normapolles) have been found in Upper Cretaceous floral material of Sweden by Skarby (1981) and Friis (1981, 1985b). Friis (1984) feels that the flowers are referable to the Juglandales–Myricales alliance but probably not to an extant family. *Trudopollis*, in particular, was obtained from one fossil flower, *Manningia*, and *Plicapollis* from another, *Caryanthus*.

C. TETRADS

Ericipites

Ericipites-like pollen tetrads were described from Eocene mimosoid (Leguminosae) inflorescences by Crepet & Dilcher (1977).

Monosulcate

Monocolpopollenites

Pollen referable to *Monocolpopollenites tranquillus* (Pot.) Thomson & Pflug was found in palm flowers by Schaarschmidt & Wilde (1986) in the Eocene of Messel, Germany.

Information is building up regarding "Cenophytic" (Cretaceous plus Cenozoic) fossil flowers and their pollen, but as yet not many of the data are applicable to dispersed fossil pollen.

CRETACEOUS–PRESENT SPORES/POLLEN RANGES

Unfortunately, relatively little is available in the literature regarding spores/pollen ranges in the "Cenophytic". Some broad generalizations have already been noted and can be seen in earlier figures: first appearance of tricolpate pollen in Neocomian–Cenomanian (depending on latitude), first Normapolles in Cenomanian (southern Laurasia), disappearance of many *Aquilapollenites* spp. at the end of the Maestrichtian (and most of the rest of "Aquila" at the end of the early Paleocene), end of Normapolles in late Eocene to early Oligocene, etc. It should be emphasized, however, that the spores/pollen correlations in the Cenozoic are dependent on local climatic conditions more than on well-documented, widespread episodes of extinction (Hochuli personal communication 1981). For example, Figure 14.9 shows that the last appearance of Normapolles (= *Plicapollis*) and the first appearance of *Tsuga* (= *Zonalapollenites*) are rather different in different parts of Europe, and this perhaps shows a gradual adjustment of vegetation to changing climates. Figure 14.9c also shows some Paleocene pollen zones from western North America. The zones shown are Oppel zones, determined by tops (ends) and bottoms (beginnings) of some pollen taxa. (Concurrent range zones are very similar but emphasize the mutual occurrence in each zone of a specified suite of fossils.) Figure 14.10 shows some stratigraphic zones for the Cenozoic of Australia. The zones displayed are assemblage zones, of which a particular taxon is selected as typical. Note that the appearance of *Acacia* and *Eucalyptus* pollen, supertypical of modern Australia, is a Neogene phenomenon.

PALEOGENE–NEOGENE DINOFLAGELLATES

Dinoflagellates continued rapid diversification in the Paleogene and are hence very important stratigraphically, especially in marine sediments. As always, in some, especially nearshore, marine sediments, they provide an invaluable link between the critical marine animal fossil stratigraphic indicators and spores/pollen from land. (There are also freshwater dinoflagellate cysts, but they are relatively spotty in occurrence.) A number of dinoflagellate cyst taxa terminate at or near the end of the Cretaceous (*Dorocysta, Triblastula, Dinogymnium*, etc.), but many other taxa "sail" unabashedly across the boundary (*Oligosphaeridium, Cordosphaeridium, Leptodinium*, etc.), and it is evident that the terminal

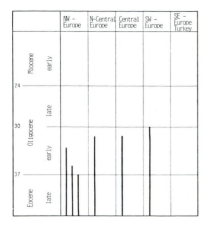

a. Normapolles: latest occurrence

in Paleogene

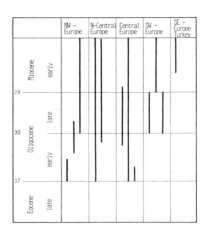

b. Zonalasporites spp. (Tsuga) :

first occurrence in Paleogene

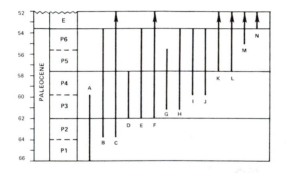

c. Paleocene pollen zones, western North America

Figure 14.9 Examples of Cenozoic pollen ranges in the Northern Hemisphere. (a) Normapolles "tops" (last occurrences). The extra bars for northwest Europe represent data from different sources. Clearly, however, the top is early Oligocene. (b) *Zonalasporites* (= *Tsuga*) "bottoms" (first occurrences). As in (a), the multiple bars in most segments represent data from different sources. It is evident that *Tsuga* moved into northwest Europe earlier than into southern Europe. (c) Pollen zones for the Paleocene of western North America based on juglandaceous pollen. The zones are mostly based on first ("bottoms") and last ("tops") occurrences of included taxa. Zones so delimited are referred to in a general sense as Oppel zones (similar to concurrent range zones). The taxa are: A, *Momipites leffingwellii* Nichols & Ott; B, *Momipites waltmanensis* Nichols & Ott; C, *Momipites wyomingensis* Nichols & Ott; D, *Momipites actinus* Nichols & Ott; E, *Caryapollenites prodromus* Nichols & Ott; F, *Momipites anellus* Nichols & Ott; G, *Momipites triorbicularis* (Leffingwell) Nichols; H, *Momipites ventifluminis* Nichols & Ott; I, *Caryapollenites wodehousei* Nichols & Ott; J, *Caryapollenites imparalis* Nichols & Ott; K, *Caryapollenites veripites* (Wilson & Webster) Nichols & Ott; L, *Caryapollenites inelegans* Nichols & Ott; M, *Juglans-Pterocarya* type; N, *Platycarya platycaryoides* (Roche) Frederiksen & Christopher.

313

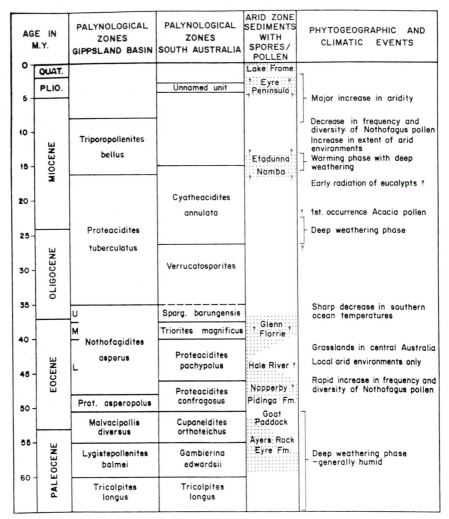

AGE IN M.Y.		PALYNOLOGICAL ZONES GIPPSLAND BASIN	PALYNOLOGICAL ZONES SOUTH AUSTRALIA	ARID ZONE SEDIMENTS WITH SPORES/ POLLEN	PHYTOGEOGRAPHIC AND CLIMATIC EVENTS
0		QUAT.		Lake Frome	
		PLIO.	Unnamed unit	? Eyre ? Peninsula ?	Major increase in aridity
5					
10	MIOCENE	Triporopollenites bellus			Decrease in frequency and diversity of Nothofagus pollen Increase in extent of arid environments
15				? Etadunna ? Namba ?	Warming phase with deep weathering
					Early radiation of eucalypts ?
20			Cyatheacidites annulata		1st. occurrence Acacia pollen
25	OLIGOCENE	Proteacidites tuberculatus		?	Deep weathering phase
30			Verrucatosporites		?
35		U	Sparg. barungensis		Sharp decrease in southern ocean temperatures
	EOCENE	M	Triorites magnificus	? Glenn ? Florrie	
40		Nothofagidites asperus			Grasslands in central Australia
		L	Proteacidites pachypolus	Hale River ?	Local arid environments only
45			Proteacidites confragosus	Napperby ? Pidinga Fm.	Rapid increase in frequency and diversity of Nothofagus pollen
50		Prot. asperopolus		Goat Paddock	
	PALEOCENE	Malvacipollis diversus	Cupaneidites orthoteichus		
55		Lygistepollenites balmei	Gambierina edwardsii	Ayers Rock Eyre Fm.	Deep weathering phase –generally humid
60		Tricolpites longus	Tricolpites longus		

Figure 14.10 Palynological stratigraphic zones for Cenozoic of Gippsland Basin (Victoria) and South Australia, Australia. This zonation dramatizes the very important point (see Fig. 14.9) that in the Cenozoic pollen/spore zonation applies only locally. Note early Oligocene cooling, Miocene warming, and the late Neogene major aridity increase, reflecting mostly Australia's northern drift toward the equator.

Cretaceous "event" was not as cruel to dinoflagellates as to the Coccolitho-phoridae. The picture seems to be more like that for pollen and spore taxa. One wonders if the sporopollenin coat which the cysts had in common with spores/pollen might have had something to do with their survival. Figure 14.11 illustrates some typical Paleogene dinoflagellate cysts. Very useful range charts for dinoflagellate cyst genera from Triassic to present have been published by Wilson & Clowes (1980). Various authors present range charts for Cenozoic taxa in Wrenn *et al.* (1986).

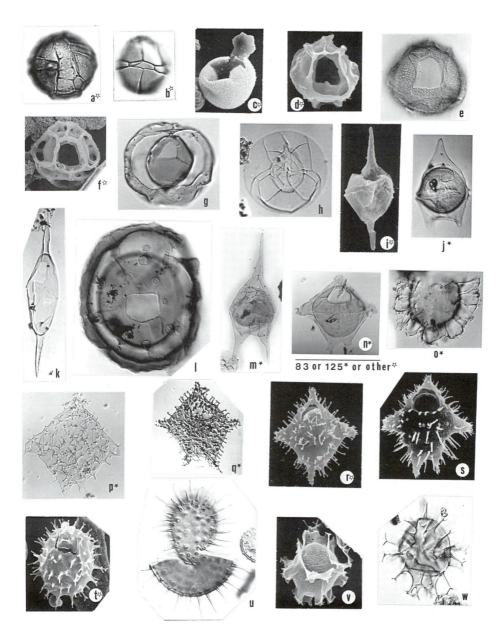

Figure 14.11 Some Cenozoic dinoflagellate cysts from North America and the eastern North Atlantic, in interference contrast light microscopy (I.C.) and S.E.M. The North Atlantic samples are from D.S.D.P. Leg 81, Rockall Plateau, and are designated "D.S.D.P." Magnification in micrometers shown by bar under (n). Sizes for specimens for which "other" is indicated are given in the caption. Note that for each specimen, not only the surface shown, but the aspect from which that surface is seen is indicated. This is a recommended practice for photos of dinoflagellate cysts, because a ventral surface seen from the ventral side is a mirror image of a ventral surface seen by focusing through from the dorsal side, and this confuses perception of plate arrangement. For

315

S.E.M., however, the view is only exterior, and this is not a problem. (a) *Impagidinium californiense* Damassa. Proximate cyst, I.C., ventral view of ventral surface, maximum dimension 43 μm, early–middle Eocene, Alaska. (b) *Impagidinium patulum* (Wall) Stover & Evitt. Proximate cyst, I.C., ventral view of dorsal surface, maximum dimension 70 μm, late Miocene, D.S.D.P. (c) *Kallosphaeridium brevibarbatum* De Coninck. Proximate cyst, S.E.M., oblique apical view showing the apical archeopyle with the multiplate operculum attached at the parasulcus, maximum dimension 41 μm, Paleocene–Eocene, Virginia. (d) *Pentadinium polypodum* Edwards. Proximochorate cyst, S.E.M., dorsal view showing clearly the precingular archeopyle, maximum dimension 65 μm, middle Eocene, Alabama. (e) *Pentadinium favatum* Edwards. Proximate cavate cyst, I.C., ventral view of dorsal surface, (see (d)), middle Eocene, Alabama. (f) *Impagidinium aquaductum* (Piasecki) Lentin & Williams. Proximate cyst, S.E.M., dorsal view showing precingular archeopyle and parasutural crests which form an open network reflecting original tabulation, maximum dimension 47 μm, middle Miocene, D.S.D.P. (g) *Invertocysta lacrymosa* Edwards. Strangely cavate cyst with the relatively dwarfed endocyst and the precingular archeopyle showing within, I.C., ventral view of dorsal surface, late Miocene, D.S.D.P. (h) *Invertocysta tabulata* Edwards. Cavate cyst as in (g), I.C., dorsal view of ventral surface, middle Miocene, D.S.D.P. (i) *Biconidinium longissimum* Islam. Proximate cyst with long apical and antapical horns, S.E.M., oblique ventral view, maximum dimension 96 μm, early Eocene, Virginia. (j) *Deflandrea phosphoritica* Eisenack, cavate cyst with intercalary archeopyle outline showing, I.C., dorsal view of dorsal surface, late Paleocene, Virginia. (k) *Palaeocystodinium golzowense* Alberti. Cavate cyst with the periphragm continuing into apical and antapical horns, intercalary plate comprising the operculum showing through from dorsal surface in this mid-focus ventral view, I.C., middle Miocene, D.S.D.P. (l) *Tuberculodinium vancampoae* (Rossignol) Wall. Odd cavate cyst with the antapical archeopyle showing in this antapical view, I.C., late Miocene, D.S.D.P. (m) *Ceratiopsis* sp. Proximate cavate cyst with thin-walled periphragm, I.C., ventral view of ventral surface, Paleocene, Alabama. (n) *Rhombodinium* sp. proximate cavate cyst with faint indications of tabulation showing, intercalary archeopyle, I.C., dorsal view of dorsal surface, lower Eocene, Virginia. (o) *Glaphrocysta* sp. Chorate cyst with processes united at the level of original theca, showing the apical archeopyle, involving several plates, I.C., dorsal view of ventral surface, Paleocene, Virginia. (p) *Gochtodinium* sp. Chorate cyst, intercalary archeopyle showing in this dorsal view of dorsal surface, I.C., early Oligocene, Alabama. (q) *Wetzeliella hampdenensis* Wilson. Characteristically angular chorate cyst with processes of medium length, the level of the grapnel ends indicating original thecal wall position, intercalary archeopyle showing, I.C., early to middle Eocene, Alaska. (r) *Wetzeliella* sp. (See information for (q).) Endocyst showing behind the partially detached operculum, dorsal view, S.E.M., maximum dimension 111 μm, early Eocene, Virginia. (s) *Wetzeliella varielongituda* Williams & Downie. Information basically as for (q), intercalary archeopyle in both pericyst and endocyst showing in dorsal view, S.E.M., early Eocene, Virginia. (t) *Apectodinium homomorphum* (Deflandre & Cookson) Lentin & Williams. Cyst with intercalary archeopyle, little or no other evidence of tabulation, S.E.M., maximum dimension 70 μm, late Paleocene–early Eocene, Virginia. (u) *Lingulodinium machaerophorum* (Deflandre & Cookson) Wall. Skolochorate cyst with long processes, with an archeopyle involving the entire epicyst attached at the sulcus, I.C., middle Miocene, D.S.D.P. (v) *Hafniasphaera goodmanii* Edwards. Skolochorate cyst, dorsal view showing the endophragm within, S.E.M., early Eocene, Maryland. (w) *Spiniferites mirabilis* (Rossignol) Sarjeant. Skolochorate cyst, a generic name introduced by Ehrenberg over a century ago, dorsal view of dorsal surface, I.C., late Miocene, D.S.D.P.

CHAPTER FIFTEEN
Neogene palynology

INTRODUCTION

The Neogene period began about 23 million years ago with the Miocene. As used in this book, it extends to present, consisting of Miocene, Pliocene and Pleistocene (which includes the present interglacial, sometimes called the "Holocene"). As usually defined, the Neogene consists of Miocene and Pliocene only, with the "Quaternary" (Pleistocene and "Holocene") being separate. Much of the middle part of the Miocene was warmer than any time since the Eocene, but the general temperature decline of the Cenozoic was reasserted about 15 million years ago. The expansion of temperate deciduous trees, grasses, composites, and other herbaceous dicots, and of conifers at high altitudes and latitudes, that began about then at middle latitudes became the signature of the later Neogene. By 20 million years ago practically all angiosperm remains, including pollen, are referable to extant families, and this datum ties in (mnemonically at least) with the beginning of the Neogene (about 24 million years ago). About 10 million years ago, the level of close to 100% extant genera was attained (see Fig. 15.1). This time also marks the formation of major Antarctic ice (glaciation may have commenced as early as 35 million yr B.P.) and the initiation in America and Eurasia of widespread steppe vegetation, dominated by grasses and by shrubby composites and chenopods. I informally refer to this 10-million-year period as the "Ultimogene" (Traverse 1982). Its inception is based on the presence of practically 100% extant plant genera. We recognize as fossil pollen primarily genera, not species, of angiosperms. (In some cases, as grasses and sedges, we tend to identify only the *family*). Thus, although the level of almost 100% extant species is not attained (as recognized from megafossils by Reid (1920); see Fig. 15.1) until about the beginning of the Pleistocene (earlier in some locations; see Leopold 1967), about 1.8 million years ago, the techniques applied in Pleistocene palynological studies are also frequently applicable in the "Ultimogene" because the studies are really based on generic identification. The use of the data is primarily for paleoclimatic and paleoecological reconstruction. As one moves back in the pre-"Ultimogene", this procedure is riskier, because the assumption of paleoecological significance of pollen data is less supportable the greater the proportion of extinct and exotic forms. Thus, the "Ultimogene" is really palynologically very different from pre-"Ultimogene", and it is not surprising that Pleistocene palynology (especially present interglacial Holocene = "post-glacial" = Flandrian palynology) has always been quite different in approach from study of older sediments. The year 1916 is often reckoned as the beginning of palynology, with the publication of Von

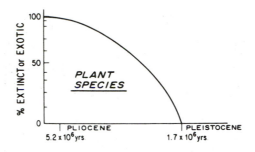

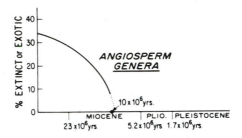

Figure 15.1 Percentage of extinct or exotic plant *species* in late Neogene time and of extinct or exotic angiosperm *genera* in the late Paleogene and the Neogene of North America. The data are all derived from megafossil floras. Jan Muller (in Traverse 1982) put the level of 100% modern families at about 20 million years ago. As pollen can routinely be recognized only to genus of extant plants, this means that the same sorts of interpretations can be made from pollen analytical data for sediments 10 million years old as for those only 10 thousand years old.

Post's (1916) post-glacial studies. The truth is that since that time "pollen analysis" (or "pollen statistics"), as this branch of palynology has sometimes been called, has always steered an independent course from "paleopalynology proper" (pre-Pleistocene palynology), and has been more connected to plant ecology, vegetational history, archeology, and paleoclimatology than to geology. "Pollen analysts" in this sense tend to be basically botanists whereas most, especially applied, pre-Pleistocene palynologists tend to be geologically oriented. I am saying here that the normal boundary between the two approaches is really best set at about 10 million years ago with the beginning of the "Ultimogene" (about 100% modern plant genera), not at 1.8 million years ago with the beginning of the Pleistocene, or at 10,000 years ago with the inception of the "Holocene" or present interglacial. (It is sometimes useful to group the Paleocene and Eocene informally. The Oligocene and the early and middle Miocene, up to the onset of major cooling about 10 million years ago, can also be informally grouped for discussion of floral evolution.)

One dramatic demonstration of the profound effect of late Cenozoic cooling is in western North America, where continuing orogeny and resultant rain-shadow development and other changes in climate were added to worldwide depression of temperature. Figure 15.2 shows such a case, the present extent of

318

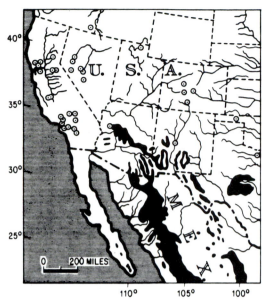

Figure 15.2 Present-day "Sierra Madrean" woodland (blackened areas) vs. records of related Tertiary vegetation (circles) in southwestern North America. This retreat is a characteristic result of cooling in the Neogene in higher latitudes, plus orogenic effects. Axelrod & Raven (1985) present more information on the Madrean flora.

"Sierra Madrean" woodland, as contrasted with much more northerly records from earlier in the Cenozoic.

Some characteristic Mio-Pliocene spores/pollen from England are displayed in Figure 15.3. All of these forms are still encountered in various parts of the Northern Hemisphere, but Pleistocene glaciation has eradicated many of the taxa from present-day Europe. Note also the use of form-generic names. In my opinion this is quite unnecessary, and even not helpful, in the "Ultimogene", for reasons already explained. (Palynologists working with Miocene–Pliocene floras customarily use many such names: see the profusely illustrated work of Nagy 1985.)

PALYNOLOGICALLY SIGNIFICANT STRATIGRAPHIC BOUNDARIES IN THE "ULTIMOGENE"

The boundaries between the Miocene and the Pliocene and between the Pliocene and Pleistocene have to be established on the basis of extension to other parts of the world of data from the type-section areas in France and Italy. Oxygen-isotope ($\delta^{18}O$) data, radiometric dating, and magnetostratigraphy extended from marine cores to type sections, plus micropaleontological data (foraminifera and nannofossils, mostly), have helped make this possible. It is a stratigraphic, not a conceptual, problem. The Miocene/Pliocene boundary has

319

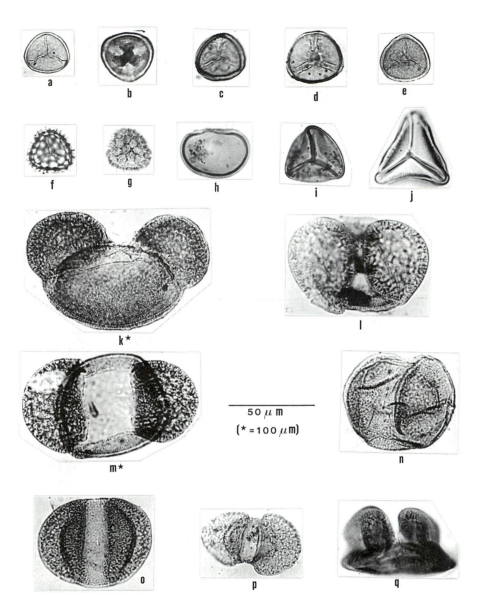

Figure 15.3 Some characteristic late Miocene to Pliocene spores/pollen. These specimens are from clays in limestone sinkholes in Derbyshire, England. Similar sporomorphs occur in mid to late Neogene sediments all over the Northern Hemisphere. As explained in the text, forms such as these, which are about five million years old, can be assumed to belong to extant genera. It does not therefore seem to me that the obligate use of form-generic names makes much sense, but it is conventional with most pre-Quaternary palynologists even in the Neogene. Magnification indicated by bars between (m) and (n) and under (y). (a) *Stereisporites stereoides* (Potonié & Venkatachala) Pflug (see also (b)–(e)). Proximal view. Spores of this form-genus are common in some Cenozoic sediments. These are derived from the *Sphagnum* sort of

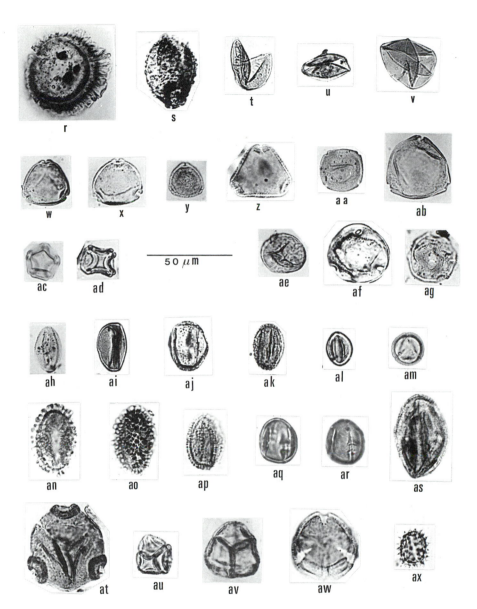

moss, being one of the relatively few bryophyte spores with enough sporopollenin to be preserved. (b) *Stereisporites wehningensis* Krutzsch. Proximal view. The laesura has thick labiae. (c) *Stereisporites germanicus* Krutzsch. Proximal view. (d) *Stereisporites granistereoides* Krutzsch. Proximal view. A form with well-marked laesural labiae. (e) *Stereisporites magnoides* Krutzsch. Proximal view. (f) *Lycopodium* sp. Proximal view, high focus, showing somewhat sinuous laesura. The extant generic name is used for such Cenozoic spores more commonly than are other extant generic names, perhaps from the unproven view that the genus has not rapidly evolved in the Cenozoic, and because reference to *Lycopodium* seems very sure. (g) Same as (f). Distal view. (h) *Laevigatosporites haardtii* (Potonié & Venitz) Thomson & Pflug. Lateral view. Such Sa0

321

fern spores have a Paleozoic to present range! (i) *Leiotriletes wolffii* Krutzsch. Proximal view, as (h). A kind of fern spore having a not very distinctive form. (j) *Gleicheniidites senonicus* Ross. Proximal view. A distinctive fern spore which should eventually be referable to a particular extant genus. (k) *Abies* sp. Lateral view. Conifer pollen is often dominant in Miocene–Pliocene sediments of the Northern Hemisphere. (l) *Pinus* sp. ("sylvestris type"). Distal view. (m) *Keteleeria* sp. Distal view. (n) *Picea* sp. Distal view. (o) *Pinus* sp. ("haploxylon type"). Distal view. (p) "*Podocarpoidites libellus*" R. Potonié. Distal view, mid-focus. Podocarpaceous conifers are now confined to the Southern Hemisphere. (q) *Cedrus* sp. Lateral view. Distinction of *Cedrus* and *Pinus* pollen in sediments containing both is very difficult. (r) *Tsuga* sp. (diversifolia section). Shows that this extant conifer pollen is Pv1. (s) *Sciadopitys* sp. Shows the peculiar hollow verrucae. This taxodiaceous genus is now confined to east Asia but in the Miocene-Pliocene was widely distributed in the Northern Hemisphere. (t) *Inaperturopollenites hiatus* (Potonié) Thomson & Pflug. The clamshell-like opening is characteristic of extant *Taxodium* pollen. (u) *Cryptomeria* sp. Lateral view. *Sequoia* pollen is very similar. Note the bent papilla on top. (v) *Graminidites media* Cookson. The thin-walled baggy nature of such grass pollen yields folded walls on collapse. Note the single annulate pore. There is no doubt that this is a grass pollen, but genera of grasses are very difficult to separate by the pollen. Use of form-taxa names for fossil grass pollen is an alternative to saying just "grass pollen" or "Gramineae". (w) *Porocolpopollenites rotundus* (Potonié) Thomson & Pflug. Polar view, with short colpae as well as pores. (x) *Trivestibulopollenites betuloides* Thomson & Pflug. Polar view. A triporate–vestibulate form, which could also easily be referred to *Porocolpopollenites*. (y) *Porocolpopollenites vestibulum* (Potonié) Thomson & Pflug. Polar view. (z) *Porocolpopollenites* sp. Polar view. Such pollen is very similar to modern *Symplocos* pollen and could be referred to, e.g., the form-genus *Symplocoipollenites*. (aa) *Carpinus* sp. Polar view. P03 is a more common *Carpinus* form, and pollen of the near-relative *Ostrya* is not easily distinguished from it. (ab) *Myrica* sp. Polar view. (ac) *Alnus* sp. Polar view (see also (ad)). Note characteristic arci. (ad) *Alnus* sp. Polar view (see also (ac)). (ae) *Ulmus* sp. Rugulate stephanoporate. (af) *Carya* sp. A triporate with the pores characteristically located off the equator in one hemisphere. *Carya* was common in the Mio-Pliocene of Europe but did not survive the Pleistocene there. (ag) *Liquidambar* sp. A reticulate P0x with characteristic large pores with distinctively sculptured pore membranes. (ah) *Tricolpopollenites reniformis* Thomson & Pflug. Equatorial view. There are hundreds of different extant angiosperm genera making such pollen. (ai) *Tricolpopollenites microhenrici* (Potonié) Thomson & Pflug. Equatorial view. (aj), (ak) *Hedera* sp. Equatorial view. "English ivy" is a sensitive indicator of climatic fluctuations in the late Neogene. (al), (am) (al) *Tricolpopollenites ipilensis* Pacltová. (al) Equatorial view; (am) polar view. (It is worth emphasizing how different these are; even professional palynologists sometimes have difficulty recognizing that polar and equatorial views, e.g., of tricolporate forms, go together.) (an), (ao) *Tricolporopollenites iliacus* (Potonié) Thomson & Pflug. Equatorial view. (an) Mid-focus; (ao) high focus. Clavate sculpture is identical to that of extant *Ilex* pollen. (ap) *Tricolporopollenites margaritatus* (Potonié) Thomson & Pflug. Equatorial view. Clavate sculpture finer than that of (an), (ao). (aq), (ar) *Tetracolporopollenites sapotoides* Thomson & Pflug. Equatorial view. (aq) Mid-focus; (ar) high focus. This Pd4 pollen, despite the specific name, is not like that of sapotaceous genera. (as) *Tricolporopollenites edmundii* (Potonié) Thomson & Pflug. Equatorial view. (at) *Corsinipollenites maii* Krutzsch. Polar view. Certainly onagraceous, probably with careful study referable to an extant genus. (au) *Empetrum* sp. Mid-focus of tetrahedral tetrad. Such Ericaceae–Empetraceae-like pollen is very common in Neogene sediments. (av) Ericaceae (?*Rhododendron*). Mid-focus of tetrahedral tetrad. (aw) Ericaceae (?*Erica*). High focus of tetrahedral tetrad. (ax) *Compositoipollenites rizophorus* Potonié. Equatorial view. Echinate Pc3 pollen certainly referable to the Compositae plays an increasingly important role through the Neogene.

been established and extended this way, with an absolute date of about 5.2 million years ago. The base Calabrian of Italy, the undoubted bottom Pleistocene, has been linked with the worldwide Olduvai magnetic event. The basal Pleistocene is thus set at about 1.7–1.8 million years ago. Figure 15.4 shows how these pieces of stratigraphic information were applied (Hsü & Giovanoli 1979) to studies of D.S.D.P. cores in the Black Sea. The "steppe/forest index" (*SFI*) is based on calculations of the ratio between pollen of major steppe indicators and major forest indicators as follows:

$$SFI = \frac{(Artemisia + \text{Chenopodiaceae} + \text{Amaranthaceae pollen})}{(\text{the above} + Pinus + Cedrus + Picea\text{–}Abies + Quercus + Alnus + \text{Ulmaceae (and other tree genera) pollen}} \times 100$$

The cold peaks shown by the *SFI* curve are clearly glacial events, which I have called alpha, beta, and gamma. This is because steppes in the Northern Hemisphere dominated by *Artemisia* and "Cheno–Ams" are located in cool, dry areas. The Olduvai magnetic event occurs in the middle of the alpha glacial, indicating that European glaciers extend back at least to the Pliocene, and that there was in the Black Sea drainage absolutely no early, non-glacial

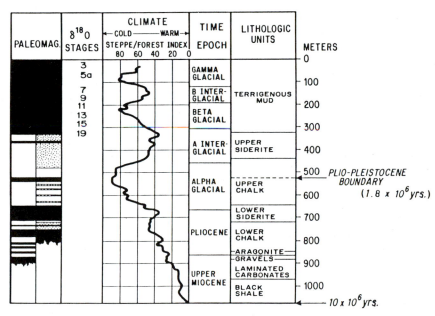

Figure 15.4 The last 10 million years in a continuous core record from the Black Sea (D.S.D.P. leg 42B, holes 380–380A). The steppe/forest index (*SFI*) curve represents a ratio of *Artemisia* plus *Chenopodiaceae* plus *Gramineae* to total pollen. The higher the *SFI* curve, the colder and drier the conditions indicated. There are three peaks of coldness called alpha, beta, and gamma to avoid unwarranted correlation with glacial periods shown in Figure 15.7. Interglacials are labeled A and B for the same reason (the current interglacial is C, not labeled). "Paleomag" on the left refers to paleomagnetic normal (black) and reversed (white) readings, interpreted to the left and measured to the right.

Pleistocene (as mentioned by Birks & Birks (1980) and by others). Oxygen-isotope data support strongly the concept of late Pliocene pronounced cooling, including presumed development of continental ice sheets in Europe and North America beginning about 3.2 million years ago, but that was not the beginning of the Pleistocene, when it is defined the only way it logically can be – stratigraphically. (It should be emphasized that the curve in Figure 15.4 is a smoothed-out curve, having the effect of grouping smaller fluctuations. Oxygen-isotope work suggests as many as 19 glaciations in the Pleistocene.)

MIOCENE–PLIOCENE PALYNOLOGY

General remarks

Inasmuch as almost all of the spores/pollen found in rocks of the "Ultimogene" (the last 10 million years) are determinable to extant genera (in some cases only to families, relatively rarely to species), it is not really very helpful that palynologists sometimes feel compelled to use artificial generic names (form-generic names) for the fossils studied, even though this *is* necessary for most forms pre-"Ultimogene". A compromise approach is that of Meon-Vilain (1970) for the late Miocene–Lower Pliocene of France, in which the reader is informed that "*Polyporopollenites stellatus*" is *Pterocarya*. "*Tsugaepollenites*", however, is not identified as *Tsuga*, probably because that was thought to be obvious. "*Quercoidites*", on the other hand, refers not only to *Quercus* pollen but in part to things such as *Q. henrici* and *Q. microhenrici*, which are likely fagalean but probably not *Quercus*, and may be forms for which a form-generic name would be appropriate. (It is likely, however, that the producing genera are extant *somewhere* in the world.) There is no reason, however, why form-generic names have to be coined just because it is the current vogue to do so. It is quite all right, for example, to refer chenopodiaceous pollen to "Chenopodiaceae" or to "Cheno-Am", without feeling compelled to create a form-genus for this concept. Menke's (1976) approach for Pliocene pollen seems very reasonable: pollen recognized as certainly rubiaceous but not for sure belonging to a particular genus is referred to as "Rubiaceae". Pollen known to belong to *Myriophyllum* is so listed. Pollen probably referable to *Ulmus* but perhaps running over into *Zelkova* is "*Ulmus-habitus*" without a new taxon being formally named. (Quaternary palynologists often use "*Ambrosia*-type", "cf. *Ambrosia*", etc., to show varying degrees of closeness of match; see Birks & Birks (1980, "nomenclature of fossils").) Some forms not recognized as to extant taxon are "Pollen-6024a" and the like. In just a few instances form-generic names are used.

Givulescu's (1962) studies of Romanian Mio-Pliocene megafossil plants illustrate a couple of the palynological problems. The floral lists include many lauraceous forms. These will not produce preservable pollen if modern examples tell us anything. On my property in Pennsylvania, *Sassafras albidum* (Nutt.) Nees and *Lindera benzoin* (L.) Blume are abundant, the latter dominant in the understory, but the pollen record would be *completely* blank for them. On the other hand, Givulescu reports many representatives of the Fagaceae, and these would be the source of "*Quercoidites*" and "*Cupuliferoipollenites*", forms that are characteristic of latest Miocene sediments in North America–Eurasia.

Not only do we have in the spores/pollen flora evidence of plants referable to modern genera, but in a broad way it is reasonable to draw paleoecological conclusions from them. However, it is obviously pressing the significance of a single genus too far, when it is known that various extant species have quite different ecological requirements. Furthermore, the stratigraphic use of palynological data in the whole Neogene is difficult, because there are very few real extinctions, only local extinctions and migrations. Thus the significance of certain "tops" (last occurrences) in northwest Europe per Van der Hammen *et al.* (1971) cannot be directly expanded to include the Black Sea drainage. The use, for example, of *Engelhardia* pollen as a Miocene–Pliocene transition indicator is worrisome, because *Engelhardia* still survives in several parts of the world, although not in Europe where its local extinction *can* be so used. Another example is *Pterocarya*, whose disappearance in northwest Europe is a useful palynological indicator for the arrival of later Pleistocene time. It still persists in the Caspian–Black Sea drainage and cannot be so used there. Van der Hammen *et al.* (1971) and Leopold (1967) give lists of local extinctions that are useful as local stratigraphic markers. Particular plant assemblages are indicative of some of the interglacials. *Carpinus* is common in the Eemian (= Ipswichian) of northern Europe.

General vegetational trends represented by palynofloras in "Ultimogene" time

In the Northern Hemisphere, the temperate deciduous hardwood forest was expanding while the paleotropical (= "mastixioid") flora was retreating. This gives the present-day Eurasian–American vegetation its character, as such taxa as the Gramineae, the Compositae, the Chenopodiaceae, *Acer, Alnus, Betula, Carya, Pterocarya, Ulmus, Pinus, Abies, Picea, Sciadopitys*, and *Tsuga* come into importance or dominance.

Secondly, outside of the areas affected by these trends, floras also were changing in distribution. Figure 15.2 shows from Axelrod's (1958) megafossil flora work in southwestern North America the related migrations of "Sierra Madrean" and "Lagunan" woodlands in response not only to cooling but to aridity caused by orogeny and epeirogeny. Van der Hammen *et al.*'s (1973) study of cores from the Colombian cordilleras dramatizes that, on top of the late Miocene cooling, mountain building can continuously produce new plant associations. Truswell & Harris (1982) have shown palynologically that arid-adapted plants expanded their range from Eocene onward in Australia, with grasslands developing in central Australia by late Eocene. The trend toward xerophytic vegetation accelerated in the Neogene, with *Acacia* appearing, eucalypts expanding their range, and *Nothofagus* declining (see Fig. 14.10). All of this indicates the Australian plate's northward movement in the Cenozoic.

Thirdly, in the Neogene, angiosperms of middle latitudes moved strongly toward deciduous and herbaceous habits:

(a) *Deciduous habit.* The coming to dominance of this character of trees and shrubs is linked to the progressive cooling of the Neogene. *Alnus, Acer,*

Ulmus, Fagus, Castanea, and *Carya* are all deciduous. *Quercus* is instructive in that it seems to show its ancestry in the Paleogene by ranging from evergreen (*Q. virginiana* L.) to irregularly or incompletely deciduous (*Q. nigra* L.), to completely deciduous (*Q. alba* L.), but even the white oak sometimes shows poor abscission in the fall.

(b) *Herbaceous habit.* This character among both monocots and dicots was an obvious response to the Neogene climatic collapse and the expansion of deserts, semiarid regions, and seasonality. The Gramineae and Compositae are primary examples of the advancing tide of herbs. In both families there are (more primitive) members that are woody plants confined to the tropics and subtropics. Pollen of both groups occurs in abundance first at the Oligocene–Miocene transition in Euramerica, but does not become really common until late Neogene. (Monoporate pollen very similar to grass pollen occurs in the not closely related, much smaller, monocot family Restionaceae, and this monoporate type has been identified in Paleocene and even latest Cretaceous rocks; see Medus (1982). Because restionaceous P01 pollen is scrobiculate and grass pollen mostly scabrate, they can be distinguished.) The Chenopodiaceae and Cyperaceae have a similar history. The chenopods are prevailingly herbaceous but have many woody members such as *Atriplex*. They are almost strictly a Pliocene and later phenomenon, though periporate (P0x) pollen that could be related occurs much earlier. Cyperaceous pollen is also primarily a Pliocene–Quaternary phenomenon, though 10My (Miocene) cyperaceous pollen is reported by Barnosky (1984). Unlike the other families just mentioned, there are no woody family members.

During the latest Miocene and Pliocene of mid-latitude America and of Eurasia, the more warmth-demanding hardwoods gradually withdrew. *Tsuga, Nyssa, Juglans, Castanea, Carya,* and *Liquidambar* retreat southward in North America and disappear in Europe. *Quercoidites* spp., such as *Q. henrici* and *Q. microhenrici,* and *Cyrillaceaepollenites* spp., such as *C. megaexactus,* disappear in Europe and perhaps are even extinct. Planderová (1972), speaking of central Europe, notes that the following families are much more significant in the Pliocene than in the Miocene (they first appeared much earlier): Gramineae, Chenopodiaceae, Umbelliferae, Oenotheraceae (Onagraceae), Rhamnaceae, Ericaceae. She also points out that Taxodiaceae, Nyssaceae, Myricaceae, and the genus *Engelhardia* are practically gone by latest Miocene. *Abies* becomes important at about the same time, along with Compositae and Gramineae. In the early Pliocene, *Acer, Betula,* and *Alnus* begin to appear in larger percentages. The early Pliocene is very rich in taxa, as new forms come in while some of the older ones are still present.

Table 15.1 shows the stratigraphic picture for some important palynological indicators, from latest Miocene to Pleistocene in northwest Europe and central Europe.

Benda (1971) studied the Miocene–Pleistocene of southwest Anatolia, and the results for Mio-Pliocene are interesting to show the possibilities and difficulties of geographically extending correlation based on Neogene pollen records (see Fig. 15.5). Black Sea D.S.D.P. cores I have studied tie in fairly

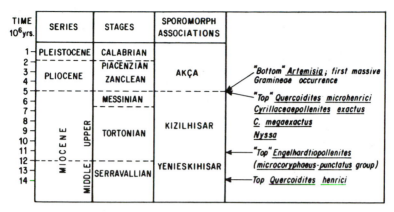

TIME 10⁶ yrs.	SERIES	STAGES	SPOROMORPH ASSOCIATIONS
1	PLEISTOCENE	CALABRIAN	
2		PIACENZIAN	
3	PLIOCENE		AKÇA
4		ZANCLEAN	
5			
6		MESSINIAN	
7			
8	UPPER		KIZILHISAR
9		TORTONIAN	
10			
11			
12			
13	MIDDLE	SERRAVALLIAN	YENIESKIHISAR
14			

(MIOCENE)

"Bottom" *Artemisia*; first massive Gramineae occurrence

"Top" *Quercoidites microhenrici*
Cyrillaceaepollenites exactus
C. megaexactus
Nyssa

"Top" *Engelhardtiopollenites*
(*microcoryphaeus-punctatus* group)

Top *Quercoidites henrici*

Figure 15.5 Miocene–Pliocene sporomorph associations from western Turkey. Note the influx of *Artemisia* and grasses, two of the principal steppe indicators beginning in the Pliocene after the end (top) of many of the Miocene warmth-loving woody plants.

well with Benda's horizons, as do the European records shown in Table 15.1: latitudinal differences were not as sharp in the Mio-Pliocene as since. However, Benda found a "top" (last occurrence) for *Arecipites* (= Palmae) pollen in the earliest Miocene, whereas I found rather abundant palm pollen in the latest Miocene of the D.S.D.P. cores. Van der Hammen *et al.* (1971) have palm pollen well up in the Pliocene of northwest Europe, which seems discordant. Palms are, however, a tricky business, as a few occur in southernmost Europe today!

The latest Miocene did not produce really cold climates even quite near the Arctic Circle. Hopkins *et al.* (1971) analyzed a flora radiometrically dated as somewhat more than 5.7 million years old, i.e., Messinian or latest Miocene, from Seward Peninsula, Alaska, latitude 65° N. The overall pollen flora includes no tundra elements. Brady & Martin (1978) report that much of Antarctica was glaciated at the beginning of the late Miocene, but that in places a rather luxuriant vegetation of Proteaceae, Fagaceae (*Nothofagus* spp.), Podocarpaceae, and ferns still existed. Mercer (1973) says that even west Antarctica was glaciated by at least 3.5 million years ago. The east Antarctic ice sheet is considerably older; various authors would have it in place by about 14 million years ago (mid-Miocene) and the west Antarctic sheet beginning perhaps as early as 9 million years ago (late Miocene), but some geologists favor considerably older, early Miocene, Antarctic glaciation. Oceanographic evidence mostly supports an approximate 10 million years ago initiation of glaciation and worldwide major cooling (see Kerr 1982). Various Soviet scientists have mentioned that extensive steppe vegetation developed in central Asia and Siberia during the late Miocene, but whether this was a modern steppe flora is not certain. Leopold (1984) and Leopold & Wright (1985) have established that in parts of western North America *Artemisia* steppe was present by late Miocene, but Leopold notes that major expansion of grasslands and steppe west of the Rockies is a Pliocene–Pleistocene phenomenon, which would agree with palynological observations for south–central Europe

Table 15.1 Stratigraphic occurrence of important floral elements in the "Ultimogene" (late Neogene) of northwestern and central Europe (names from both lists modified for ease of comparison).

Northwest Europe	Germany
"Top" mid-Pleistocene *Carya* *Castanea* *Juglans* *Ostrya* *Pterocarya* *Tsuga*	
"Top" early Pleistocene *Fagus* *Phellodendron*	"Top" early Pleistocene *Carya* *Eucommia* *Fagus* *Juglans* *Pterocarya* *Tsuga*
"Top" about end of Pliocene *Aesculus* *Cupuliferoidaepollenites fallax* *Liquidambar* *Nyssa* *Sciadopitys* *Sequoia* *Zelkova*	"Top" in or at end of Pliocene *Castanea* Cupressaceae (= *Inaperturopollenites dubius*) *Cupuliferoidaepollenites fallax* *C. quisqualis* *Cyrillaceaepollenites exactus* *Liquidambar* *Nyssa* *Platanus*

"Top" about mid-Pliocene
Eleagnus
Palmae
Symplocos
Tricolporopollenites edmundii

"Top" about end of Miocene
Cupuliferoipollenites villensis
Cyrillaceaepollenites exactus
Engelhardtioipollenites
Quercoidites henrici
Q. microhenrici
Rhoipites pseudocingulum

"Top" in or at end of Pliocene, cont.
Taxodium
Tricolporopollenites edmundii
Sciadopitys
Sequoia–Metasequoia–Glyptostrobus

"Top" about end of Miocene
Araliaceoipollenites euphorii
Betulaceoipollenites bituitus
Engelhardtioipollenites punctatus
Palmae
Quercoidites henrici
Q. microhenrici
Triatriopollenites myricoides

(Traverse 1982). Leopold's studies also established from modern pollen deposition that the pollen content of lake and alluvial sediments does indicate the broad vegetational type in the original area of deposition.

Most paleobotanists and palynologists who have worked in the tropical Cenozoic believe that in the lower latitudes there has been no massive change of floras at the generic level since pre-Messinian Miocene. Even at higher latitudes in Eurasia–America, profound changes of the vegetation in place in the Miocene did not begin until late Pliocene, with the arrival of continental ice sheets.

Thus, the use of pollen floras for broad-scale stratigraphy in the Miocene–Pliocene–Pleistocene is not practicable, though, within a selected area, trends may be observed and used. Van de Weerd (1979), for example, has cautioned well:

> Uppermost Miocene and Lower Pliocene associations are rather similar and cannot be clearly separated. The morphotype distributions of *Pinus* do not provide an unambiguous boundary . . . pollen associations within a basin are stable over long periods. Gradual changes in the frequencies of pollen within lithological units cannot be detected. Marked differences in pollen associations are due to tectonic events . . . boundaries present in one basin may be absent in other basins.

The Pliocene record

The Pliocene palynoflora of Eurasia–North America was characterized especially by conifers, though angiospermous pollen is numerically dominant (except in mountainous or high-latitude areas). For example, in the Black Sea D.S.D.P. cores, when one works downward through the Pleistocene one is struck by the sudden abundance and diversity of conifers in the Pliocene: *Tsuga* spp., *Abies* spp., *Cedrus, Podocarpus, Pinus* spp., plus many species of Taxodiaceae such as *Taxodium, Sequoia,* and *Sciadopitys,* and Cupressaceae–Taxaceae forms. Presumably this indicates widespread cool but not yet arid climate. The first massive cooling of the late Neogene occurred in the late Pliocene, perhaps about 2.5 million years ago, whereas because of dating of the type sections the Pleistocene does not begin until about 1.8 million years ago. Suc (1980) studied sections in France and Spain that were mostly Pliocene and noted that the bases of his sections are dominated by Taxodiaceae. The middle of the sections shows a replacement of taxodiads by abietineans. Toward the top of the Pliocene, xerophytic plants are present in great numbers: grasses, chenopods, *Artemisia,* in brief, steppe-like indicators. In the Mediterranean area, *Pinus*–Taxodiaceae–Cupressaceae–*Sciadopitys* are abundant in early Pliocene, followed by *Pinus* and other conifers, then Podocarpaceae, then *Pinus–Sciadopitys–*Compositae at the Pliocene/Pleistocene boundary (Sauvage 1979, Sauvage & Sebrier 1977, studying sections in Greece). The first interglacial was characterized by *Tsuga–Sciadopitys–Carya–Pterocarya.* I also observed this flora in the Black Sea D.S.D.P. cores, after the "alpha" glaciation. Bertolani Marchetti *et al.* (1979) observed that in northern Italy *Sciadopitys* terminated near the end of the Pliocene (see Fig. 15.6).

330

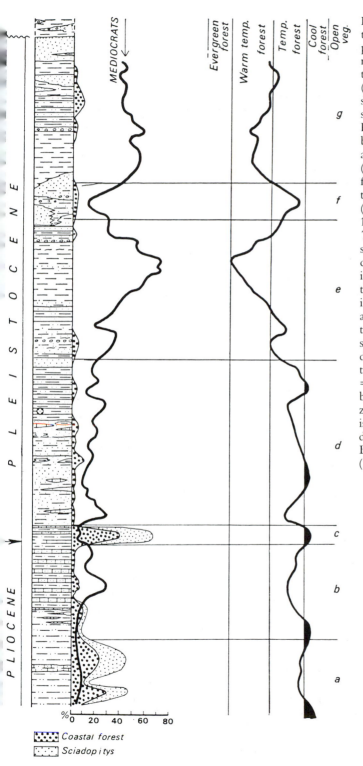

Figure 15.6 Late Pliocene to (ca.) first half of Pleistocene pollen record, Stirone River, northern Italy. Note especially the termination of *Sciadopitys* (today there is only one species, confined to Japan) soon after the end of the Pliocene, and the diminution but periodic minor reappearance of *Sequoia–Taxodium* ("coastal forest", now confined to North America) at that time. The mediocrats (see "mesocratic" in Figure 15.8) curve refers to *Quercus*, *Tilia*, *Ulmus*, and other tree/shrubs characteristic of the climatic-optimum part of interglacials. The curve on the right is a climatic curve, indicating cooler to the right, as shown by the vegetation types at the top. The patterns shown on the left of the diagram are standard indications of sediment type: dots = sand and silt; lines = shale; blocks = limestone; etc. The zones a–g indicate characteristic vegetational complexes described in detail by Bertolani Marchetti *et al.* (1979).

Palynological writers have interpreted a "preglacial Pleistocene" ("Pre-Tiglian") and have drawn conclusions from this concept. This is no longer tenable, as we have seen. The "Tiglian" of the Netherlands sections belongs in the Pliocene (pre-1.8 million years). The "Tiglian warming" perhaps shows in the D.S.D.P. record just before the "alpha" glaciation (see Fig. 15.4). The Pliocene/Pleistocene boundary depends on sections in Italy for which there is marine fossil control. These sections have not been studied palynologically until comparatively recently. It would be fortuitous if a universally present and marked Pliocene/Pleistocene palynological boundary existed. Thus, the termination of *Sciadopitys* (Taxodiaceae) pollen near the boundary in Italy shown in Figure 15.6 is not to be taken as a demonstration of an extendable datum. Nevertheless, there are observable worldwide palynological effects of the late Pliocene cooling. For example, great expansion of *Artemisia* and Gramineae pollen in the U.S.A. Gulf Coast coordinate with other signs of the event, e.g., mottled clays. Suc (1980) observed steppe–forest alternation in the Plio-Pleistocene of France. This and related phenomena are widespread, at least in the Northern Hemisphere. Boulter (1971a, b) has shown that, given the knowledge of where such sediment occurs, even a Neogene deposit for which stratigraphic evidence is not helpful can be palynologically dated. Boulter analyzed "pockets" of clay from sinkholes in Carboniferous limestone in Derbyshire, England, and dated the palynoflora as Miocene/Pliocene boundary (5.3 Ma). See Figure 15.3 for examples of this flora.

PLEISTOCENE PALYNOLOGY

General remarks

The Pleistocene is a time of "catastrophes", that is, a time of comparatively rapid climatic changes. The steppe/forest curve in Figure 15.4 is a smoothed-out, running-average curve, and even this curve is based on very widely spaced samples. The oscillations are really much more numerous. One can observe the same sort of rapid Pleistocene oscillations on a small scale in places in Scotland, where birch or pine forests have been killed and covered by blanket peats since the climatic optimum of a few thousand years ago (human clearing of forest has probably also played a role). The violent and frequent swings of climate in the Pleistocene are shown by detailed steppe/forest index curves in Black Sea cores (Traverse 1978a) and by non-tree pollen vs. tree pollen (arboreal pollen) (NAP/AP) fluctuations in many other places. In areas such as Africa and southwest U.S.A. and Australia, oscillations of dry vs. moist conditions rather than temperature changes are the important matters, whereas at high latitudes and elevations temperature swings have been important.

The rapidity of climatic swings (some quite brief) in the late Cenozoic can apparently be shown by palynological analysis of closely spaced dark and light laminations of Black Sea D.S.D.P. cores, as shown in Table 15.2. The swings in steppe/forest pollen index and accompanying chemico-physical measurements indicate very rapid changes, in the range of a century. Based on oxygen-isotope data, the light layers represent colder episodes than the dark layers.

Table 15.2 "Light"–"dark" cycle pairs from Deep Sea Drilling Project (leg 42B, 1975) holes 379A and 380–380A, showing the regularly found association in "light" samples between high *SFI*, high $CaCO_3$, low SiO_2, and less negative $\delta^{18}O$ isotope values; compared with "dark" samples with lower *SFI*, lower $CaCO_3$, higher SiO_2, and more negative $\delta^{18}O$ isotope values. "Light" samples apparently represent the cooler part of a cycle and "dark" samples warmer parts of a cycle.

Sample	Pairs I–III, 380A:51:3						Pair IV, 379A:60:2	
Pair (cycles)	I		II		III		IV★	
Color	Light	Dark	Light	Dark	Light	Dark	Light	Dark
Depth (cm)	76–77	75–76	110–111	109–110	116–117	115–116	10–12	6–9
SFI	54	10	70	9	72	36	55	9
$CaCO_3$†	92	46	94	81	89	53	67	45
SiO_2†	7	17	6	14	6	30	20	33
$\delta^{18}O$‡	−2.20	−3.32	−2.72	−3.25	−2.82	−3.03	−5.80	−6.19

★This pair was reported in Traverse (1978b) and a sample residue remained for recent x-ray and isotope analysis.
†X-ray analysis by J. Pika, E.T.H., Zürich. Values expressed in percentage of total crystalline inorganic matter.
‡Isotope analysis by J. Pika, E.T.H., Zürich. The more negative values indicate warmer. Presumably the explanation for the high readings for pair IV is that, if a pair is deposited during a generally warmer period, both "light" and "dark" will be more negative than are sediments laid down during a generally colder period. The darker layer of a "pair" is almost always more negative than the lighter layer. Values are expressed as relative enrichment of ^{18}O against the "PDB" standard, which is taken as 0 on the scale. ("PDB" refers to Peedee Formation belemnites, the University of Chicago standard.)

The precise causes of the climatic swings are not known, and it must be mentioned that some palynologists (Davis & Botkin 1985) are of the opinion that, at least for cool temperate forests, short-term climatic changes on the scale of a few hundred years are not picked up by the pollen record, partly because of lag time in vegetational response. There is also a sampling problem in getting samples from very short intervals. Palynologists seldom sample sediments centimeter by centimeter; 10 cm or much wider spacing is more common.

The conventional divisions of the "Quaternary" and Pleistocene are shown in Figure 15.7. As mentioned earlier, I would prefer to discard the term "Quaternary", and use Pleistocene as the final epoch of the Neogene, both of which would run to the present. The "Holocene" for the last 10,000 years is also distasteful to me, implying that we no longer are in the Pleistocene. I would call it "present interglacial", or some neutral term such as "post-W" (= post-Wisconsinian, post-Würm). (In western Europe there is some vogue to use "Flandrian" for the last 10,000 years, but I now realize that getting it accepted in this sense worldwide would be impossible. I also realize that abandoning Quaternary as equal to Pleistocene plus "Holocene" is not something that can be accomplished here.) Originally, the Pleistocene was envisioned as one great glacial time, by Louis Agassiz and others. Later scientists such as Geikie (see Charlesworth 1957) showed that there were multiple large glaciations ("polyglacialism" – see discussion of the history by West (1985)). How many has been much disputed. Geikie suggested six. Penck & Brückner (1909), working on alpine sections of Germany and Switzerland, claimed four major alpine glaciations and named them Günz, Mindel, Riss, and Würm for superimposed terraces of Danube tributaries, which are listed alphabetically. In America for a time there was a vogue for five great glacial episodes: Nebraskan, Kansan, Illinoian, Iowan, and Wisconsinian. Under the influence of the European chronology, the Iowan was suppressed as a separate mega-glaciation, and a four-glaciation Pleistocene came to be widely accepted: Nebraskan, Kansan, Illinoian, and Wisconsinian. As can be seen in Figure 15.7, the Riss and Würm together are now thought in Europe to be equivalent to the Wisconsinian, and the Danubian is equivalent to the Nebraskan, the Günz to the Kansan, and the Mindel to the Illinoian. Whereas in Europe the intervening warmer times have been traditionally labelled by the preceding and following Alpine glacials (thus, Günz–Mindel, etc.), in North America separate names are used for the interglacials: Aftonian, Yarmouthian, and Sangamonian. In Europe the interglacials in areas away from the Alps are commonly given distinctive names: Ipswichian, Hoxnian, etc. The truth seems to be that there were multiple cold phases, alternating with multiple warmer phases. In my work with Black Sea D.S.D.P. cores, I concluded that the most reasonable grouping of multiple oscillations was into three great glacial times, which I called alpha, beta, and gamma, to avoid confusion with the classical names (see Fig. 15.4). This broad grouping is at least as compatible with oxygen-isotope data from marine cores as a four-phase or a six-phase model.

GLACIAL – INTERGLACIAL SUBDIVISIONS

CLIMATIC REGIME	NORTH AMERICA	BRITISH ISLES AND NORTHWEST EUROPE	CENTRAL EUROPE
Interglacial	Holocene ("Post Glacial")	Flandrian	Flandrian
Glacial	Wisconsinian	Devensian–Weichselian	Würm
Interglacial	Sangamonian	Ipswichian=Eemian Interglacial / Wolstonian–Saalian	Riss–Würm Interglacial / Riss
Glacial	Illinoian	Hoxnian–Holsteinian	Mindel–Riss
Interglacial	Yarmouthian	Anglian–Elsterian	Mindel
Glacial	Kansan	Cromerian	Günz–Mindel
Interglacial	Aftonian	Menapian	Günz
Glacial	Nebraskan	Waalian	Danubian–Günz
		Eburonian	Danubian
Interglacial	"pre-Nebraskan"	Tiglian	Biberian–Danubian
Glacial	"pre-Nebraskan"	pre-Tiglian	Biberian
			(Transitional Beds)
		Reuverian	

ca. 10⁴ years — PLEISTOCENE
1.8×10⁶ years — PLIOCENE

Figure 15.7 Pleistocene subdivisions as used in North America and Europe.

Palynology of glacial-interglacial cycles

Our best information centers on studies of the last mega-glaciation (Riss + Würm = Saalian + Weichselian = Wisconsinian) and of the interglacial that preceded it (Sangamonian in North America = Mindel–Riss = Holsteinian in Europe). The terminal Wisconsinian (about 18,000 yr B.P.) seems to have been the coldest time of all (Peterson *et al.* 1979), archetypal Pleistocene. Godwin (1975), working on both of the last two European interglacials, the Ipswichian – from the American point of view, a Wisconsinian interstadial – and the Hoxnian (= Holsteinian = Sangamonian of North America) in Britain, has prepared a palynological model for the glacial–interglacial transitions (see Fig. 15.8). Figure 15.9 gives a pollen diagram for the Ipswichian interglacial or interstadial, and Figure 15.10 for the Hoxnian and for what we so far have experienced of the present, Flandrian (= Holocene), interglacial. Obviously the interglacials are rather similar to each other in comparable areas. Van der Hammen *et al.* (1971) have also diagrammed these events. Their presentation emphasizes the impact of temperature on moisture. It gets cold first, then dry, in a glacial; and warm first, then wet, in an interglacial.

That it is very difficult to transfer such a model directly to other parts of the world is shown by Figure 15.8b. Heusser (1977a) finds a number of cold–warm oscillations during the last 100,000 years of the Washington State area, northwest U.S.A. The exact relation of these to the Würm–Wisconsinian, etc., is unclear. The oscillations are obviously numerous, and the great glaciations are groupings of closely spaced cold times. It is also well-known that sedimentary factors (associations of palynofloras with sediment types) have a considerable bearing on a spore/pollen "signature" (the shape of a pollen analytical curve). It is therefore unreasonable to expect to find exactly the same sequences in peat from a peat bog, in silt from the center of a large lake, and in a core from offshore clayey silt, even in the same general area.

Baker (1986) describes the pollen and plant megafossil record of the last glacial–interglacial cycle from a site in Yellowstone National Park, Wyoming. The peak warm period assumed to represent the Sangamonian interglacial is represented by a flora dominated by *Pseudotsuga–Pinus* forest, and a climate considerably warmer than any Holocene climate is suggested.

The sorts of pollen on which the palynological records discussed above are based represent a relatively small list, of mostly wind-pollinated extant plants, primarily those from mid-latitudes of the Northern Hemisphere. Figures 15.11 (mostly gymnosperms) and 15.12 (angiosperms) present illustrations of a number of such forms, comprising a large percentage of all the important taxa.

Glacials vs. interglacials and water budgets: "pluvials"?

This is a very important and vexing question to which palynology has made some contributions. The long and the short of it is that despite exceptions and over large areas cold means dry, and warm means wet. It was long believed that the reverse was generally true. That is to say that glacials were associated

336

THE INTERGLACIAL CYCLE

Characteristics of	CRYOCRATIC	PROTOCRATIC	MESOCRATIC	TELOCRATIC
			Mean temperature	
Climate	Cold	Warm	Thermal maximum	Cooling
Soils	immature. unstable, base-rich	fixed but transitional	brown earths	podsols and blanket-bog
Vegetation	open herb and low shrub	park-tundra to light wood	closed deciduous forest	coniferous woodland and acidic heath
Floristic elements	arctic and alpine	residual arctic-alpine; steppe and S. European; weeds and ruderals.	woodland plants and thermophiles	recession of thermophiles

a.

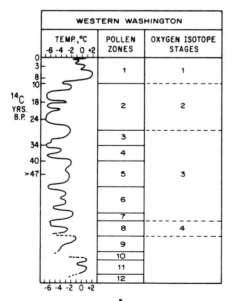

b.

Figure 15.8 Glacial–interglacial "cycles" in perspective. (a) The formal interglacial cycle of vegetation–soil–climate alterations indicated by pollen analytical data of northwest Europe. Note that mesocratic is equivalent to a climatic optimum. (b) Oxygen-isotope data and ^{14}C dated core from western Washington State, U.S.A., correlated to pollen zones, with temperature estimate curve. The approximately 80,000 year record is characterized by frequent, even rather dramatic, shifts. Putting the Pleistocene record into a three or four glaciation–interglacial framework represents very considerable smoothing of the record. In fact, there have been hundreds of smaller climatic changes.

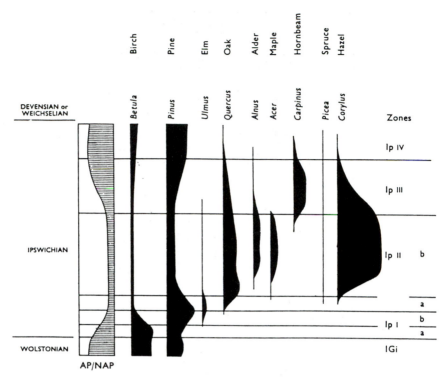

Figure 15.9 Composite pollen diagram for Ipswichian interglacial (see Fig. 15.7) of eastern England representing vegetational changes during stages of this interglacial. (An interglacial contemporaneous with the Ipswichian is not recognized as a separate interglacial in North America – it would be a late Wisconsinian interstadial.) Compare with Figure 15.10 for evidence that glacial–interglacial cycles are repetitive (in one geographic area) and that the present (Flandrian) interglacial is "normal".

outside of the glacial areas with wet "pluvials". This does seem to be true at some places. But in Africa, for example, the colder times seem to have been prevailingly dry periods when the Sahara has advanced. Street & Grove (1979) have shown that, in the tropics generally, high lake levels were interglacial phenomena due to increased monsoons. However, in North America, the argument about pluvials still goes on. Wells (1979) and Benson (1978) believe that lakes such as Lahontan and Bonneville in the central west of the U.S.A. had high stands during glacials ("pluvials"). This could, however, be due to increase in effective moisture because of lower temperature. Brakenridge (1978) showed that temperature factors alone could account for all vegetation displacement in the late Pleistocene of the western U.S.A. and also for the higher lake levels, because of lowered evaporation, not greater precipitation (see Fig. 15.13). Spaulding *et al.* (1983) emphasize that the full glacial climate of the southwest U.S.A. *was* moist – effectively moist compared to the Holocene, because of lower temperatures and prevalence of winter precipitation in contrast to Holocene summer precipitation, but possibly also with higher

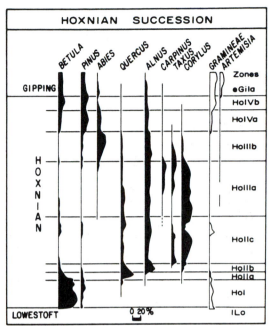

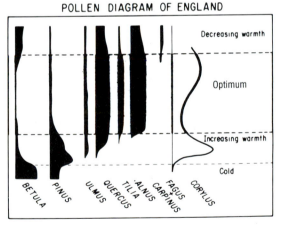

Figure 15.10 Two more interglacial pollen diagrams from England, for comparison with Figure 15.9. (a) The Hoxnian diagram is for the interglacial known as Sangamonian in North America and Mindel–Riss in central Europe. (b) The Flandrian interglacial is less satisfactorily also known as "post-glacial" or Holocene. Compare with Figure 15.9 to show that glacials–interglacials have a repetitive character, e.g., *Betula* peaks in late glacial time, minimizes during the climatic optimum, etc. There are small characteristic differences too, e.g., *Carpinus* has a somewhat different response to local circumstances in each of these three interglacials.

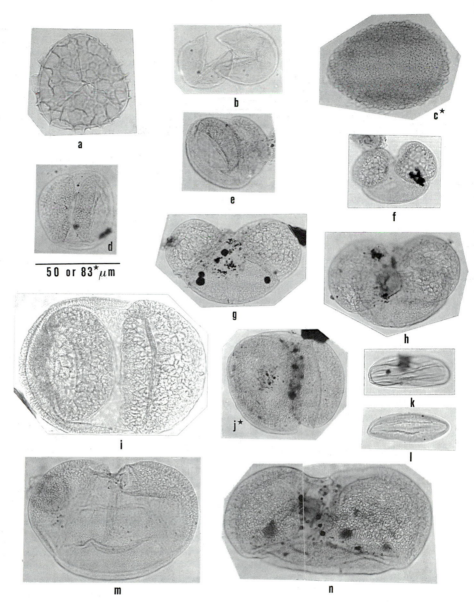

Figure 15.11 Some significant sporomorphs (mostly gymnosperm pollen) from the Northern Hemisphere Pleistocene, including the present interglacial (= Holocene). Some of the specimens are from a Holocene peat from New Brunswick, Canada (N.B.); others are from Pleistocene levels in D.S.D.P. hole 380 in the Black Sea (D.S.D.P.). The specimens from the New Brunswick peat have expanded somewhat since preparation. Magnification indicated by bar under (d). (a) *Lycopodium* sp., proximal view (N.B.). A frequent constituent of Pleistocene samples, known amounts of *Lycopodium* spores are also added to samples by some palynologists for use in calculation of the concentration of spores and pollen. Fern spores are also frequently

340

annual precipitation than today, especially in some areas. That the situation is very complicated is demonstrated by Jansen *et al.* (1986) in a multipronged study of ocean cores, indicating that a swing to more glacial conditions in the Northern Hemisphere 300,000–400,000 yr B.P. was correlated with a trend to more interglacial conditions in the Southern Hemisphere!

In Australia Dodson (1977) shows that *Casuarina* plus *Eucalyptus* forests have greatly expanded in the Holocene of coastal Australia, whereas very dry open woodland dominated the same areas during full glacial time, 11,000–26,000 yr B.P.

Van Zinderen Bakker's (1976) palynologically based diagrams for the African cold vs. warm fluctuations clearly show that warmer times are prevailingly moister (see Fig. 15.14). Flenley's (1979) extensive survey of the African equatorial rainforest in the Pleistocene supports the idea generated from palynological and other evidence that the time from 26,000 to 12,000 yr B.P. was cooler *and* drier than at present. On the other hand, a thorough review of tropical pollen analysis in Africa, as well as in South America and other continents, by Livingstone & Van der Hammen (1978) shows that the interpretation of the as yet rather limited number of data points is difficult. It is very clear, however, that tropical climates are and have been very unstable.

---

encountered in Pleistocene work. (b) *Taxodium* sp. (swamp cypress). Two specimens in lateral view (D.S.D.P.) showing the "clam-shell" appearance usually presented by the split-open fossil pollen. (Other taxodiads as well as some Cupressaceae pollen sometimes look much the same.) (c) *Tsuga* sp. (hemlock). A monosaccate conifer fossil pollen with very characteristically ropy-ruffled exine (N.B.). (d) ?*Cedrus* sp., distal view (D.S.D.P.). In practice the differentiation of specimens of *Cedrus* from those of *Pinus*, where both occur, is so difficult that it is probably better to lump them as *Pinus/Cedrus*. With rare exceptions, *Pinus* pollen will be much more abundant. (e) ?*Cedrus* sp., distal–lateral view (D.S.D.P.). See comments under (d). (f) *Pinus* sp., lateral view (D.S.D.P.). *Pinus* is a large genus, and various people have demonstrated the possibility of separating the pollen as to species or groups of species, on the basis of the morphology. However, in work with sediments such as those in the Black Sea or the Gulf of Mexico, with streams contributing sediment and pollen from a very large area, it is only practicable to count "*Pinus*". (g) *Pinus* sp., lateral view (D.S.D.P.). The black spots are pyrite (probably marcasite variety) crystals, a product of sulfur-bacterial activity and a feature of pollen deposited in a reducing environment. (h) *Pinus* sp., distal–lateral view (D.S.D.P.). See (f) and (g). (i) *Abies* sp., distal view (D.S.D.P.). Two sorts of giant bisaccates occur commonly in Pleistocene sediments – *Abies* (fir) and *Picea* (spruce). The two genera of conifer trees are somewhat similar phenotypically. *Abies* pollen looks more like a giant pine pollen grain than does *Picea*, however. See (m) and (n). (j) *Abies* sp., distal view (D.S.D.P.). Photo at lower magnification than most others on this plate. Note, as in (i), the pine-like appearance. (k) *Ephedra* sp., proximo-distal view (D.S.D.P.). A psilate form. *Ephedra* pollen is often found in the same samples as *Artemisia*, but is never as abundant. It is a polyplicate (= taeniate) pollen grain with a long fossil history. See (l). (l) *Ephedra* sp., proximo-distal view (D.S.D.P.). A species with fossulate sculpturing in addition to being polyplicate. (m) *Picea* sp., lateral view (D.S.D.P.). A giant conifer bisaccate pollen with sacci that blend into the corpus and more or less continue its outline in lateral view. See *Abies*, (i) and (j). (n) *Picea* sp., lateral view of somewhat flattened specimen (D.S.D.P.). See (m).

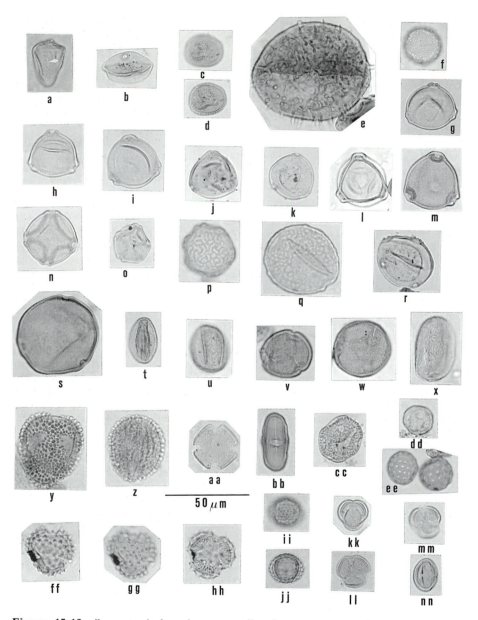

Figure 15.12 Some typical angiosperm pollen forms encountered in the Northern Hemisphere Pleistocene, and the present (Holocene = Flandrian) interglacial. Some of the specimens are from a Flandrian peat deposit in New Brunswick, Canada (N.B.), some are modern pollen from water samples from Trinity River and Trinity Bay, Texas, U.S.A. (T.R.), and some are from Pleistocene levels in D.S.D.P. hole 380, Black Sea (D.S.D.P.). N.B. forms have expanded somewhat in preparation. Magnification indicated by bar under (aa). (a) Cyperaceae (sedge). Lateral view of the typically pear-shaped pollen, showing an ulcus at the top and one on the surface (arrow)

342

(T.R.). (b) Gramineae (grass). Obliquely distal view showing the annulate pore with operculum, and the characteristic folded nature of most fossil grass pollen (D.S.D.P.) (c), (d) *Typha* sp. (cattail), or the closely related *Sparganium*. Two levels of focus (D.S.D.P.). *Typha* pollen is mono-ulcerate. (e) *Nuphar* sp. (pond-lily). Monosulcate echinate pollen, distal view (N.B.). (f) *Sagittaria* sp. (arrowhead). A weakly periporate echinate pollen of a common inhabitant of wet places (D.S.D.P.). (g) *Carpinus* sp., or *Ostrya* sp. (ironwood). Two very common betulaceous tree genera, polar view (D.S.D.P.). Separation of the many triporate pollen types in Pleistocene pollen analysis is very difficult. *Carpinus* and *Ostrya* pollen are so similar that best practice is to include them under "*Carpinus/Ostrya*". The pore structure is relatively small and protrudes relatively little from the perimeter. (h) *Myrica* sp. (wax myrtle). Triporate pollen of a common shrub genus, polar view (N.B.). See comments under (g). The pore structure is relatively heavy but does not have a well-developed vestibulum as does *Betula*. In practice, *Myrica* spp. pollen are difficult to separate from *Corylus* spp. (i) *Myrica* sp. (N.B.). See comments under (g) and (h). (j) *Corylus* sp. (hazelnut). Polar view (D.S.D.P.). Very similar to *Myrica*, pore structure somewhat less pronounced, annulus not as heavy. See (g) and (h). (k) *Corylus* sp. Polar view (D.S.D.P.). See (g), (h), and (j). (l) *Betula* sp. (birch). Polar view of triporate pollen of a very common Pleistocene tree (N.B.). The heavy structure of the pore apparatus is characteristic, as is the large vestibulum between pore and os. The triangular figure connecting the pores represents an exine band at the equator, indicating the shape and size of the grain before KOH treatment and mounting in glycerin jelly caused swelling poleward. (m) *Betula* sp. See (l). Probably a different species from that of (n), though from the same preparation (N.B.). (n) *Alnus* sp. (alder). Polar view of 4-stephanoporate pollen of a very common shrub and tree genus (N.B.). The thickened bands (arci) connecting pore structures are characteristic. See (o). (o) *Alnus* sp. A 5-stephanoporate form (D.S.D.P.). See (n). The specimens (n) and (o) were probably originally about the same size, but different sedimentary history and processing technique has swollen specimen (n). (p) *Ulmus* sp. (elm). Polar view of 6-stephanoporate form of this common tree pollen, in high focus to demonstrate characteristic rugulate sculpture (T.R.). (q) *Ulmus* sp. A 5-stephanoporate form (N.B.). See (p). (r) *Pterocarya* sp. Polar view of 5-stephanoporate pollen of an Old World judlandaceous tree common in Pleistocene samples of Black Sea–Mediterranean area (D.S.D.P.). (s) *Carya* sp. (hickory). Polar view of P03 pollen of this very common Pleistocene tree genus (D.S.D.P.). (t) *Quercus* sp. (oak). Equatorial view of verrucate Pc0 pollen of this very common Pleistocene tree genus (D.S.D.P.). *Quercus* is a very large genus, and pollen of at least some of the species can be distinguished by combination of light microscopy and S.E.M. In many Holocene sediments, *Quercus* and *Pinus* combined are 50% or more of the palynoflora. (u) *Quercus* sp. A different form (D.S.D.P.). See (t). (v), (w) *Fagus* sp. (beech). Polar and obliquely equatorial views, respectively, of Pc3 pollen of a common Pleistocene tree genus, showing the evenly scabrate sculpture and spherical shape (D.S.D.P.). (x) *Acer* sp. (maple). Equatorial view of Pc0 pollen of a common Pleistocene tree; high focus to show the striate pattern of sculpture (N.B.). Pollen of some other plants, especially in the Rosaceae, has similar sculpture. (y), (z) *Ilex* sp. (holly). Polar and equatorial views, respectively, of Pc0 pollen of a common tree and shrub of Pleistocene, with characteristic clavate sculpture (N.B.). (aa) *Fraxinus* sp. (ash). Polar view of Pd0, reticulate pollen of a common Pleistocene tree (T.R.). (ab) *Umbelliferae* (carrot family) pollen. Equatorial view of characteristic perprolate Pc3 pollen (T.R.). Herbaceous umbellifers are abundant contributors of pollen to Pleistocene sediments. Pollen of many umbelliferous genera are very similar. (ac) Caryophyllaceae (pink family). Reticulate P0x pollen of herbaceous dicot (D.S.D.P.). Many members of the family make similar pollen. (ad) Amaranthaceae (amaranth family). Micropitted P0x pollen of herbaceous dicot (D.S.D.P.). The amaranths and chenopods (ae) form a complex of periporate pollen which many palynologists lump in counting as "Cheno-Ams". (ae)

While the period 20,000–12,500 yr B.P. was, for example, in general drier than the present in the tropics, there are places that were exceptions. Rossignol-Strick & Duzer (1979a) show, in a palynological study of a core from offshore Senegal (northwest Africa), very dry circumstances at the 18,000 yr B.P. glacial maximum and very moist during the mid-Holocene (which has sometimes been called the hypsithermal, or thermal maximum). That these generalizations only work as a broad view is shown, however, by Sowunmi's (1981) work in Nigeria, where very complex oscillations of moisture-loving and xerophilic vegetation were measured palynologically. The picture agrees in general with glaciations at higher latitudes being correlated with dry conditions at low latitudes, partly because of marine regression, resulting from lower sea levels during glaciations. Primarily, however, the low-latitude dryness has to do with changes in atmospheric circulation brought on by glaciation: shifts in position of trade winds, etc.

Palynological information on late Pleistocene chronology and vegetational history

Woillard's (1978) work with the long Grande Pile peat core in France (Fig. 15.15) is very important as the most complete, continuous, non-marine palynological record in western Europe. The commencement of the Eemian (= Riss–Würm = Wisconsinian interstadial) at about 130,000 yr B.P., as indicated by arboreal pollen (AP), looks good. The end of the Eemian is at the break toward colder conditions at about 125,000 yr B.P. This trend agrees well with Shackleton & Matthews (1977) 125,000 yr B.P. level for Barbados, based on oxygen-isotope studies. In any event, Woillard's data and the oxygen-isotope data agree well with the Black Sea data in indicating one large, last interglacial (Riss–Würm = Sangamonian = "gamma"). Wijmstra's (1969) diagram for a long core from Macedonia looks remarkably like the Grand Pile core, with a prolonged glacial period, apparently a "gamma" (Black Sea), with

---

Chenopodiaceae (beet family). Microreticulate P0x pollen of abundant herbaceous dicot family (D.S.D.P.). Chenopodiaceous genera are difficult to separate on pollen characters. See also comment under (ad). (af), (ag) Compositae (aster family). Long-spined sort, polar view, high to mid-focus and high focus, respectively (D.S.D.P.). This huge family of herbs and shrubs is ubiquitous in the Pleistocene. It is difficult to separate the many genera on pollen characters in routine analysis – most of them are Pc3–echinate. (ah) Compositae. Long-spined type, mid-focus, showing the clearly expressed columellate structure (D.S.D.P.). See comments under (af) and (ag). (ai), (aj) *Ambrosia* sp./*Iva* sp. (ragweed, Compositae). Equatorial views, high focus and mid-focus, respectively (T.R.). This herb complex is the bête noire of pollinosis, and because it is a weed also is abundant in Holocene sediments where human habitation was a factor. (ak)–(an) *Artemisia* sp. (sagebrush, Compositae). Oblique–polar, polar mid-focus, polar high focus, and equatorial views, respectively, of short-spined Pc3 Compositae pollen (D.S.D.P.). Some species ("sagebrush") of this genus thrive especially in cold, dry environments such as large areas of the modern Rocky Mountains. It is characteristically associated with chenopod pollen in pollen floras sedimented from steppe areas. (There are *Artemisia* spp. with rather different ecological significance.)

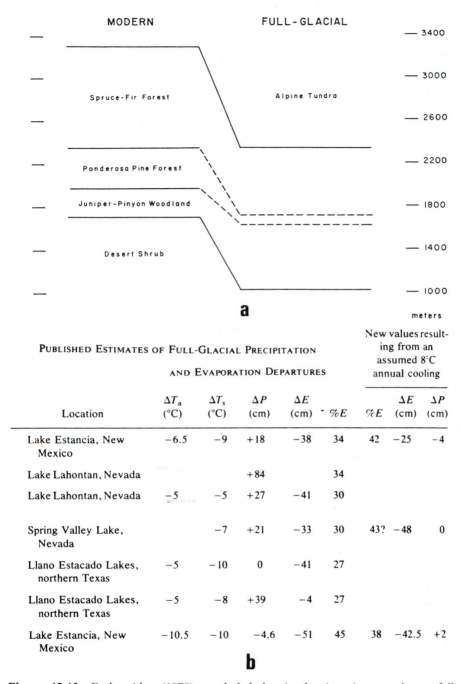

MODERN   FULL-GLACIAL

— 3400

— 3000

Spruce-Fir Forest   Alpine Tundra

— 2600

— 2200

Ponderosa Pine Forest

Juniper-Pinyon Woodland

— 1800

Desert Shrub

— 1400

— 1000

a

meters

	ΔT_a (°C)	ΔT_s (°C)	ΔP (cm)	ΔE (cm)	%E	%E	ΔE (cm)	ΔP (cm)

PUBLISHED ESTIMATES OF FULL-GLACIAL PRECIPITATION AND EVAPORATION DEPARTURES — New values resulting from an assumed 8°C annual cooling

Location	ΔT_a (°C)	ΔT_s (°C)	ΔP (cm)	ΔE (cm)	· %E	%E	ΔE (cm)	ΔP (cm)
Lake Estancia, New Mexico	−6.5	−9	+18	−38	34	42	−25	−4
Lake Lahontan, Nevada			+84		34			
Lake Lahontan, Nevada	−5	−5	+27	−41	30			
Spring Valley Lake, Nevada		−7	+21	−33	30	43?	−48	0
Llano Estacado Lakes, northern Texas	−5	−10	0	−41	27			
Llano Estacado Lakes, northern Texas	−5	−8	+39	−4	27			
Lake Estancia, New Mexico	−10.5	−10	−4.6	−51	45	38	−42.5	+2

b

Figure 15.13 Brakenridge (1978) concluded that in the American southwest full glacial conditions caused displacement to lower altitudes of vegetation communities, as shown in (a). The displacement was estimated from pollen and pack rat (megafossils from middens) data. Broken lines indicate displacements based on only two sites.

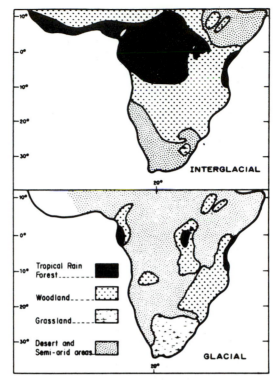

Figure 15.14 Generalized hypothetical vegetation maps of Africa south of the Sahara during glacial and interglacial maxima, based partly on palynological data. Glacial times were prevailingly drier. (On the basis of more recent information, Van Zinderen Bakker would slightly decrease both the tropical rainforest area and the desert area for the interglacial, while increasing the woodland area concomitantly.)

interstadials toward the bottom, identified by increased AP. Delcourt (1979) has analyzed cores from Tennessee, hundreds of kilometers south of the glacial limit, representing the last 25,000 years. The glacial maximum at 19,000–16,300 yr B.P. is represented in Tennessee by dominance of *Picea*, *Abies*, and *Pinus banksiana* pollen, probably indicating cold–dry conditions. Mixed mesophytes began to appear in the pollen record at 16,300 yr B.P., and this sort of forest was replaced largely by oak–hickory forest in the mid-Holocene.

Heusser (1977b) has investigated a variety of sites in central and northwestern North America where records going back from 30,000 to more than 50,000 yr

---

Brakenridge also concluded (b) that full glacials were *not* "pluvials", but that the observed increase in volume of southwestern lakes during glacials resulted from decreased evaporation. ΔT_a and ΔT_s are changes in annual and summer temperatures, respectively; ΔP is change in annual precipitation; ΔE is change in annual evaporation; $\%E = \Delta E/$modern E. Although previous estimates had claimed much elevated precipitation during glaciations, Brakenridge's recalculations show precipitation to have been little different from present levels.

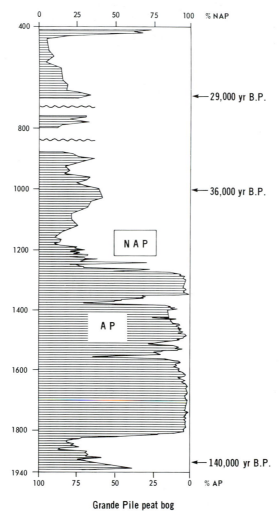

Figure 15.15 Perhaps the longest, most nearly complete late Pleistocene pollen record in existence, Woillard's (1978) study of the Grande Pile peat bog, Vosges Mountains, northeastern France. The NAP (non-tree pollen) vs. AP (tree pollen) curve shows the last major interglacial (Eemian = Sangamonian), which Woillard subdivides into Eemian plus St Germain I and II, but it clearly seems to be one interglacial with minor colder interludes. In any event, the AP dominance is clear. The collapse to the last full glaciation was dramatic, as forests largely gave way to steppe, tundra, and grasslands in periglacial areas and NAP greatly increased. The sensitivity of NAP/AP to glacial–interglacial conditions is characteristic all over the northern parts of the Northern Hemisphere.

B.P. are obtainable. In northwest Alaska and most other sites, evidence of a pre-40,000 yr B.P. interstadial or interglacial shows clearly. During fully glacial times, Illinois apparently somewhat resembled present-day northwest Alaska. As is discussed later, in connection with Figure 17.6, Delcourt (1979) was able to show, by comparison of surface sample pollen floras with palynofloras from cores, that Tennessee at 19,000 yr B.P. had vegetation resembling today's vegetation in south–central Manitoba, Canada.

CHAPTER SIXTEEN

Holocene interglacial palynology

INTRODUCTION

Most palynologists and others concerned with study of the (relatively) glacier-free time in which we live agree that, aside from human disturbance of Earth systems, it is presumptuous to speak of it as a completely different epoch, the "Holocene". Also, it is too soon to presume that the Pleistocene is over, and thus "post-glacial" is likewise inappropriate. In westernmost Europe there is a trend to speak of the "Flandrian" as the timespan of about 10,000 years since the last major retreat of continental ice sheets. However, this term cannot readily be transferred to other parts of the world. A term such as "present interglacial" has problems too, as we do not know for sure that another glacial time follows. For the present we seem to be obliged to follow convention and say "Holocene" for the last approximately 10,000 years. Holocene palynology has always had different approaches from palynology of older sediments, as we have seen. One can practically neglect the presence of extinct or grossly exotic species. Thus, floral studies based on known present plant associations and their ecological requirements, and rather precise paleoclimatological deductions, are possible. In the pre-Holocene Neogene, pre-10,000 years ago, the same sort of approach is possible, but the *SFI* curve in Figure 15.4 is an illustration of differences. The *SFI* curve is too crude for the Holocene, where more precise analytical methods are possible. On the other hand, the *SFI* curve is also progressively less good as one works back to the early Pliocene and into the Miocene, because plant communities with no modern analog were dominant in the Black Sea drainage, e.g., *Artemisia* and grasses were no longer found. Palynological count data in the pre-Holocene can be mathematically analyzed with multivariate analytical techniques to pick up associations of taxa that one might not recognize in terms of modern plant associations. This is a better approach than to attempt to reconstruct pre-Holocene forest communities from the pollen analytical results, as is possible in the Holocene, based on modern analogs. In the pre-Holocene we may be obliged to use fungal spores, acritarchs, algal colonies, wood and cuticle fragments, and especially dino-flagellate cysts to tell us things about temperature, salinity, and pH. The paleoecological approach in the Holocene has to do largely with forest history, plant-association history, and other related ecological matters. For the most part, peats and lake sediments, prevailingly autochthonous, are studied. In the pre-Holocene, we often study allochthonous sediments (though peats and lake sediments are also investigated), and we therefore look at a broader, regional picture for our environmental trends. In Holocene palynology the normal approach is very careful study of pollen rain from the potential source plant communities and their ecological requirements. Precise pollen analytical counts of very closely spaced samples from cored sediments, and the plotting of these analyses in rather standardized "pollen diagrams", is employed, in conjunction with ^{14}C dates (see examples later in this chapter). A good example of a broad-scale pre-Holocene approach is the work of Heusser *et al.* (1980), in which modern pollen "rain" of surface samples from Alaska to California was studied, and the known broad climatic indications were applied mathematically to pollen analytical data from cores obtained in Washington State, to ascertain probable temperature and precipitation at various levels in the cores representing about the last 80,000 years – well back into the Pleistocene.

HOLOCENE PALYNOLOGICAL METHODS

Pre-Holocene, Neogene palynologists mostly work with sedimentary materials obtained with conventional geologic methods: outcrop samples, oil-well cuttings, cored intervals, side-wall cores, and the like. Holocene (and late Pleistocene) palynologists are mostly interested in lake, swamp, and bog deposits reflecting the history of the local vegetation, and they sample them in a variety of ways quite different from conventional geologic methods. The devices used include the Hiller and other sediment (especially peat) samplers (Fig. 16.1a & b). Hiller and similar samplers are devices with a chamber on the end for taking a plug or core of soft sediment from a carefully measured depth. Extension rods enable reinsertion of the device into the sampling hole and repeated sampling to depths of 5 m or more. More commonly used now, where possible, because a continuous relatively undisturbed section of core is obtained, are various sorts of piston cores (Fig. 16.1 c–e). A tripod is usually used in pulling the core; chain hoists are sometimes used if the sediment is compact, but usually the corer can be pushed in and pulled out by hand, using rope or wire for attachment to the tripod. The coring device consists basically of tubing, usually aluminum, with a strengthening bit-like device on the end to penetrate the sediment, and a rubber piston inside. The tube sits on top of the sediment and maintains a partial vacuum behind the core when it is pulled, to prevent loss of the core. There are many variations on these themes, but the basic idea is the same. Piston corers can be operated in shallow water either by standing on the bottom or by using a raft or boat. When working in water, some sort of casing is necessary. The collected cores are sealed in the field, or may be extruded, described, and wrapped there. In the laboratory they can be subsampled by sawing open the sections of tubing lengthwise, at which time samples for other purposes can be obtained, the cores described, and photography of the relatively fresh sediments accomplished. Samples for palynology are processed in various versions of the standard methods given in the Appendix, though the prevalence of peat in the samples has caused Holocene palynologists to emphasize KOH cooking and acetolysis, both aimed at destruction of the abundant cellulose and cellulose derivatives in peaty sediment, and HF digestion where siliceous minerals are abundant. The

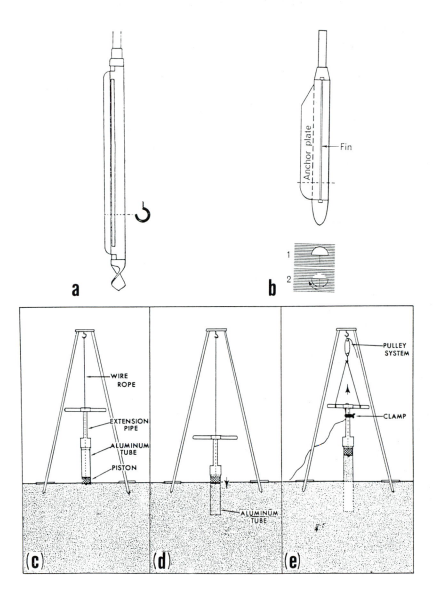

Figure 16.1 The three most common sediment samplers in use by late Pleistocene palynologists. All are really "generic" and have variants. (a) The "business end" of a Hiller peat borer, much used for sampling of fibrous peats. At the tip there is a sharp auger device which penetrates the peat as the auger is twisted into it, using a handle attached to the end of screw-together rods or pipe sections. The sampling chamber, shown also in section at right, is a cylinder within a cylinder, with an outer, flanged cutting edge, such that when the direction of turning is reversed to counterclockwise, a sample of peat is cut by the blade and taken in. Turning in the opposite direction closes the chamber, and the sampler and sample can be withdrawn. Most versions have a sample chamber about 50 cm long, and many meters of peat can be cored quite

351

completed slides are analyzed by counting the spores and pollen and calculating spores/pollen percentages and/or concentrations per gram of sediment, or annual pollen influx per area (cm^2) of sedimentary surface (for which rate of sedimentation must be known). Sometimes known amounts of a foreign pollen (most commonly *Lycopodium* spores, *Eucalyptus* pollen, or polystyrene spherules) are added to each sample before processing, so that pollen present in the preparation can be expressed as a ratio to this foreign pollen, especially as a means of determining the abundance of fossil pollen per unit of original sediment sample. The most common approach by far is to express the amounts of various sorts of spores/pollen as percentages of either total pollen or of a "pollen sum" from which certain forms, such as aquatics, are excluded. The idea is to get a number that relates to the composition of the vegetation that produced the pollen. This concern for relationship to vegetation composition has greatly influenced Holocene and late Pleistocene palynology because many palynologists in this specialty are ecologically based, and many sorts of pollen are well-known to be over- or underrepresented. However, pollen analytical data that do not directly relate to forest composition can nevertheless yield very valuable information about climatic and other changes. For example, some pollen may be reworked from older

---

accurately by repeated sampling. Unfortunately, the sample is somewhat contorted and smashed laterally, and may be contaminated superficially, though the vertical integrity is preserved. Various scientific supply houses sell these, or they can be made up by a machine shop. (b) The "Russian" peat sampler (sometimes called Macaulay sampler), which collects a 50 cm sample as does the Hiller, but has the advantage of not compressing the sample laterally during the sampling operation. This sampler is especially good for peat and sand. The sampler consists of a 50 cm × 5 cm half-cylinder, fixed to the sampling head, which is rotated 180° when the desired depth in the peat is reached. This encloses a half-cylinder of peat against the central anchor plate which remains stationary as the half-cylinder is rotated (see sectional drawings 1 and 2). The half-cylinder sample is bisected by a fin plate at right angles to the anchor plate. The sample is then easy to remove and is usually enclosed in plastic sheeting in the field. Unfortunately the only way to obtain such a sampler is to have one made by a machine shop. (c)–(e) One version of the piston coring device, often called a Livingstone piston corer or Livingstone sampler, after D. A. Livingstone, who first used such sampling devices. This sort of sampler is poor for fibrous peat but excellent for non-fibrous peat and mud. The operation sequence is: (c) preparatory to coring; (d) coring tube pushed into ground while piston is held stationary with respect to ground by wire fixed to tripod; (e) core withdrawn from ground while piston is held stationary with respect to core tube by clamping wire rope to extension tube pipe. The sample is brought to the surface in sections of aluminum tubing, which can be sealed, frozen if desired. The tubes can be sawed in half later in the laboratory for interval sampling. In other versions, the sampling tube for the piston is steel and permanent, and the sample is extruded from it after collection into a plastic tube for storage and study. The accuracy of piston samplers is in general better than that of Hiller-type samplers, though compression of sample and loss of various intervals occurs with piston devices. Some of my students have used locally available materials such as irrigation pipes for collecting tubes and trees cut down at a site for the tripod, to approximate the set-up shown here, in places such as remote areas of Honduras and Montana.

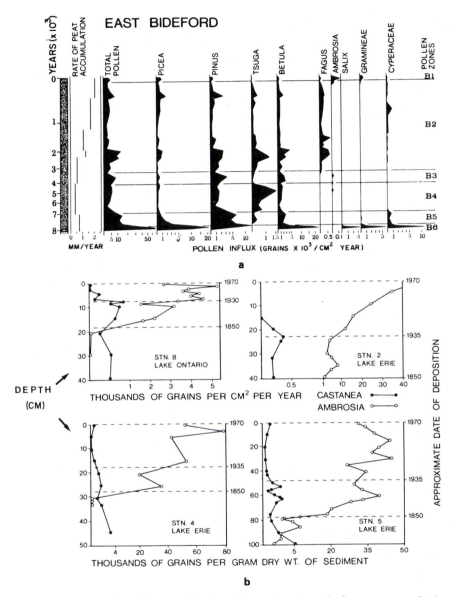

Figure 16.2 Examples of the use of pollen influx ratios instead of percentages of either total palynomorphs or of a selected pollen sum. (a) East Bideford, Prince Edward Island, Canada, diagram on pollen influx basis, for comparison with same data diagrammed on percentage basis in Figure 16.7. This method requires ^{14}C or other precise dating, so that the annual amount of sediment accumulation can be measured (see "rate of peat accumulation"). Then the number of grains per area (cm^2) per year can be calculated. The method avoids the problems of (1) overabundant forms swamping out the percentage of other forms and (2) the effects of changing sedimentation rates on calculations of concentrations per gram of sediment. (b) Use of per gram or per volume

353

sediments and indicate from their presence erosion in the basin of deposition caused by high precipitation levels.

In many branches of palynology it has been pointed out that percentage data have a built-in bias, because the percentages must total 100%, and the percentage of pollen A therefore influences the percentage of pollen B. Palynomorphs per gram of sediment, or as a ratio to an added "constant", or per area per year (pollen influx), or per volume of sediment have been suggested as alternatives. The subject is more fully discussed in the next chapter. Figure 16.2b shows an application of both pollen influx and pollen concentration methods to a problem of change in a very short timeframe.

PRESENTATION OF "POLLEN ANALYSIS" DATA

Symbolic presentation of data for the various pollen types was developed quite early in the history of late glacial/postglacial pollen analysis. Jessen (1920) used the term "pollen spectrum", and pollen analysts thus developed a "spectral" diagram, presented in circular form. Each of the kinds of tree pollen counted was presented as a fraction of the circle; consequently a form with 10% would get 36° on the circle. The taxa were presented in the same order and in the same pattern, to make the diagrams easier to compare. However, it was not practicable to display more than a few taxa, and these "pie" diagrams are no longer used. The use of circles of different sizes on maps to demonstrate the relative abundance of a certain type of pollen or pollen associations at various localities is a related idea.

Pollen analytical results lend themselves to representation in diagrams in which the depths of the samples are displayed on the ordinate axis, and the percentages or other indication of amount of pollen are shown on the abscissa. Frequently the concentrations are expressed logarithmically or to different scales. An early idea was to connect the points and identify them by using symbols for each sort of spores/pollen, e.g., solid circles for *Pinus*. Figure 16.3 shows a modern pollen diagram employing such symbols for a few taxa. The system of symbols originally introduced by Erdtman is shown to the right.

The line-and-symbol diagrams are sometimes difficult to read. When more than five or six taxa are on the diagram, it is practically impossible. The symbols are not always consistently used. Hence, line-and-symbol diagrams were mostly displaced by "sawblade" diagrams, of which Figures 16.6, 16.7 and 16.8 are examples. As can be seen from the figures, such diagrams made it possible to display not only the sorts of tree pollen (= arboreal pollen = AP) but also non-tree pollen (NAP) and aquatic pollen (AqP). Some authors have used bar graphs in the same way. Because it could be argued that the points

---

and influx pollen concentration to measure the near-disappearance of *Castanea* (chestnut) and the great expansion of *Ambrosia* (ragweed) in the time of human population explosion in the eastern Great Lakes area of North America. Note that in this particular case, as in many, concentration per gram of sediment, a much less expensive procedure, yields acceptable results. The methods of calculation are explained in the text.

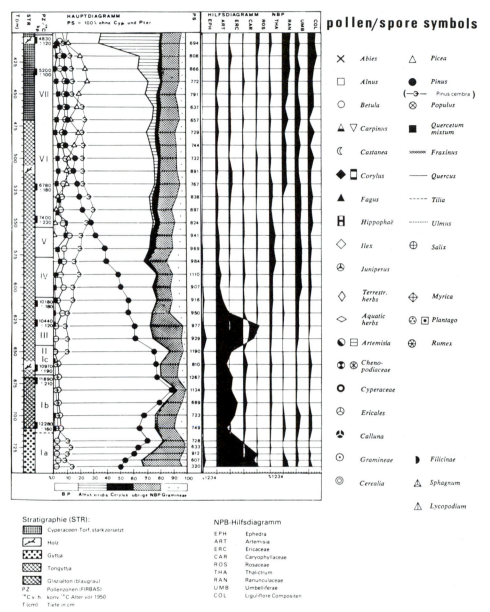

pollen/spore symbols

✕	*Abies*	△	*Picea*
☐	*Alnus*	●	*Pinus*
		(—⊙—)	Pinus cembra)
○	*Betula*	⊗	*Populus*
△ ▽	*Carpinus*	■	*Quercetum mixtum*
₵	*Castanea*	xxxxxx	*Fraxinus*
◆ ☐	*Corylus*	——	*Quercus*
▲	*Fagus*	-----	*Tilia*
⏚	*Hippophaë*	········	*Ulmus*
◇	*Ilex*	⊕	*Salix*
⊛	*Juniperus*		
◇	Terrestr. herbs	⊕	*Myrica*
◇	Aquatic herbs	⊙ ▣	*Plantago*
◕ ⊟	*Artemisia*	⊛	*Rumex*
◑ ⊛	Cheno-podiaceae		
○	Cyperaceae		
⊛	Ericales		
♠	*Calluna*		
⊙	Gramineae	▶	Filicinae
◎	Cerealia	△	*Sphagnum*
		⚠	*Lycopodium*

Stratigraphie (STR):

▦	Cyperaceen-Torf, stark zersetzt
⟋	Holz
▨	Gyttja
▨	Tongyttja
▨	Glazialton (blaugrau)
PZ	Pollenzonen (FIRBAS)
¹⁴C v. h.	konv ¹⁴C-Alter vor 1950
T (cm)	Tiefe in cm

NPB-Hilfsdiagramm

EPH	Ephedra
ART	Artemisia
ERC	Ericaceae
CAR	Caryophyllaceae
ROS	Rosaceae
THA	Thalictrum
RAN	Ranunculaceae
UMB	Umbelliferae
COL	Liguliflore Compositen

Figure 16.3 The use of symbols for pollen and spore types in Pleistocene palynological diagrams. To the right, the symbols commonly used in Europe, with one parenthetical additional type, *Pinus cembra*, because it occurs in the example of an analysis from the German language literature, shown to the left. The example is a late glacial to hypsithermal record from Sass de la Golp, southern Switzerland. Explanation of German captions: STR = stratigraphy; PZ = pollen zones (Firbas zones – see ·Fig. 16.4); Hauptdiagramm = main diagram; PS = pollen sum; Hilfsdiagramm =

355

measured really represent separate "events", not a connected curve, this has some conceptual advantages. However, "sawblades" are easier to read and are the customary method. In the late glacial to Flandrian diagram shown in Figure 16.3, the main diagram uses symbols, and the various supplementary diagrams use sawblades. I feel strongly that most late glacial/Holocene pollen analytical diagrams could be improved by standardization and simplification. Some units are shown for which the counts are probably too small to be statistically significant. Such counts should be listed in a table. In fact, I believe that the great majority of publications would be improved by giving much of the raw data (counts) in tables and diagramming only the major or most significant taxa preferably by composite summary, and smoothed-out curves. Unfortunately, the trend has been in the other direction: diagrams with sawblades for taxa not even mentioned in the text, diagrams so large they have to be presented as fold-outs. The purpose of diagrams is exclusively visual, and if the purpose is really just to list the numbers, an appendix or table is better. Unfortunately, the literature is full of pollen analytical fold-out diagrams, some with as many as six folds. A few simple diagrams of the major points would be more useful, more readable, and more likely to be consulted and understood.

Another problem with glacial–Holocene palynological diagrams is that the taxa are often given in no standardized order, but according to "dealer's choice". It is not possible to assume that curves for *Alnus, Abies, Fagus, Salix,* etc., will always be found in the same part of the diagram, as might seem reasonable. One prominent Quaternary palynologist wrote me:

> You asked if there is any convention in the ordering of genera in Quaternary pollen diagrams. Authors vary greatly, and there is no generally accepted discipline. The Minnesota practice, which I . . . think best is to place genera in order of time of appearance, so the order will differ between diagrams. It is also usual to divide the diagram into sections for trees/shrubs/woody vines/herbs (including the pteridophytes, usually grouped together) and obligate aquatics . . . The modern practice is to have a percentage pollen sum calculated on everything, excluding only obligate aquatics such as water-lilies. Pollen sums based on trees alone are increasingly unusual. It is also unusual now to exclude the pteridophytes from the pollen sum. (W. A. Watts personal communication 1980, quoted with permission.)

---

supplementary diagram; NBP = NAP; BP = AP; übrige NBP = other NAP; Cyperaceen-Torf, stark zersetzt = much decomposed sedge peat; Holz = wood; Tongyttja = clayey gyttja; Glazialton (blaugrau) = glacial clay, blue-gray; ^{14}C v.h. konv. (^{14}C-Alter vor 1950) = ^{14}C age before 1950 (= yr B.P.); Tiefe in cm = depth in cm; Compositen = Compositae. Note high *Artemisia* in the late glacial (before 10,000 yr B.P.) and the decline of *Pinus* in the Holocene. The use of the symbols is not common outside of Europe.

This quotation also makes the point that the basis for calculation varies somewhat from author to author. In addition to the pollen sum and percentage methods, one must be alert for less common calculations based on neither, but on pollen per gram of sediment, or pollen influx based on calculations of pollen sedimentation per unit surface area or per volume of sediment.

HOLOCENE CHRONOLOGY

Typical transition to the Holocene in northern Europe features decrease of *Artemisia* and increase of *Betula, Juniperus,* and *Salix* at the end of the Weichselian (Würm). The "Alleröd interstadial", generally viewed as occurring just before the end of the last glaciation, could perhaps just as well be taken as the beginning of the Holocene, although it was followed by renewed glaciation. It should also be noted that the Holocene is in a sense a relative chronological unit, beginning at different times in different places, although we use 10,000 yr B.P. as a sort of convenient average. (In central Greenland, one could argue that it is not yet fully Holocene. Similarly, in a sense, northern Scandinavia is now about Dryas I or earlier.) The names of the stratigraphic–chronological terms usually used are shown in Figure 16.4. The divisions are based on northern Europe, where the work began, and are now time-stratigraphic, based on ^{14}C dating. In northeastern North America, where Holocene research also has a long history, these units have little applicability. The "Two Creeks interstadial" in North America, for example, is probably not directly equivalent to the Alleröd interstadial. On the other hand, Holocene diagrams from Japan, though based on quite different taxa, have the same general shape as those of Europe (see Tsukada 1957, 1958).

The pollen diagram of the British Flandrian (= Holocene) is so well-known that standardized versions of it have even been used as a logo (by the American Association of Stratigraphic Palynologists for several conventions). Figure 15.10b shows a simplified, caricature version. The diagram shows the termination of the latest glacial and the earliest Flandrian. *Betula–Pinus* dominance in the early Holocene is replaced by interglacial-type dominance of *Quercus–Alnus–Ulmus–Tilia* in the middle Holocene. This dominance characterized the warmest part of the Flandrian, the "climatic optimum" (in some of the literature called "hypsithermal"), about 9,000 to 2,000 yr B.P. (The warmest middle part of this period is sometimes called the "altithermal": about 7,500–4,000 yr B.P.) During the last 2,000 years it has not been so equable in western Europe, and birch, for example, has re-expanded its range in northern Europe. Some northern forest lands have reverted to scrub. In various places, all or part of this post-hypsithermal is known as the "little ice age".

Figure 16.5 shows the application of Firbas (F) and Overbeck (O) schemes in Germany in particular, with the forest equivalents given in Figure 16.5b. The Bölling interstadial is omitted in many diagrams. The original German has been left on the figure as the German terms were given great currency by Firbas and by Overbeck, and because, from Von Post (1916) until after 1945, Scandinavian palynologists frequently wrote in German, so students will

Years B.P.	BLYTT–SERNANDER ET AL.	MONTELIUS	FIRBAS 1949		OVERBECK–SCHNEIDER 1938	JESSEN–IVERSEN 1935–1941	GODWIN 1956	
Present			X	Nachwärmezeit	XII	IX	VIII	POSTGLACIAL
1,000	Subatlanticum	Iron Age			XI			
2,000			IX		X			
3,000	Subboreal	Bronze Age		IX		VIII	VIIb	
4,000		Neolithicum	VIII	Spätewärmezeit				
5,000			VII		VIII			
6,000	Atlanticum			Mittlerewärmezeit		VII	VIIa	
7,000			VI					
8,000		Mesolithicum	Vb	Frühewärmezeit	VII	VI	VI	
9,000	Boreal		Va		VI	V	V	
10,000	Preboreal		IV	Vorwärmezeit	V	IV	IV	
11,000	Younger Dryas		III	Jüngere Dryas	IV	III	III	LATE GLACIAL
12,000	Allerød		II	Allerød	III	II	II	
	Older Dryas	Paleolithicum	Ic					
13,000	Bølling		Ib	Ältere–Dryaszeit	II	I	I	
14,000	Pleniglacial		Ia					
15,000					I			

Figure 16.4 Various schemes for pollen zonation of the late glacial (Weichselian) and post-glacial or Holocene (= Flandrian = present interglacial) of Europe, along with the vegetation zones of Blytt–Sernander, and the prehistorical designations of Montelius. Compare with Figure 15.7.

358

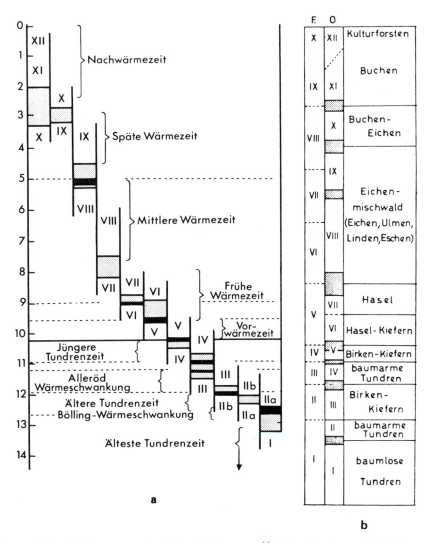

a

b

Figure 16.5 (a) Summary of relationship between ^{14}C dates (scale on the left, in years B.P.) and Overbeck Holocene subdivisions for western Europe from a series of overlapping European records. The stippled areas between Roman numerals represent range of ages, black bands signify zones of especially numerous dates. (B) Firbas ("F") and Overbeck ("O") zonation compared, for Germany, with the vegetational history indicated by pollen analyses (see Fig. 16.4). Original German terminology retained because much of it has been widely used in Holocene palynology. Nachwärmezeit = Sub-Atlantic, or post-climatic optimum; Späte Wärmezeit = Subboreal; Mittlere Wärmezeit = Atlantic; Frühe Wärmezeit = Boreal; Vorwärmezeit = Preboreal; Jüngere Tundrenzeit = younger tundra period (later Sub-Arctic); Alleröd Wärmeschwankung = Alleröd warm oscillation; Ältere Tundrenzeit = older tundra period (earlier Sub-Arctic); Bölling Wärmeschwankung = Bölling warm oscillation; Älteste Tundrenzeit = oldest tundra period (Arctic); Kulturforsten = cultivated forests; Buchen = *Fagus*

359

encounter the terms. The numbers applied to the late glacial and interglacial "times" comprise a classification named for Blytt and Sernander, Scandinavian botanists–geologists of about a century ago, who noticed evidence for post-glacial floral change and suggested in outline a model for the post-glacial fluctuations upon which Von Post, Firbas, Jessen, Overbeck, and Godwin, among others, later built. Firbas' 10 "times" have been mostly replaced by Overbeck's 12 "times". Others have used still different numbers. This is unfortunate because now "IX" in some papers does not mean the same as "IX" in other contributions.

The "chronology" applies only in northern and central Europe, and even there it is time-transgressive. Parts of central Europe are now in the Sub-Atlantic, but much of Scandinavia is not, if *Fagus* is the signature. (One can define the Firbas, etc., zones radiometrically with arbitrary dates, and then they are not, of course, time-transgressive.) Figure 16.6 shows a typical sawblade-style pollen diagram for Germany. The vegetational history can vary considerably even in a rather small region, especially if there are considerable altitudinal differences between sites studied.

The palynologically–botanically based chronology has been very useful as a reference, and before radiocarbon even for dating. In northern Europe it was long ago observed that at about the beginning of the Iron Age, 500–600 B.C., the nature of the peat in raised bogs changed abruptly – a "Grenz Horizont" – what is probably the same thing shows in some places as the lower boundary of Godwin zone VIII (= bottom of Firbas IX) (see Fig. 16.4; obviously at other times elsewhere). Another example of direct application of the European Flandrian (Holocene) chronology is the demonstration of "isostatic rebound" in Oslofjord, Norway, by pollen analysis, as is discussed below (see Fig. 16.13). When Von Post introduced pollen analysis as a practical tool in 1916 he considered it primarily of geochronological usefulness (Faegri 1974). However, with the advent of direct, absolute ^{14}C radiometry, this aspect of pollen analysis (in Holocene palynology) has taken a back seat to the use of the art for analysis of vegetational change.

An attempt has been made in eastern North America to establish a Blytt–Sernander type division of the Holocene. A classification analogous to the Blytt–Sernander one was introduced by Deevey (1949) (see Fig. 16.7a). Largely because North America is too large and too diverse for such a scheme to work over any very large area, Deevey's or similar schematic subdivisions have not been widely adopted. The pollen zones are clearly local and time-transgressive. "C3", for example, cannot be applied as a dateline in both Ohio and Nova Scotia. It does seem to be true that a three-fold division of the "post-glacial" in the northeast U.S.A. is a reality. Figure 16.7b shows that, when Holocene pollen zones are used in North America, they are usually described in such a manner as to make clear that the zones are purely local in significance.

(beech); Eichen = *Quercus* (oak); Eichenmischwald = *Quercetum mixtum* (mixed oak forest); Ulmen = *Ulmus* (elm); Eschen = *Fraxinus* (ash); Hasel = *Corylus* (hazel); Kiefern = *Pinus* (pine); Birken = *Betula* (birch); baumarme Tundren = tree-poor tundra; baumlose Tundren = treeless tundra.

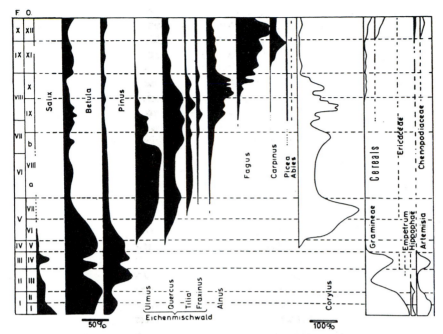

Figure 16.6 Schematic latest Pleistocene pollen diagram from near Göttingen, Germany. Black curves = AP; white curves = NAP. Only the most significant NAP are shown. See Figures 16.4 and 16.5 for orientation. F = Firbas, O = Overbeck. Eichenmischwald = *Quercetum mixtum* (mixed oak forest). Note that this association is the "signature" of the hypsithermal in Europe. Getreide = cereal grasses.

SOME CHARACTERISTIC HOLOCENE POLLEN ANALYSES

Europe In a "classic" British Flandrian pollen analysis, of the sort on which the caricature in Figure 15.10b is based, there is a *Corylus* maximum in VI, the *Betula* decline beginning in VI, and the *Ulmus* decline at VIIA–VIIB. (Numbers are in the Godwin sequence, per Figure 16.4.) The diagram lacks the continental *Fagus–Carpinus* "signature" of the Sub-Atlantic (see Fig. 16.6). It has been shown (Birks 1973) that this diagram cannot be directly applied outside of England, not even in Scotland. Figure 16.6 is a characteristic continental diagram. Note especially the climatic optimum (Wärmezeit) and the *Ulmus* decline (later than in England). Note also the grain grasses as evidence of human activity in the Sub-Atlantic. Swiss lake diagrams show abundant steppe pollen (*Artemisia*–Chenopodiaceae–grasses) in the late glacial, and sometimes *Juglans* pollen to show introduction by Roman settlers of walnuts about 2,000 B.P., an interesting example of human activity. In the Mediterranean very different Holocene diagrams are obtained. Mixed oak forest is present from the beginning of the Holocene. *Pistacia*, not seen in central or northern Europe, is a characteristic form. The decline of steppe pollen begins much earlier in this area than farther north.

361

Pollen zone	Pollen description	Zone Age (years B.P.)
C-3	Return of spruce and fir	0 ± 80 - 1600 ± 100
C-2	Hemlock minimum; pine, beech, oak, and birch maxima	1600 ± 100 - 4000 ± 120
C-1	Hemlock maximum; pine minimum	4000 ± 120 - 7250 ± 175
B	Pine maximum at about 8500 ± 180 years B.P.	7250 ± 175 - 9300 ± 200
A	Spruce and fir	9300 ± 200 - 10,500 ± 200

a

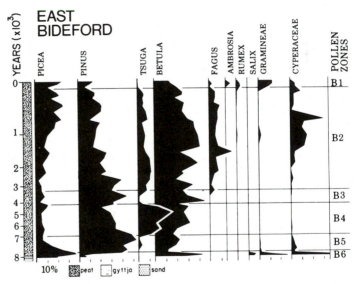

b

Figure 16.7 Holocene palynological zonation in North America. (a) An attempt was made to establish a Blytt–Sernander type division of the "post-glacial" for North America, and this has been applied by various palynologists since. Such a zonation by McDowell *et al.* (1971) for Bugbee Bog, Vermont, is shown here, along with radiocarbon dates. The A–B–C1–C2–C3 division came originally from Deevey (1949). The hypsithermal (climatic optimum), consisting of approximately B–C2, was originally published by Deevey & Flint (1957). The hypsithermal was originally defined as having time significance, but the climatic signal is actually time-transgressive, occurring at different times in different localities. The scheme is now mostly of historical significance. (b) Other palynologists recognize only local pollen zones and thus use arbitrary letters, partly to avoid confusion with the Deevey zones. In this diagram by Anderson (1980), of a site in Prince Edward Island, Canada, the "B" notation for the zones refers only to the site name (Bideford).

362

Students should consult the very useful series of pollen maps of Europe by Huntley & Birks (1983), covering the last 13,000 years. These maps show for Europe the history of vegetational changes as revealed by pollen analysis, in a most dramatic way.

North America   This continent is too vast and vegetation too variable to produce the sort of common thread that pollen diagrams for northwest Europe have had. This is certainly one reason why North American Holocene studies have lagged behind European, though research was already under way in the 1920s. During the 1930s and 1940s "friends of pollen analysis" published a *Pollen and Spore Circular*, mostly under the auspices of Paul Sears at Yale, later at Oberlin. Many scientists, especially Pleistocene geologists, saw the potential importance of pollen analysis in America and encouraged it. Kirk Bryan, geomorphologist at Harvard, and R. F. Flint, Pleistocene geologist at Yale, are examples. (It was in the *Pollen and Spore Circular* that Hyde & Williams (1944) introduced the word "palynology".) The thinness of the coverage is dramatized by the presence in Pennsylvania (about 650 km wide) of only a handful of Holocene pollen analyses.

The pollen diagram for a locality in Prince Edward Island in the Atlantic provinces of Canada (Fig. 16.7b) shows that replacement of *Picea* by *Pinus*, characteristic of northeastern U.S.A. diagrams, has only partially occurred in Prince Edward Island, and the subsequent rise in oak pollen seen in the presumed climatic optimum of New England has not happened at all this far north. This sort of phenomenon is called a migration lag.

The fact that a climatic optimum warmer than present existed in the Holocene of northeastern U.S.A. is dramatically demonstrated by Davis *et al.* (1980) from studies in the White Mountains of New Hampshire, where pollen and megafossils show that *Pinus strobus* and *Tsuga canadensis* trees grew at elevations hundreds of meters higher than their present limits for thousands of years of the Holocene.

In Appalachian parts of North America the glaciations caused withdrawal of vegetation types southward into a somewhat larger continent, as world sea level fell. Figure 16.8 presents a West Virginia diagram by Watts (1979); he has published pollen diagrams from a number of previously poorly covered areas south of the glaciated parts of eastern North America. As would be expected, events are well ahead of those in New England. In all of these diagrams the biggest changes in vegetation coincide with the end of the Wisconsinian glaciation. (This sort of change is, however, not synchronous across the U.S.A.) Nothing else approaches this in magnitude. In South Carolina there is practically no *Picea* record. *Nyssa* and *Liquidambar* invaded South Carolina about 9,500 yr B.P. The refugium was probably well down in Florida (Watts 1980). The change in vegetation occasioned in north–central North America by retreat and disappearance of the glacial ice was profound and affected Pleistocene animal populations directly (Whitehead *et al.* 1982).

In the western U.S.A., various diagrams show that we are in a position to map the migration of some taxa during the Holocene. Figure 16.9 shows that in New Mexico a big change occurred at the end of glacial time with massive

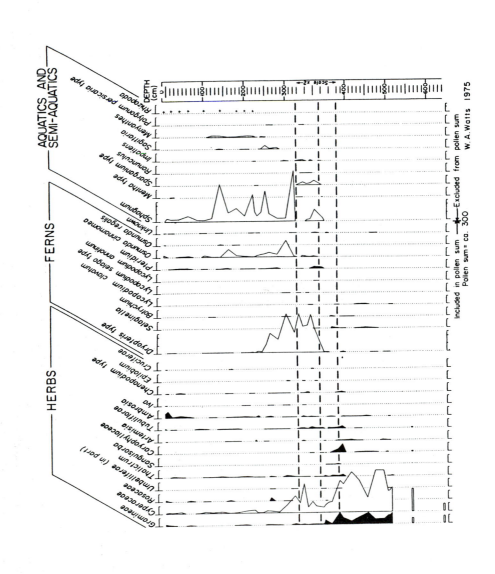

AQUATICS AND SEMI-AQUATICS

FERNS

HERBS

DEPTH (cm)

W. A. Watts 1975

Included in pollen sum ⟶ Excluded from pollen sum
Pollen sum = ca. 300

Scale x2

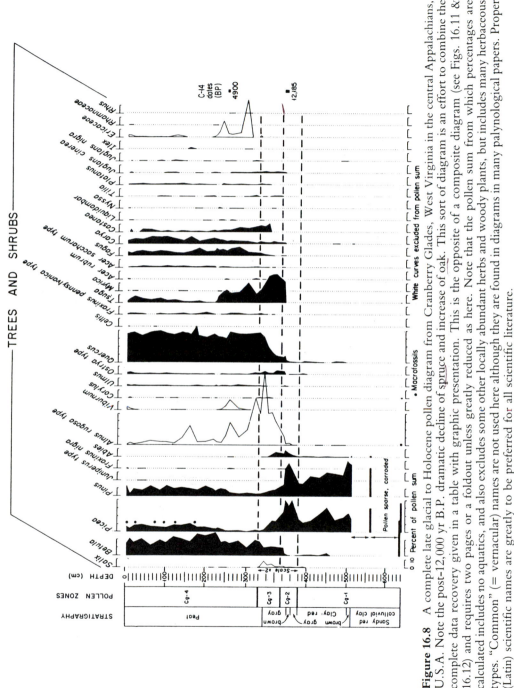

Figure 16.8 A complete late glacial to Holocene pollen diagram from Cranberry Glades, West Virginia in the central Appalachians, U.S.A. Note the post–12,000 yr B.P. dramatic decline of spruce and increase of oak. This sort of diagram is an effort to combine the complete data recovery given in a table with graphic presentation. This is the opposite of a composite diagram (see Figs. 16.11 & 16.12) and requires two pages or a foldout unless greatly reduced as here. Note that the pollen sum from which percentages are calculated includes no aquatics, and also excludes some other locally abundant herbs and woody plants, but includes many herbaceous types. "Common" (= vernacular) names are not used here although they are found in diagrams in many palynological papers. Proper (Latin) scientific names are greatly to be preferred for all scientific literature.

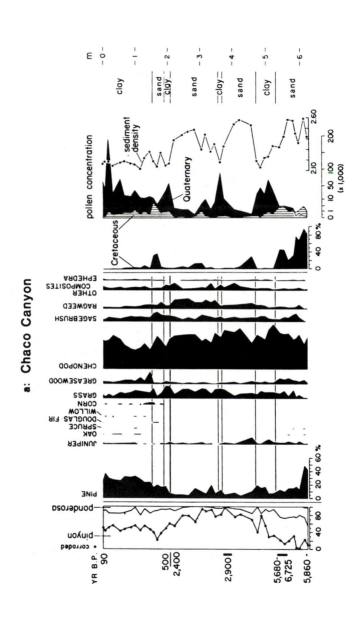

a: Chaco Canyon

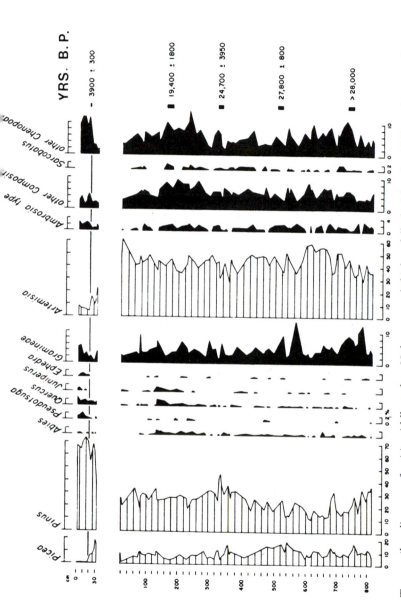

Figure 16.9 Two pollen diagrams for (a) middle to late Holocene, and (b) last full glacial and middle Holocene, of northwestern New Mexico, U.S.A. Note the importance of Gramineae (= grass) and Chenopodiaceae plus *Artemisia* (= sagebrush), the signature of Northern Hemisphere steppes, in the record right down to the present, though more abundant in the full glacial. These diagrams also illustrate several other points. Note the separation of pine species into *P. ponderosa*, *P. edulis*, and corroded pine at the left in (a); and also the plotting of Cretaceous reworked (also called secondary) palynomorphs at the right. Pollen concentration represents grains per gram of sediment. The ages indicated at the left would have to be more closely spaced to permit calculation of pollen influx accurately, but the Chaco Canyon record is from fluvial sections, for which pollen influx is not as significant as for lakes, ponds, and bogs. Students should note from these and other diagrams that Pleistocene palynologists are not consistent in the use of Latin versus vernacular plant names ("Douglas fir" = *Pseudotsuga*, etc.), nor is there any firm convention as to the order of presentation of taxa, e.g., *Quercus* (oak) is before *Juniperus* (juniper) in (b) but not in (a). Clearly Pleistocene palynologists tend to plot the taxa as the "sawblade" graphs fit best on the diagram. Some sort of convention would probably be desirable.

decline of *Artemisia* ("sagebrush") and grasses. *Pinus* greatly expands post-11,000 yr B.P. and has been more than 80% of all tree pollen for thousands of years. In some western diagrams ash-falls are very significant as marker horizons for dating purposes. Figure 16.9a also demonstrates the very frequent use of vernacular names for taxa on diagrams. In scientific literature this should probably be avoided, though in presenting talks, review papers, or newspaper articles to laymen, one can scarcely get around it. "Sagebrush" is a good example of the technical problems: although "sagebrush" is usually restricted to various species of *Artemisia*, "sage" commonly means dozens of species in at least four different genera, in different families! In Figure 16.9 "sagebrush" means *Artemisia*, "ragweed" is *Ambrosia et al.*, greasewood is *Sarcobatus*, and the other vernacular names are unequivocal. In the southwestern U.S.A. and Mexico generally, the comparative dearth of coreable sediments (natural lakes or wetlands) has hampered the rapid expansion of the database, though much work has been done, even, for example, on pack rat middens. Pollen studies in the southwest U.S.A. are especially important because they tie in with archeological and dendrochronological studies, as has been abundantly demonstrated by the work of Bryant and coworkers in Texas and vicinity (see Bryant & Holloway 1985).

From the western U.S.A., Figure 16.10 presents pollen analytical data from California and Colorado. The decline of *Artemisia* and Gramineae post-7,000 yr B.P. plus expansion of *Abies* is notable in the California diagram, which also illustrates the value of showing macrofossil data in association with pollen stratigraphy. Charcoal influx presumably indicates human activity as well as natural fires. The diagram from Colorado is given to show an interesting graphical technique, stressing concomitant AP expansion and *Picea* decline in the post-7,000 yr B.P. part of the record.

Figure 16.11 presents diagrams from Iowa in the central U.S.A. The Lake Okoboji diagram illustrates use of key taxa to typify "stratigraphic zones" in the record. Here NAP *expands* in post-glacial time! This reflects prairie expansion, with increasing grass and composite pollen. The generalized composite diagram to the right is also shown for Iowa, to illustrate this very useful method of dramatizing the data. Such diagrams are perhaps better than detailed diagrams, with the supporting data to be presented in tables. The overall Holocene vegetational history of the north–central U.S.A. is now reasonably well understood from many pollen analyses coupled with radiocarbon dates (Webb *et al.* 1983).

Figure 16.12 illustrates diagrams from north–central and northwestern North America. In the Yukon and Alaska, steppe taxa decline post-9,000 yr B.P., and most tree genera (*Picea, Pinus, Betula, Alnus*) greatly expand soon thereafter. *Picea* declines in Saskatchewan post-8,000 yr B.P., however. The summers became too hot and total effective precipitation too low, presumably. Figure 16.12c is also an excellent illustration of the desirability of generalized, smoothed-out, composite summary diagrams that make it easier for the reader to grasp the main points. (More detailed information can be made available in tabular form.)

a: BALSAM MEADOW, CALIFORNIA

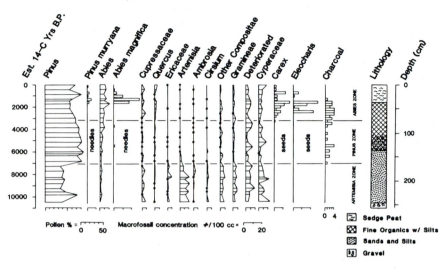

Pollen % = [0 — 50]   Macrofossil concentration #/100 cc = [0 — 20]

☒ Sedge Peat
☒ Fine Organics w/ Silts
☒ Sands and Silts
☒ Gravel

b: POUDRE PASS BOG, COLORADO

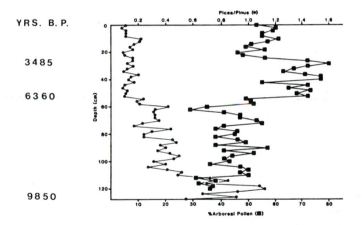

Figure 16.10 Pollen analytical data from the western U.S.A., also illustrating some graphical techniques. (a) In this diagram from California, note the decline of *Artemisia* and Gramineae and increase in *Pinus* since 6,500 yr B.P. Also note that megafossil (= macrofossil) elements are plotted and are much more abundant in the last few thousand years, also reflecting change in the abundance of vegetation. Charcoal abundance indicates forest fires and is probably related to human activities. The pollen percentages in this case are based not on total pollen but on a *pollen sum* of all non-aquatic plants. (b) This diagram from Colorado shows another technique for presenting AP vs. NAP, plus the relative abundance within AP of *Picea* vs. *Pinus*. The relative decrease in *Picea* probably bespeaks warming.

369

APPLICATIONS OF HOLOCENE PALYNOLOGY

One of the first applications of pollen analysis was to archeology. The Iron Age was linked to the beginning of the Sub-Atlantic (Firbas zone IX) of central Europe. The onset of agriculture was marked by the first abundant occurrence of pollen of human-introduced weeds. The spreading of heath in Scotland from clearing and grazing activities (Birks 1973), the clearing of forests for field crops, and the coming in of cereal grains and walnuts are all shown in pollen analyses in Europe. Van Geel (1972) was able to show the advent of field cropping to a section of western Germany about 700 B.C. and *Secale* pollen commencing about 50 B.C. A pollen study of the sediments associated with a Neanderthal burial is credited with showing that an individual who died some 50,000 years ago was buried in or on a bed of flowers (Leroi–Gourhan 1975). Palynologists have studied the stomach contents of a late glacial musk ox (Benninghoff & Hibbard 1961) and of burial vases (Rue 1982), with interesting and significant results.

Pollen analysis continues to be important to archeology, but the problems encountered are often considerable. For example, sediments from cave sites are

a: LAKE OKOBOJI, IOWA

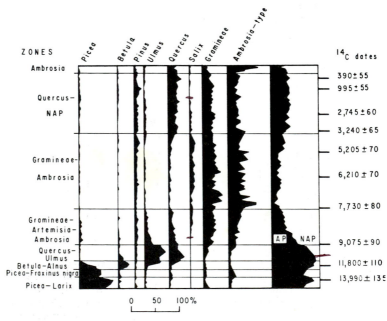

Figure 16.11 Diagram of Wisconsinian glacial to Holocene time, Iowa, central U.S.A. (a) A diagram from a large lake in Iowa, showing the use of dominant genera as names for pollen zones. Obviously the diagram is of selected forms only, as several of the typifying genera for zones (*Artemisia*, *Fraxinus*, and others) are not on the diagram. The increase of NAP from about 9,000 to 3,000 yr B.P. clearly represents increase in

370

b: IOWA—THE LAST 30,000 YEARS
A GENERALIZED COMPOSITE POLLEN DIAGRAM

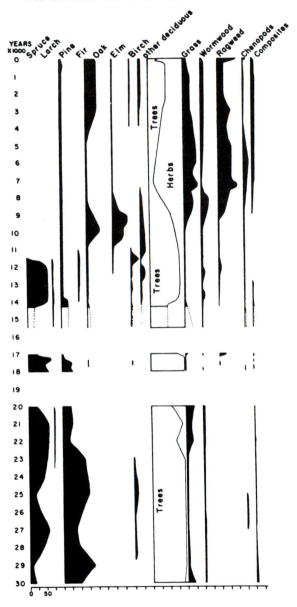

Gramineae and Compositae, reflecting the expansion of prairie. (b) A composite, generalized, smoothed-out diagram for the past 30,000 years, based in part on the same information as in (a). Generalized diagrams have much to recommend them, and they can now be computer-generated from the pollen counts. Purely objective, all-inclusive, diagrams tend to be printed too small to read, or they appear as bulky foldouts.

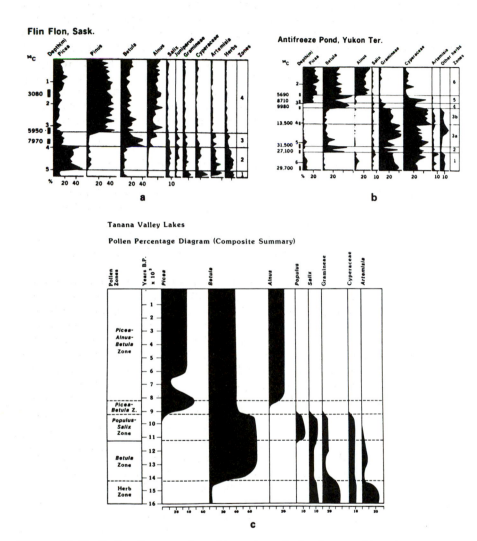

Figure 16.12 Pollen diagrams from (a) northern Saskatchewan, (b) Yukon Territory in northwestern Canada, and (c) central Alaska. Diagrams (a) and (b) are clearly of selected pollen, while diagram (c) is a composite summary, smoothed-out diagram (see caption for Fig. 16.11). Note that the Yukon and Alaska diagrams are very similar for the 16,000 years they have in common: decline of steppe taxa such as *Artemisia* and Gramineae post-9,000 yr B.P., along with expansion a little later of *Alnus* and *Picea*. The Saskatchewan diagram shows decline of steppe taxa and NAP generally and expansion of *Pinus* 2,000 years later (at about 6,000 yr B.P.).

372

often studied, but such samples are in my experience usually very poor in spores/pollen, I would guess because of the restricted air circulation in caves, plus post-depositional oxidation and disturbance of sediment. Furthermore, the pollen that is found is often introduced or badly mixed up by sedimentary and biogenic processes in the cave. However, I would say on experience with one deposit of bat guano in Texas that cores of such deposits are quite promising. Samples from other sorts of archeological digs (soils, garbage dumps) are often very much weathered and for that reason pollen-poor or (usually) barren. Various techniques may be employed to increase the number of palynomorphs recovered, especially starting out with very large samples, then using physical dispersion techniques, and float–sink, panning, and screening methods. In these ways it is sometimes possible to overcome the worst problem of archeological palynology, which is the derivation of conclusions from the study of very small numbers of palynomorphs. A special case is the study of pollen in fossil human feces. These coprolites are often comparatively rich in pollen, and their study can be very revealing as to the diet of the humans who made them, and the climatic conditions at the time of production (see Martin & Sharrock 1964, Bryant 1974a, b). Dimbleby (1985) has summarized the available information about archeological palynology in a text that describes the possibilities and pitfalls.

Holocene palynology has contributed to an understanding of the autecology of certain plants. For example, Davis (1980) has called attention to the dramatic drop in pollen counts of *Tsuga canadensis* (= hemlock) pollen in eastern North America 4,000–5,000 yr B.P. She believes that this is best explained by massive attack on *Tsuga* by a plant disease. There is a model in historic time: the near-extinction by disease of *Castanea dentata* (= chestnut) in the eastern U.S.A. since 1900. Figure 16.2 shows palynological evidence for the disappearance from its former range of this once dominant tree.

A direct contribution of Holocene pollen analysis to tectonics is the palynological evidence for "isostatic rebound", the elastic rise in the Holocene of land masses once covered and weighed down by ice (see Fig. 16.13). Hafsten (1956) shows such a case from Oslofjord in Norway. Sediment samples recognized as originally deposited at sea level at various localities, now found at various elevations above present sea level, were dated palynologically according to Jessen's (see Fig. 16.4) numbering scheme for the Holocene. The data show that since the early Preboreal (Jessen IV), about 10,000 yr B.P., isostatic rebound of 220 m (= 2.2 cm/yr) has occurred.

Bernabo & Webb (1977) demonstrated the possibility from the analyses available of constructing isopollen (= "isopoll") lines for selected genera and plant associations at various times during the late glacial and Holocene. Sensitive ecological markers, such as the conifer–hardwood/prairie border, can reveal much about what would happen in certain areas if temperature and/or moisture factors changed a little.

The *Handbook of Holocene palaeoecology and palaeohydrology*, edited by Berglund (1986), contains very useful sections on all of the field and laboratory techniques, as well as on statistical (numerical) methods and other matters relating to Holocene palynology.

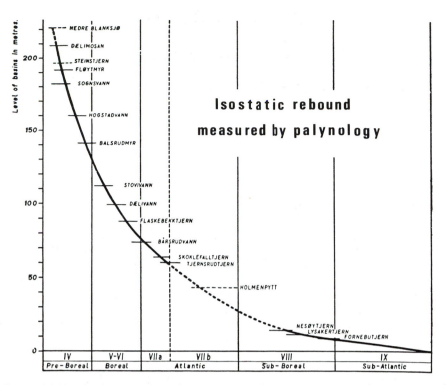

Figure 16.13 An interesting application by Hafsten (1956) of palynologically based chronology that had been correlated with varved-clay studies before common application of radiocarbon dating. (Hafsten has more recently applied radiocarbon dating to these localities, but there are technical difficulties that have made absolute dating so far impossible.) Each of the named levels refers to a small basin in the Oslofjord area, southeastern Norway. Hafsten found the marine-lake contact in cores of the sediment in each basin and then placed the pollen flora of that interval in the appropriate pollen analysis zone. (Hafsten used Jessen's scheme for southern Scandinavia; see Figure 16.4.) The present elevation of the basins is shown on the left in meters. The 220 m elevation at Nedre Blanksjo, with a pollen flora of early Preboreal (Firbas IV; see Fig. 16.4) age (10,000 yr B.P.) indicates isostatic rebound of 2.2 cm/yr.

CHAPTER SEVENTEEN

Production, dispersal, and sedimentation of spores/pollen

INTRODUCTION

The production and subsequent distribution of spores/pollen has been much studied from modern models, and the information transferred back to older sediments. In order to understand sedimentation of palynomorphs in the fossil record, this is the most obvious approach, but it is not without shortcomings. For example, the organic productivity of angiosperms is much greater than that of older groups of embryophytes, and it is very likely that this includes microspore abundance. It is also probably too facile to presume that extinct forms of pteridophyte spores and gymnosperm pollen were transported and preserved in the same way that ragweed pollen is. Nevertheless, the actualistic approach yields information that is applicable to older sediments.

SPORES/POLLEN AS SEDIMENTARY PARTICLES

As produced by the source plants, spores/pollen are solid bodies of low specific gravity. They have a resistant, organic outer coat or shell, an inner cellulosic wall, and a protoplasmic interior which is destroyed rather quickly by bacteria, fungi, oxidation, and hydrolysis upon falling to the ground or into water. A spore/pollen grain ready for fossilization is therefore a more or less hollow, more or less spherical body consisting of the sporopolleninous outer shell of the grain. Such particles, along with other sedimented plant products, are sometimes referred to as sapropel if they are a major constituent of the sediment. More specifically, various plant products including spore/pollen exines in sediment are sometimes termed "type III kerogen" (Batten 1981a).

Production of spores/pollen

It is now axiomatic that wind-pollinated seed plants, and some spore-producing plants such as *Lycopodium, Pteridium,* and *Sphagnum,* produce prodigious quantities of spores/pollen. Straka (1975), in discussing a series of measurements in Darmstadt, West Germany, says that a single measured cubic meter of space yielded during the whole year 12.5 million spores/pollen! In east Texas I harvested the pollen from a few branches of *Pinus taeda* and extrapolated to show that an uncrowded 15 m tree could produce about 5 liters of pollen per year. A modest 100 hectare woodlot of such trees would

375

produce in the spring something like 250,000 liters of pine pollen – a large railroad carload. I once measured the pollen output of single *Zea mays* plants as of the order of 25 cm³ per plant. At this rate a corn field could therefore produce about 500 liters of *Zea* pollen per hectare. *Lycopodium* produces isospores in such large amounts that they have long been harvested commercially and sold by the kilogram for various purposes. Textbooks have frequently quoted rather old data originally from Pohl (1937a, b) on pollen productivity of plants. The data are certainly all right as to order of magnitude, and are displayed here in Table 17.1. Some species produce much more spores/pollen than others, and a large wind–pollinated tree produces far more pollen per year than a herb, even if the herb is one like *Rumex*, with huge productivity per inflorescence. I would "guesstimate" that the average hectare of woodland in eastern North America produces at least 3,000 liters of pollen per year, though the amount will vary greatly depending on specific composition of the forest. It is obvious that wind–pollinated trees are the major producers. *Acer* (maple), *Pyrus* (apple), and *Tilia* (basswood, linden), all partly or wholly insect–pollinated, are not in a pollen–productivity class with anemophilous genera such as *Alnus* (alder), *Corylus* (hazel), and *Pinus* (see Table 17.1). Some zoophilous plants such as the Orchidaceae and certain Asclepiadaceae produce large pollinia in which all the pollen of one pollen chamber is shed as a cemented-together unit. Such plants are so adapted to pollination by animal vectors that they are very economical in pollen production, and thus their occurrence as fossil sporomorphs is minimal! The production of flowering plant pollen is a seasonal phenomenon, with certain plants favoring cool, moist weather for flower and pollen maturation (such as spring in Switzerland – see Fig. 17.1), while others favor drier, warmer conditions. In areas with alternation of dry and wet seasons, most but not all angiosperms flower in the moist season.

Preservability in sediment

The propensity of spores/pollen to be preserved in sediments depends largely on the amount of sporopollenin in the exine, which is partly a function of the thickness of the exine and of the division of the exine between endexine and ektexine (the more ektexine, the more sporopollenin), and partly due to other factors. It is clear that *Equisetum* spores and *Populus* pollen are not likely to be as strongly represented as sedimentary particles as other exines because of relatively low sporopollenin content and therefore low preservability. *Pinus*, on the other hand, is not only abundantly produced and relatively buoyant, but is also very rich in sporopollenin. The character of the sporopollenin in individual taxa is also apparently significant. *Beta vulgaris* (and presumably other chenopodiaceous pollen, such as *Atriplex* and *Salicornia*) has moderately high sporopollenin content (about 17% – see Table 3.2) and the exines are extraordinarily durable. I have cooked a sample of *Beta* pollen for many hours in successive treatments of HCl, KOH, and various organic solvents without affecting the integrity of the exines. On the other hand, the family Lauraceae as a whole is characterized by very low sporopollenin content in the exine. Acetolysis of pollen of this family produces only fragments of very thin exine.

Table 17.1 Pollen production, per flower (or male cone) and inflorescence (or group of male cones on one branchlet). The numbers are rounded to nearest thousand. *Vallisneria* is an aquatic plant, as an illustration of low pollen productivity. Some plants produce relatively small flowers but large inflorescences, e.g., *Polygonum*. The super-producers per branchlet on the list, indicated by asterisks, are either large trees or else shrubs and herbs of sorts that grow in dense stands, so that the immensity of the productivity is relatively much greater than indicated.

Species	Pollen production per flower (or male cone)	Pollen production per inflorescence
Vallisneria spiralis	70	140
Polygonum bistorta★	6,000	2,860,000
Sanguisorba officinalis	11,000	–
Fagus sylvatica	12,000	174,000
Calluna vulgaris	18,000	–
Betula verrucosa★	20,000	5,453,000
Fraxinus excelsior	25,000	1,606,000
Carpinus betula	28,000	–
Quercus robur	41,000	555,000
Tilia cordata	44,000	200,000
Secale cereale★	57,000	4,241,000
Pinus sylvestris★	158,000	5,770,000
Aesculus hippocastanum	180,000	765,000
Picea excelsa	590,000	–
Pinus nigra★	1,480,000	–
Populus canadensis★	–	5,800,000
Alnus glutinosa★	–	4,445,000

★Super-producers.

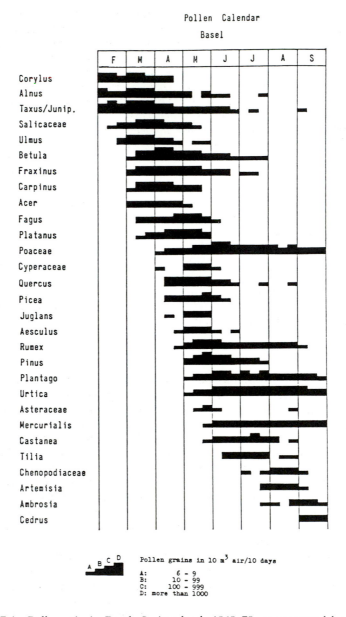

Figure 17.1 Pollen rain in Basel, Switzerland, 1969–79, as measured by collecting pollen with a Burkhard pollen trap located on a building in that city. The collections also include considerable fungal spores and industrial dust particles, not graphed here. These continue through the fall and winter months, October–January, but relatively small amounts of pollen are trapped then. Note that readings of more than 1,000 pollen grains of one taxon per cubic meter of air occur, e.g., *Betula* in April!

378

Had this "lauraceous tendency" ever become prevalent in the angiosperms, Cenozoic palynology would be even more of a dinoflagellate story than it is!

As is stressed elsewhere, the sedimentary situation into which palynomorphs are deposited also has very great control over the likelihood of preservation: acid environments preserve better than alkaline, reducing environments better than oxidizing, and quiet sedimentary situations better than very energetic ones. The exclusion of oxygen is especially important. However, even some seemingly unlikely sites of deposition such as soils can preserve pollen, though preservation in soils is usually very poor, and thus the records are probably biased toward resistant types. However, in soils, downwashing and redistribution by vectors such as earthworms must be allowed for. Spores/pollen spectra of soils often permit drawing of conclusions about vegetational history during soil genesis (Dimbleby 1961).

However, Cushing (1966) has shown that preservability of spores/pollen in sediment is the result of complex factors, not just sporopollenin content alone. Furthermore, the destruction of exines follows various pathways in various situations: degraded exines in which the structure is generally altered are common in silty sediments, whereas superficial corrosion (pitting, etc.) is more likely to be encountered in peat, and crumpling is characteristic of copropels (caused by processing in animal guts). Also, Cushing's studies showed that in some situations a seemingly delicate pollen such as *Populus* can outlast apparently tough forms such as *Alnus*.

"Pollen rain"

This refers to pollen sedimentation from the air. Pine and other pollen occurs in sediment at the Mid-Atlantic Ridge and other places far removed from the source. Erdtman's (1954) vacuum-cleaner experiments on board a passenger ship in the Atlantic were enough to show this, though the observation has often been repeated in other ways. Maher (1964), for example, showed that *Ephedra* and *Sarcobatus* pollen were transported many hundreds of kilometers from western North America to the Great Lakes region. Scott & Van Zinderen Bakker (1985) report that on Marion Island, southern Indian Ocean, exotic pollen comprises more than 1% of pollen spectra from surface samples. Most of the forms come about 2,000 km from southern Africa, but some are from more distant South America. They are transported by the prevailing winds and perhaps by ocean currents.

As spores/pollen are silt-sized or very fine sand-sized particles (see Fig. 17.2) and are low in specific gravity, they obviously can get into the upper atmosphere when the air is turbulent, and there is certainly every reason to assume that some such grains can and do circle the Earth in the same way as volcanic dust. However, most pollen falls out of the air very near the producing plant. This is evident from the sinking rates reported in Table 17.2. (Recently published data for three species of saccate conifer pollen, based on sophisticated modern techniques, yielded an average rate of 2.7 cm/s (Niklas 1984).) An average sinking rate of 3 cm/s is nearly 2 m/min, and it is quite clear both from this theoretical consideration and from many observations that

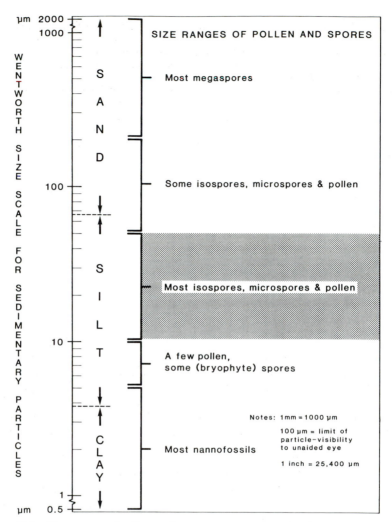

Figure 17.2 Size ranges of pollen and spores in comparison to clastic particles, per the Wentworth scale, plotted logarithmically. Most spores and pollen are silt sized, although a few are fine sand sized, and megaspores are practically all sand sized. As the specific gravity of sporopollenin (about 1.4) is less than that of mineral clastic particles (about 2.5), and because palynomorphs are not solid, they will tend to sort in sedimentation in a mineral class of somewhat smaller particles than themselves (see Stanley 1969).

most pollen, even pollen from a large tree, reaches the ground a few tens of meters from the tree unless the air is turbulent. Clearly pollen from tall trees will spread farther laterally while settling than pollen from shrubs and herbs, other things being equal. More solid data based on marked pollen would be desirable, but I think it is clear from many studies already made that at least 95% of all pollen has normally settled down well within a kilometer of the

Table 17.2 Sinking speed in air, measured and predicted from Stokes' law, of various wind-pollinated pollen and *Lycopodium* spores. There are some surprises. Grass pollen (*Zea, Secale*), despite being wind-pollinated, is quite dense and sinks rapidly. *Lycopodium* spores and *Taxus* pollen, rather dense-looking, are very "light".

	Sinking speed (cm/s)	
Species	Measured	Calculated from Stokes' law
Zea mays	24–30	13
Abies alba	12–39	–
Larix spp.	10–22	–
Picea spp.	6–9	6
Secale cereale	6–9	–
Fagus sylvatica	6	5
Pinus spp.	3–5	3
Corylus avellana	2–3	2
Alnus spp.	2–3	2
Cannabis sativa	2–3	2
Taxus baccata	1–2	1
Lycopodium spores	2	–

source plant. On theoretical grounds, Tauber (1965) calculated that most pollen, even from an elevated source, would be on the ground within 2,700 m. Tsukada (1982) reports that 90% of pollen of *Pseudotsuga menziesii* (northwest U.S.A.) falls within 100 m of the source tree. Various palynologists have pointed out that, in a forest, the canopy of trees is a controlling factor. Thus, the autumn leaf fall provides one peak in "pollen rain" that is entirely secondary – pollen trapped on leaves finally reaches the ground and streamflows (Loeb 1984). Figure 17.3a shows Tauber's (1967) model and Figure 17.3b presents Jacobson & Bradshaw's (1981) model for the general scheme of the early stages in pollen's journey from tree to sediment.

One reason why one intuitively tends to doubt the generalization that most pollen drops near the source vegetation is that pollen is found in well-reported aerobiological counts of spores/pollen in the air in all sorts of places, even within a city. (See Figure 17.1 for a pollen "calendar" for a location in western Europe. Note that 10 m^3 is a lot of air, and that the counts were made for 10-day segments of time.) The anecdotal impression of pollen abundance in the air is intensified by the incredible sensitivity of human sufferers from pollinosis to even tiny amounts of pollen. Five or even fewer pollen grains of an offending species can make a sensitive person ill. Thus, sensitive persons may pick up enough pollen from an open window of an urban building to make them ill, but the amount of pollen is very small. Mandrioli *et al.* (1982) report that they collected at various atmospheric levels up to 800 m above sea level in northern Italy significant amounts (up to 12 grains/m^3) of tree (*Quercus farnetto, Fagus* sp., *Ostrya carpinifolia*) pollen that, judging from the meteorological conditions and species composition, apparently had crossed the Adriatic from Yugoslavia. But even these amounts are not large when compared to the total

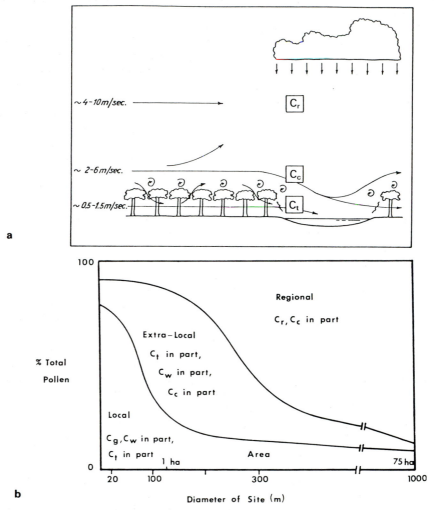

Figure 17.3 Local distribution of pollen and spores by wind and water. (a) Tauber's (1967) model for pollen dispersal by wind in a forested area. Pollen is shown as moved by three routes, of which the total transport is a composite: C_r = pollen brought down by rain, C_c = pollen carried above the forest canopy, and C_t = pollen carried through the trunk space in the forest. When passing over a lake the air currents bend down. However, most pollen reaching a lake does so by the lower route of C_t. (b) Jacobson & Bradshaw (1981) presented a more complete model by including C_w (= surface runoff) and C_g, a gravity component for pollen dropped directly at the site of deposition. For other symbols see (a). "Regional" refers to pollen from more than several hundred meters from the basin of deposition; "Extra-local" means pollen from plants 20 m to several hundred meters from the basin; and "Local" indicates plants from within 20 m. In both models, the activities of streams entering the basin are not taken into account.

production of spores/pollen, much of which is eventually washed into streams and into basins of deposition.

One of the best illustrations of airborne pollen versus water-borne pollen in the literature is Muller's (1959) study of the Orinoco Delta. *Podocarpus* pollen, a bisaccate, was present in the sediment as demonstrably airborne, but was numerically overwhelmed by that reaching the basin of deposition in the streams. Thus, in measuring the "pollen rain", one must be aware that a number of conflicting mechanisms are at work.

Various devices are used to trap and measure airborne pollen. The typical apparatus is a sophisticated mechanical device for passing air over sticky slides. These devices are often located on tops of buildings. Pollen rain has also been measured over long terms from stationary traps on the ground, or the spore/pollen content of moss polsters and other surface litter may be analyzed, as may surface sediments from lakes. Palynologists have also measured the pollen in the atmosphere at various levels using pollen traps. From a geological point of view, however, perhaps the most significant data are from studies of spores/pollen in the water of streams of various sizes, and we have so far very few data of this sort.

The specific gravity of whole pollen is a tricky matter. Wodehouse (1935) and others have shown that the protoplasmic content of spores/pollen picks up moisture from the atmosphere. The specific gravity of *Zea mays* pollen, given in Firbas (1949) as 0.35, will approach 1.0 if the pollen is moist enough. The specific gravity of pollen is also dependent on air spaces within the grain. Bisaccates such as *Pinus* have much trapped air in the sacci. The specific gravity of pollen exine itself is well above unity – about 1.4 – so fossil exines devoid of air and protoplasmic contents sink in water but float in a $ZnCl_2$ or $ZnBr_2$ solution with specific gravity about 2.0; this is the basis of one of the laboratory techniques for concentrating fossil palynomorphs. Some sorts of pollen, especially bisaccates, will float for long periods in water. Most *Pinus silvestris* pollen grains I put in large beakers of water were still floating a year later. Those still floating when mounted in water on a microslide showed air bubbles filling the sacci. The grains that had sunk did not show these air bubbles; the sacci were "wet". When the exines are completely filled with water, the specific gravity of sporopollenin governs, and the exines sink. (See also later field evidence for long-term floating of pollen in the Bahamas.)

Barriers to transportation of spores/pollen in the air are many, even if there are strong winds and the air is turbulent. Figure 17.3a shows that pollen well above the tree canopy of a forest travels nearly an order of magnitude faster in the wind than does pollen near the ground, where it is impeded by tree trunks and shrubs. Once spores/pollen reach the ground, further movement is mostly by water, except in open areas such as grassland or desert, where repeated deflation is likely. Trees that are wind-pollinated apparently have had to adapt to the exigencies of too little pollen reaching the area of the female gametophyte. Niklas (1984) has shown that cones of coniferous trees are constructed to direct incidental air currents in order to maximize the probability of pollen capture.

It is precisely because spores/pollen mostly sink to the ground quickly that the pollen rain of an area tends to be characteristic, that is, rather constant from year to year. ("Iatropalynology", the application of medical methods to the alleviation of pollinosis, assumes the relative predictability of spores/pollen in the air in a given area.) Studies of pollen in surface sediment have shown that several collecting stations in about the same latitude and altitude will yield a rather similar pollen "signature", if the vegetation is similar.

However, the extent to which pollen rain in surface sediment presents a picture from which the source vegetation can be "reconstructed" is a complex matter. For one thing, there is "outside" pollen, despite the prevailingly local character of the pollen rain. "Outside pollen" ranges from 100% on Antarctic or Greenland glaciers to near zero in a wooded peat swamp. McAndrews (1984) showed that ice from a glacier on Devon Island, Arctic Canada, contains pollen from sources more than 1,000 km away. When the ice melts, the pollen can be reworked into glacially derived sediments. On the Isle of Skye, Birks (1973) has shown that surface samples from some environments yield mostly local pollen, e.g., a woodland that produces plenty of pollen itself and impedes penetration of pollen rain from outside. Sub-alpine grassland, scrub, and alpine vegetation have as much as 40% outside pollen. Mack & Bryant (1974) note that in steppe communities of the Columbia Basin, Washington State, U.S.A., pine pollen may equal 50% of pollen in surface samples, though collected 10 km from pine forests. Mack et al. (1978) showed that, although in general the pollen flora of surface samples from Washington and Idaho could be related, with corrective factors, to the surrounding vegetation, samples from grass–dominated areas in the sub-alpine zone contain much tree pollen from nearby forests. Markgraf (1974) has shown that areas above the timberline in mountains produce very little pollen – at least an order of magnitude less than lower in the same mountains – and thus the pollen that comes in with regional winds is relatively prevalent above the timberline, as in marine sediments from far offshore (see also Fig. 17.4).

The altitudinal diminution in pollen rain is matched by a latitudinal decrease in Arctic areas, from Boreal forest to tundra to ice field, something like an order of magnitude in each case (see Birks et al. 1975). Jacobson & Bradshaw (1981) have shown that when pollen is deposited in a small basin, such as a lake with no inflowing streams, the proportion of autochthonous (locally produced) pollen varies inversely with the size of the basin. The effective "pollen rain" of an area is a product of three factors:

(a) what is produced locally (some species are much overrepresented),
(b) what is preserved (no Sassafras, little Populus),
(c) what comes in from outside.

Heusser (1978) has shown that off the Pacific coast of North America the general pollen rain is 100–1,000 grains/m^2 per year.

A study by Grindrod (1985) of mangrove vegetation on a prograding shore in Queensland, Australia, demonstrated that pollen analysis can be employed to create models for plant successions even in such highly mobile, unstable environments.

384

a

Dominant trees
in vicinity of
sample

Percent pollen in sample

	Pinus	Picea	Fagus	Carpinus	Quercus
Pinus	85.7	1.2	0.5	1.1	1.9
Picea	54.0	29.0	---	---	1.5
Fagus	51.1	2.1	22.0	1.3	6.1
Carpinus	68.0	1.0	1.0	13.0	2.0
Quercus	68.7	3.0	1.0	1.3	9.3

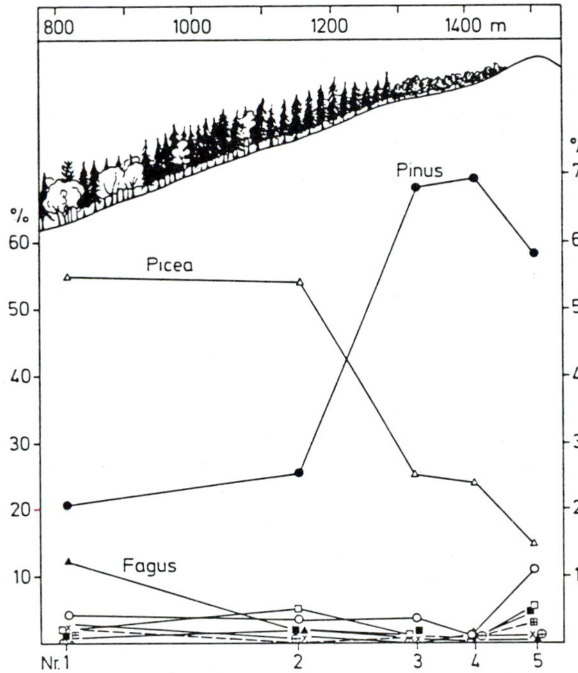

b

Figure 17.4 Pollen rain as a reflection of standing vegetation. (a) Pollen in surface samples from a locality in Germany. The data here and from many other places have shown that pollen of a given tree taxon is most abundant in surface sediment from areas where the taxon is dominant. Thus, *Carpinus* pollen is most abundant in soil from *Carpinus*-dominated woods. However, wind-pollinated taxa which produce large amounts of pollen, such as *Pinus* in this case, tend to overwhelm pollen of other taxa, even in their own areas. Samples do not total 100%, because of minor constituents not listed. (b) Pollen rain in comparison with forest composition in a mountainous area of central Europe. Sample No. 1 came from a *Picea-Fagus* forest, No. 2 from *Picea* forest, No. 3 from a boundary zone for dwarf *Pinus*, No. 4 from the dwarf *Pinus* zone, and No. 5 from above the timberline. *Pinus* is abundant (20%) at Nos. 1 and 2, where there are no pine trees. At No. 5, easily airborne *Pinus* dominates in the total absence of trees, but other pollen from the mixed forest are also blown in to a smaller extent (symbols are those given in Fig. 16.3). The relationship of vegetation to pollen rain is clearly complex.

385

VEGETATIONAL ANALYSIS FROM POLLEN ANALYTICAL RESULTS, "R VALUES", ETC.

The principal paleoecological purpose for study of the pollen rain that produced a palynoflora is to relate it to the source vegetation, partly with the hope of relating a known fossil pollen spectrum to the unknown vegetation that produced it. Tracing the changing composition or distribution of vegetation through time enables one to draw conclusions about the past climate and environmental changes. For example, maps can be produced based on isopolls (see Fig. 17.5) showing lines connecting points with the same percentage of spores/pollen of particular taxa. Clearly, despite over- and underrepresentation of taxa and other detracting factors, these isopoll maps do really represent changes in vegetation and thus of climates. Even a single taxon can show much. Iversen (1944), for example, studied *Hedera*, showing the presence or absence of freezing weather; and *Hippophae* pollen in northwestern Europe (Straka 1975) indicates pioneer colonization of deglaciated areas. However, the aim of pollen rain investigations is usually to seek a model for interpretation of pollen diagrams, but over-, under-, and non-representation prevent direct translation of even Holocene pollen analyses into vegetation reconstructions. Cross (1984) has shown that in an arid environment, for example, pollen floras of surface samples reflect the vegetation very poorly. In an investigated part of Baja California, surface samples had *Opuntia* pollen where no *Opuntia* plants were present, whereas *Lycium* and *Larrea*, the dominant shrubs, were not represented by pollen in the surface samples. In tropical rainforest areas, surface samples are dominated by wind–pollinated plants, such as grasses, composites, and chenopods, with very small amounts of (animal–pollinated) pollen from the dominant trees (Rue 1986). Early in the history of Holocene pollen analysis, correction factors were suggested. For example, it was proposed (Faegri & Iversen, 1975; the idea goes back at least to the 1940s) that in northwest Europe *Betula* and *Corylus* counts should be divided by 4 to correct for overrepresentation of birch and hazel pollen – one unit of *Betula* or *Corylus* plants is interpreted as producing four units of pollen. Davis (1965, 1973) has published sophisticated arithmetic methods for calculating analogous ratios: R_m is the ratio determined from comparison of the pollen rain with the actual vegetation. Davis' R_m is based on the same idea as the division by 4 for *Betula* and *Corylus*, as cited above. An R_m of 4 means that the fossil pollen percentage is divided by 4 to get the corresponding density (percentage) of the taxon in the vegetation. Davis (1963a, b) showed that the use of an "R ratio" to correct the fossil pollen counts in a subsurface sample yields a modified "percentage" which is close to the presumed composition of the fossil (i.e., theoretical) forest. However, it is clear that Davis' method would not help if the forest had large amounts of *Persea, Lindera* or *Sassafras*, of which no pollen is preserved, and probably would not work for forests rich in *Populus* or *Liriodendron*, of which relatively little pollen would be found. It also would not work in tropical rainforest areas with enormous numbers of species represented by very few individual trees. However, Davis has shown that the method does work in a small and

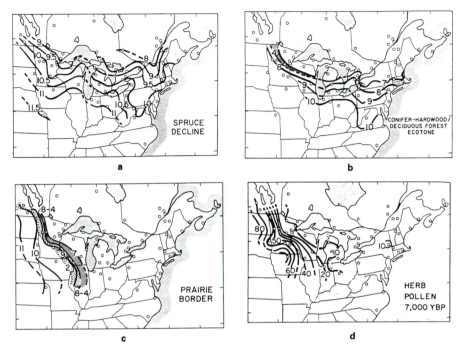

Figure 17.5 Extrapolation of pollen analytical data for graphical interpretation of vegetation distribution in the North American Holocene. Preparation of such maps is based on isopollen lines (see Fig. 17.6). (a) *Picea* (spruce) decline. The lines are isochrones showing how many thousands of years B.P. *Picea* pollen analytical values fell below 15%. (b) Conifer/hardwood deciduous forest ecotone (= transition between the two vegetational types), showing where this zone was 10,000, 9,000, 8,000, and 7,000 yr B.P., based on pollen analyses. For Minnesota, the situation is confused by the prairie expanding eastward post-8,000 yr B.P., then moving westward again after 4,000 yr B.P. (c) The location of the prairie border is shown by isochrones for 11,000, 10,000, 9,000, 8,000, and 7,000 yr B.P., based on herb (= NAP) isopolls. Shaded areas refer to the region over which the prairie retreated after reaching maximum Holocene extent at 7,000 yr B.P. (d) Isopolls can show the concentration of high counts for given pollen types at a selected time, in this case for herb (= NAP) pollen at 7,000 yr B.P. This shows that the prairie border in (c) is based on the 30% herbaceous pollen isopoll.

thoroughly studied environment. Janssen (1967) demonstrated that R values can vary tremendously even in the same general area, with the same vegetational patterns throughout, e.g., *Betula* from 0.5 to 15. The R ratios tend to be locally too large when the taxon really is rare to absent in the actual very local vegetation, and too small when the taxon is correspondingly locally abundant. Parsons & Prentice (1981) have examined the mathematics of the R value method in detail and have suggested computer-based multivariate and other statistical methods, designed to counter the problems encountered in application of R value techniques for vegetation reconstruction from pollen data.

Webb *et al.* (1981) use regression analytical techniques with a rather simple formula for converting pollen analysis data to estimates of relative abundances of taxa in forests. The basic formula is

$$p_{ij} = v_{ij}r_j + p_{oj}$$

where p_{ij} is the percentage of taxon j at site i in the pollen assemblage, v_{ij} is the percent abundance of taxon j at site i in the forest vegetation, r_j is a representation coefficient for species j (comes from pollen productivity), and p_{oj} is the percentage of taxon j at site i in the pollen assemblage, but derived from outside the local area (regional background pollen influx). (The equation can be further corrected by factors for error in measurements.)

In a study of 1,684 modern pollen (surface sample) samples in comparison to forest-inventory summaries for eastern North America, Delcourt & Delcourt (1985b) state that taxon calibration based on known pollen production and dispersal data are the best available means for reconstructing forest history. Pollen percentages (P_a) are transformed into preliminary vegetation values (V_a) using geometric-mean linear regression equations:

$$V_a = \frac{P_a - P_0}{r} \text{ where } P_a \geqslant P_0$$

where P_0 is the y-intercept of the regression line and r is the slope of the regression line. The resulting V_a values must be corrected based on the sum of all V_a values at a specified time, in order to represent the paleovegetation values based on 100% recalculated vegetation (V_c):

$$V_c = \frac{V_a}{\Sigma V_a} \times 100$$

Solomon & Webb (1985), noting that pollen data integrate information from 50–3,000 km^2 around the basin of deposition, have discussed the subject of modeling of vegetation on the basis of pollen analyses and use of forest-stand simulation models (JABOWA, FORET). Computer-based manipulation of the two sorts of data offer potential for an integrated approach.

Overpeck et al. (1985) have demonstrated how dissimilarity coefficients, measuring the difference between multivariate samples, can be used to identify modern analogs for fossil pollen samples. The mathematics of this and other analog techniques based on multivariate analysis are complex and should be studied in the original papers or in Birks & Gordon (1985), a specialist text on mathematical approaches in pollen analysis.

Delcourt & Delcourt (1985a) have shown that data on relative abundance of different sorts of pollen in present-day surface sediment of eastern North America can be compared directly with forestry data on the abundance of trees of the same species in forests close to the surface-sediment sources (see Fig. 17.6a–e). From such and other plant distribution data it is possible (Fig. 17.6f, g; Delcourt 1979) to plot areas in eastern North America where the present-day vegetation best matches the environmental requirements of dominant species at various levels of a late Pleistocene core in Tennessee. The level dated 19,000 yr B.P. ("19"), for example, finds its closest analog today in Manitoba.

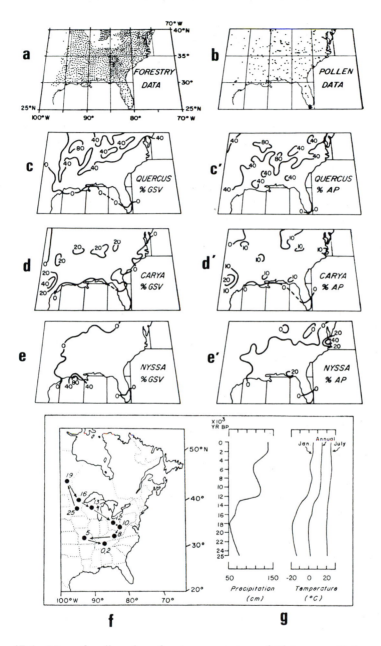

Figure 17.6 Use of pollen data for interpretation of Pleistocene–Holocene plant distribution and paleoclimates. (a)–(e) The relationship between present-day forestry data on relative dominance of tree taxa in commercial forests, and the contribution of pollen from these taxa to surface sediment: (a) and (b) indicate geographical sources for data; (c) shows the isophytes of *Quercus* (oak) abundance as a percentage of GSV (growing stock volume), from forestry data; (c') shows isopolls for the pollen

389

POLLEN PER GRAM, POLLEN INFLUX, ETC.

From the beginning of pollen analysis the most frequently used method of data presentation has been by percentage of each important taxon. Often the total palynoflora to which the percentages are related is modified from the actual total counted by eliminating certain forms – the modified total is the "pollen sum". The present trend is to use pollen sums that include nearly all land plant spores/pollen. In any event, there are clearly difficulties with the percentage approach. For example, if the palynoflora consists overwhelmingly of *Pinus* and *Quercus* pollen, an increase in *Quercus* pollen will cause a decrease in *Pinus* percentage, even if the absolute numbers of *Pinus* remain about the same. A number of suggestions have been made to correct for this difficulty. One is to express spore/pollen concentration as a ratio to a common fossil that is not usually part of the spore/pollen analysis, such as fungal spores, but fungal spores are not themselves constant.

As noted in the previous chapter, foreign spores/pollen (*Lycopodium* spores, etc.) can be added to samples before processing, and the fossil palynomorphs can then be expressed as a ratio to the added spores. This technique yields not only such a ratio, but by simple calculation also gives the absolute number of fossil palynomorphs in the original sample. If the weight of the original sediment sample was measured, the concentration of palynomorphs per gram of sediment sample can then easily be calculated (Stockmarr 1971). This Stockmarr technique depends on tablets of *Lycopodium* spores held together by $CaCO_3$ and water-soluble organics. Five tablets contain 105,000 ± 3,000 spores. If five tablets are added to the sediment sample before processing, then

$$\frac{105,000 \times \text{palynomorphs counted on a slide}}{\text{number of } Lycopodium \text{ spores counted}} = \frac{\text{number of palynomorphs}}{\text{in sediment sample}}$$

---

percentage, as a percentage of total AP (tree pollen) in surface sediment; (d) and (d′) present comparable data for *Carya* (hickory), and (e) and (e′) for *Nyssa* (tupelo, black gum). Oak and hickory pollen show a close relationship between isophyte and isopoll data, as is the case for most wind-pollinated tree genera, which allows quantitative calibration from pollen data for reconstructing changes in forest dominance in time. However, *Nyssa*'s abundance in the forests and the percentage of AP shown by its pollen in surface sediment are rather widely different. This is presumably a product of the local nature of production and the durability of *Nyssa* pollen, as well as the rather local abundance of *Nyssa* trees, primarily in swamp-forest situations, and the similarly patchy distribution of surface pollen samples. (f) and (g) Applications of the known environmental requirements of plants dominant at various levels of sediment cored at Anderson Pond, Tennessee: (f) represents location of modern pollen samples in North America with composition equivalent to pollen samples from various levels in the cores, from 25,000 yr B.P. ("25") to just before present ("0"); (g) shows the precipitation and temperature changes inferred from weather-station data at the location of modern analogs for the fossil-pollen assemblages.

The method has the advantage that not all the palynomorphs on a slide need to be counted to make the calculation. However, in our laboratory we use a more direct method involving no additives:

(1) Weigh the original sample.
(2) Weigh the final maceration residue plus mountant (glycerin jelly in our laboratory).
(3) Weigh the slide and coverslip before and after adding the drop of residue and mountant.
(4) Count fossils on the slide (or a carefully measured fraction, if it is very dense) and calculate fossils per gram as follows:

$$X = \frac{BD}{CA}$$

where X = microfossils per gram of sediment, A = grams of rock sample, B = grams of maceration residue plus mountant, C = grams of residue plus glycerin jelly on slide, D = microfossils counted on whole slide.

As is discussed elsewhere, the total amount of spores/pollen per gram in a sample is interesting from a sedimentological point of view. The amount of a particular pollen type per gram is in some ways more significant than the percentage, as it is related to a non–pollen datum and is not affected by fluctuations in the amounts of other pollen types.

However, spores/pollen per gram *is* affected by fluctuation in the sedimentation rate, and, from the fact that total pollen per gram of sediment in the Great Bahama Bank is of the order of a 1,000/g or so, in silts from a river delta of the order of 50,000/g or so, and in silts from the Gulf of Mexico of the order of 10,000/g, it is obvious that spores/pollen per gram cannot be directly compared from one environment to another. Even in one place, if sedimentation of mineral matter goes up significantly, the amount of pollen per gram of sediment must go down if pollen delivery to the basin of deposition remains constant. To compensate for the difficulty of differing sedimentation rates, Davis (1967) developed methods for calculating a "pollen accumulation rate" or "pollen influx" (Davis *et al.* 1973). In effect this is done by correcting data for pollen/volume of sediment by a factor depending on sedimentation rate, so that the data are expressed as pollen/cm^2 (surface) per year. Obviously, this elegant method depends on ^{14}C or other dates in sufficient number to ascertain that sedimentation rate has remained constant, or to what degree and when it has fluctuated. Pollen influx, then, is the amount of total spores/pollen, or of a particular spore/pollen type, falling on a unit area of the basin of sedimentation per year. The calculation is as follows:

pollen influx (*PI*)

$$= \text{spores/pollen per cm}^2 = \frac{\text{spores/pollen per cm}^3}{\text{years for deposition of 1 cm (vertical) of sediment}}$$

The amount of spores/pollen per cm³ of sediment can be calculated by correcting data for spores/pollen per gram by the specific gravity of the sediment. It can be calculated directly, of course, by adding known amounts of foreign spores/pollen to measured volume, as in the Stockmarr method described above.

Figure 16.2 shows an application of the pollen influx approach to a problem in Lake Erie. This illustration also shows that concentration per gram of sediment, a less complicated and less expensive procedure (no ^{14}C dates needed), gives acceptable results, at least in some cases. Where sedimentation rates are very variable, however, there will clearly be a problem.

Pollen influx studies have been used other than in connection with efforts to reconstruct vegetation. If the average pollen influx in an area is known, this estimate of the number of spores/pollen incorporated into sediments per unit surface area per unit time (usually per cm² per year) can be used to detect very rapid changes in sedimentation rate, such as occur in an ash-fall, or to estimate the length of time during which the ash fell (see Mehringer *et al.* 1977). (The pollen deposition rate (*PDR*) of some authors is a measurement essentially equivalent to *PI*.)

Another approach that can be objective is the use of ratios between counts of various taxa, to show regional paleoclimatic trends. Firbas *et al.* (1939) long ago showed that the ratio between *Picea* and *Fagus* pollen in surface and subsurface samples from the Oberharz region of Germany can be revealing. Samples from a number of different locations in Flandrian zone IX, the "beech time", yielded consistent *Picea/Fagus* ratios of 0.3–0.4, whereas surface samples from the same areas show the present pollen rain to yield a *Picea/Fagus* ratio of 7.0–14.0. Firbas related the sets of ratios to the respective climatic regimes.

A ratio such as

$$\frac{\text{spores/pollen}}{\text{total palynomorphs}}$$

can yield an indication of the relative contribution of land-based flora to total palynoflora, and

$$\frac{\text{phytoplankton}}{\text{total palynoflora}}$$

can indicate the "marine influence", if the dinocysts and acritarchs are known to be marine. Manum (1976a, b) shows, in a study of palynomorphs in D.S.D.P. cores from the Norwegian Greenland Sea, that the ratio between marine and non-marine palynomorphs can be displayed graphically to advantage.

Birks & Birks (1980) have shown very clearly that sophisticated mathematical methods such as factor analysis, cluster analysis, principal-components analysis, and other matrix-algebra-based, computer-programmed techniques offer great possibilities for detecting from the data the existence of significant patterns. For example, cluster analysis dendrograms can show unsuspected similarity between samples, as can principal-components analysis, but principal-components analysis can also reveal clearly associations between many more types of pollen than can be conveniently shown in a dendrogram. The example in Figure 17.7 of the application of principal-components analysis shows that pollen spectra sometimes sort out on the basis of linkage of high and low analytical values, related to environment. Figure 17.8 shows an interesting application of factor analysis, demonstrating linkage between groups of taxa and geographic distribution. It would be almost impossible to make the necessary calculations of such association without this mathematical tool. All of these multivariate mathematical methods demonstrate association, but not what the association means. The methods have also been applied to older sediments, and the older they are, the more difficult it is to interpret the displayed associations. Palynologists in practice are best advised to visit statisticians and computer specialists at a university or elsewhere for help in the necessary programming and analysis of their data.

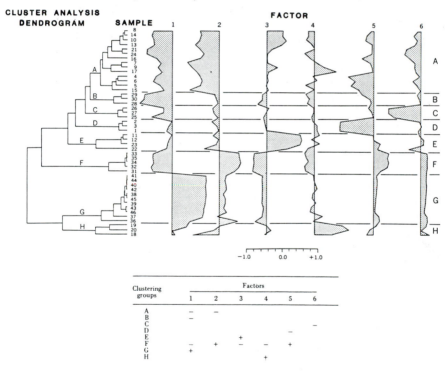

Figure 17.7 Palynological application of principal-components and cluster analyses. These two multivariate statistical methods are used to detect and describe patterns in pollen data. If there are *m* samples and *n* pollen types, the data can be regarded as a set of *m* points in *n*-dimensional space. Cluster analysis produces a dendogram or clustering tree in which the samples are grouped with other samples, on the basis of similarity to each other. Principal-components analysis produces a set of variates that are linear combinations of the data from the pollen samples, and are uncorrelated to each other.

SEDIMENTATION OF SPORES/POLLEN AND OTHER PALYNOMORPHS

Spores/pollen exines (and other palynomorphs such as dinoflagellate cysts and fungal spores) when occurring in water are technically, from their size, silt or very fine sand particles (see Fig. 17.2), which behave in water according to the same principles as govern other clasts. The low specific gravity (about 1.4) of palynomorphs, the fact that they contain internal space, their tendency after initial stages of preservation to be tiny disks rather than spheres, and other factors give them somewhat different settling characteristics from mineral particles of the same maximum size (specific gravity equal to or greater than 2.4). As is true for sedimentary particles generally, the settling speed in water basically follows Stokes law, a formula devised for expressing the settling rates of spherical bodies (which palynomorphs usually are not):

$$w = \left(\frac{(\rho_s - \rho)g}{18\mu} \right) d^2$$

where $\rho_s - \rho$ is the density difference between the particle (ρ_s) and the fluid (ρ), g is the acceleration due to gravity, μ is the dynamic viscosity of the fluid, d is the diameter of the particle, and w is the settling or fall velocity (from Blatt *et al.* 1972).

Palynomorphs are more or less hollow and variously wrinkled as soon as the protoplasts are destroyed (or after excystment for dinocysts). They also often have external processes or other sculpturing and are usually more or less flattened to tiny disks. Laboratory measurements I have made show that speeds of sinking in water for acetolyzed (substantially like fossil pollen, except not flattened) modern pollen are generally slower than the calculated speed, often 50% slower. According to Muller (1959) an average pollen exine should sink at the rate of 4 cm/h, but Muller's measured rate is a bit higher, of

---

The method is useful in reducing the dimensionality of data sets and in clarifying patterns in the data. These and other statistical methods are applied to a palynologist's data with the aid of computer programs, and the average palynologist is best advised to read Birks & Gordon (1985) and papers that use multivariate analytical techniques, and then seek help and guidance from the computer science and statistics departments of his institution. The pollen analytical data presented here are from Osgood Swamp, California, representing about the last 12,000 years. The sample numbers (1–46) are for samples ranging from surface (1) to 440 cm (46), but the samples are not presented in stratigraphic sequence, rather in the order of their clustering, though most of the clusters are themselves stratigraphic subsets. The clustering is on the basis of pollen types in common. The principal-components analysis depends on a series of complex pollen analytical factors. Factor 1, for example, is related to high *Artemisia* values and low *Pinus*, *Abies*, and *Isöetes* values. This factor is highly positive at the glacial/post-glacial interface of cluster G, highly negative in clusters representing the higher levels of the core.

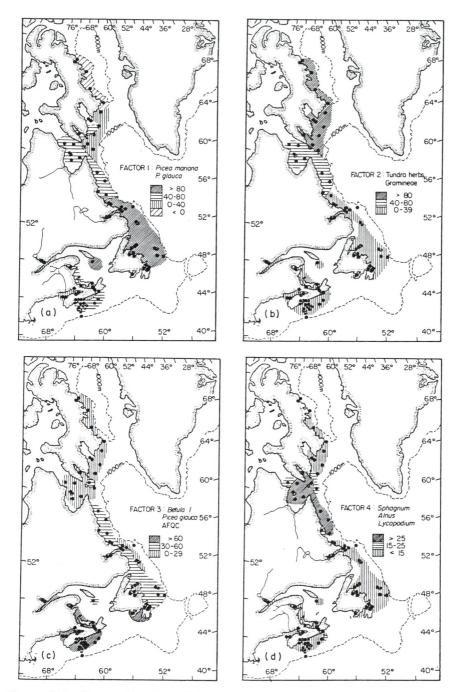

Figure 17.8 Geographic plots of factor analysis data for pollen in recent marine sediments, eastern Canada. Factor analysis is a multivariate technique in which large

the order of 7–17 cm/h, about like that for fine silt (4–8 cm/h). Stanley (1965) observed that palynomorphs usually sort out in a mineral silt fraction one class smaller than would be expected from the palynomorph size, which is another way of saying that they do not settle as fast as mineral particles of the same size. The relative proportion of various classes of palynomorphs (spores/pollen, dinoflagellates, acritarchs) in sediments is a function of distance from shore and depth of water, as can be seen in Figure 17.9.

The Hoffmeister patent statement

Until the mid-1950s most palynologists assumed that those spores/pollen which enter the fossil record owe their distribution primarily to the atmosphere, especially to wind patterns. Thus, spores/pollen should as easily be found in sands as in silts, inasmuch as they would fall out of the air into all sorts of depositional environments. It turns out that only palynofloras from sediment of small ponds, soils, and swamp-generated peat contain mostly autochtonous spores/pollen that dropped out of the air in the neighborhood of the site of deposition. Most fossil spores/pollen in sediments have been transported, sometimes very long distances, by streams and by ocean currents. The spore/pollen flora of the Mississippi River Delta, for example, contains spores/pollen from the vegetation of North Dakota, Pennsylvania, Minnesota, and all areas between those states and Louisiana, carried by the various tributaries of the Mississippi. There are also reworked sporomorphs from Paleozoic to Neogene, and variously sorted and reworked, as well as more or less contemporary, dinoflagellate cysts, some formed in the present Gulf of Mexico; in short, a pot pourri, the understanding of which depends partly on sedimentology. Peck (1973) has shown that even in a very small catchment basin in Yorkshire about 97% of the spores/pollen in the sediment reaches its destination in the small stream and runoff, not in the air. As late as the mid-1950s, the prevailing idea still was that pollen and spores reach sites of deposition mostly by air.

W. S. Hoffmeister, of the oil company now called Exxon (= Standard Oil of New Jersey), startled the palynological world in 1954 by obtaining a U.S. patent for virtually the whole field of paleopalynology, especially as it applies to paleoecologŷ (Hoffmeister 1954). Later (1955) Hoffmeister dedicated his patent to the public, and explained (to me) that his only purpose was to prevent others from patenting the method and preventing its free use.

Of much more palynological significance than the patent itself was the announcement it made that Hoffmeister's palynological group in Tulsa had discovered that spores/pollen are distributed in sediments according to

---

matrices of variable data are simplified by grouping the taxa into coherent assemblages which covary in a similar fashion. Factor 4, for example, is a sub-Arctic assemblage, with a geographic distribution favoring the 56°–60° latitudes. The four factors displayed here, and two others (5: *Pinus banksiana* + *Tsuga canadensis*; 6: Gramineae + *Picea mariana*) account for 95% of the variance in the samples studied. Note that a taxon can occur in more than one factor, e.g., *Picea glauca* in factors 1 and 3.

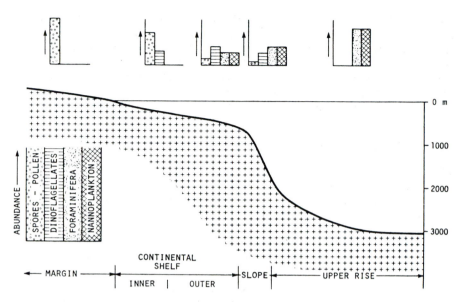

Figure 17.9 Distribution of various microfossil groups in sediments of marine or near-marine sedimentary environments. Spores/pollen are shown as a constituent of all environments except deep sea (more than 3000 m), in decreasing abundance toward the open ocean. Actually, even deep-sea sediments contain some pollen, especially bisaccate conifers, but the amounts are about 50/g, only about 0.01 times as abundant as in average shelf deposits (Melia 1982). Only spores/pollen provide a link between terrestrial and truly marine environments. The marginal and inner continental shelf environments contain only palynomorphs: spores/pollen, dinoflagellate cysts, and (not shown) acritarchs. Acritarchs have about the same distribution as dinoflagellate cysts. Freshwater environments lack foraminifera and nannoplankton. Relatively exceptional freshwater dinoflagellates, plus a suite of freshwater acritarchs, are found in totally non-marine environments, along with pollen and spores.

principles that are basically sedimentological. Hoffmeister stated that the abundance of spores and pollen in sediment decreases sharply (more or less logarithmically) as one moves offshore. He also noted that there is a marked sorting among fossil spores/pollen, with smaller forms being relatively more abundant than larger forms, as one samples further offshore. Hoffmeister stated that in samples of shale:

(1) proximity of an ancient shoreline (of potential importance for oil exploration) is indicated where the ratio

$$\frac{\text{large (70–120 μm) spores/pollen}}{\text{small (20–50 μm) spores/pollen}} \simeq 0.25$$

(2) proximity of an ancient shoreline is indicated where the concentration of spores/pollen is about 7,500 per gram of sediment.

397

However, various investigators have since stressed that, while palynomorph concentration decreases with distance from shore in a general way, far offshore, however, more important factors are sediment size and bottom morphology. Palynomorphs are much more abundant in sediment offshore from stream mouths than where no such drainage is nearby. Heusser (1983) and Heusser & Balsam (1977) have shown that concentration tends to be lower on the continental shelf, and higher on the slope and rise. In the western North Atlantic they found 10 grains/g on the abyssal plain and as high as 230,000/g on the slope, a product of the deposition of palynomorph–rich sediment on the slope.

Muller's Orinoco Delta work

About the same time as Hoffmeister's patent statement, Jan Muller, a Royal Dutch/Shell employee, accompanied the Van Andel expedition to the Orinoco Delta of South America, sponsored by the Netherlands government. Muller's (1959) palynological work showed that the total spores/pollen of surface sediment followed in a very general way Hoffmeister's outline, but that the picture is much more complicated than the Hoffmeister patent statement said. As can be seen in Figure 17.10, the concentration of fossils per gram of sediment in the Orinoco Delta–Gulf of Paria is about 7,500/g only in a few places where the conditions are "normal". In areas of little water turbulence ("low energy"), the concentration is much greater than 7,500/g, e.g., south of the Peninsula of Paria. In "higher-energy" areas, such as just west of Trinidad, the concentration per gram is considerably less. *Rhizophora* (red mangrove) pollen is very small in size and it does generally increase in percent of total pollen as one moves offshore, though the situation is complicated by current patterns (see Fig. 17.10c). Dinoflagellate cysts (= Muller's "Hystrix"; they were not recognized as dinoflagellates at the time), of the same general size as spores/pollen, and apparently of the same approximate chemical composition, are distributed quite differently from spores/pollen in the sediment (see Fig. 17.10h). Their distribution, unlike that of land-originating spores/pollen, is a thanatocoenosis, apparently primarily a product of nutrient availability for the dinoflagellates and only secondarily to later operative sedimentological factors. Fossil fungal spores (see Fig. 17.10g), consisting of chitin, were found hardly at all in the marine environment but in enormous quantity in all parts of the delta itself. (By contrast, Wang *et al.* (1982) and others have shown that halophyte pollen, such as *Salicornia*, occurs practically exclusively offshore in the Yangtze River Delta of China.)

Muller's work established the importance of clastic sedimentation to fossil palynomorph distribution. The significance of local "energy level" (water turbulence) was demonstrated. That wind distribution plays some role even in marine sediments was shown by bisaccate pollen of *Podocarpus* (Fig. 17.10f), blown in small amounts in a northwest arc from upland source areas in Trinidad. Long-distance transport by water was shown by sedimentation offshore of very small amounts of *Alnus* pollen, originating hundreds of kilometers upstream and in a direction opposite the prevailing wind.

Palynomorph sedimentation in the Bahamas: a non-clastic model

Soon after Muller's work in the Orinoco area, Traverse & Ginsburg (1966), working for Shell Oil Company on sedimentation in the Bahamas, applied palynological techniques to that totally non-clastic environment, which also has no streams. The Great Bahama Bank is a drowned Pleistocene island with mostly very shallow water, relatively little current flow and a series of low islands (see Fig. 17.11a). The sediment is mostly of two types:

(a) carbonate "sand", consisting of shells and shell fragments of small animals, and of fecal pellets, and
(b) silt-size carbonate "mud".

The concentration of pollen is typified by pine pollen per gram of sediment (Fig. 17.11b). Very small amounts of pine pollen, or none at all, are found in samples from the sandy areas where the water is comparatively turbulent. In areas of quieter water, in the lee of Andros and Eleuthera Islands, comparatively great peaks of abundance of pine pollen were found, in the "muddy" (silt-sized) sediment. The line of demarcation between high concentrations and low (or zero) concentrations of pine pollen is very sharp, tested by very close sampling. Figure 17.11d shows the close association of pine pollen concentration with sediment type for a transect off Eleuthera Island. Relatively high pollen concentration is a phenomenon of silt-sized sediment, as would be expected. It is also worth noting that the concentration of spores/pollen per gram in carbonate mud is one to two orders of magnitude less than in most clastic silts: more or less 100–1,000/g as against more or less 1,000–100,000/g (or more). Shipboard continuous centrifrugation of water west of Andros Island showed comparatively large amounts of pine pollen in the water at localities where small amounts were found in the sediment, and conversely (Fig. 17.11c). It seems clear that pine pollen containing trapped air floats on Bahama Bank water for long periods of time, and then is sedimented out when it finally is wet and reaches a quiet water area that acts as a "pollen trap". The buoyancy of bisaccate pollen in water is probably primarily responsible for its occurrence in deep-sea sediments where little else occurs, though it should be noted that Koreneva (1964), on whose work much of this information depends, emphasizes that some sorts of fern spores are apparently also buoyant (air trapped in the perispore?) and occur in sediment in remote oceanic locations. Davis (1968, 1973) has demonstrated a similar "pollen-trap effect" in a small lake with no stream influx. Funkhouser & Evitt (1959) published a laboratory technique for concentrating and cleaning palynomorph preparations by agitation in watch-glasses, and Tschudy (1960) once designed an apparatus for separating spores/pollen types by vibration induced in a perforated tube. Both of these laboratory techniques depend on the same flotation property of spores/pollen as observed in the field in the Bahamas. More recently, Wang *et al.* (1982) have shown in the Yangtze River Delta area that pine pollen as a percentage of total pollen is higher in most of their offshore stations than on land, bespeaking the ready transportability of pine pollen.

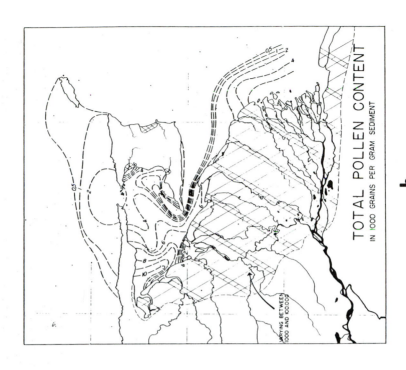

TOTAL POLLEN CONTENT
IN 1000 GRAINS PER GRAM SEDIMENT

VARYING BETWEEN
1000 AND 100000

b

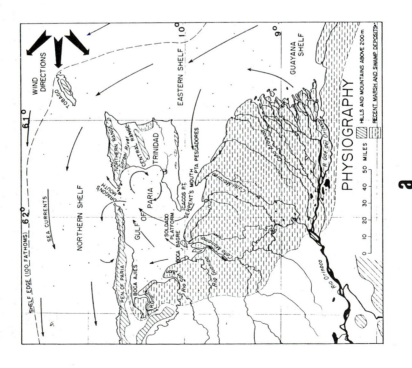

WIND
DIRECTIONS

TOBAGO

SHELF EDGE (100 FATHOMS) 62° 61°

SEA CURRENTS

NORTHERN SHELF

PEN OF PARIA

BOCA AJIES

Rio Sa

Rio Guanipa

Northern Range
Coura River

NORTHERN
RANGE

DRAGONS
MOUTH

GULF
OF
PARIA

SOLDADO
PLATFORM

BOCA BAGRE

Rio Morichal

CENTRAL RANGE

TRINIDAD

CACOS PT
SERPENT'S MOUTH
PTA PESCADORES

EASTERN SHELF 10°

9°

GUAYANA
SHELF

Caño Mocorro

Caño Macareo

R Gde del Orinoco

Rio Orinoco

PHYSIOGRAPHY

0 10 20 30 40 50 MILES

HILLS AND MOUNTAINS ABOVE 200 m

RECENT MARSH AND SWAMP DEPOSITS

a

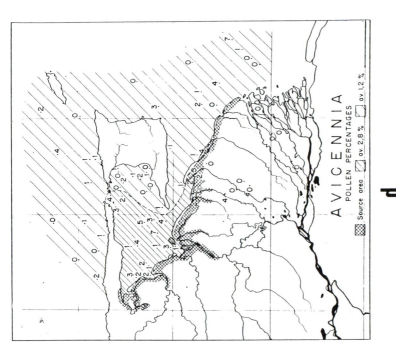

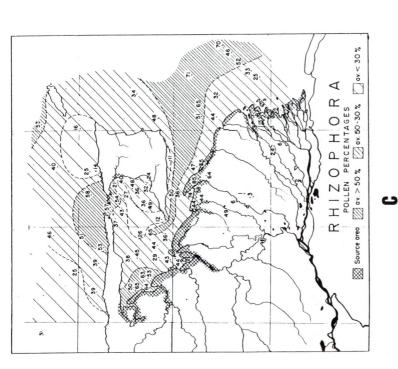

Figure 17.10 Muller's pioneering study of palynomorph sedimentation in the Orinoco Delta region demonstrated that in a clastic sedimentary environment palynomorphs behave primarily like water-borne, fine silt particles, wind distribution being of secondary importance except locally and in special cases. (a) General layout of Orinoco Delta in northern South America, about 10° north of the Equator. Note prevailing eastern winds and sea currents. (b) Total pollen content in grains per gram of sediment. Note that there is a general decrease in amount offshore, but that water turbulence and current are very important. The highest concentrations are in quiet embayments and on the western Gulf of Paria. Hoffmeister's "7,500 per gram" for shoreline is approximately correct only for some places along the delta. Off Trinidad and the Peninsula of Paria, the concentrations are much less. (c) *Rhizophora* (red mangrove), a very small, smooth, thin-walled pollen grain, is carried easily by the water, actually increasing in percentage in some far offshore locations. (d) *Avicennia* (black mangrove), an insect-pollinated, moderately heavily sculptured, medium-sized pollen grain is never abundant and

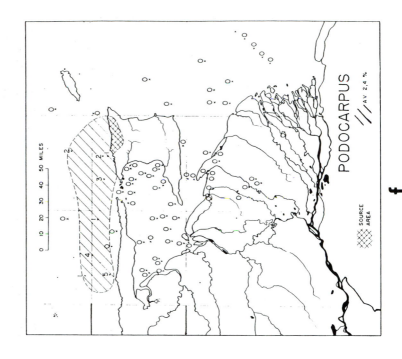

PODOCARPUS

AV 2,4 %

0 10 20 30 40 50 MILES

SOURCE AREA

f

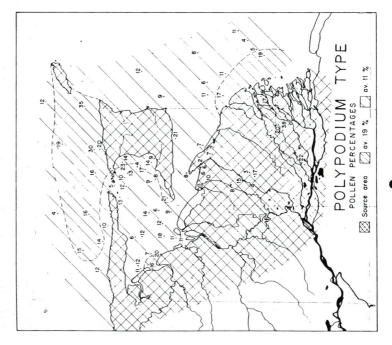

POLYPODIUM TYPE

POLLEN PERCENTAGES

Source area   ov 19 %   ov 11 %

e

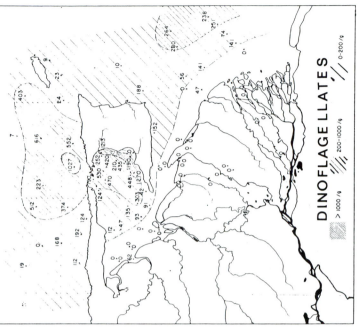

DINOFLAGELLATES

▨ >1000 /g   ▥ 200-1000/g   ⫽ 0-200/g

h

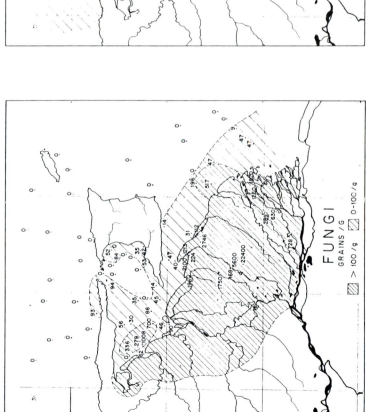

FUNGI

GRAINS / G

▨ > 100 /g   ⫽ 0-100/g

g

mostly drops out of the water close to sources of the producing plants. (e) *Polypodium*-type fern spores, although heavy-walled, are apparently buoyant and are found in sizable percentages far offshore, in agreement with observations for some fossil fern spores. (f) *Podocarpus*, a buoyant bisaccate conifer pollen grain, is an exception to the general indication for water transport in the Orinoco area. The small percentages of *Podocarpus* pollen encountered west of Trinidad clearly are dropped there as wind-borne grains. (g) Fungal spores are a deltaic phenomenon, and the sharp dropout of fungal spores offshore is dramatic, presumably related to the specific gravity of the spores. (h) Dinoflagellate cysts show a distribution related partly to sedimentary factors (note that concentration in the Gulf of Paria along the west side of Trinidad parallels pollen concentration there). However, dinoflagellate distribution is also related to nutrient availability, not sedimentation – see sizable numbers at some offshore locations, and the lack of specimens near the influx of fresh water.

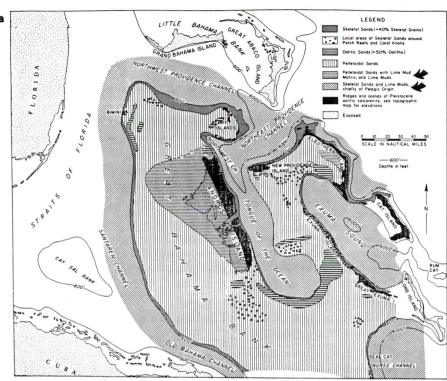

a

LEGEND

Skeletal Sands (>40% Skeletal Grains)

Local areas of Skeletal Sands around Patch Reefs and Coral Knolls

Oolitic Sands (>50% Ooliths)

Pelletoidal Sands

Pelletoidal Sands with Lime Mud Matrix, and Lime Muds

Skeletal Sands and Lime Muds, chiefly of Pelagic Origin

Ridges and islands of Pleistocene oolitic calcarenite, see topographic map for elevations

Exposed

SCALE IN NAUTICAL MILES
0 10 20 30 40 50

— 600 —
Depths in feet

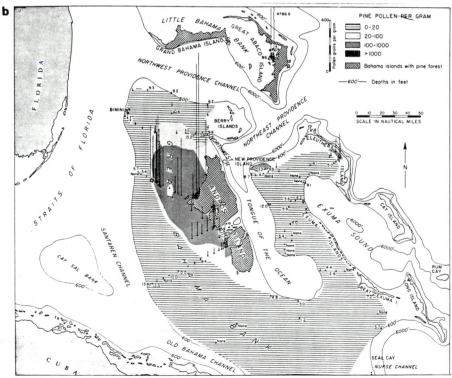

b

PINE POLLEN PER GRAM

0-20

20-100

100-1000

>1000

Bahama islands with pine forest

— 600' —   Depths in feet

SCALE IN NAUTICAL MILES
0 10 20 30 40 50

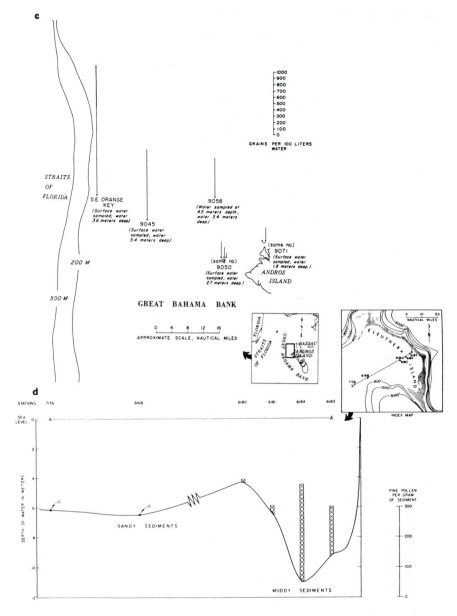

Figure 17.11 Palynomorph sedimentation on the Great Bahama Bank, a carbonate environment with no streams and no clastic sedimentation. The Bank is a drowned Pleistocene island with mostly very shallow water, over which currents drift in a generally northwestward direction. Prevailing winds are also toward the northwest. Palynomorph sedimentation here thus has provided an interesting comparison with sedimentation in clastic environments (see Figs. 17.9 & 17.10). (a) Sedimentary types of surface sediment. Note especially (arrows in legend) the lime mud areas in the lee of Andros and Eleuthera Islands, and the general prevalence of pelletoidal sand over much of the Bank. (b) Pine pollen per gram of sediment. Note the prevalence of high values in the lee of large islands, closely agreeing with location of lime mud. Sediment from areas of pelletoidal sands is practically devoid of pine pollen. Evidently, pine pollen

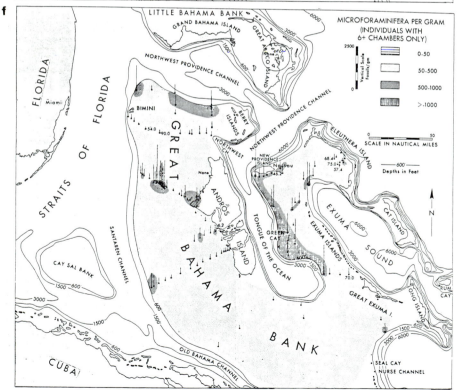

Dinoflagellate cysts (Fig. 17.11e) and foraminiferal chitinous inner tests ("microforaminifera"; Fig. 17.11f) found in the Bahamas sediment represent thanatocoenoses of remains which are not primarily distributed according to sedimentary environment, but apparently in response to biological factors, mostly the availability of nutrients to the living organisms, such as in upwelling areas near the Tongue of the Ocean. Melia (1984) has made a similar observation about distribution of dinoflagellate cysts and microforaminifera off the coast of West Africa.

Farley (1982) has pointed out that the relative paucity of spores/pollen in my study of the non-clastic, carbonate sediment of the Bahamas, in contrast to sediments of clastic environments, is a confirmation of the water-transport source of most fossil sporomorphs, which travel in water as clastic particles. If air transport were of major importance, Bahamas surface sediment would be as rich as is ordinary silt formed in such an environment, which it is not, by at least an order of magnitude. That the distribution of palynomorphs in sediment is primarily a product of sea currents in the basin of deposition has been repeatedly demonstrated, but Melia (1984), working in northwest Africa, noted that spores/pollen distribution in sediment there reflected both oceanic transport mechanisms and the relatively strong offshore African wind patterns. Onshore vegetational complexes were also traceable in the marine sediments.

Further contributions to spore/pollen sedimentation

Cross *et al.* (1966) and coworkers have shown that palynomorph sedimentation in the Gulf of California agrees in general with patterns suggested by the above models (see Fig. 17.12). They showed in addition the importance of bottom morphology. Lower Gulf fine sediments are comparatively rich in palynomorphs, and upper Gulf sediments are coarser and relatively poor in palynomorphs. But the Gulf of California sediment contains considerable reworked palynomorphs from older sediment, and these are comparatively abundant. Some of the concentrations of palynomorphs are influenced by nearness to mouths of source streams. Smaller variations represent turbulence

---

floats for considerable periods over the Bank, eventually being trapped in areas of relatively low energy. (c) Pollen in the water off Andros Island has a reciprocal distribution to that in the sediment: the areas with high pollen in sediment have low pollen in the water and vice versa (see (b)). (d) A traverse off Eleuthera specifically confirms the relationship of pollen highs to mud, and pollen lows to sand, as would be expected from the silt-to-finest-sand size of pine pollen. Because of relatively low specific gravity (about 1.4), and because it is not solid, pollen sorts with mineral particles (specific gravity about 2.5) a bit smaller than pollen (see Stanley 1969). (e) Dinoflagellate cysts display a distribution that is partly a result of sedimentary factors (some were encountered in the water studies of (C)) and partly a product of nutrient availability. (f) Foraminiferal chitinous inner tests ("microforaminifera") show a distribution entirely different from palynomorphs proper, presumably because the foram fossils are a thanatocoenosis related primarily to factors encouraging the development of foraminiferal populations.

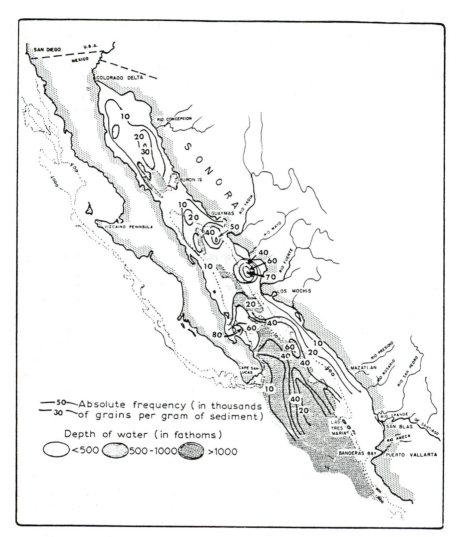

Figure 17.12 Palynomorph sedimentation in the Gulf of California. In this complex of clastic sedimentary environments in a semiarid climatic zone, the sedimentation of pollen and spores is correspondingly complex. However, in general, palynomorphs are more abundant in offshore silts and near delta mouths and in some submarine channels. Coarse sediments nearshore are poor in palynomorphs, but the transition to silt beyond these sands can be abrupt. Most forms decrease in absolute abundance with distance from shore, but pine and mangrove increase in relative abundance.

patterns, e.g., near the tip of Baja California. Pine pollen from trees on shore is transported to the Gulf of California by water currents, against the direction of the wind.

Stanley (1965), working on U.S. Atlantic coastal shelf deposits, and Heusser (1978), investigating the U.S.–Canada Pacific Northwest, also have shown that palynomorph concentration on a continental shelf reflects the presence of streams, among other factors. In this connection, Mudie's (1982) work demonstrates for eastern Canada that the estuaries of major rivers yield sediment with proportionately higher palynomorph levels than is true for areas with only small streams. Chowdhury (1982) shows that, in the German Bay (southeast North Sea), marine current systems play a large role in distribution of palynomorphs, though the rivers (Elbe, Weser, Ems, Rhine) are the major source of pollen. In sequences of riverine sediments such as the Catskill Magnafacies, New York–Pennsylvania, overbank deposits are barren or nearly so of palynomorphs, whereas channel sediments may be richly productive (Traverse et al. 1984). Elsik & Jarzen (1984) present similar results from recent overbank deposits of Zambia.

Koreneva (1971) found very low concentrations of spores/pollen in surface sediment of the Mediterranean, mostly less than 10/g (Melia (1984) found less than 50/g in deep ocean basins). The concentrations seem in general to agree with the concept of pollen settling out in quiet areas or catchment basins (such as the northern Adriatic), and for apparent concentration of palynomorphs to be greatest where general mineral sedimentation rate is lowest. Koreneva (1966) also has demonstrated that spores/pollen of land plants occur in small amounts hundreds of kilometers from the source vegetation in the Pacific, and the types found occur there primarily because of the physical characteristics of the particular spores/pollen types. Koreneva has also made interesting observations in the Sea of Okhotsk and elsewhere that link the concentration of spores/pollen in sediment to the particle size of the enclosing sediment, and to the characteristics of the ocean floor: depressions act as pollen traps, as in the Bahamas (see Fig. 17.11d). Darrell (1973) and Darrell & Hart (1970) have shown by statistical analysis of sporomorph complexes of Mississippi River Delta sediments that various onshore depositional environments (marshes, levees, etc.) differ markedly in palynomorph concentration from offshore environments of the same system (mouth bars, prodelta). The same studies showed significant differences in pollen spectra between the various onshore environments, but not between various offshore environments. Surprisingly, 87% of the palynomorph taxa identified were sedimented independently of environment, just as likely to be sedimented in one as in another. The other 13% were facies-dependent and their preferred sites of deposition were:

(1) reworked forms – levees, channels and offshore,
(2) bisaccates – offshore,
(3) tree pollen – channels, levees and offshore,
(4) marsh pollen – marshes.

Groot & Groot (1971) and, in much more detail, Heusser (1983) and Sarro et al. (1984) have pointed out that the spore/pollen content of sediment depends to a considerable extent on the productivity of vegetation in nearby source continents, as well as on distance from that source.

The relative abundance of reworked forms increases with distance from source continent. Stanley (1969) has in this connection cautioned against too facile interpretation of past continental climates from pollen spectra of deep-sea cores obtained far from shore, especially because of reworking. Heusser's figures well illustrate the total situation (see Fig. 17.13). Total pollen concentration is affected by stream systems, as shown by lobes of higher concentration associated with major streams of eastern North America. Concentration is also affected by bottom geometry – note the outer continental shelf lows, *vis-à-vis* slope analyses (maximum value 23,000 grains/g). Note also the general control of vegetation types in the nearby continents, as demonstrated by factor analysis. On a much smaller scale, in Lake Turkana, Kenya, Vincèns (1984) has shown that deltaic sediments contain pollen derived from the regional montane vegetation, whereas the strictly lacustrine sediments yielded mostly pollen from the local steppe vegetation.

Wang *et al*. (1982), working in the area of the Yangtze River Delta, China, have shown that the amount of palynomorphs per unit of sediment varies with the environment of deposition, as Muller found in the Orinoco Delta. Wang *et al*. also emphasize that sediments from the mouth of the Yangtze had lower values than either upstream or offshore stations. I would attribute this to turbulence and would expect that measurements of the water itself would reveal high values at the mouth.

Williams (1971) has demonstrated that in marine environments generally the distribution of dinoflagellate cysts is a product primarily of the organic productivity of the superficial water (see Fig. 17.14). This result agrees well with Traverse's and Muller's earlier studies of limited areas (see Figs. 17.11e and 17.10h). Dinoflagellate cysts offer some possibilities for paleoecologic interpretation that spores/pollen in marine sediments do not. The source for dinoflagellates is limited as to range of required environmental characters. Dinoflagellate cysts can be directly tied to oceanic temperature, nutrient supply, and currents. Various authors, e.g., Scott (1982), have suggested that smooth cysts (dinoflagellate and acritarch) are characteristic of nearshore, high-energy, turbid environments, whereas spiny cysts suggest deeper, low-energy, cleaner water. Dinoflagellate presence or absence can sometimes be used to differentiate marine from non-marine sediments, but this is tricky because freshwater or brackish-water dinoflagellates can be very abundant in non-marine sediments.

Reworked sporomorphs often demonstrate the effectiveness of long-distance transport in rivers; e.g., Stanley (1969) showed that Cretaceous sporomorphs in Pleistocene sediments of the Gulf of Mexico probably travelled over 2,000 km from the Great Plains, by way of fluviatile transport. (Reworking can be a major factor in palynomorph sedimentation. Chowdhury (1982) notes that up to 30% of the surface spores/pollen in German North Sea sediments are redeposited forms, as old as Carboniferous.) Bonny (1976, 1978) has shown that, even in small lakes, as much as 85% of the pollen in the sediment reaches its destination by the stream(s) feeding the lake. On the other

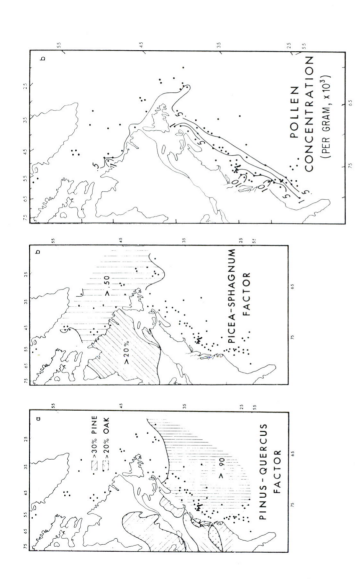

Figure 17.13 Pollen concentration in sediments off the east coast of North America, showing maxima on the continental slope, and offshore from stream systems with maximum sediment load. The two diagrams to the right are based on Q-mode factor analysis of pollen concentration. The *Pinus–Quercus* factor accounts for 67% of the variance in the data set, and especially dominates in the sediment east of onshore dominance of trees of these genera (>30% pine and >20% oak in the onshore vegetation is shown by the patterns indicated toward the top; >.90 refers to factor loadings for pollen in the sediment). The *Picea–Sphagnum* factor comprises 19% of the variance, and is similarly associated with vegetation sources to the west (>20% pattern is for vegetation; >.50 refers to the factor loading for this subset).

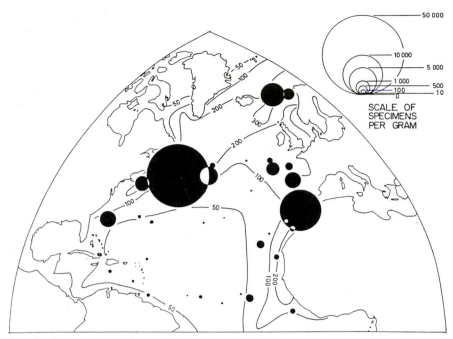

Figure 17.14 Dinoflagellate cyst concentration in marine sediments (see also Figs. 17.10 & 17.11). Dinoflagellate cysts are not abundant in situations far from continental margins, but they tend to abound in regions in continental shelf areas where, presumably, circumstances favor both dinoflagellate populations in general and encystment in particular. In Black Sea cores, I have studied a few, probably brackish-water, Pleistocene samples containing several *million* dinoflagellate cysts per gram!

hand, Hooghiemstra & Agwu (1986) showed that, in the Atlantic off northwest Africa, the trade winds and African Easterly Jet (Saharan Air Layer) were the main suppliers of pollen to the sediment, despite the generally greater significance of water transport.

SORTING OF SPORES/POLLEN PER SEDIMENT TYPE

Earlier in this chapter I have discussed the fact that palynomorphs are silt- and finest sand-sized clasts and are mostly fellow travellers of fine silt, as they sort out with slightly smaller mineral particles. Well-sorted claystones and well-sorted coarse sandstones will not ordinarily contain more than a trace of fossil palynomorphs. Carbonate sediments are prevailingly non-clastic and most of them are palynomorph-poor, although some can and do contain beautifully preserved palynomorphs (Scott *et al.* 1985b). Coal balls of Europe and North America are also calcareous sediments, in which palynomorphs are frequently very well-preserved. Permineralizations such as these examples are, however, very atypical sedimentary rock. Coals are a special case. Most were formed in warm temperate to subtropical swamps. Many contain much spores/pollen (if attrital or coal macerals predominate), nearly all of which were autochthonous. Vitrinitic coals, consisting largely of wood and other tissues, are often very poor in spores/pollen, however. Coal-derived palynofloras tend to be less diverse than those from associated clastic sediments, which derive sporomorphs from a much broader area.

Various studies have shown that when one considers broad characters of palynofloras, not the individual taxa, palynofloras of quite different age can be grouped according to sediment type (see Fig. 17.15b). The general composition of the palynoflora (proportion of monosaccates and bisaccates – conifers, fern spores, etc.) from a Lower Jurassic marine sediment is often more like that of an Upper Jurassic marine sediment than it is like that of a Lower Jurassic siltstone. In many instances, the interpretation of a fossil palynoflora depends on sedimentological environment rather than on autecology of the source plants. For example, as pointed out by Chaloner & Muir (1968), prevalence of saccate pollen in marine Carboniferous sediment usually demonstrates not dominant Pv2 producers near the basin of deposition, but quite the opposite: deposition at considerable distance from shore because of greater amounts of bisaccate pollen being transported by air and water to that site. Chaloner & Muir called this the "Neves effect", from the work of R. Neves, which they interpreted as supporting the theory (see Fig. 17.15a). Extant models certainly suggest that this could be possible. For example, high percentages of *Rhizophora* pollen in the Gulf of Paria (Fig. 17.10c) are explained, not by nearby mangrove vegetation, but by deposition at some distance from shore, where the comparatively great transportability of tiny *Rhizophora* pollen is operative. Figure 17.16 presents a model for the Neves effect and related phenomena.

Interpretation of fossil palynofloras must always take sedimentary factors into consideration. For example, the occurrence in latest Triassic–early Jurassic (Rhaeto-Liassic) sediments of layers dominated by tremendous quantities of *Corollina* (= *Classopollis*) pollen obviously means presence somewhere of large numbers of cheirolepidiaceous shrubs and/or trees that produced this pollen type. (Cheirolepidiaceae is the family of Mesozoic conifers which includes *Hirmeriella*, formerly called *Cheirolepis*.) However, whether the pollen dominance was because the *Corollina*-producing plants were coastal organisms like mangrove or cypress today, or whether the plants were upland shrubs/trees and the pollen blooms resulted from sedimentary sorting, can be debated. It has been noted (Hughes 1976), for example, that *Corollina* pollen usually is an order of magnitude more abundant in shales than in associated coals. In some Carboniferous coal-bearing sequences it has been observed by Smith (1962) and others that *Densosporites* spores (lycopod-derived) trend toward dominance at the top of coal beds and in the roof shales (see Phillips *et al.* 1974, Chaloner & Muir 1968). Scott & King (1981) show a similar sequence with lycopod-derived megaspores such as *Zonalessporites*, which are megaspores of densospore (a microspore) producers. Such a situation, presumably edaphically controlled, would appear to be plant-successional, because coal palynofloras are practically 100% authochthonous. Densospore producers perhaps succeeded swamp trees as the water deepened and became more marine influenced. On the other hand, the situation could theoretically be

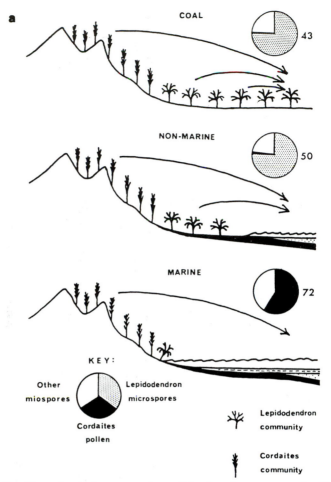

Figure 17.15 Two classic demonstrations of the fact that spores and pollen produced contemporaneously are selectively sorted by sedimentation factors into different environments. (a) The "Neves effect": saccate pollen (*Florinites*) of *Cordaites* trees are carried by both wind and water into marine environments, whereas microspores of *Lepidodendron* (*Lycospora*) are produced in swamps and are not so transported out of the swamps. Some other taxa of miospores are produced in other environments and do float reasonably well, and are carried seaward. In the coal swamps the locally produced *Lepidodendron* spores completely overwhelm the relatively few *Florinites* pollen grains that come in. In the non-marine, non-swamp situation there are more *Florinites* but not proportionately so many as offshore. On the other hand, Hughes (1976) has demonstrated that some Mesozoic coals produced in conifer swamps are dominated by bisaccate conifer pollen, whereas another conifer pollen, *Classopollis* (= *Corollina*), behaves as the "Neves effect" would predict: it is dominant in contemporaneous shales and limestones produced offshore from the coal swamp. Pollen and spores of coal are more a reflection of the local vegetation than of sedimentary factors. (b) Spore assemblages of British Jurassic and Lower Cretaceous rocks of different lithology. The

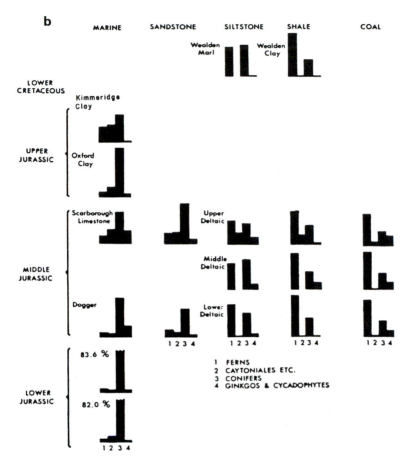

general palynological composition clearly depends heavily on the environment of deposition, as reflected by lithology, even over a large timespan. Sorting according to sedimentological effects is obviously at least partly responsible, although spores and pollen produced in a coal swamp reflect the immediately local vegetation and have little chance of being carried out of the swamp.

sedimentological, with densospores perhaps coming in with invading marine water. Chaloner & Muir called the plant-successional model the "Smith effect", from the work of A. H. V. Smith, which demonstrated it. (A regionally climatically controlled succession, such as the typical Euramerican Holocene pollen diagram, Chaloner & Muir called the "Von Post effect".) The sedimentological "natural history" of spores/pollen is an advantage to their biostratigraphic use because they themselves are relatively free of purely local phenomena such as water temperature or salinity (coastal plants such as mangroves are clearly exceptions) but the sedimentological history of palynomorphs makes them sensitive to depositional facies.

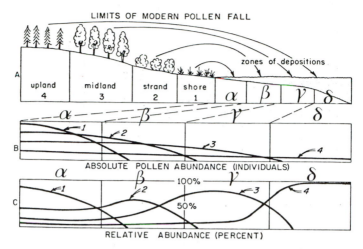

Figure 17.16 Neves effect (see Fig. 17.15) and related phenomena graphically displayed for a hypothetical coastal environment. Coniferous vegetation in upland zone 4 produces pollen that arrives in about equal small amounts (= absolute abundance) in depositional zones α, β, γ, and δ. This is somewhat less true of midland vegetational zone 3, because the pollen in this zone tends not to be as buoyant as conifer pollen, but is produced copiously per unit area and so starts out abundant in depositional zone α but tails off at the beginning of depositional zone δ. Strand (2) and shore (1) vegetational zones produce abundant but (in the case displayed!) typically heavy pollen which is therefore abundant only in the relatively near vicinity of production (depositional zones α and β). Note that when plotted by percentage, the upland (buoyant, gymnosperm) pollen attains 100% in depositional zone δ, though its *absolute* amount remains approximately constant from zone α to zone δ. However, compare with information in Fig. 17.10c for an example of shoreline vegetation (red mangrove) which produces small, buoyant pollen which would show quite a different distribution (more like an upland coniferous pollen in this example). Each depositional situation is a little different, but the basic analysis of the sedimentological factors follows the same "laws".

Biological "sorting": marine palynomorphs

Some palynomorphs originate in the water of the basin of deposition, e.g., those derived from marine animals or protists, such as chitinozoans, scolecodonts, and microforams, as well as acritarchs and dinoflagellate cysts, which are mostly derived from marine plants or protists. For fossils that originate in the water, one must be aware of ecological factors affecting production and dispersal. In the modern Gulf of Mexico–Caribbean area, for example, microforams are abundant in the calcareous sediment of the Bahamas and Yucatan Peninsula areas, but are rare in the clastic sediments of the northern Gulf, e.g., the Mississippi Delta or Trinity–Galveston Bay. As already observed, it has been suggested that spiny acritarchs and dinocysts are more common in deeper water, and relatively non-ornamented forms are more common in shallower, higher-energy situations where there may also be more brackish water. In the Bahamas, high concentrations of dinoflagellate cysts and microforams generally are strongly correlated with the presence of

Plate 1 Spores/pollen exine coloration with geothermal maturation ("coalification"). "Munsell prod. no." refers to Munsell color standards, a series of opaque lacquer films on cast-coated paper. *TAI* = thermal alteration index, a scale based on both color of spores/pollen and on vitrinite reflectance measured by reflected light under oil. Fluorescence refers to the tendency of exines to fluoresce (to produce visible light when exposed to ultraviolet light). Exinite is the coal petrological maceral consisting of exines; vitrinite is a maceral of woody or similar plant tissue origin.

ORGANIC THERMAL MATURITY	COLOR OF FOSSIL SPORES/POLLEN	MUNSELL PROD. NO.	TAI = 1-5	VITRINITE REFLECTANCE	COAL RANK	EXINITE FLUORESCENCE: AMOUNT AND COLOR
IMMATURE		17,391	1	0.2%	Peat	High to Medium; blue-green
		20,520	1+	0.3%	Lignite	High to Medium; green-white
		19,688	2-			
		14,253	2	0.5%	Subbituminous	High to Medium; white-yellow
		13,800	2+		Bituminous, High Vol. C	
MATURE MAIN PHASE OF LIQUID PETROLEUM GENERATION		12,424	3-	.9%	Bituminous, High Vol. B	High to Low; yellow
		15,816	3		Bituminous, High Vol. A	
		17,209	3+	1.3%	Bituminous, Medium Vol.	Low: dark yellow to orange-brown
DRY GAS OR BARREN		15,814A	4-	2.0%	Bituminous, Low Vol.	No Fluorescence of Spores/Pollen Exines
		19,365	4	2.5%	Semi-anthracite	
	BLACK & DEFORMED		(5)		Anthracite	

nutrient-rich water. (Melia (1984) has made a similar observation in Africa.) For phytoplankton (acritarchs and dinocysts primarily), it has also been suggested that low taxon diversity indicates nearshore deposition. However, Dorning (1981), in a study of British Silurian sediments, notes a peak in generic diversity of phytoplankton in open shelf environments with less diversity both toward the shore and toward deeper water. He also found that more highly ornamented forms dominate in shelf areas, whereas less ornamented forms and sphaeromorphs were dominant in nearshore localities and to some extent also in deeper-water environments beyond the shelf. In an extensive systematic study of dinoflagellate cysts in the North and South Atlantic, Wall *et al.* (1977) found that the distribution of fossil cysts (a thanatocoenosis) parallels reasonably well the distribution of the corresponding living dinoflagellate species. They also noted that there are two major trends in distribution, one correlative with distance offshore (environmental), the other with latitude (climatic). Both species diversity and cyst density tend to increase seaward, as a result primarily of sedimentological factors. Biologically, dinoflagellates that provide fossil cysts are as a class adapted to unstable environments in shallow-water situations along continental margins, and around oceanic island groups, prevailingly in tropical regions (Wall *et al.* 1977). Individual species, however, tend to be adapted to the most stable sectors of these unstable environments.

Classifications of organic particles in sediment

Palynologists have long noted that maceration residues prepared from sedimentary rocks contain a wide range of organic particles in addition to palynomorphs. This is especially true of marine shales. Manum (1976a) suggested the term "palynodebris" for such particles, whether carbonized or not. Some examples of palynodebris are shown in Figures 17.17 and 17.19. Several palynologists have made an effort to put study of such particles on a systematic basis, with the hope that such study could contribute to understanding of the total sedimentological picture. As Bujak *et al.* (1977) stress, the range of kinds of organic particles in sedimentary rocks also controls to some extent the production of hydrocarbons from the rock during thermal maturation. Bujak *et al.* term the four principal kinds of organic particles:

(a) amorphogen (amorphous: structureless organic matter),
(b) phryogen (non-woody plant material, including palynomorphs),
(c) hylogen (from woody material),
(d) melanogen (opaque organic matter).

Amorphogen and phryogen are more likely to produce liquid hydrocarbons in time than is hylogen, and melanogen is least likely to be productive. Manum & Throndsen (1978) found that, in the Spitsbergen Tertiary, phryogen is usually dominant in marginal marine sediments, amorphogen is high in offshore marine sediments, and hylogen is most common in non-marine deposits (melanogen was not common in any samples). Those particles which originally were fragments of plants are sometimes called "phytoclasts".

417

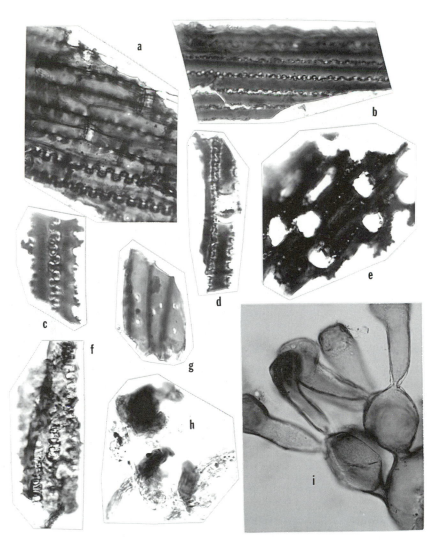

Figure 17.17 Commonly encountered palynodebris from various Neogene sediments. Most of this palynodebris consists of plant fragments ("phytoclasts") and would fall in Habib's tracheid and cuticle category of the exinitic and tracheal facies, and in Masran & Pocock's structural terrestrial (telinite) and biodegraded terrestrial categories. There are also many more or less amorphous sorts of palynodebris (see Fig. 17.18). Magnification shown by bar under (l). (a)–(f) Fragments of seed plant cuticle, Pliocene sediment of Black Sea: (e) shows a quite carbonized cuticular fragment in which entire stomatal apparatuses are missing, leaving lacunae, and (f) is a very much degraded fragment, still recognizable as cuticular. (g) Seed plant wood fragment showing portions of several tracheids with bordered pits, Pliocene sediment of Black Sea. (h) Characteristic degraded, indeterminable, organic debris of palynological preparations, Pliocene sediment of Black Sea. (i) Probably chitinous animal (or fungal?) material, Recent sediment of Black Sea. (k), (l) Portions of vessel elements with pitting, Pleistocene

418

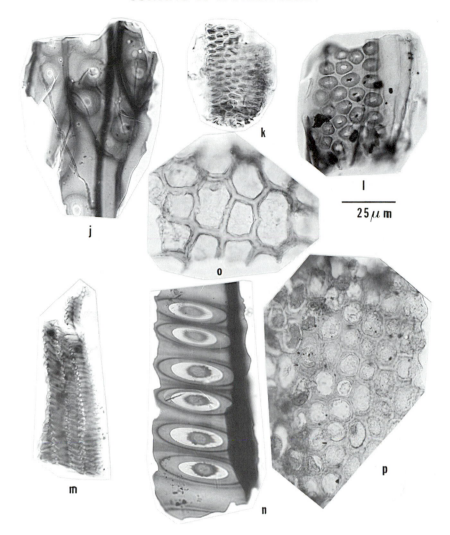

sediment of Black Sea. (m) Plant tracheary material, much degraded, Pleistocene sediment of Black Sea. (n) Portion of one coniferous tracheid with bordered pits, Pleistocene sediment of Black Sea. (o), (p) Portions of sheets of probably chitinous cellular tissue, perhaps animal, Recent sediment of Gulf of Mexico.

Masran & Pocock (1981) devised a classification for phytoclasts (Fig. 17.18a) based largely on botanical and coal petrological classification of particulate palynodebris. The classification emphasizes the sorts of original tissues represented, and the categories of coal macerals these become. The classification depends partly on experimental disaggregation of plant material, producing artificial "phytoclasts". Masran (1984) has demonstrated the usefulness of the method in a study of North Atlantic Jurassic–Cretaceous sediment from D.S.D.P. cores. The organic sediments could be classified as marine or non–

a. Masran & Pocock:

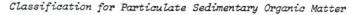

Classification for Particulate Sedimentary Organic Matter

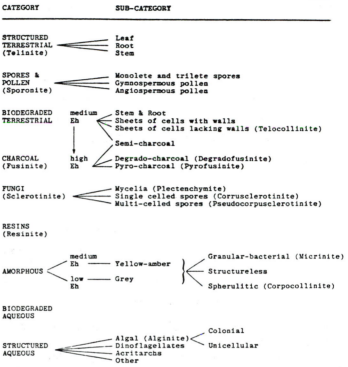

CATEGORY	SUB-CATEGORY

STRUCTURED TERRESTRIAL (Telinite) — Leaf, Root, Stem

SPORES & POLLEN (Sporonite) — Monolete and trilete spores, Gymnospermous pollen, Angiospermous pollen

BIODEGRADED TERRESTRIAL medium Eh — Stem & Root, Sheets of cells with walls, Sheets of cells lacking walls (Telocollinite), Semi-charcoal

CHARCOAL (Fusinite) high Eh — Degrado-charcoal (Degradofusinite), Pyro-charcoal (Pyrofusinite)

FUNGI (Sclerotinite) — Mycelia (Plectenchymite), Single celled spores (Corrusclerotinite), Multi-celled spores (Pseudocorpusclerotinite)

RESINS (Resinite)

AMORPHOUS — medium Eh — Yellow-amber, low Eh — Grey — Granular-bacterial (Micrinite), Structureless, Spherulitic (Corpocollinite)

BIODEGRADED AQUEOUS

STRUCTURED AQUEOUS — Algal (Alginite) — Colonial, Unicellular, Dinoflagellates, Acritarchs, Other

Figure 17.18 Two classifications of "palynodebris", the dispersed organic matter found in palynological preparations, including palynomorphs but other organic (mostly plant) matter as well. (a) Masran & Pocock's (1981) classification of palynodebris ties in with coal petrologic maceral categories (given in parentheses) and is primarily descriptive. (b) Habib's (1979, 1982a, b) classification of associations of palynodebris as "palynofacies" encountered at levels in sediment corresponding to various distances from land is paleoecological, dividing all palynodebris into: (1) the exinitic facies (nearest to shore), including almost no dinoflagellates; (2) the tracheal facies, dominated by bisaccates and other, more or less sorted pollen likely to be transported long distances, plus dinoflagellate cysts; (3) the shredded xenomorphic facies, dominated by shredded organic debris of uncertain origin, plus dinoflagellates and varying amounts of pollen; (4) the micrinitic facies, representing far offshore black clays with much black, carbonized plant debris and few palynomorphs; and (5) the globular xenomorphic facies (farthest offshore) with very few palynomorphs, and globular organic debris of marine origin probably zoöplanktonic fecal pellets. The example used by Habib was Cretaceous, hence the mention of *Classopollis* (= *Corollina*) pollen. The abundance of palynodebris fragments decreases from the exinitic (100,000/g) to globular xenomorphic (500/g), providing another parameter for identification of the constituent facies.

420

b. Habib Classification Of Palynodebris:

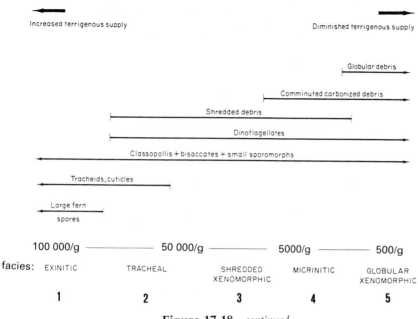

Figure 17.18 *continued*

marine in origin and conclusions could be drawn about the post-depositional history of the sediment.

Habib (1979, 1982a) has classified the organic particles in palynological residues from a somewhat different point of view. The overall character of the palynologic residue from a given sample is termed "palynofacies" (see Fig. 17.18b), and separated into five categories: (1) exinitic, (2) tracheal, (3) shredded xenomorphic, (4) micrinitic, and (5) globular xenomorphic.

Exinitic and tracheal residues typically occur in sediments that were rapidly deposited. They are comparatively rich in total organic matter. Micrinitic residues are typical of more slowly deposited sediments. Habib's system thus classifies each sample as either 1, 2, 3 or 4. A micrinitic sample in the Habib scheme might be 20% melanogen, 40% amorphogen, and 20% phryogen in the Bujak *et al.* scheme described above.

Batten (see Fig. 17.19) has proposed a series of "palynofacies" or assemblage types, which are varying complexes of certain palynomorphs and certain categories of "palynodebris". Batten's palynofacies will apply, of course, only to a rather limited area and to closely associated stratigraphic units, although generalized classes of his facies could be worked out. Combaz (1964) originally defined "palynofacies" for reference to the general aspect of palynological preparations. (The term "palynofacies" is unfortunately sometimes also used to refer to suites of palynomorph taxa associated with particular depositional paleoenvironments, without reference to palynodebris.)

421

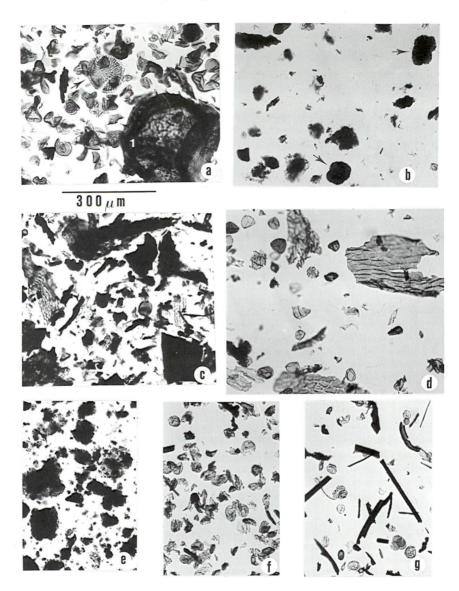

300 μm

Figure 17.19 "Palynofacies" of palynologic assemblage types, as varying complexes of palynomorphs and "palynodebris". This concept was introduced by Batten (1973, 1980, 1981a, 1982), and is useful in understanding the general origin of sediment and its propensity to be associated with hydrocarbons. In Batten's (1973) formally described assemblage types, each type has a key palynological character, which is usually a palynomorph but may be a kind of palynodebris ("brown wood") or a general palynological character ("diverse miospore content"). Illustrated here are various palynofacies in Batten's sense, as seen under the microscope at low power. Magnification indicated by bar under (a). (a) Facies from channel-fill siltstone, containing abundant trilete spores and megaspores: *Minerisporites* sp. (lower right,

422

Caratini *et al.* (1983) have emphasized the significance of the agents and conditions of transportation to the assortment of palynodebris found in a sediment and have introduced useful methods for diagrammatically representing the color and size range of the particles found. Scott & Collinson (1978) have urged the use of S.E.M. in systematic study of such organic particles in sediment.

Leopold *et al.* (1982) have shown by parallel studies of lignin derivatives and pollen in an 11 m (about 13,000 yr B.P.) core in Lake Washington that late Pleistocene lake sediments deposited when the lignin derivatives indicate a treeless area contain a pollen spectrum (mostly *Pinus*) suggesting forest! This seems to mean that the chemistry of the sediments reflects the immediate environs of the lake, whereas the pine pollen arrived from some distance, probably by air. Sediments from small and moderate-sized lakes are indeed an exception to the general rule that most palynomorphs in sediments got there by water transport.

AMOUNTS OF PALYNOMORPHS IN SEDIMENTS

The concentration of palynomorphs in sediments varies from near zero in well-sorted sandstones (even if the sandstone contains much black organic matter – most often sand-sized particles of vitrinite) and claystones ("dirty", that is poorly sorted, claystones and sandstones are often rich in palynomorphs) to 5 million(!) per gram of sediment (or rarely even more) in some sorts of coal and some organic marine sediments (about 5 million dinoflagellates per gram in some cores of Black Sea sediment; see Fig. 17.20). An average figure for a productive siltstone is about 10,000/g. Farley (personal communication) calculates mathematically that, with perfect packing, a 30 μm spore with walls 2 μm thick could theoretically attain a density of about 170×10^6/g, about an order of magnitude more than is ever observed in nature.

---

indicated by white "1"), and *Schizosporis reticulatus* Cookson & Dettmann (upper left, indicated by arrow). Lower Cretaceous, England. (b) Facies mainly consisting of fluffy amorphous organic matter, probably of algal origin, and *Botryococcus* colonies (arrows), probably from a freshwater mudstone. These colonies are thick-walled, hence appear dense. Lower Cretaceous, England. (c) Facies from a non-marine channel-fill siltstone dominated by abundant vascular plant remains, woody and membraneous, with a few miospores. Lower Cretaceous, England. (d) Facies with abundant plant cuticle and *Densoisporites* sp., non-marine gray shale. Mid-Jurassic, North Sea. (e) Facies from a marine "black shale" dominated by amorphous organic matter, a sort of facies that is likely to be associated with hydrocarbon generation, given sufficient thermal maturation. Upper Cretaceous, Helgoland, Germany. (f) Facies from a mudstone containing numerous specimens of a single species of peridinioid dinoflagellate cyst. The sediment was deposited in a brackish-water mudplain environment. Lower Cretaceous, England. (g) Facies dominated by fusinite–inertinite (carbonized plant remains), and conifer pollen, especially *Corollina* (= *Classopollis*), of which two tetrads are visible here. From a calcareous mudstone deposited in a basin-margin environment. Uppermost Jurassic, England.

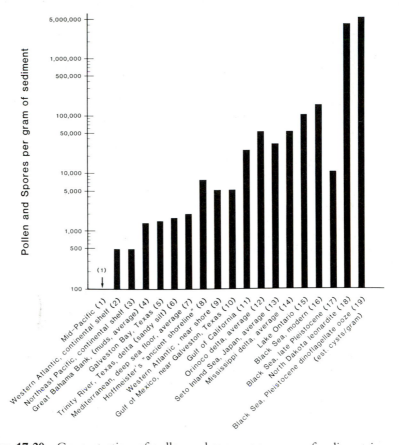

Figure 17.20 Concentration of pollen and spores per gram of sediment in various situations, plotted logarithmically. Deep oceanic sediments are practically barren. Nearshore sediments range around 1,000–5,000/g, being higher near river discharges. Lakes and inland seas are about an order of magnitude higher. Comparison of Black Sea modern sediment with late Pleistocene sediment from the same locations (see data bars 16 and 17) is instructive. With current high stand of the Black Sea, rivers are bringing relatively less mineral sediment than in the Pleistocene when water level was lower and the streams rejuvenated. Thus, if one assumes delivery of spores/pollen to the feeder streams to have remained relatively constant, we can explain the fact that the numbers of palynomorphs per gram of sediment are much higher now than in the late Pleistocene. The maximum concentration of palynomorphs per gram, found in some fossil "oozes" and in some coals, is in the 5,000,000/g range. Sources for data: (1) Koreneva (1964); (2) Stanley (1965); (3) Heusser & Balsam (1977); (4) Traverse & Ginsburg (1966); (5) and (6) Traverse, unpublished data; (7) Rossignol-Strick (1973); (8) Hoffmeister (1954); (9) Heusser (1983), Mudie (1982); (10) Traverse, unpublished data; (11) Cross *et al.* (1966); (12) Muller (1959); (13) Matsushita (1982); (14) Darrell & Hart (1970); (15) McAndrews & Power (1973); (16) Traverse (1974b); (17) Traverse (1974a); (18) Traverse *et al.* (1961); (19) Traverse, unpublished data.

Concentrations of 100,000/g are common in some deltaic shales–siltstones: Habib (1982a) estimates 40,000–100,000/g for carbonaceous clays and silty clays. Bahamas Bank calcareous silts ("muds"), on the other hand, contain of the order of 1,000/g or less. Koreneva (1971) found only a few hundred or less per gram in most Mediterranean surface sediments. Melia (1982, 1984) reported only 2,000/g in ocean sediment off the west coast of Africa and only 50/g in deep ocean basins. (Most limestones contain very few palynomorphs; dolomites are invariably barren, probably because of secondary recrystallization effects.) Spores/pollen have been obtained from bedded salts (Klaus 1955), but saltdome salt, presumably transported, is normally barren, however. Spores/pollen have been reported from petroleum (Jiang & Yang 1980, Yang & Jiang 1981), but the concentration is very low, less than 0.1/g (!). Because oxidation destroys sporopollenin and chitin, soils and redbed sediment, regardless of particle size, are usually barren. Weathered outcrops are also likely to be barren, even if the same rocks are productive from cored material nearby. Kuyl *et al.* (1955) reported tropical areas in which outcrops were weathered and barren to a depth of over 10 m. (Weathering is one source of difficulty for palynology of archeological materials, which are often soils or soil-like. The concentration of spores/pollen is usually abysmally low. Special preparation techniques start with samples much larger than normal, but much archeological palynology seems to depend on very small numbers of palynomorphs.)

Outcrop samples of shale are frequently barren or have only poorly preserved palynomorphs because of oxidation–weathering, whereas cores of the same horizon yield abundant, well-preserved palynomorphs. Because of the carbonization of chitin and sporopollenin by temperatures greater than 200°C, palynofloras that will permit study are not obtained from rocks cooked by lava flows or intrusions. Similarly, the geothermal gradient results in very carbonized (dark) palynomorphs in deeply buried sediments and no well-preserved spores/pollen occur below about 5,500 m. Metamorphism of all kinds progressively destroys palynomorphs, so that low-volatile bituminous and anthracitic coals are barren, while closely associated high-volatile bituminous coal has abundant spores/pollen. Evidence of folding and faulting such as slickensides are nearly always contrary indicators, as also are evidence of cleat, cleavage, induration, and recrystallization in rocks.

SPORES/POLLEN IN WATER

Not nearly enough work has been done on the behavior of spores/pollen as sedimentary particles in water, especially in major streams. (Unfortunately, we are running out of undammed major streams to study!) Figure 17.21 presents the sort of data that are currently available. Federova (1952) studied the Volga River and showed that the amount of spores/pollen in the water varied from about 23,000 to about 45,000 per 100 liters of water. Her laboratory technique consisted of allowing the pollen to settle from the water by gravity. She did not study the water of different levels in the river. Groot & Groot (1966) ·used continuous centrifugation and measured 50,000–800,000 spores/pollen per 100 liters of water in the estuary of the Delaware River.

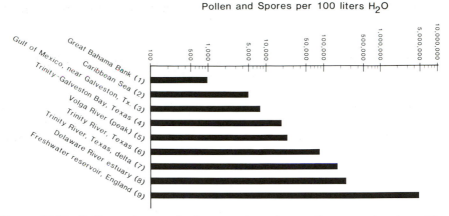

Figure 17.21 Pollen and spores in the water in various sedimentary situations. The concentration of palynomorphs per volume of water varies considerably from season to season and in response to storms and other factors. Therefore, all of the data with the exception of (5) are estimated averages for multiple measurements. Pollen and spores are expressed per 100 liters, as the numbers so generated are then similar to those for palynomorphs per gram of sediment (see Fig. 17.20). The data are plotted logarithmically. Water from mid-ocean localities would presumably contain at least an order of magnitude less per 100 liters than even the water of Great Bahama Bank. The reading of 5,000,000 per 100 liters of water for a small reservoir in England (9) is an indication of the high density that can be obtained in water with limited influx, closely surrounded by pollen-producing vegetation. Sources for data: (1) Traverse & Ginsburg (1966); (2) Farley (1987); (3) and (4) Traverse, unpublished data; (5) Federova (1952); (6) and (7) Traverse, unpublished data; (8) Groot (1966); (9) Peck (1973).

Using shipboard continuous centrifugation on the Great Bahama Bank, Traverse & Ginsburg (1966) found pollen (mostly *Pinus*) of the order of 1,000 per 100 liters of water. In 1960–62 I studied the Trinity River, Texas, near its mouth. The Trinity River at the time was relatively unaffected by dams and culture generally. I sampled at various stations in the lower river, the delta, and in Trinity–Galveston Bay on a monthly basis for about a year. Using a pumping apparatus mounted in a small boat, I sampled surface, mid-depth, and near-bottom water. The entire sample (20 liters) was evaporated in the laboratory to obtain a sludge which was then centrifuged, and processed by ordinary palynological techniques. Practically 100% recovery was assured by this technique. Figure 17.21 shows the average figures obtained: about 90,000 per 100 liters of water in the river some miles upstream, about 180,000 per 100 liters in delta river water, about 20,000 per 100 liters in Trinity Bay, and about 8,000 per 100 liters in the Gulf of Mexico off Galveston. There was much seasonal variation, depending on a variety of factors, especially flowering seasons and rainfall. The range of values for all depths and all stations was 10,000–500,000 per 100 liters of water. One of the interesting sidelights of this work was the frequent appearance of *Engelhardia* pollen, obviously reworked from Tertiary rocks far upstream, probably from the Eocene, about 300 km northwest. Frequent appearance of this reworked form seemed to be directly related to heavy precipitation in upstream areas.

REWORKING, RECYCLING, "STRATIGRAPHIC LEAK" OF PALYNOMORPHS

The splendid toughness of chitin and sporopollenin makes for some perhaps unexpected problems and opportunities in paleopalynology. Weathering and erosion of siltstones release massive amounts of spores/pollen by simple disaggregation of the sediments. Sediment samples from off the Mississippi River Delta contain not only spores/pollen from extant vegetation of the whole Ohio River–Missouri River–Mississippi River drainage but a grand assortment of reworked spores/pollen and other palynomorphs from Devonian on up. Kemp (1972) has noted the presence in sediment of the West Ice Shelf, east Antarctica area, of abundant spores/pollen and microplankton of Permian, Cretaceous, and Paleogene age.

Collinson *et al.* (1985) have described the reworking of Paleozoic and Mesozoic megaspores into Paleocene deposits of southern England, along with Cenozoic megaspores belonging to the time of deposition. Obviously the sediment included a fraction with clastics in the sand size range, and the megaspores of various ages were sorted into this fraction.

It is easy to recognize a reworked Pennsylvanian *Densosporites* in a modern sediment as recycled, but a *Carya* pollen grain reworked from an interglacial terrace deposit or from the Pliocene is a different matter. Spores/pollen are capable of repeated "recycling", though more than one cycle lessens the relative abundance of reworked fossils and also corrodes the fossils more, so that they are easily recognizable as redeposited. Wilson (1976) documents a case from the Pennsylvanian of Oklahoma in which recycled forms of Ordovician–Mississippian ages, and from both marine and non-marine environments, are regularly present. As Muir (1967) has pointed out, spores weathered and eroded from sedimentary rock are usually (but not always!) more carbonized than fresh palynomorphs and hence may be chemically more resistant to oxidation during processing, but also much more brittle. The percentage of reworked fossils is usually smaller than that of contemporaneously produced spores/pollen. Reworking in peat – hence in coals – is practically non-existent; nearly all of the spores/pollen come from the vegetation where the peat was produced. The palynoflora is therefore characteristically autochthonous. Some interesting studies have been made of spores carried within pieces of coal into younger sediments. For example, Bartlett (1929) studied megaspores from pieces of Pennsylvanian coal reworked into Pleistocene glacial drift in Michigan.

Muller (1959) found in the recent sediments of the Orinoco Delta that reworked forms are normally a small percentage of total palynomorphs in shelf sediment, but may be very high in delta levees. However, reworked spores/pollen can sometimes be a high percentage of the total palynoflora. I once studied a sample of siltstone from upper levels of sediments of the Gulf of Mexico which was more than 90% reworked: the palynomorphs were

Paleogene. The sample probably happened to consist mostly of a pebble of Paleogene shale saltated into the recent sediment. Reworked palynomorphs have provided clues as to direction of stream transport and sediment source, e.g., for a Cretaceous conglomerate, determined from Mississippian spores/pollen in certain of the conglomeratic clasts. Needham *et al.* (1969) reported use of reworked Carboniferous palynomorphs as tracers of sedimentation patterns in the northwest Atlantic. In the Black Sea reworked spores/pollen are much more abundant in surface sediment formed during the last glaciation than they are now, presumably because streams rejuvenated by glacial drop of sea level delivered far more sediment derived from weathered and eroded sedimentary rock than is now the case (Traverse 1974a; see Figs. 17.22a, b).

When palynomorphs of discordant age occur in sediments, the usual situation is reworking of older material into younger: the discordant forms are the older. However, sometimes one finds discordant forms that are younger than the associated rock. This is called "stratigraphic leak" and can be caused in several ways.

(a) Artificial contamination: For example (Traverse *et al.* 1961), drilling muds are frequently composed partially of partly oxidized coals such as the naturally occurring leonardite (Paleocene, North and South Dakota). Drill cores and well cuttings where such "mud" was used will be coated with, and cracks partly filled with, mud, which may be difficult to remove before processing. Thus a Triassic core can be heavily charged with Paleocene spores/pollen, as leonardite, for example, may contain 5 million spores/pollen per gram! In addition to drilling-mud additives, the circulating drilling mud also includes a mixture of particles from horizons stratigraphically above the drill bit, and these can also occur in drilled samples, just as does the drilling-mud additive. These are reasons why sidewall cores are preferable to drill cuttings for paleopalynology.

(b) Natural percolation of weathered-out spores/pollen from superficial rock.

(c) Deposition in karst topography: An example of this sort of stratigraphic leak is the Independence Shale of Iowa (Urban 1971), in which Mississippian and reworked Devonian spores were deposited in Mississippian time in caves in Devonian limestone, producing shales with a partially Mississippian palynoflora, occurring within the Devonian limestone sequence. Another good example is a Jurassic palynoflora found in Carboniferous limestones of Ireland (Higgs & Beese 1986), indicating karst formation in the older limestone, in Jurassic time.

Discordant palynofloral elements may result from mass movement of sediment, of which the best example is the intrusion of saltdomes into sediment above. Often large amounts of sediment are thereby intruded. Ecke & Löffler (1985), for example, describe a situation in Germany in which Zechstein (Upper Permian) evaporite penetrated and mixed with Röt (Lower Triassic) evaporite, resulting in a mixture of Permian with Triassic palynomorphs. Similarly, a melange rock can contain blocks of sedimentary rock of quite disparate age picked up in the course of the tectonic processes that produced the Franciscan Melange of California, for example (Traverse 1972).

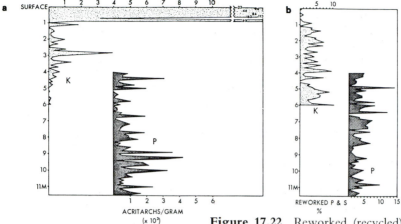

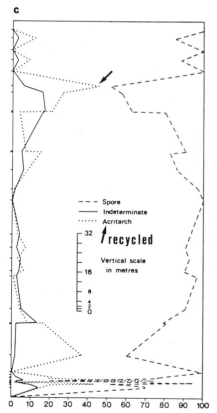

Figure 17.22 Reworked (recycled) palynomorphs, because of the toughness of sporopollenin and chitin, are a very frequent occurrence in paleopalynology. At times, reworked palynomorphs even are revealing as to the source of sediment in a rock, or in other ways. (a), (b) Palynomorph counts from overlapping sediment cores (K and P) from the Black Sea, representing about the last 25,000 years of sedimentation. As the Black Sea now is connected to the Mediterranean because of Holocene high sea level, amounts of palynomorphs per gram of sediment are very high – less sediment increases the concentration of fossil palynomorphs per gram of sediment (see acritarchs per gram in (a)). Before the present high sea level was reached, the streams in the basin were more actively bringing in sediment, and the large amount of sediment caused palynomorphs per gram to be less than is now the case. At the same time, as shown in (b), the percentage of reworked forms has markedly dropped in the late Holocene, because lower erosional rates are coupled with lower amounts of recycled palynomorphs. (c) In a study of non-marine Lower Devonian spores in England, Richardson & Rasul (1978) found abundant Ordovician and Silurian acritarchs, clearly reworked on the basis of both age and original environment. Because of differences in carbonization level, it was demonstrable that the Devonian streams contributing to the sedimentation of the Lower Old Red Sandstone studied were eroding marine source rocks of at least four different ages and geothermal history. Note towards the top (arrow) a level at which reworked acritarchs are about 50% of all counts.

429

In both this case and in a conglomerate, careful sampling can separate the rock fragments producing disparate palynofloras, so the palynofloras are not really mixed as they are when reworking represents weathered and eroded material.

Detection of reworked forms varies from exceedingly easy to very difficult. The detection depends fundamentally on incongruence.

Preservation discordance   Reworked palynomorphs are usually (but not always!) more poorly preserved than the palynomorphs that came into the basin of deposition from contemporaneous vegetation. This can be picked up from corroded, ragged or thin walls, or a different natural color. Stanley (1966) noted that reworked palynomorphs often stain differently and proposed the practical use of this detection method: he used safranin-O and showed that reworked forms could stain either more or less intensely, and showed that reworked forms have ektexine vs. endexine staining characteristics differing from those of the non-reworked forms. Presumably this is a reflection of oxidation–alteration during weathering. Staining often works, but it is tricky to apply with certainty, especially where the problem of detection is most severe, with say Pliocene into Pleistocene. Fluorescence microscopy can sometimes demonstrate a difference in fluorescence level between "native" and reworked palynomorphs. Drilling-mud palynomorphs – "artificially reworked" – are often much better preserved than the "native" fossils. For example, abundant late Cretaceous spores/pollen in Triassic well cuttings may stand out because the "native" Triassic forms are carbonized darkish brown whereas the probable drilling-mud forms are yellowish and much better preserved. Richardson & Rasul (1978) were able to use carbonization differences to indicate different sources for Ordovician–Silurian acritarchs reworked into Devonian sediment (see Fig. 17.22c). Legault & Norris (1982) found that Devonian spores/pollen reworked into Cretaceous sediments were more damaged by the experience than were Devonian acritarchs associated with them. Whether this is a universal situation is not known. On the other hand, reworked dinoflagellate cysts in some Cenozoic Arctic palynofloras are reported to be sometimes better preserved than the "native" palynomorphs.

Stratigraphic or ecological discordance   Pennsylvanian or Devonian spores/pollen in a Cretaceous sediment will usually stand out like a sore thumb. However, for some generalized forms, e.g., psilate or scabrate trilete spores, this is not the case. Marine dinoflagellates in a non-marine sediment is another case of reworking which is easy to detect. In other instances, e.g., well-preserved recycled Cretaceous spores in the Neogene of Black Sea cores, it is much trickier. There are examples of palynologists being fooled by such a combination into suggesting an intermediate age! It should also be emphasized that reworking can be complicated by multiple episodes of recycling. Legault & Norris (1982) studied a Pleistocene–Cretaceous–Devonian sequence in which Devonian palynomorphs were reworked into the Cretaceous, and, later on, Cretaceous and Devonian together reworked into Pleistocene.

CHAPTER EIGHTEEN

Some factors affecting practical applications of paleopalynology

POST-DEPOSITIONAL ALTERATION OF PALYNOMORPHS: THERMAL MATURATION (= "CARBONIZATION")

Although sporopollenin (and chitin) is a very tough material, it is not indestructible. Post-depositional oxidation (weathering) can corrode or even destroy palynomorphs. Furthermore, post-depositional heating causes chemical changes. These are of the same sort as affect organic matter generally, e.g., in coal beds. Just as the coal series proceeds from peat to anthracite by grades, with loss of H and O and concomitant enrichment of C and molecular condensation, the same occurs with dispersed sporopollenin, though apparently not as fast as it does with other organic substances (see Fig. 18.1a). Dispersed organic matter taken as a whole, however, coalifies at about the same rate as associated coal beds. The process of coalification of dispersed organic matter in sedimentary rock by thermal alteration can be called catagenesis (Dow 1977). If one wishes more precision, one may follow Hayes *et al.* (1983) in referring to thermal change up to 50°C as diagenesis (R_0 up to 0.5), that in the 50°–150°C range as catagenesis (R_0:0.5–2), that in the 150°–250°C range as metagenesis (R_0:2 –4), and that above 250° as metamorphism (R_0 above 4). R_0 is a measure of reflectance (see Glossary).

An interesting example of the principle of coalification of dispersed organic matter is the demonstration by Hatcher *et al.* (1962) that the organic matter in coal balls has the same rank as the enclosing coal. Such dispersed insoluble organic matter is often called "Kerogen", and Durand (1980) is a good source of information about its study. Spore/pollen exines in sub-bituminous coal behave more like exines prepared from living plants than the wood from the same coals resembles wood from a lumber yard, as to elasticity and staining properties. The principal observed change in spores/pollen exines along the carbonization–coalification route is change of color in transmitted light. Fresh exines of modern plants are pale yellowish to almost colorless. If exines are heated, e.g., by deep burial or proximity of the enclosing sediment to a lava flow, the color intensifies from yellow to orange to brown, dark brown, and ultimately black. Experiments by McIntire (1972) showed that temperature in excess of 200°C is the primary factor, along with length of time of exposure to the elevated temperature. Elevated pressure alone, not accompanied by heat, does not cause carbonization. Piérart (1980) found that heating in air at 150°C caused carbonization, whereas heating in a nitrogen environment required

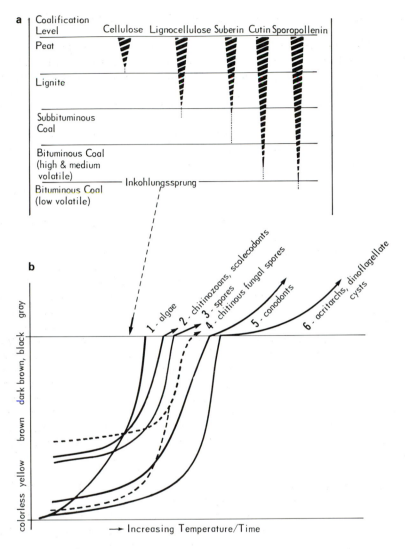

Figure 18.1 Alteration of sporopollenin and other organic substances with coalification by geothermal alteration (= maturation) (see Plate 1). Note that (a) reads *down* to more mature, (b) reads *up* to more mature. (a) Persistence of chemically distinct materials with coalification level. "Inkohlungssprung" is German for "coalification leap", referring to a quantum change between medium-volatile and low volatile bituminous coal. Sporopollenin retains some of its distinctive properties (color, elasticity, refractive index) right up to that point. (b) Trends in alteration in color of various kinds of palynomorphs with geothermal maturation; "Inkohlungssprung" at top. (1) Algal remains start colorless, are relatively quickly altered in color. (2) Chitinozoans and scolecodonts, composed of pseudochitin and chitin respectively, start dark yellow and darken at first slowly, then rather rapidly to the "Inkohlungssprung". (3) Spores/pollen exines start yellow, alter to brown and black, about the same as chitinozoans and

432

higher temperatures. Carbonized sporopollenin is more oxidation-resistant than raw sporopollenin.

Naturally, the darkest that exines can be and still be subject to study by transmitted light is dark brown. When enclosed in coal, exines reach this point at about the "Inkohlungssprung", the point in the natural carbonization process at which a quantum jump in molecular organization of coal occurs (see Fig. 18.1). It occurs in coal in the low-volatile bituminous coal range at about 75% fixed carbon (= about 25% volatile matter).

As is shown in Figure 18.1b palynomorphs consisting of compounds other than sporopollenin respond to geothermal maturation somewhat differently. Fungal spores, for example, start either lighter or somewhat darker (pale brown) than spores/pollen (usually pale yellow) but are a bit more resistant to change than even sporopollenin. Acritarchs and dinoflagellate cysts start almost colorless and require more temperature exposure to darken. In other words, when spores/pollen exines are already dark brown, acritarchs may be still relatively light in color.

Inasmuch as exine enclosed in a shale is a "mini-coal" consisting of 100% exinite, study of the color of the exines by transmitted light, or of their reflectance by reflected light, can reveal the thermal history and state of the enclosing rock, a matter of considerable importance to hydrocarbon exploration. Some decades ago, White (1915, 1935) showed that in the Appalachian coal field there is a direct relation between coal rank and occurrence of hydrocarbons: anthracite fields do not have any, high-volatile bituminous fields may have liquid petroleum, and low-volatile bituminous fields have gas but no liquid hydrocarbons. The first proposals to use state of carbonization of dispersed organic material in rocks to predict hydrocarbon potential of the rocks came from palynologists who could do it by color (Gutjahr 1966; he eventually used light transmission as measured by a photometer). This is still done, although the field of "thermal maturation" of dispersed organic matter is now mostly a separate endeavor of "organic petrologists" who measure the reflectance of specially prepared dispersed organic matter, more often of vitrinite particles than of exinite. However, the color of palynomorphs in preparations is still useful as a measure of carbonization. Batten (1980), for example, suggests the seven-point scale given in Table 18.1.

This may be compared with Pearson's color chart (Plate 1), in which the color is directly related to the numerical thermal alteration index (*TAI*). Batten (1981a) has stressed that normal oxidative methods used in laboratory

---

scolecodonts. (4) Chitinous-walled fungal spores start practically colorless to brownish yellow to light brown and, according to unpublished observations of W. C. Elsik, retain the original color longer than spores/pollen before darkening to black. As is true for (2), (3), (5), and (6), they presumably undergo a further, slower change in character beyond the "Inkohlungssprung". (5) Organic matter within conodonts, phosphatic animal microfossils, also undergo color change with geothermal maturation. (6) Dinoflagellate cysts and acritarchs require the greatest length of time to darken with geothermal maturation. In other words, samples with very dark spores/pollen may contain dinoflagellate cysts in studiable condition.

Table 18.1 Batten's (1980) scale of palynomorph colors.

Observed color of palynomorph	Significance
(1) colorless, pale yellow, yellowish orange	chemical change negligible; organic matter immature, having no source potential for hydrocarbon
(2) yellow	some chemical change, but organic matter still immature
(3) light brownish yellow, yellowish orange	some chemical change, marginally mature but not likely to have potential as a commercial source
(4) light medium brown	mature, active volatilization, oil generation
(5) dark brown	mature, production of wet gas and condensate, transition to dry gas phase
(6) very dark brown-black	overmature; source potential for dry gas
(7) black (opaque)	traces of dry gas only

procedures do not lighten the color of sporomorphs enough to invalidate the use of color of processed specimens as a geothermal indicator. Thus, carefully recorded color observations by the working palynologist can be used for reconnaissance of the geothermal history of the rock in a wildcat oil well. Bujak *et al.* (1977) have pointed out that various categories of organic matter in sedimentary rock do not all follow the same geothermal maturation course (see also Fig. 18.1).

The course of carbonization or thermal maturation can also be observed by use of fluorescence microscopy of both exinite (spores/pollen in coal) and vitrinite (wood and bark tissues in coal) (see Plate 1). The particles in question are placed in an ultraviolet light beam, and the fluorescent light emitted is studied. Teichmüller & Ottenjahn (1977) have shown that fluorescence microscopy very sensitively shows diagenetic changes of spore walls and other organic substances in the context of oil formation. Teichmüller & Durand (1983) demonstrate that fluorescence microscopy reveals changes in the macerals of coal, related to bitumen formation during coalification. Van Gijzel (1981) reports that the principal stage of petroleum generation ($R_{oil} = 0.50$–0.85) is characterized by double peaks in exinite fluorescence spectra, and the "oil death line" ($R_{oil} = 1.15$–1.35) is reached when fluorescence does not occur.

Carbonization can be one of the banes of palynologists' existence. Because of the geothermal gradient alone, sedimentary rock buried more than about 5,500 m does not yield palynofloras of recognizable spores/pollen. (The process of carbonization by geothermal gradient follows "Hilt's law" (Dow 1977).) Carbonized spores/pollen are difficult to study even when they can be successfully separated from the enclosing rock. Furthermore, as the principles of maceration depend in part on the differential reaction of sporopollenin to oxidation compared to other organic matter present, highly carbonized spores when mixed with other organic matter are difficult or impossible to prepare. Spores/pollen cannot be macerated out of low-volatile bituminous to anthracite coals, even though presumably equivalent coal beds of medium to high-volatile constitution yield abundant fossils. Furthermore, although other preservation problems can be partially overcome, carbonization due to deep burial or tectonic activity cannot. It was long felt that the Triassic–Jurassic rocks of the Hartford–Springfield Basin (Newark Supergroup) would not yield palynofloras, because the typical lithology is reddish sandstone and shale: too oxidized. However, productive grayish shales do exist, though the productive shales are not "typical" lithology for the area (Cornet & Traverse 1975). On the other hand, samples of sediments buried by 10,000 m or more of rock, such as from deep wells now being drilled, will never be palynologically productive. Sediments from part of a mountain chain which has been subject to much tectonism are very unlikely to be productive. For example, Triassic sediments from the Alberta Rockies are in my experience in this category.

"MARGINAL PALYNOLOGY"

Most of the photomicrographs of fossil palynomorphs in this book and in practically all palynological publications are of "super specimens" illustrating the morphological features of the taxa concerned. Most of the specimens encountered in practical palynology are not so well preserved. They may be variously torn, compressed, crumpled or folded (see Fig. 17.19). They may be more or less corroded by oxidation or fungal–microbial attack. They may be riddled by pyrite or other crystals, or rarely even by molds of coccoliths (Batten 1985)! Sometimes, in extreme cases of corrosion, the palynomorphs are mere "ghosts". Worst of all, sometimes the palynomorphs may be far along the diagenesis–carbonization pathway and are just black silhouettes. Nevertheless, many of the sorts of miserable palynomorphs just mentioned can be recognized, at least to genus or group, e.g., "circumpolloid", and this sort of identification can be the basis of informative palynological study. I call this kind of work "marginal palynology" (Traverse 1972). An example is the Danville–Dan River Basin (Newark Supergroup, Virginia and North Carolina), in the shales of which there are few palynomorphs and these are highly carbonized, apparently because of higher-than-normal heat flow in the basin. It is probable that oxidizing conditions penecontemporaneous with deposition caused sparse and corroded palynomorphs even before depositional diagenesis. Pyrite precipitation occurred after deposition, further ruining preservation. Nevertheless Robbins (1982) showed that gross study of these spores/pollen "wrecks" was possible and permitted approximate dating of the sediments (see Fig. 18.2).

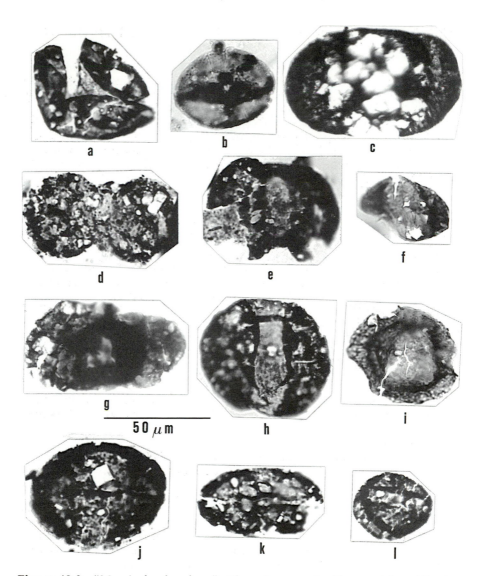

a
b
c
d
e
f
g
50 µm
h
i
j
k
l

Figure 18.2 "Marginal palynology". The palynomorphs illustrated in publications are usually well-preserved, well-oriented specimens. However, even in good preparations many specimens are folded, crushed, or corroded and nevertheless often are identifiable by a palynologist who is familiar with well-preserved specimens of the same taxon. Furthermore, the palynologist is often called on to work with poorly or very poorly preserved specimens where they are the only, or among the few, fossils present. Often, after a "foothold" is attained to suggest the approximate age level, even execrable, badly preserved, corroded, and carbonized "wreck" specimens such as those illustrated here can be identified. The assemblage shown is from the Triassic (Karnian) sediments of the Danville–Dan River Basin of Virginia–North Carolina. The blackness (carbonization) is best explained by post-depositional heat flow in the basin, and the

PALYNOSTRATIGRAPHY (= THE USE OF PALYNOLOGY FOR STRATIGRAPHY)

The principal application of the study of fossil palynomorphs is for stratigraphic purposes, in the simplest case to propose a geologic age for an unknown sample. Palynologists are frequently asked to perform this basic stratigraphic service, especially in sequences of sedimentary rock such as the Catskill Formation (Upper Devonian) of Pennsylvania–New York, which is often described as almost "unfossiliferous", meaning no megafossils of marine organisms with a limited stratigraphic range. During D.S.D.P. work in the Black Sea, my duty as on-board palynologist was to ascertain, based on what was known from surrounding areas, when the drilling reached Miocene, as a working decision had been made to drill that deep, if possible. The method usually employed for fixing a date is to plot from the literature and/or previous investigations the ranges of the forms found. The most probable date is the date showing the best control by upper limits ("tops") and lower limits ("bottoms"), making allowances for geographic position; for example, in the Black Sea drilling, none of the forms encountered is truly extinct, but pollen of palms is not usually found in the general area above top Miocene. Beginning students, using this method, are almost always able to fix the age of an unknown sample to the nearest stage of a period.

In oil-company-oriented palynostratigraphy the operational task is to correlate sequences from one well with those from others. The purpose of this is simply orientation. If an oil pay zone is anticipated from other information just above zone "d", one can make money for his employer by recommending further drilling to reach it, or to stop drilling if correlations show that it has been passed without luck. Correlation may show that zones have so thickened

---

mineral (mostly pyrite) casts by reducing environment in the Danville–Dan River lake at the time of deposition. The specimens are brittle, and very gentle processing must be used. Magnification indicated by bar under (g). (a) *Duplicisporites granulatus* Leschik, proximal view. Note mineral casts, and torn, squashed nature of specimen. (b) *Aratrisporites saturni* (Thiergart) Mädler, proximal view of monolete, zonate microspore. (c) *Alisporites grandis* (Cookson) Dettman, distal view of large bisaccate. Open spaces are due to mineral casts and corrosion. (d) *Platysaccus* sp. Despite carbonization, multiple mineral casts, and corrosion, the small corpus and large sacci make generic identification certain. (e) Probably *Lunatisporites* sp., distal view. Obviously a bisaccate, and the generic determination is suggested by apparent taeniae. (f) *Klausipollenites schaubergeri* (Potonié & Klaus) Jansonius, distal view. Small sacci and relationship of sacci to corpus permit identification. (g) *Triadispora plicata* Klaus. Relationship of sacci and corpus, structure of corpus, and occasionally visible trilete mark allow identification. (h) *Sulcatisporites australis* (de Jersey) Dunay, distal view. Nature of sulcus and its relationship to sacci characteristic for taxon. (i) *Tulesporites briscoensis* Dunay & Fisher. Monosaccate. Despite tears and cracks, overall morphology recognizable. (j) *Ovalipollis* sp., distal view. Straight sulcus and shape suggests this taxon, despite mineral casts, carbonization and corrosion. (k) *Cycadopites* sp., distal view. Monosulcate. (l) *Paracirculina maljawkinae* Klaus, polar view, recognizable by angular folds associated with rimula.

in comparison to a standard that the goal is too deep in a new well to make reaching the pay area feasible.

Paleopalynology has some major advantages and disadvantages for doing stratigraphy. One problem is that paleopalynological assemblages are very likely to be facies-controlled. As discussed earlier, palynofloras often are a reflection of the sediment type. The apparent correlation of two Pennsylvanian coal beds by palynology, for example, is often not a true correlation at all, but merely a reflection of the similarity of coal-bed palynofloras over a considerable timespan, given similar ecological conditions: this is a typical palynofacies (in the palynostratigraphic sense of the word). The waxing and waning of spores/pollen taxa over hundreds or even thousands of feet of section may mean only environmental changes affecting the production of spores/pollen on land or changes in the environment of deposition, e.g., deeper or shallower water, affecting the palynoflora.

Correlation by palynofacies may even be satisfactory for practical work within a small area, but such correlations cannot be extended far laterally. On the other hand, as long ago pointed out by Kuyl et al. (1955), spores/pollen have big advantages for biostratigraphy over marine fossils, such as foraminifera. This advantage is that palynomorphs of various kinds occur in all sedimentary environments: spores and pollen of land plants come into sedimentary basins primarily with the sediment load of streams. Palynomorphs are found in all non-marine sediments of the requisite particle size and geochemical history. They also occur in practically all shaley marine sediments, although the same limitations as for non-marine shales apply. Dinoflagellate cysts mostly originate in marine environments. Thus palynostratigraphy can correlate levels in marine sediments with those of non-marine sediments.

Obviously, correlation by palynofloras should ideally depend on real "tops" and "bottoms" (last occurrences of and first occurrences of taxa, respectively) based not on facies (migration) but on extinction of some forms and evolution of other forms. In the time represented by some parts of the geologic column, plants were evolving rapidly, and spores/pollen are very effective for stratigraphy, e.g., the Middle and Upper Devonian. At other times, dinoflagellates were more rapidly evolving than land plants, and dinocysts are very effective for practical stratigraphy if marine sediments are available, e.g., much of the Jurassic. Palynostratigraphic zones for correlation are based on various combinations of first and last occurrences (see the section on "Data management in palynostratigraphy" which follows).

Another problem for spores/pollen-based palynostratigraphy is provincialism. Almost since their origin land plants have not been cosmopolitan, and therefore the lateral extension of correlation by spores/pollen is risky, the more so the greater the area involved. Precise transcontinental correlation is difficult, intercontinental correlation (except in very broad terms) usually impossible.

Reworking is discussed earlier. This problem can obviously lead to great difficulties with correlations in general, and with the determination of "tops" in particular. An example extending beyond the confines of practical well correlation is that of plant extinction at the end of the Cretaceous. Was there for plants an "event"? Megafossil plant compressions are not often recycled, but spores/pollen very often are. Thus, to determine whether various *Aquilapollenites* species terminate at the end of the Maestrichtian (top of the Cretaceous) or march on into the Danian (bottom of the Paleocene) requires certainty that *Aquilapollenites* occurrences in the Danian are not examples of reworked pollen. As a general rule, one should be suspicious that spores/pollen/dinocysts occurrences may be reworked when they are relatively rare specimens, occurring stratigraphically well above rather abundant counts of the same form. Careful inspection of such specimens may yield preservational hints that they are not last-ditch stragglers but reworked. Because of the durability of palynomorphs generally, "bottoms" have a theoretical tendency to be sharper than "tops", but when drilling cuttings are used, the bottoms are usually blurred by caving of upper cuttings into lower levels.

Dinoflagellate cysts are both marine and non-marine, but marine forms are vastly more common. In the pre-Devonian palynostratigraphy is only practicable with marine palynomorphs: chitinozoans, scolecodonts, and acritarchs. The possibility of correlating non-marine with marine rocks is what put palynology in business as a practical oil-company tool (Kuyl *et al.* 1955). In the Lake Maracaibo Basin (Venezuela), Shell palynologists demonstrated that they could do what foraminifera specialists could not do: correlate the extensive non-marine sections with each other, and with the associated marine sections.

DATA MANAGEMENT IN PALYNOSTRATIGRAPHY

Correlation techniques based on palynofossils are basically the same as those used for other sorts of fossils, subject to the limitations already mentioned. Because the number of specimens obtained is normally large, palynostratigraphy is ideally suited to employ statistically based, computer-managed methods. Basically, the palynostratigrapher depends on concurrent range zones, Oppel zones, or assemblage zones (where barren intervals intervene) (see Fig. 18.3). Oppel zones are based on upper and lower limits of several taxa each. Concurrent range zones are similar to Oppel zones, but these zones are determined by presence of ranging-through forms as well as first and last appearances. They are subject to the limitations mentioned: reworking which extends a top into a pseudo-top, facies-controlled distributions which show an apparent top based only on environmental factors. See the *International stratigraphic guide* (Hedberg 1976) for definitions of kinds of zones.

For some parts of the geologic column, palynostratigraphy has been practiced and results published over a long enough timespan that well-accepted, named zones exist. The Devonian–Carboniferous of parts of the world is such an example. Some palynostratigraphers in some situations use "phases" instead of zones (see Fig. 18.4), emphasizing the progressive nature of alteration of palynofloras over time, instead of the disjunct nature of zones, each distinct. Many palynostratigraphic correlations and proposed zones exist only in files and computer storage of oil companies and related enterprises. Enough of this has been published from time to time that the literature provides sufficient information for most routine palynostratigraphic work.

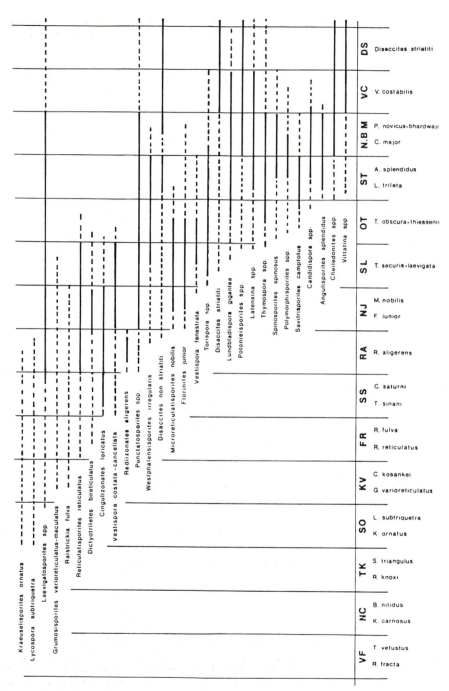

Figure 18.3 Part of a range chart of palynomorph species in the Carboniferous (zone VF is Dinantian) to lowest Permian (VC and DS are Autunian) of western Europe.

Edwards (1982a) has summarized the various biostratigraphic methods that are numerically based, easily handled by computers, and applicable to relatively finely tuned palynostratigraphy, as follows:

Probabilistic methods   The method depends on placing event A above or below event B on the basis of probability studies, e.g., binomial or trinomial functions.

Multivariate techniques   One approach is to use numerical expressions of similarities or differences between pairs of samples, usually assuming that a taxon occurs in all samples between its first and last occurrences: the "range-through" method. Multivariate analysis can be used to find groupings of samples. This analytical method also permits calculations of the degree of restriction of a taxon to a particular zone and the proportional contribution of the taxon to the zone. Principal-components analysis is a multivariate technique that can be used to find groupings of events as displayed by a series of sections. Christopher (1978) has developed a technique for preparation of "multi dimensional" stratigraphic charts based on "similarity coefficient matrix contouring". This has the effect of grouping samples with high similarity coefficient values from pairs of stratigraphic sections which, for example, then form "hills" in a three-dimensional graph. The method is an elegant way of presenting stratigraphic correlations, but the graphs are hard for the uninitiated to interpret.

Graphic methods   The basic idea of Shaw (1964) for graphic correlation has been applied to palynostratigraphy by Miller (1977). Further explanations are provided by Edwards (1984b). This stratigraphic method compiles ranges of taxa in measured sections, which can be corrected for sedimentation rate. The ranges within the sections are plotted graphically. A mathematical expression for time equivalence for levels in pairs of sections is determined. A two-axis graph is produced which can be used for correlation purposes. The curve can be fine tuned by input from additional sections beyond the first two. The Shaw–Miller graphic method uses absolute scales. Edwards (1978) suggested a different, "no-space", graph which operates on relative position only.

Relational methods   Range charts are used to generate zonation by relational matrix. The zones are based on mutual presence (never absence) of two or more taxa. The zones devised are similar to assemblage zones.

Broken lines represent reduced or discontinuous presence. The palynological zones are *concurrent range zones. Assemblage zones* are similar but emphasize the biological integrity of the group of taxa. *Oppel zones* are also similar but are always determined by first and last appearances of several taxa. The zones are given a name and symbol based on the names of characteristic taxa included; thus, DS (*Disaccites stiatiti*) for the uppermost zone.

	AEGEAN	BITHYNIAN	PELSONIAN	ILLYRIAN
Densoisporites nejburgii	○○○○○○			
Platysaccus leschikii	○○○○○○			
Alisporites cymbatus	○○○○○○			
Voltziaceaesporites heteromorphus	○○○○○○○○○○○○○○○○○○○○○			
Lunatisporites sp. div.	○○			
Aratrisporites sp. div.	○○			
Angustisulcites gorpii	○○○○○○○○○○○○○			
Kraeuselisporites hystrix	○○○○○○○○○○○○○○○○○○○○○○○○○○○○○○○○○○			
Striatoabieites aytugii	○○○			
Triadispora crassa	○○			
Angustisulcites klausii	○○			
Stellapollenites thiergartii	○○○			
Triadispora plicata	○○			
Microcachryidites sp. div.	○○○			
Concentricisporites sp. div.	○○○			
Strotersporites nov. sp.	○○○			
Distriatites insculptus	○○○			
Illinites chitonoides		○○○○○○○○○○○○○○○○○○○○○○○○○○○○○○○○○○○○○		
Perotrilites minor			○○○○○○○○○○○○○○○○○○○○○○○○○○○○	
Dyupetalum vicentinense			○○○○○○○○○○○○○○○○○○○○○○○○○○○○○○	
Tsugaepollenites oriens			○○○○○○○○○○○○○○○○○○○	
Triadispora polonica			○○○○○○○○○○○○○○○○○○○○○○○○○○○	
Cristianisporites triangulatus			○○○○○○○○○○○○○○○○○○○	
Protodiploxypinus potoniei				○○○○○○○○
Kuglerina meieri				○○○○
Uvaesporites gadensis				○○○○
Staurosaccites quadrifidus				○○
Ovalipollis pseudoalatus				○○
Circumpolles group				○○

crassa-thiergartii phase

thiergartii-vicentinense phase

Figure 18.4 Use of "phases" instead of zones in the middle Triassic. Aegean, etc., are substages of the Anisian stage. The use of phases is subtly different from that of assemblage and other zones (compare Fig. 18.3). The emphasis is on gradual development over time of successive assemblages, rather than on the complete distinctiveness of each assemblage. To underscore this, the designation of successive phases typically includes a taxon name from the previous phase. Thus, in the two phases shown here, *crassa* is *Triadispora crassa, thiergartii* is *Stellapollenites thiergartii* and *vicentinense* is *Dyupetalum vicentinense*. In practice, one uses the phases very similarly to zones.

Many palynologists now use the computer for handling of masses of palynological data by various kinds of multivariate analyses and other mathematical and graphical manipulation, not only for stratigraphy but also for paleoecological and even for morphological studies. Boulter & Hubbard (1982), for example, used cluster analysis and principal-components analysis in an effort to reconstruct from their palynological data the vegetational patterns of Paleogene sediments in southern Britain. Although this approach has state-of-the-art importance, it is beyond the scope of this book to present the application of multivariate analytical techniques such as cluster analysis to palynological problems, and the subject is changing too fast for such directions to be practicable. Most academically based palynologists will be best advised to turn to colleagues who are expert in mathematics, computer science, and statistics for help with this sort of problem, as well as with those related to data storage and retrieval. See Green (1983) for an example of a computer program for paleopalynology (POLSTA) which aids in the handling of pollen analytical data; there are many other programs being used. It is worth emphasizing that unwary palynologists using the dendrograms generated by cluster analysis, or the information from principal-components analysis, etc., without really understanding the limitations of the various methods, may find "relationships" that are not real, or overlook relationships that are obscured by the misuse of these powerful mathematical tools. For the average student, study of range charts from the literature, showing zones, and correlation of them with informal zones generated from the sections studied, is the most useful approach, until one needs help from statistical-mathematical consultants.

The use of the computer, both mainframe and personal, has already had great impact in palynology, not only for preparation of distribution maps and graphs (e.g., of cluster analysis, etc.; see Huntley & Birks 1983), but also for data storage and management in all aspects of palynology (see Barss *et al.* 1984, re BIOSTRAT). This subfield of paleopalynology is changing so rapidly that it is best to refer readers to the appropriate personnel, e.g., in the palynological laboratories of the national geological surveys, for information about the latest status of software for palynological data management.

Most educational institutions and research laboratories have people technically able to advise palynologists on statistical tests to use for a variety of purposes. For example, t tests can be used to determine if morphological features of palynomorphs are different enough to merit recognition; χ^2 tests may be used to express variance between samples, and therefore to test whether significant boundaries exist or whether the pollen counts of two analysts differ significantly (Dudgeon 1985, Gray & Guennel 1961, Mosimann 1965).

Palynological laboratory techniques

This appendix is not intended to provide an atlas of all palynological techniques used by various laboratories. Rather, it provides simple directions for processing various sorts of samples, as we do it in our laboratory at Penn State, along with comments about techniques used elsewhere, and some related matters. The emphasis is on techniques that can be adapted easily to laboratory facilities even of small colleges. The reader would also find it useful to consult Doher (1980). This 30-page publication about palynomorph processing contains directions for various preparation techniques, such as use of vinylite AYAF mounting medium. Another valuable publication on techniques is Phipps & Playford (1984). Kummel & Raup's (1965) *Handbook of paleontological techniques* contains a number of chapters about palynology. The chapter by Gray summarizes the then current maceration methods, most of which are still in use. Well-described techniques in Evitt (1984), intended for preparation of dinoflagellate cysts, are also applicable to spores/pollen.

EXTANT SPORES/POLLEN

Introduction

People who work with fossil spores/pollen should study extant forms, at least to get ideas on identification. When studying post-Cretaceous materials, modern forms are obviously important for suggesting possible botanical relationships. By late Neogene time practically all of the forms encountered are from extant genera, and collections of modern spores/pollen provide the raw material for all identification efforts.

It is possible to study "raw" pollen, and indeed aerobiologists routinely do this: atmospheric pollen is trapped, often directly on a slide to which Vaseline (petrolatum), glycerin or some other sticky mountant has been applied, and the spores/pollen are studied without treatment. It is possible to recognize pollen so prepared, especially as the aerobiologist has to recognize only a relatively few taxa that occur over and over. However, to determine the exine features with any degree of satisfaction, this is not a good approach, as untreated exines have oil droplets and other intercalary inclusions, or external coatings, and furthermore the protoplast and the intine of the spores/pollen make microscopic study of the exine very difficult. The adherent oils and included lipids and the whole protoplasm and intine need to be removed if the exine is to be properly studied. Also, the exine alone is what a paleopalynologist studies, so the reference materials need to be, as it were, artificially fossilized.

One way this can be done is to boil anthers, sporangia, or even small flowers in 10% KOH. The cellulose is hydrolyzed, and protoplasts lyzed. The resulting exines are, however, practically colorless and must be stained for proper light microscopy (safranin-O, basic fuchsin or other red stains are good, because optical systems are usually corrected for green light). G. Erdtman (with help from his chemist brother) long ago introduced the practice of acetolyzing sporulating/flowering material in a mixture of nine parts acetic anhydride and one part concentrated sulfuric acid. The procedure has been retained by palynologists since, with little change (Traverse 1955). The reaction is as follows:

$$(C_6H_{10}O_5) + 3(CH_3CO)_2O \xrightarrow[\text{catalyst}]{H_2SO_4 \text{ as}} (C_6H_7O_5)(CH_3CO)_3 + 3CH_3COOH$$

| cellulose | acetic anhydride | cellulose triacetate | acetic acid |

(For structure of cellulose, see Chapter 3: cellulose is really a long-chain compound, with thousands of repeating units.) The sulfuric acid is a catalyst and also a desiccating agent. Sporopollenin of most pollen comes through relatively unscathed, at least at the level of magnification usually used, but the color is altered from almost colorless to a yellow, amber or orange color depending on length of treatment and thickness of exine. Some modern pollen, e.g., Malvaceae, are characteristically thick-walled. It is impossible to prolong acetolysis long enough to remove the cellulose of anthers and other flower parts of these plants without getting the exines very dark brown, sometimes almost black. These grains should therefore be bleached with sodium hypochlorite (use laundry bleach) after acetolysis. Some grains are very pale even after acetolysis. These can be stained (we use basic fuchsin or safranin-O; basic fuchsin is more specific for sporopollenin but is trickier to use). In practice, staining and bleaching are seldom necessary. If acetolysis is done carefully, 95% of the forms processed will come out an acceptable yellow to orange color. Apparently, exines of at least some sorts of pollen are considerably modified in microstructure by acetolysis when it is combined with other treatments, as demonstrated by Aubran (1977) for the use of acetolysis, plus potassium permanganate fixation (of cycadeaceous pollen), and by Hafsten (1959) for acetolysis combined with HF and bleaching (oxidation). In my experience, this is not a serious problem for practical palynology, where one seldom uses these combinations of methods.

It is very important to understand that reference spores/pollen should not ordinarily be prepared from flowers, anthers or pollen collected by palynologists from the field into envelopes or vials, although it is possible to do this, and there are sometimes adequate reasons. (Flowers, anthers, or even pollen itself can be collected into vials of glacial acetic acid. Glacial acetic dries the material, which must be water-free for acetolysis.) In the laboratory, the glacial acetic acid can be removed by centrifugation, the acetolysis can be done on the glacial acetic acid-dried material (acetone will do the same thing but is very volatile and very flammable). However, by far the best procedure (see Traverse 1965) is to take the trouble to make pressed, dried plant specimens using conventional botanical techniques. These specimens will match up with field notes and numbers, and they can be used to prepare standard herbarium sheets. These are "voucher sheets", and the identity of the plants producing the spores/pollen can be checked by a plant taxonomist. Study of the pollen morphology has shown me that specimens of *Nyssa sylvatica* were marked "*Maclura pomifera*" in several prominent herbaria. This dramatizes the point: if university herbarium curators make mistakes in identification of material, the average palynologist should not depend on his own field identifications.

Preparation of spores/pollen for microscopic study from extant plant material

Note: This assumes pressed–dried plants. Fresh flowers, etc., must be first dried: glacial acetic acid or acetone are good drying agents for this purpose because one can go directly from a thorough soak (overnight is best) in these to acetolysis mixture after centrifuging off the drying liquid. Heating the glacial acetic acid and sample will expedite the reaction. If acetone is used, do not heat, because acetone is too volatile and flammable. Freeze-drying is another possibility, and it is also possible to dry plant material very quickly by a very brief (2–3 min) treatment in a microwave oven (Hall 1981, Bacci & Palandri 1985). Investigations in our laboratory show that microwave drying does not harm all pollen exines in a flower.

Outside the "wet laboratory" (a) Isolate sporiferous/polleniferous material as much as possible, dissecting off leaves, stems, peduncles, sepals, petals, etc. Ideally, just sporangia/anthers are best (easy with *Magnolia*, *Lilium*, and some ferns), but for plants with tiny flowers, such as Umbelliferae, or ferns with well-protected sporangia, whole flowers or pieces of frond with many sporangia are best, although one should take care to remove as much extraneous tissue as possible. Work on a piece of white paper, using needles, scalpel, and fine scissors (Fig. A.1). A dissecting microscope (or at least a hand lens) is very handy. The best material is several florets just on the point of opening but not yet open. Younger florets will often contain immature, atypical pollen, while older, open florets may have shed much or all of the pollen. Where large flowers such as *Camellia* are the subject, one collects just anthers, but may wish to avoid opening a flower and thus damaging the specimen too much. In such cases, and with fern sori and gymnosperm male cones, always carefully study the anthers, sporangia or pollen sacs with a dissecting microscope or hand lens, to be sure that abundant spores/pollen are present. The spore/pollen-bearing material that is difficult to pick up can be easily poured from the sheet of paper mentioned above into a collecting envelope. Use as little material as possible to get a good preparation; this makes for cleaner slides. If the material is from a working herbarium, the curator will welcome palynological predation more cheerfully if one is very conservative: anthers only where possible, as few florets as possible, etc.

  (b) Record data immediately on the collecting envelope; one can get too little information but never too much. Get the original collector's name and field number, locality information, and the herbarium where the voucher sheet resides. If it is one's own specimen or if the herbarium curator wants it done,

a   **b**

Figure A.1 Preparation of modern spores/pollen for the reference collection. (a) The only sensible way to do this is from herbarium sheets. The sheet, either in the palynologist's own collection or in a herbarium, provides a "voucher" for the pollen preparation. The palyniferous material should be removed carefully with scissors, forceps, and/or small knives. In most cases two or three whole flowers should be collected into envelopes with full data on identity recorded (printed envelopes as shown encourage getting all data each time). The data side of the envelope can be glued to a file card as a record. (b) A very simple setup, as shown here, works satisfactorily, though aluminum heating blocks per Traverse (1955) have the advantage of being water-free. The operator is shown grinding a flower through a brass screen (40 × 40 mesh) with thumb and/or finger, through a funnel into a centrifuge tube. The tube is heated in a water bath (a saucepan is fine; it can be fitted with a rack as shown). A single sample can be run in a centrifuge tube held in place in a beaker of water by a piece of aluminum foil with a hole in it (see arrow). This method has the advantage of preventing droplets of water from getting into the acetolysis mix.

annotate the sheet from which material is taken. Remember that all of this preliminary work should be done outside the processing laboratory because contamination, though seldom a serious problem, can become so if spores/pollen are willy-nilly introduced to the dust load of the "wet" laboratory. I favor preserving the data side of the collecting envelope and mounting it on an index card as the file record of the spore/pollen preparation. However, I do not assign a spore/pollen collection number to the preparation until it has proven productive. If it does prove productive, a serial number is assigned to the preparation. These numbers are recorded serially in a book. The completed slides are filed by family, then by genus and species; the book is organized by serial number, and the card files with the original collection envelope panels are filed by genus. So one can find a particular preparation in the collection by: (1) number, (2) genus (+ species), or (3) family (+ genus and species) (see Fig. A.2).

In the "wet laboratory": acetolysis   (c) Prepare the acetolysis mixture of nine parts acetic anhydride and one part concentrated sulfuric acid. Add the acid to the anhydride in a hood (acetic anhydride is very hydrophilic and fumes

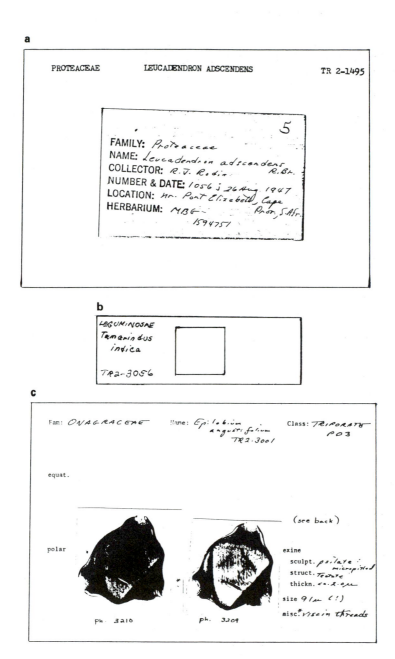

a

PROTEACEAE   LEUCADENDRON ADSCENDENS   TR 2-1495

5

FAMILY: *Proteaceae*
NAME: *Leucadendron adscendens*
COLLECTOR: *R. J. Rodin.*   *R. Br.*
NUMBER & DATE: *1056 ; 26 Aug. 1947*
LOCATION: *Nr. Port Elizabeth, Cape*
HERBARIUM: *MBG -*   *Prov., S.Afr.*
                    1594751

b

LEGUMINOSAE
Tamarindus
indica

TR2-3056

c

Fam: *ONAGRACEAE*   Name: *Epilobium*   Class: *TRIPORATE*
                              angustifolium   *PO3*
                              TR2-3001

equat.

                                                    (see back)

polar   exine
                                                    sculpt. *psilate ;*
                                                             micropitted
                                                    struct. *tectate*
                                                    thickn. *<± 2. 0 µ*
                                                    size *9 / µ (!)*
                                                    misc.* *visein threads*

ph. 3210   ph. 3209

Figure A.2 Simple data management for modern spores/pollen preparations at Penn State. (a) A file of 5 inch × 8 inch cards is maintained, on which the panels from the collecting envelopes used for herbarium materials are fastened. These cards are filed by *genus*, providing a generic–specific index for the collection. (b) Photoduplicate of slide from the collection, demonstrating results of writing with India ink directly on the .slide. These slides are filed by *family*, then by *genus* and *species*, providing a family index

449

will go for the moisture in your mucous membranes with uncomfortable results). The mixing reaction is exothermic. Do not stopper the mixture until it has "settled down". Be certain that all utensils used in preparing the mixture are absolutely dry, as even a little moisture inadvertently introduced will turn the mixture immediately dark brown and render it useless. The correct color for the completed mixture is pale yellow. Each preparation will take about 12 cm^3 of mixture. It is best to calculate how much will be needed and to make approximately that much, discarding the excess in a hooded sink with plenty of running water, or into a "decant bottle" which is periodically dumped in an approved place and manner; check with safety officers of your institution. Usually the decant bottle can be dumped in a hooded sink if a great excess of water is admixed, but local conditions may prevent this. In our laboratory, we fill our decant bottles half-full of saturated calcium chloride solution, in order to convert HF to relatively harmless CaF_2. Acetolysis mixture can be stored in a refrigerator in a stoppered bottle for later (but not too much later) use. It will change with time from the original pale yellow through orange and eventually to brown, decreasing in effectiveness as it darkens. It is better not to store the stuff beyond the time required for one session of pollen preparation, but on occasion I have left it in a refrigerator for a week and found it still yellowish and usable. On other occasions even 24 h was too long.

(d) For each envelope of sporiferous/polleniferous material to be processed, prepare a 15 ml high-quality, high-temperature-resistant glass centrifuge tube, to which a laboratory number has been affixed (I prefer the special laboratory tapes for this purpose). This number should be recorded in a working laboratory book, along with the species name and more information, if more than one preparation of a species is to be made. Stand the tube in a tube rack and put a small glass funnel in the tube. On the top of the funnel put a slightly cupped square of 40 × 40 (0.420 mm) mesh brass screening. Pour the contents of one of the collecting envelopes on the screen and rub the anthers, etc., through with a thumb or fingertip. Note that the screen must be thoroughly cleaned from previous use, or the resulting preparation will have contaminants! I favor cleaning the screen by brushing with a toothbrush, followed by flaming in a gas flame (hold square with a longish pair of channel-lock pliers). However, the resultant preparation will then be contaminated with abundant black flecks of charcoal, unless the screen is thoroughly rid of them by vigorous knocking against the edge of the bench or other surface.

(e) Wash polleniferous material from the screen and sides of funnel with an acetone wash bottle (do not work near a flame!). Put the funnel aside on a labeled paper towel. Centrifuge sample for 3 min (top speed in a clinical

---

to the collection. (c) "Morphocard": one of these (5 inch × 8 inch) cards is made for each preparation in the collection, including photomicrographs and/or drawings and filed by morphologic type (triporate, P03 in this case). In the file each morphologic type is subdivided according to sculpture (micropitted in this case). This file is very useful in identification of unknown Neogene pollen. Not shown, filed by preparation number, are vials containing the residues from which further slides can be made. The vials fit into holes drilled into pieces of pine board.

centrifuge, or 2,500 r.p.m. in a tachometer-controlled machine) and discard acetone into running water in a hooded sink. Replace funnel in tube and, now working in hood, carefully add acetolysis mixture to tube. Fill tube to within about 2 cm of the top. Equip tube with narrow stirring rod (about 3 mm: heavier stirrers sometimes break tubes). Put tube with stirring rod in a small beaker of boiling water on a hot plate in hood. If more than 1–2 tubes are to be processed, I favor using a rack with holes for the tubes that fits into a saucepan (Fig. A.1). (See Traverse (1965) for description of the virtues of aluminum heating blocks: no problem with constantly refilling water baths, no spoiling of acetolysis mixture by water condensate, more readily controlled temperature, e.g., at higher elevations where water boils at too low a temperature. Water baths are more likely to be available and are all right if properly monitored.)

(f) Heat at 100°C for 10–12 min (usually 12 min; if working at higher elevations, the water bath will not be 100°C and the time must be adjusted upwards). Stir (gently) every 3–4 min, leaving stirring rod in the tube. (Pure samples of pollen can be acetolyzed for shorter periods of time, but if much cellulose is present, the long acetolysis is needed.) Acetolysis mixture turns dark brown if the procedure is working. Remove stirring rod and centrifuge sample in the same manner as for acetone removal, except that this step must be accomplished in a hood. Decant into hooded sink or decant bottle, depending on your laboratory circumstances. Never decant in open sink, as the acetolysis mixture fumes are very toxic to people! (Note about decanting: Centrifuge tubes with a well-packed residue should be decanted in a quick, smooth, inverting movement. Novices make problems for themselves by gradual decanting, as this causes the liquid to bite into the sediment at the bottom. However, I would advise anybody who is having a decanting problem and wants to avoid redoing a lot of work to decant centrifuge tubes into a beaker first, and then empty the beaker if all went well.)

(g) Add distilled water to the tube. (The semidistilled water in laboratory taps labeled as "distilled" is fine. So is water from a dehumidifier. My experience is that it is usually unwise to use local tap water, which is a source of contaminants and sometimes of distressing calcium salts.) Stir thoroughly; I prefer to stopper the tube and shake it, but prior loosening with a thin stirring rod may be necessary (be careful not to break the tube!). Centrifuge again. Decant into hooded sink. Repeat washings followed by centrifugation until no acetolysis mixture remains. Usually this is three washings, unless there was too much plant material. If one is in doubt, after the third washing it is safe, and the best test, to taste with the tip of the tongue. After the first wash it is all right to work in an unhooded centrifuge and to decant in an open sink. After the last decantation keep the tube in the inverted position and put it mouth down on a paper towel in a tube rack to drain for about 10 min.

(h) Label a vial to receive the residue and label desired number of slides which should be placed along with coverslips on a warming table set at a temperature of 38–40°C. Add warm (about 50°C) glycerin jelly (see section on mounting media in "Notes" that follow) to the residue, the amount depending on the amount of residue and the density of spores/pollen desired on the slides; start with about five times as much mounting medium as you have residue, and adjust if necessary after inspecting the first slide. To add glycerin jelly, I use a 10 ml pipette from which I have cut off a section of the tip, as the original tips are too narrow, and soon clog with jelly. Stir thoroughly with a narrow, warmed (hold under hot water tap) stirring rod, but avoid introducing air bubbles. Keep the tube warm by frequently holding under hot water tap. Make desired slides by inserting a short glass tube (2 mm diameter) with the end stopped by a finger, about half-way to the bottom of the well-mixed jelly residue, then releasing the finger to admit a few drops of mix. A drop of mix is put on the slide and covered by a coverslip (working on a warming table). I favor placing the drop in the center of the slide, as the centered preparation is more conveniently studied microscopically than is a preparation which extends to one edge or the other. Leave slides on warming table at least 24 h and not more than 72 h to cure (= lose water, mostly). Then clean the slide and label. Peel-off labels are acceptable, but in my opinion the best way to label is to clean the end of the slide to be labeled with 50% alcohol (rubbing alcohol is satisfactory) to remove all traces of finger oiliness and then write the label directly on the glass with an India ink pen. In a minute or two the label can be painted over with colorless fingernail polish. The coverslip can be carefully ringed with the polish at the same time. Use a "bead" that just gets up onto the coverslip. It is not possible to study specimens critically that lie under the ring. If such a specimen is later encountered, use a little acetone on a folded corner of tissue to remove the acetate film in that area, and polish the area with another tissue moistened with saliva (or water for the squeamish – saliva is better). After studying or photographing the specimen, renew the ring. The ringing greatly prolongs the life of the slide. Oxidation of specimens on the slide is hastened by oxygen from the atmosphere, including that in bubbles on the slide.

Notes

Contamination problems Although some palynologists use elaborate precautions against atmospheric and other pollen contamination in their laboratories, my experience is that this is hardly ever a problem. The pollen prepared from plant material or from sediment samples should completely overwhelm isolated contaminants for one thing, or the slide should not be used (atmospheric contaminant pollen will seldom exceed 1–2 per slide even when conditions are relatively bad, with many people coming and going in a flowering season). Spores/pollen falling into the mounting medium or otherwise introduced after processing contain protoplasts and thus stick out prominently in the preparation. This is just not a serious problem, though I advocate a moderate degree of sensible caution, e.g., no flowering material in the wet laboratory. Vials, beakers, tubes, and other specimen containers should be kept covered when left overnight or longer. A much more serious source of contamination in my experience is insufficiently cleaned glassware, especially centrifuge tubes. Plastic tubes are more likely to offend than glass because of the presence of numerous scratches. Phipps & Playford (1984) advise against using mortar and pestle for crushing, as their surfaces may be pitted and cause contamination. This is true, but very minimally true, if the mortar and pestle are carefully cleaned after each use. As noted above, a stray contaminant grain now and again really does little harm, as it will be overwhelmed by the palynomorphs that belong in the sample. Results should never depend on very rare specimens for just the reason that contamination cannot be completely excluded. Nevertheless, the Phipps & Playford method of crushing between two aluminum pie plates which are then discarded is certainly elegant.

Coverslip thickness, upside-down slide curing, and coverslip sandwiches Sometimes one has difficulty with spores/pollen lying too far below the coverslip to be studied with the very short working distance of oil immersion lenses. Undestroyed chunks of plant tissue or foreign matter exacerbate the problem. In the first place, never use coverslips thicker than no. 1 (some people use the thinner no. 0, but they are very fragile). It is possible to cure slides upside down, using coins or wooden strips on the warming table, under the ends of the slides, or slides can be cured upside down in racks in a warming oven. The spores/pollen will mostly sink in the liquid glycerin jelly to come to rest against the coverslip and will stay there if the slides are also cooled upside down while the glycerin jelly jells. (Another way to keep specimens near the plane of the underside of the coverslip is described below under double mounting.) Where really critical microscopy is planned, however, the best idea is to make coverslip "sandwiches", with the residue between two no. 1 coverslips. These can be fastened to a slide with bits of sticky tape. A specimen can even be studied and photographed from both sides in this manner. (I also use this technique for S.E.M. work: see below.)

What to do if a centrifuge tube breaks Glass centrifuge tubes, even the best ones, do break. Yet for acetolysis work, they are best: centrifuging works better, and they are easier to clean. (Both glass and plastic centrifuge tubes can be cleaned of possible contaminants with strong oxidants such as sodium hypochlorite. Stand the tubes in a beaker of such solution.) Usually a preparation from a centrifuge tube that breaks can be saved, though it is better to start over if there is plenty of material. First, the centrifuge shields in which the tubes are spun must be clean: wash and brush them out frequently. If a tube breaks, dump the contents, rubber cushion, broken glass, and all, into a beaker. Use needles and forceps to remove the cushion and glass shards. If the sample was in acetolysis mixture or other strong chemical, dilute with enough water in the beaker to make this possible. If necessary, strain through a small plastic sieve to separate small glass shards. Centrifuge to recover the residue and proceed as if nothing had happened. Obviously, it is tempting to use plastic tubes all the time.

Non-centrifugation Occasionally, for reasons I do not understand, an acetolyzed residue will display repulsion of particles and will refuse to "go down". It may be necessary to basify the mixture with some 10% potassium hydroxide after which the residue will pack down normally. The alkali can then be washed out by a couple of water changes and centrifugations. (C. Barnosky (personal communication) suggests adding of ethyl alcohol to reduce the specific gravity of the liquid as a way of encouraging centrifugation in difficult cases.)

About mounting media   Spores/pollen are mounted on microscope slides in a wide variety of transparent media. Some have even advocated corn syrup or various self-prepared plant gums. Today there are only a few preferred mountants. The most common is glycerin jelly. Canada balsam, a natural resin from coniferous trees, is popular. It is soluble in xylene. Its index of refraction is good for pollen study (see Table A.1). The ideal index of refraction should be a little different from that of sporopollenin but not too different. Glass has an index of refraction of 1.54. For example, water, at 1.34, is too different from sporopollenin, and specimens in water show too much contrast and a very dark outline. Glycerin jelly, a little below, and Canada balsam, a little above sporopollenin's *RI* of 1.48 are both good. However, specimens must be run through alcohol and xylene changes before they can be mixed with balsam, and the liquid balsam remains liquid (especially in the center of a preparation) until the xylene gradually evaporates, sometimes after many months. In time the originally almost colorless balsam turns dark, but balsam preparations are good for many decades.

Table A.1   Refractive indices (*RI*) of some palynological mounting media. As has been frequently explained (Berglund *et al.* 1959), palynomorphs give poor definition in transmitted light if mounted in a medium with refractive index either too different from or too similar to sporopollenin. An *RI* moderately above or moderately below that of the specimens is best. For example, water mounting gives very harsh contrast and a black outline. Further, refractive index too close to that of sporopollenin gives poor, even blurred, definition without sufficient contrast. The relatively good definition of sporomorphs in Canada balsam would be difficult to explain if the *RI* for acetolyzed or fossil sporopollenin were as high as Christensen (1954) and Jones (1984) have reported for fresh sporopollenin. Polyvinyl alcohol, the usual primary mountant in double-mounting techniques, also gives good resolution at *RI* =1.55. That glycerin jelly, Canada balsam, and PVA all give very good resolution with palynomorphs tends to support the measurement of 1.48 for acetolyzed or fossil sporopollenin.

Substance	RI
air	1.00
water	1.34
silicone oil	1.4
glycerin jelly	1.43
AYAF (vinylite)	1.46
glycerol (glycerin)	1.47
sporopollenin, acetolyzed or fossil	1.48*
Canada balsam	1.53
quartz glass	1.54
Lakeside 70†	1.54
polyvinyl alcohol (PVA), as solid film	1.55

* *RI* = 1.55–1.62 for fresh sporopollenin, according to Christensen (1954) and Jones (1984).
† Used as adhesive for rock thin sectioning.

Although I have tried to find a satisfactory synthetic substance, I keep coming back to glycerin jelly. Despite the fact that it is not really permanent (most of my 40-year-old preparations are spoiled, apparently by autoxidation of the sporopollenin), it is fairly durable: a well-sealed preparation will stay in good condition at least 10 years. Furthermore, glycerin jelly has an ideal refractive index, and is perfect for photomicrography. It is thermoplastic at a lowish temperature (about 45°C, depending on how it is made) so that residues can simply be stored in it, melted when needed, and new slides made. It is water-soluble, so that no extra steps are needed to go from final water washes of a residue, before combining with mountant. As an alternative storage method, residues can be stored in water containing a biostatic compound, and slides made by mixing glycerin jelly with the stored residue. As glycerin jelly is water-soluble, a residue may be recovered from it for further processing at any time. Because of thermoplasticity the mountant in the vicinity of a specimen can be touched with a warm instrument to melt the mountant locally and permit turning of a specimen. Because the constituents can be purchased almost anywhere, one is independent of scientific suppliers, though very good glycerin jelly can also be purchased from them.

Many palynologists who study and count late Quaternary pollen find it essential to be able to turn over their specimens during counting to verify morphological features. They therefore prefer to use silicone oil, as it remains liquid, does not evaporate, and has a satisfactory index of refraction. It is true, however, that specimens will wander, especially if the slides are not stored horizontally. Glycerin (= glycerol) can be used in a similar fashion, but the refractive index is less satisfactory. (Tacking down the coverslips with nail polish retards but does not prevent wandering.) Silicone oil can be purchased from: Accumetric, P.O. Box 843, Elizabethtown, KY 42701, U.S.A. It is manufactured by Dow Corning Corp. as: 200 Fluid Dimethylpolysiloxane (viscosity: 2,000 centistokes). This information is from Dr. Cathy W. Barnosky, Carnegie Museum of Natural History, Pittsburgh, who uses the following modification of Faegri & Iversen's (1975) method for mounting fossil pollen residues in silicone oil:

(1)   Wash with water.
(2)   Wash with a few drops of water and 95% ethanol.
(3)   Wash with 99% ethanol; stain with safranin-O or fuchsin if desired.
(4)   Wash with t-butyl alcohol.
(5)   Add about 1 ml t-butyl alcohol, transfer to small vial, add silicone oil, leave for evaporation for about 24 h. The vials may be kept for future use.
(6)   Add the amount of silicone oil needed for optimal concentration of the pollen.

In making the slides the smallest possible amount of liquid should be used. The droplet spreads under the coverslip very slowly. For fossil slides small coverslips (18×18 mm) are to be recommended. Reference slides of recent pollen may be sealed with nail polish. (Paraffin is preferred by some, as there is some deterioration over a period of several years caused by a reaction between nail polish and the silicone oil.)

"Double mounting"   Many laboratories use a technique for slide preparation that results in specimens closely adhering to coverslips, using two mountants: a thin film in which the palynomorphs are enclosed and a second mountant, the function of which is to fasten the coverslip to the slide. The following is an example of double mounting used in the palynological laboratory of the Geological Institute, Swiss Federal Technical Institute ("ETH"), Zürich, adapted from procedures followed in palynological laboratories of the Geological Survey of Canada:

Ingredients   Polyvinyl alcohol solution: 50 g polyvinyl alcohol in 500 cm^3 distilled water. Mix and heat until solution clears – *do not boil*. Stir constantly, otherwise solution will form crystals. After PVA has dissolved, filter through filter paper and add a few drops 37% formalin to prevent fungal growth. Store at room temperature. (It should be rather viscous, like syrup.)

Epo-fix (tradename for an epoxy resin manufactured by Buehler Co., 2120 Greenwood Street, Evanston, IL 60432, U.S.A.): mix four parts Epo-fix with one part Epo-fix hardener. Mix only small amount needed, as it hardens within 30 min.

Procedure
(1) Drain finished residue completely by inverting tube on paper towel.
(2) Add few drops distilled water, one drop phenol (5% solution) to residue in test tube.
(3) Prepare coverslips to receive residue by affixing tiny numbered bits of press–on label to one side of coverslip, and inverting, so that unlabeled side will receive residue.
(4) Using capillary pipette, put one drop PVA on coverslip (two drops if coverslip larger than 22 mm square).
(5) Using another capillary pipette, mix residue thoroughly with its few drops distilled water and drop of phenol (see step (2)). Put one drop of residue on coverslip – mix thoroughly with PVA, and spread evenly, using side of pipette, being sure to get to edge of slip, and to eliminate areas of thick residue. Hold slip by edges with fingers while doing this. Set aside on sheet of white paper overnight. (It is desirable to cover all prepared coverslips with a box lid or similar item to prevent dust collection or other contamination.) Be certain all coverslips are labeled (see step (3)).
(6) Proceed to next residue, using another pipette, of course. Discard pipettes after use. You can use the same pipette in PVA for all samples being mounted at one time, provided you do not touch any of the residues with this pipette, i.e., always mix on coverslip with residue pipette. (Note: The easiest way to manage the PVA is to pour amount you intend to use for the job (25–50 cm^3, perhaps) into a very small (100 cm^3) beaker. As long as you are certain you have not contaminated the PVA, you may return unused portion to your primary container.)
(7) When job is done for the day (and it is best to save up residues until you have eight or ten to do), put a large box lid or other protection over the coverslips you have just coated with residue, as mentioned in step (5).
(8) Next day, or when convenient, mount coverslips on slides as follows:

(a) Use adhesive such as Epo-fix, mixing just the amount you need.

(b) Using a capillary pipette, wooden stick or other disposable utensil, put two drops of Epo-fix on slide and lower coverslip, residue side down, onto slide, using the same method (with needle, lowering coverslip slowly) that is used when mounting coverslips with glycerin jelly.

(c) Set aside to harden completely (several days) before removing the tiny labels and affixing permanent labels. It is not necessary or desirable to ring the coverslips with fingernail polish, as is usually done with glycerin jelly preparations.

(9) Storing of residue still remaining in test tubes should be done as part of the first day's procedure, when all coverslips have been coated with that day's samples. Leftover residue may be stored upright in small glass vials with screw tops (taped shut for additional protection). An extra drop of phenol may be added. Residue liquid may evaporate, and this should be checked frequently.

Bleaching   Some spores/pollen are thick-walled or otherwise end up too dark (e.g., some Malvaceae almost black, as mentioned earlier) in an acetolysis long enough to disintegrate the cellulosic tissues. Sometimes preparations are accidentally overacetolyzed, and the pollen is rendered too dark. In these cases the preparation can be bleached after acetolysis, as follows:

Add to the washed residue in a 15 ml centrifuge tube enough laundry bleach (ca. 5% sodium hypochlorite solution) to fill the tube about one-third full. Reaction time and concentration of solution determines degree of bleaching. About 2–3 min is usual. After this time, fill the tube with distilled water and centrifuge. If desired, the reaction can be stopped by adding 5% KOH. Water wash the residue until clean (no trace of bleach odor or taste in water; usually two or three water changes).

Staining   People who prepare spores/pollen by potassium hydroxide cooking usually stain. KOH-treated specimens look as if they had been acetolyzed and then bleached. Some sorts of pollen, e.g., sedges such as *Juncus*, are so sensitive to acetolysis that KOH boiling is preferred. For these specimens, for the relatively few things that come through acetolysis very pale, or for grains that were overbleached, staining is very helpful.

To a few drops of residue in neutral water, add a drop of basic fuchsin stain (saturated water solution of the stain, using 0.5 cm^3 ethyl alcohol per 100 cm^3 water). Centrifuge and add mountant, etc. Also frequently used is safranin-O, but it is not a good stain in combination with glycerin jelly. Proceed as for basic fuchsin, except that safranin-O is normally dissolved 1:100 in 50% ethyl alcohol. Allow to stand for 1–2 min after stirring thoroughly. Then centrifuge and follow with one water wash. Safranin-O is a more general stain than basic fuchsin and will make nearly all organic matter on the slide reddish. Red stains are preferred, because optical systems are said to be corrected for green light, and it is felt that definition is best with reddish–orange color. Malachite green and Bismarck brown have also had some vogue as pollen stains. Addition of a little basic fuchsin to the glycerin jelly will counteract the tendency of the mountant to destain the preparation when basic fuchsin is used.

FOSSIL PALYNOMORPHS

General instruction

Collection of suitable rock samples for paleopalynology requires that the collector understands the basic natural history of palynomorphs, discussed at various places in this book. Spores/pollen exines, other sporopolleninous palynomorphs, and chitinous/pseudochitinous palynomorphs occur in sedimentary rocks as organic silt-sized particles, subject to certain constraints. They are sensitive to high pH over long periods and are thus not usually common in limestones. They are sensitive to oxidation even for short times, and are thus not usually found in redbeds or deeply weathered rocks. They are sensitive to heat alteration (carbonization = coalification, darkening of color, ultimately becoming opaque and otherwise devoid of sporopolleninous–chitinous character), thus not occurring in rock that is or has been deeply buried (5,000 m or so) or much metamorphosed (not extractable from slates or anthracites), or subjected to heat from lava flows or intrusions. They are sensitive to recrystallization processes, thus are never found in dolomites, and seldom in heavily cemented, indurated rock. As to size, they are silt to very fine sand particles and thus are not found in well-sorted claystones and coarse sandstones.

The field collector should therefore look for "fudgy" siltstones (see below under field methods). All in all, cores are the best sample source when available. Fresh roadcuts are just about as good and offer more possibility to study the geological environment, collect megafossils, etc. Well cuttings are poor because they are often contaminated with drilling mud containing both caved microfossils from up-well, and usually additive pollen from the drilling mud. Also, the bagged sample is only generally referable to the listed depth. Sidewall cores made by shooting into the wall of the hole are excellent but seldom available. Pieces of rock from museums are sometimes very good, usually associated with good collection data, and frequently tie in with important research projects. However, they are subject to bias because they were almost never collected with palynology in mind. For example, a palynologist once reported the near absence of palynomorphs in rocks of a certain age in North America, based on study of museum fossil matrixes. The rocks were collected mostly for fossil vertebrates and vertebrate footprints and were mostly redbed coarse shales and sandstones. Non-redbed siltstones from most of the same areas are productive. Sometimes great perseverance is required: Leopold & Wright (1985) report that they processed over 450(!) rock samples to obtain 35 which were productive, some apparently marginally so.

The dispersed palynomorphs in sedimentary rocks are nano-coals, and would constitute a fossil fuel if they, instead of mineral clasts, predominated in the rock. The principal of palynological maceration is to separate out these nano-coals from the mineral clasts, and then further treat them for microscopy as necessary. The initial treatment of a prevailingly mineral sample will ordinarily be with hydrofluoric acid to break down siliceous minerals. Concentration of the palynomorphs may then be required, for partial elimination of other "nano-coals", such as wood, leaf cuticles, and amorphous organic matter. The most commonly needed additional treatment is oxidation to break up massive organic matter to release included palynomorphs, or to lighten very dark palynomorphs. This oxidation procedure must be applied as little as possible and with great caution because sporopollenin/chitin are quite oxidation-sensitive themselves. The oxidation makes humic acids available; these must then be removed in basic solution. Sometimes pH 8 is enough. In other cases acetone or even hot 10% potassium hydroxide must be used. If the sample itself is already carbonaceous – a peat, coal or very carbonaceous shale – the initial treatment will have the purpose of structural breakdown of the organic mass by oxidation, which is then followed by hydrofluoric acid to remove the mineral matter remaining. Figure A.3 shows the normal sequence of treatments for a sedimentary rock other than such a coal. Figure A.4 shows the sequence of treatments for oxidation of coaly material, whether from coal or from an acid-treatment residue of carbonaceous sedimentary material generally. The oxidation procedure produces artificially "regenerated humic acids", which are soluble in alkali, as are normally occurring humic acids or naturally (by weathering, etc.) regenerated humic acids.

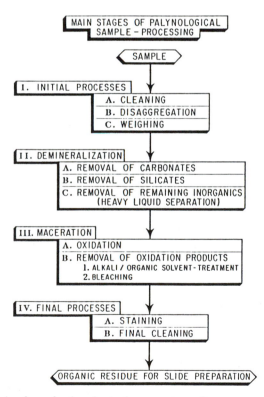

Figure A.3 Basic plan of palynological processing of an average sedimentary rock. Coals, and some other special sorts of rock, require different sequences of operations.

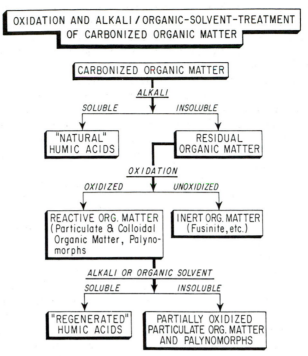

Figure A.4 Reaction scheme for oxidation (with Schulze's reagent or other oxidant) of carbonaceous matter, in order to release the sporomorphs, which are (slightly) more resistant to oxidation than other coal macerals.

In my experience, almost every suite of samples presents some unique problems, and the following directions are only a general guide. Modifications will constantly be necessitated by observations "in the kitchen", and indeed a kitchen is a better analogy to a paleopalynological laboratory than is a geochemistry laboratory. Different samples will behave quite differently under the same treatment. Further, there are usually alternative methods that one can use, as required by circumstances. For example, because HF was not allowed in the laboratory on the Glomar Challenger, I used wet physical disaggregation of the shales in an electric blender, followed by alternate cooking in 10% hydrochloric acid and detergent (Calgon®). This freed enough palynomorphs so that they could then be floated off from the mineral clasts in a heavy liquid solution.

Some sample will contain a great deal of very fine particulate matter, and the major preparation problem will be to disperse the fine particles with a dispersing agent such as Darvan® or sodium pyrophosphate (see Cwynar et al. 1979), followed by screening on very fine screen that will pass the fine particles but not the palynomorphs.

Hardly anything is absolutely essential to palynological processing. Good processing can be and has been accomplished without most of what we regard as routine. For example, without a hood one can macerate in plastic milk jugs out of doors (well away from people!), carefully decanting hydrofluoric acid into containers with very large volumes of water and disposing of this with even larger volumes of water (the amount of HF is small enough for disposal in pits well away from habitation). Few of the other chemicals are particularly dangerous and if diluted sufficiently can go in any sewer. Instead of an electric centrifuge, a hand centrifuge can be used. If enough time is available, repeated decantation will work without centrifugation at all. Batten & Morrison (1983) have shown that membrane dialysis can be used to rid a sample of acid without centrifuging or decanting. The equipment of modern kitchens suffices to do nearly all the necessary laboratory operations. Add a supply of plastic bottles, plastic refrigerator dishes, a couple of thermometers, a simple balance and a hand centrifuge, and one is in business.

I use the chart displayed in Figure A.5, and associated instructions, to teach new laboratory assistants and students how to process rock samples of unknown character. As has been emphasized, this is only a general guide and must be modified in one way or another for almost all sets of samples. For example, if concentration of palynomorphs is very low one may use massive samples (hundreds of grams) and physical disintegration in a large container, followed by large-scale "swirling" to get off a silty fraction with the palynomorphs which can then be more conventionally processed. Some pollen-bearing samples seem more or less impossible. Chitinous insect parts make up most of the organic matter of bat guano. After demineralization, it is next to impossible to separate the matted mass of insect mouthparts, etc., from the spores/pollen. Many of the mouthparts are in the same size range as spores/pollen, preventing screening. Chemical techniques that destroy chitin (oxidation) also destroy sporopollenin. There is plenty of pollen in Baltic and other amber, but no technique I have tried will dissolve the amber and leave the pollen. True oil shale, such as the Mahogany Ledge of the Green River Oil Shale, is also well-nigh impossible. The "kerogen" of the oil shale holds it together against all maceration processes I have tried.

Maceration and slide preparation

To supplement the following schedule, consult the references mentioned at the beginning of this appendix. I can only add the important advice to use common sense and observe carefully what is going on. Always keep careful notes of each process used with new samples; how long, what concentration, what results (color observed, etc.) for each step. When problems arise, these notes are very helpful.

Some additional cautions

(1) Make careful notes of steps performed as to the amounts of chemicals, time, temperature, etc. Use a permanently bound notebook. Keep all notes about the sample together by allowing sufficient space in the notebook. Such notekeeping is of paramount importance.

(2) Many rocks will require different and/or additional procedures. Remember that the basic principle (Fig. A.3) is to separate first of all the total organic residue, then to separate the pollen and spores from the other organic

BASIC PRINCIPLE: | ORIGINAL SAMPLE | → | CONCENTRATE ORGANICS | → | CONCENTRATE POLLEN, SPORES, LEAF CUTICLE, ETC. |

↓ REMOVE INORGANICS   ↓ REMOVE UNWANTED ORGANICS

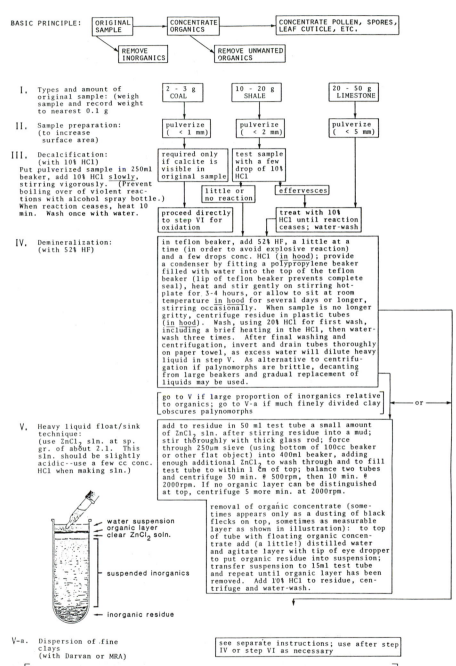

I. Types and amount of original sample: (weigh sample and record weight to nearest 0.1 g

| 2 - 3 g COAL | 10 - 20 g SHALE | 20 - 50 g LIMESTONE |

II. Sample preparation: (to increase surface area)

| pulverize (< 1 mm) | pulverize (< 2 mm) | pulverize (< 5 mm) |

III. Decalcification: (with 10% HCl) Put pulverized sample in 250ml beaker, add 10% HCl slowly, stirring vigorously. (Prevent boiling over of violent reactions with alcohol spray bottle.) When reaction ceases, heat 10 min. Wash once with water.

required only if calcite is visible in original sample

test sample with a few drop of 10% HCl

little or no reaction   effervesces

proceed directly to step VI for oxidation

treat with 10% HCl until reaction ceases; water-wash

IV. Demineralization: (with 52% HF)

in teflon beaker, add 52% HF, a little at a time (in order to avoid explosive reaction) and a few drops conc. HCl (in hood); provide a condenser by fitting a polypropylene beaker filled with water into the top of the teflon beaker (lip of teflon beaker prevents complete seal), heat and stir gently on stirring hot-plate for 3-4 hours, or allow to sit at room temperature in hood for several days or longer, stirring occasionally. When sample is no longer gritty, centrifuge residue in plastic tubes (in hood). Wash, using 20% HCl for first wash, including a brief heating in the HCl, then water-wash three times. After final washing and centrifugation, invert and drain tubes thoroughly on paper towel, as excess water will dilute heavy liquid in step V. As alternative to centrifu-gation if palynomorphs are brittle, decanting from large beakers and gradual replacement of liquids may be used.

go to V if large proportion of inorganics relative to organics; go to V-a if much finely divided clay obscures palynomorphs

— or →

V. Heavy liquid float/sink technique: (use ZnCl₂ sln. at sp. gr. of about 2.1. This sln. should be slightly acidic--use a few cc conc. HCl when making sln.)

add to residue in 50 ml test tube a small amount of ZnCl₂ sln. after stirring residue into a mud; stir thoroughly with thick glass rod; force through 250μm sieve (using bottom of 100cc beaker or other flat object) into 400ml beaker, adding enough additional ZnCl₂ to wash through and to fill test tube to within 1 cm of top; balance two tubes and centrifuge 30 min. @ 500rpm, then 10 min. @ 2000rpm. If no organic layer can be distinguished at top, centrifuge 5 more min. at 2000rpm.

removal of organic concentrate (some-times appears only as a dusting of black flecks on top, sometimes as measurable layer as shown in illustration): to top of tube with floating organic concen-trate add (a little!) distilled water and agitate layer with tip of eye dropper to put organic residue into suspension; transfer suspension to 15ml test tube and repeat until organic layer has been removed. Add 10% HCl to residue, cen-trifuge and water-wash.

water suspension
organic layer
clear ZnCl₂ soln.

suspended inorganics

inorganic residue

V-a. Dispersion of fine clays (with Darvan or MRA)

see separate instructions; use after step IV or step VI as necessary

Note: if palynomorphs are sufficiently concentrated, and light brown or lighter in color, after V or V-a, go directly to IX. If too much organic matter is present or if palynomorphs are too dark, go to VI first. If too much mineral matter is present, go to VIII first.

VI. Oxidation:
(with Schulze's sln.:
sat. K$_2$ClO$_3$+conc.
HNO$_3$)

Note: water wash
residue immediately
after Schulze treat-
ment (3 washes); then
proceed to step VII.

treat with "Schulze's sln." as follows: mix with residue
(in test tube or beaker, depending on volume of residue:
about 3x as much sat. K$_2$ClO$_3$ as you have residue; now
working in hood, add about as much conc.HNO$_3$ as you have
K$_2$ClO$_3$ to make the proportion 50/50. Note:[3] varying proportions of these two components, e.g. 70/30, also generally
give good results. Heat in water bath (tubes) or on hot-
plate (beakers) gently, stirring occasionally: 2-45 min. for
shale residues, 15-90 min. for lignites, 3-5 hrs. for
bituminous coals; proper oxidation is the point at which the
pollen and spores are comparatively light and transparent;
check residue (in water!) frequently during oxidation to
avoid over-processing.

Note re coals: before oxidizing, add a little 10% KOH to a tiny fragment of
the coal to assist in determining how much oxidation to employ. In some few
instances of low rank, already oxidized (weathered) coal, only KOH and no
Schulze treatment is necessary. At the other extreme, there will be no reaction
with KOH, indicating considerable oxidation (up to 5 hours) is necessary.

VII. Removal of humic acids:
(with 5% KOH, dilute
NH$_4$OH, 20% K$_2$CO$_3$, or
acetone); this normally
follows VI, but see
instructions.

test a drop of oxidized residue (in water suspension) first
with a drop of 5% KOH on a microscope slide, and examine under
scope (always being careful to spill any of residue on
lens or stage; a special scope should be kept in wet lab for
this purpose, always covered when not in use to prevent lens
clouding from acid fumes); if sporopollenin is destroyed or
damaged by KOH, test additional drops of residue with weaker
reagents listed left here; if sporopollenin is still damaged or
dissolved, OMIT THIS STEP, but do not return tested drops to
remaining residue; if 5% KOH does not dissolve humic or tracheid
debris, treat another drop with a drop of acetone; if little
reaction occurs, repeat step VI, followed again by step VII;
if sporopollenin is not damaged, but humic debris dissolved,
return drop to remaining residue and treat all of residue with
the appropriate reagent (KOH, etc.) by heating for (usually) 5-
10 min. (no longer!; sometimes merely the addition, centrifuging and
decanting of KOH will be sufficient without heating--only experience
with a particular sort of sample will tell); centrifuge and wash
residue with distilled water until no further discoloration of
solution occurs.

VIII. Optional additional
demineralization:
ZnCl$_2$ float/sink
technique, or 30 min.
HF treatment

if residue still contains too much clay or silt, wash
and centrifuge residue several times to insure removal
of all traces of base, then run a ZnCl$_2$ float/sink procedure in a 15ml tube, or treat the residue briefly in
HF, heating in a plastic 15ml tube in a water bath in
a 400ml beaker (secure tube in beaker with foil lid).

IX. Staining (optional)
and making slides:
(remember to weigh
slides, etc. if cal-
culating spores/pollen
per g)

(use basic fuchsin or safranin-O: for basic fuchsin, first
add a drop of dil. NH$_4$OH, then several drops conc.basic
fuchsin water sln. to residue, shake or mix, centrifuge,
decant liquid, drain; for safranin-O, use eye-dropper of
conc.stain (in water), stir thoroughly with residue, let
stand 1 min., add water to fill tube 1/2 full, centrifuge,
water-wash once, decant liquid, drain.)

add to stained or unstained drained residue in test tube
enough glycerin jelly (preferably stained, if residue is stained) to insure a high concentration of palynomorphs on slide; this is best ascertained by examining
residue in water until suitable concentration is noted, and
then (after, of course, washing and draining residue!)
adding the same amount of glycerin jelly; transfer thoroughly
stirred residue to storage vial (weigh before and after introducing residue, if spores/pollen per g is being calculated).

using small glass tube and precleaned, labelled slides
(weighed if pollen/g calculations will be made; also
weigh coverslip), place two or three drops of well-mixed
residue on center of slide (on warming table at 38-40°C);
touch jelly drop with edge of coverslip and gently lower
slip so that jelly drop is centered in middle of coverslip
(a bent needle is useful for this); release coverslip and
allow jelly to flow to margins before attempting to adjust
coverslip; allow slides to cure for 24-48 hrs. on warming
table (if desired, slides may be cured upside down by supporting the edges of the slide with coins or thin strips
of wood; this makes fossils lie close to coverslip); ring
coverslip with colorless fingernail polish to seal and
prolong life of preparation.

Figure A.5 Flowsheet for paleopalynological processing. Processing of rock samples for palynology is a sort of commonsense "kitchen chemistry" and is not easy to summarize in such a diagrammatic fashion. Practically every group of samples will pose different problems, and the student will need to respond to these by varying the procedures. Nevertheless, a thorough understanding of the general rationale will greatly help in determining what to do when problems arise.

matter. Avoid "heroics" of processing, and overprocessing to produce the "perfect slide". Follow progress of a maceration by using a rough scope in the wet laboratory to look at a drop of the maceration in progress (never from HF without thorough water washing first!). When palynomorphs are easily studiable, it is best to stop processing and begin studying. When in doubt about the wisdom of applying an additional procedure, it is best to divide the residue and experimentally further process only a part of it. Too much processing often destroys the palynoflora, or selectively destroys it, so that it is no longer representative. (On the other hand, there are limits as to how "dirty" a preparation can be and still be studiable!)

Some other pointers:

(a) Never apply a float–sink procedure after HF digestion, if the residue is richly organic. Everything will tend to float, and a grand mess will result.

(b) Never "store" a residue (overnight, over a weekend) in an oxidizing (e.g., nitric) acid, or in a strong alkali (e.g., potassium hydroxide mixture). However, it is all right to leave partly processed residue at any stage in water, or even in a non-oxidizing acid (HCl, HF) for days or weeks.

(c) Do not use Schulze's mixture or other oxidizing technique on a residue that is already partly alkali-soluble – test a bit of washed residue with 10% KOH – a dark coffee color will be imparted to the solution if the residue is in part alkali-soluble.

(3) Label everything – especially things left overnight.

(4) Always balance the centrifuge carefully.

(5) Use distilled water to prevent contamination and because tap water is too "hard".

(6) Hoods must have exhaust fans turned on all the time. HF is very bad for humans, as well as for equipment.

(7) Do not put HF beakers on stirring warm plates until reaction has ceased at room temperature. Otherwise the reaction may become too violent when the temperature is raised.

(8) If a reaction becomes too violent, cut it with alcohol from a spray bottle (see Fig. A.6a). This will knock down the reaction without diluting the components.

(9) If it is not convenient to remain in the laboratory until a reaction is complete, the total treatment may be administered in increments – it never hurts palynomorphs to sit in HF, HCl or water, as mentioned above.

(10) Always decant acids, etc., into decant bottles in fume hoods. When the bottles are about three-quarters full they should be carefully and slowly poured down hooded sinks, using plenty of running water. (In our laboratory we use 5 liter plastic, narrow-mouthed decant bottles which are prepared for use by being filled with 2 liters of calcium chloride solution, which serves to dilute the various acids poured into it, and to convert HF to CaF_2.)

(11) Clean equipment after use. You may use it next!

Calculation of pollen/spores per gram of sample This is simple if the vial is very well-mixed before placing part of its contents on a slide. But you must remember to do some weighing as you go along.

(1) Weigh the original (dry) sample.

(2) Weigh the vial in which you store your residue, both before and after putting the residue in it. If the residue represents only one-half, one-fourth, or some other increment of the original dry sample, this must also be noted.

(3) Weigh the slide and coverslip which will receive a drop of the residue both before and after putting the residue on the slide. A microanalytical balance is needed for this step.

(4) Use the equation

$$X = \frac{BD}{CA}$$

outlined in the previous chapter to calculate the number of palynomorphs per gram of sample; where X = number of microfossils per gram, A = grams of sediment sample, B = total grams of maceration residue plus glycerin jelly, C = grams of residue plus glycerin jelly on slide, and D = number of microfossils on slide. Obviously, this also requires that all the specimens on the slide be counted. One should therefore aim at a slide that has less than 1,000 specimens, as counting more than 1,000 is too time-consuming. If necessary, a line may be ruled in indelible ink, connecting opposite corners of the coverslip, and only half of the slide counted, and the counts multiplied by 2 before running the calculation per above equation.

Additional comments on the gravity separation technique Gravity separation methods (see Figs. A.6b–d) are commonly used in palynology to improve concentration of organics for spore/pollen analysis. Many methods or techniques are applicable, but the $ZnCl_2$ method has many advantages and only a few disadvantages. The advantages include:

(1) Relative safety. (No dangerous fumes exist compared with bromoform–alcohol mixtures of the correct density, although the latter may be used in the same way as $ZnCl_2$ or $ZnBr_2$ solutions. Bromoform is expensive and very poisonous. Also, residues must be water-free before going into bromoform–alcohol. Separation is accomplished with a separating funnel.)

(2) At this writing, $ZnBr_2$ costs only 1.6 times as much as $ZnCl_2$ per gram, but $ZnBr_2$ has a specific gravity of 4.2, as against 2.9 for $ZnCl_2$, meaning that it takes 1.4 times as much $ZnCl_2$ to make up a 2.1 specific gravity solution. This makes $ZnBr_2$ almost as economical as the more viscous and more difficult to manage $ZnCl_2$. Furthermore, recycling of $ZnBr_2$ solution by dilution, centrifugation and reconcentration by evaporation is more practicable. In other words, if the relative price of $ZnBr_2$ permits, use $ZnBr_2$.

(3) Little or no chance of contamination, as the solution need not be reclaimed. This also saves time. (However, the $ZnCl_2$ can be reclaimed if this is desired. Easiest method: dilution, centrifugation, evaporation to restore specific gravity of 2.1. See comments *re* $ZnBr_2$ under (2). Reclamation works more easily with $ZnBr_2$.)

(4) Stock solution does not deteriorate if properly stored, but one should always check specific gravity before use!

(5) The desired specific gravity of 2.0–2.2 is readily adjustable by addition

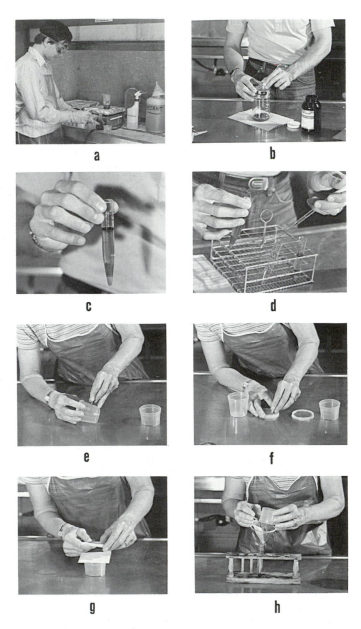

a   b

c   d

e   f

g   h

Figure A.6 A few pointers about "wet laboratory" processing of palynological samples. (a) People working with HF and other corrosive chemicals *must* wear goggles (ordinary eyeglasses are marginally acceptable) and *should* wear gloves. A plastic apron is a good idea to protect clothing (or one can wear only old clothing in the laboratory). Also shown is the decant bottle (right) into which spent HF and other acids are poured for later disposal. The bottle is kept charged with CaCl$_2$ solution which converts HF to

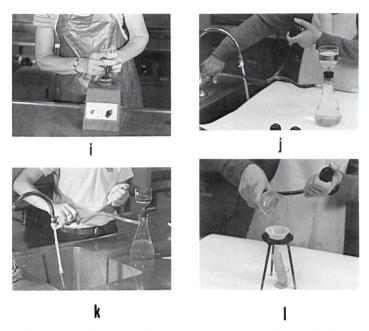

i   j

k   l

CaF$_2$, and dilutes the other acids. Next to the decant bottle is a plastic spray bottle containing 95% ethyl alcohol for spraying when foaming occurs after HF or HCl is added to sediment samples; the alcohol does not dilute the acid if 95% or absolute alcohol is used. On the stirring hotplate (use only low heat settings – Teflon *will* melt) is a Teflon beaker (below) with sediment in HF. Inserted in the top of the Teflon beaker is a polypropylene beaker of water, which acts as a condenser for HF. The Teflon beaker has a Teflon-coated stirring bar in it (not visible), which is rotated by the magnetic field of the stirrer. The operator is using a Teflon spatula to transfer sediment from a 50 ml plastic centrifuge tube following HCl pretreatment to a Teflon beaker for HF treatment. (b) Mixing ZnCl$_2$ (or ZnBr$_2$) solution with a processed sediment sample for heavy-liquid separation. After mixing the residue with some ZnCl$_2$, it is passed through a 250 μm brass screen which stands in a 400 ml beaker. The bottom of a small beaker is used to press the slurry through the screen, and more ZnCl$_2$ is poured in to complete the sieving. In this way solution and sediment are thoroughly mixed. (c) Separation of layers in heavy-liquid procedure after centrifugation (see Fig. A.5). Black organic fraction is shown on top, followed by a layer with suspended particles, then a packed layer of (light-colored) minerals at the bottom. (d) A disposable pasteur capillary pipette with a rubber bulb is used to remove the organic layer after a small amount of water has first been added to it and the water and organic layer carefully mixed with the tip of the pipette. This fraction is then washed by centrifugation. (e) Preparation of inexpensive filter unit for Swiss bolting cloth (= Nitex®) for filtering off clays and/or very finely divided organics. The bottom of a small plastic sample container is cut off. The unit to the right has already been cut. (f) The lid of the dish has the center part cut out. A completed lid and base are to the right. (g) Nitex bolting cloth (we use 7 μm opening) is fastened to the plastic dish base with the lid. (h) Many laboratories use either ultrasonic or suction in combination with Nitex, with the cloth supported on perforated disks or screens. However, it is possible to get most of the clay out of the sample by either lightly finger-stroking the under surface of an (inverted)

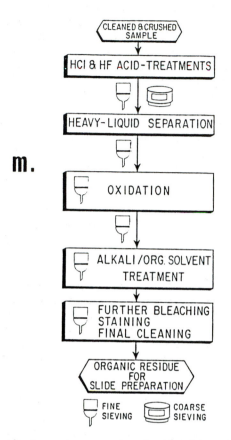

m.

CLEANED & CRUSHED SAMPLE

HCl & HF ACID-TREATMENTS

HEAVY-LIQUID SEPARATION

OXIDATION

ALKALI /ORG. SOLVENT TREATMENT

FURTHER BLEACHING
STAINING
FINAL CLEANING

ORGANIC RESIDUE
FOR
SLIDE PREPARATION

FINE SIEVING COARSE SIEVING

unit containing the residue in water, or tapping the side of the unit with a suitable rod. The operator has completed this procedure and has dumped the piece of cloth into a funnel. A water wash bottle will now be used to flush the organic residue into a centrifuge tube. (i) Electric stirring devices are very handy when multiple stirrings are necessary. The operator is capping the tube with a thumb because only water is present. If corrosive substances are present, a rubber stopper is used, but only after the reaction is *complete*; otherwise a dangerous small-scale explosion may occur. (j)–(l) Modified Reissinger apparatus (M.R.A.) in action. The glass funnel contains a shelf of fritted glass (= sintered glass) with holes rated at about 20 μm. The funnel is provided with suction (partial vacuum) from tap water. The operator has a rubber bulb in the left hand, closing the space in it to complete the partial vacuum. The fluid containing the residue is drawn down. Palynomorphs stay on the surface of a fritted glass filter, or just below it. In (k) the operator is interrupting the suction with the right hand, while pumping the bulb with the left to back-flush the filter, as it tends to clog. (Batten & Morrison (1983) have described use of a similar outfit, and Neves & Dale (1963) have shown that this sort of filtration system can be operated mechanically with an electric motor.) In (l) the operator has attached the hand bulb directly to the funnel for back-flushing the desired organics through a funnel into a centrifuge tube. (m) The multiple stages of processing at which the M.R.A. can be used are shown. See Ediger (1986a) and the text here for more information.

of water or $ZnCl_2$. The basic principle of the method is that chitin–sporopollenin have a specific gravity of about 1.4, whereas the lightest minerals have a specific gravity of about 2.5. The slow centrifugation that precedes the fast centrifugation is to permit minerals and palynomorphs to "work past each other in the traffic". An immediate fast centrifugation would cause the minerals to drag the palynomorphs down. "Slow centrifugation" can also be applied in water to separate palynomorphs by causing minerals to centrifuge out while palynomorphs are still suspended in the water. ("Short centrifugation" refers to using a short enough time for centrifugation to throw coarse minerals down, but not long enough to throw down palynomorphs, which can then be decanted off. On the other hand, the method can also be used to throw down spores but not suspended clay particles! See Darvan method below). This may, however, result in differential loss of larger palynomorphs.

(6) Processing can be maintained in a water base, whereas the bromoform–alcohol treatment requires alcohol washes both before and after the separation.

The principal disadvantage is the high viscosity of $ZnCl_2$ solution. $ZnBr_2$ solution is much less viscous. However, the viscosity of $ZnCl_2$ solution does not seem to interfere with good organic retrieval.

Addition to maceration instructions: dispersion of fine clays, screening
If standard maceration procedures are employed, one often obtains a residue which is more or less unstudiable because of a profusion of very finely dispersed particles (usually clay minerals) in the 1 μm or smaller size range. These can be removed by a variety of screening techniques. The technique below is simple and inexpensive and usually works acceptably (see also Cwynar *et al.* 1979).

Darvan® dispersant method   Use Darvan® no. 4 (from R. T. Vanderbilt Inc., Norwalk, CT 06851, U.S.A.), made up as follows: 12g Darvan, 1 liter distilled water, and few drops 37% formalin (make up in 1 liter wash bottle, mix thoroughly, keeps indefinitely). (Note: Sodium pyrophosphate can be used in the same general way (see Cwynar *et al.* 1979). Another dispersant is "Quaternary O", a high-molecular-weight quaternary ammonium surface-active agent (see Hills & Sweet 1972). This dispersant is favored by some for megaspore work.)

Procedure   (Use after HCl–HF and/or gravity separation.) If you notice that your sample is full of finely divided clay which obscures the spores/pollen, you may disperse the clay at any stage in your maceration sequence.

(1) Start with washed, drained residue in 50 ml (or 15 ml, if residue very small) test tube. (If you have more than 1 cm of residue in a tube, divide into two tubes.) Add about twice as much Darvan solution as you have residue. With stirring rod, stir up thoroughly from bottom of tube, so no residue is left sticking to sides and bottom.

(2) Agitate tube on tube mixer (such as Vortex Genie®) for 1 min to mix Darvan and residue (see Fig. A.6i).

(3) While still on mixer, fill tube with distilled water to within 2 cm of top.

(4) Centrifuge at 1,400 r.p.m. (no more!) for 1 min (no longer!). Do not use

brake to slow or stop centrifuge. Decant carefully into large beaker (check later for spores inadvertently poured off, although in our experience this practically never happens). This is called "short centrifugation" and will eliminate much clay.

(5) Wash 3–4 times, or until decant is clear, stirring carefully, centrifuging as in step (4), and decanting into the large beaker. Most of the fine mineral fraction will be poured off in this way.

(6) Sieve to separate rest of fine fraction from residue. (If residue still contains large particles, sieve first through a 210 μm brass sieve, washing through thoroughly and saving what goes through for examination.) If no large particles, skip this and go directly to sieving through 7 μm nylon screen (see Figs. A.6e–h) (Material – Nitex® industrial Swiss nylon monofilament screen fabric no.HD 3–7 super – available from Tetko Inc., Elmsford, NY 10523, U.S.A.) Wash thoroughly with warm water. Palynomorphs will be held back on the screen, and unwanted fine particles will go through.

(7) Very carefully, using a funnel (see Fig. A.6h), wash residue from cloth into 15 ml test tube(s).

(8) Centrifuge, stain if desired, add mounting medium, make slides.

(Note: Discard screen after use for one sample. If you plan to macerate other fractions of the same sample, you may retain the original screen for this after washing it thoroughly. It may be kept between paper towels (labeled!) or in an envelope.)

"M.R.A." sieving technique   This technique, using what we call in our laboratory a "modified Reissinger apparatus" (M.R.A.), based on sintered glass (= fritted glass) filters attached to funnels, can be employed at any stage in palynological processing (Ediger 1986a). "M.R.A. is particularly useful after heavy-liquid separation and after oxidation. See Figures A.6j–m for illustrations of the set-up of this apparatus, and the various steps used.

Procedure

(1) Screw discharge tube to tap and poke rubber end down sink drain.

(2) Fasten connecting tubing to outlet at top of discharge tube. Fill flask with water.

(3) Attach filter funnel firmly to top of flask. Wash filter by turning on tap and squirting a little water on the filter. Then pump (to back-flush) and close end of bulb with thumb alternately. (This also accomplishes filtering.)

(4) Pour residue into top of funnel. Add plenty of water and filter with alternating back-flushing as in step (3). (Turn on hot water tap as well as cold tap if not enough pressure has been produced, and/or pinch bulb tubing to increase pressure.)

(5) When the filtering is completed, remove funnel from apparatus and attach straight tube and bulb to it. Add a little water, pump, and pour residue into test tube(s), repeating until all residue has been removed.

(6) To clean the funnel, attach to a flask which has been set up *in a hood*, being sure the apparatus is connected to a water tap. Pour enough "chromic acid" into funnel unit to cover the surface of the sintered glass filter thoroughly. Turn on tap and the "chromic acid" will run through. When this

is accomplished, add water to rinse thoroughly. Turn off tap, remove filter, and rinse thoroughly by hand. You may have to repeat the cleaning treatment several times for especially stubborn samples.

Sintered glass filters with various ratings as to effective size of openings are available, e.g., from Schott Glass or Robo Glass, West Germany. Figure A.6m shows the way the "M.R.A." filter can be used at many stages of palynological processing. However, the technique does not work well for palynomorphs less than 20 µm in diameter. (The sintered glass filters with small enough openings clog too easily.)

Other filtering techniques　Several other filtering techniques involving more equipment are used with success in various laboratories. Neves & Dale (1963) described a filtering technique based on a sintered glass disk (porosity 2) in a Buchner funnel mounted on a pressure flask. An air pump reverses the flow of air periodically (50 s filtration flow, 5 s reversed air flow) to keep the sintered glass disk from clogging. Good results have been reported by various laboratories using variations of the method, essentially as for "M.R.A." described above. Caratini (1980) has used an ultrasonic generator in combination with nickel filters. The technique has the advantage over the M.R.A. and other sintered glass disk methods of not needing to be constantly unclogged, and over the nylon netting used in the Darvan method of being more or less permanent. However, the metal filter unit alone costs hundreds of dollars, and an ultrasonic generator must be added to this cost. Caratini (personal communication 1986) now uses a special ultrasonic generator with Swiss bolting cloth filters (= Nitex) on supporting perforated disks. Ultrasonic generators are much used in palynological laboratories, but they pose some problems, especially that they tend to break up brittle palynomorphs. If very carelessly used, they can also damage human retinas. Batten & Morrison (1983) describe the use of ultrasonics for cleaning samples in considerable detail. They use vibration of 50 kHz.

"Swirling"　This technique takes advantage of the differential response to turbulence of palynomorphs of different sizes and conformation and of the "junk" in palynological maceration residues, the same physical properties as form the basis for concentration of palynomorphs in sediments during deposition. "Swirling" involves agitation of palynomorph residues in water in a watchglass rather like panning for gold. The principle in "swirling", however, is not based on specific gravity (because sporopollenin is all pretty much the same in this respect, about specific gravity 1.4), but on size, shape, and morphology of the microfossils. The technique is very helpful in separating palynomorphs from other organic "junk" in a particular residue, without resorting to screening and heavy-liquid gravity separation. As is true of slow (or "short") centrifugation, however, there is some danger of differential loss of certain palynomorph fractions.

Procedure　Although simple, it must be learned through experience, as different samples behave in different ways and are best treated by slightly different techniques. For the neophyte it can be slow, messy, and frustrating;

but with experience one is able to "clean up" a sample in a few moments. It is especially valuable in the separation of spores from large tissue fragments of similar density. Unless great care is used, certain size groups of spores may be lost. Therefore, it is not generally recommended for quantitative work. The residue should be in a fluid (generally aqueous) carrier. To avoid equipment damage and flocculation, it should be neutral.

(1) Place a small amount of residue (usually a few drops, although amount varies somewhat with spore content, degree of "cleaning" to be done, number and size of slides to be made, etc.) in a clean 3 inch watchglass and fill two-thirds full with water. Allow to settle a few seconds.

(2) Gently swirl the contents by moving the watchglass in a circular motion, so that the center of the watchglass circumscribes a small (1–5 mm) circle about a point on the table. This causes a certain fraction of the residue to be suspended.

(3) From the suspension, decant the fraction, or pipette to another watchglass and inspect both glasses with a dissecting microscope. If one fraction contains no palynomorphs it may be discarded, and the process repeated with the palyniferous fraction to attain a desired concentration. By suitable adjustment of the swirling speed, particles of different size and general morphology from the palynomorphs can be separated from them. It is even possible sometimes to separate one kind of sporomorph from others by very critical "swirling", to make a super-clean preparation of that kind. This is seldom important as capillary tube pipetting techniques working under the microscope to select specific fossils do it more accurately. However, the most useful application of this technique is to rid a troublesome sample of unwanted minerals and large organic particles.

(4) Inspect separated fractions under the microscope. The scope must have at least a 1 inch working space. Magnification of $100\times$ or $150\times$ appears to be the best compromise between enlargement and field area. If a fraction is found to be barren (or essentially so in non–quantitative work) it is discarded.

(5) Repeat the process on the sporiferous fraction, using slightly varying swirling amplitude and vigor, until the particles heavier and lighter than the spores are removed. Material of the same density and size as the spores cannot be removed. Particles of different size from the sporomorphs can be separated owing to the difference in energy required for the microcurrents in the glass to "pick up" objects of different size. (Note: The very simplest form of swirling, letting the residue and water settle for a minute and then decanting the "discolored" water, is almost always beneficial in removing colloid-sized material. If the remaining residue is then somewhat violently swirled, the spore-bearing fraction may be decanted from the heavy and large "dregs" which escaped earlier processing (often mineral particles and heavy tissues).)

Finest degree of separation is accomplished by gently jiggling the residue into the center of the watchglass and then very gently swirling it at a very small amplitude. This generates a column of "smoke" which is of slightly different density-size characteristic from the non–rising residue. In this way separation of different genera can sometimes be made.

A number of techniques have been described which, like swirling, depend on exploiting differential physical properties of what remains after palynological basic maceration. Tschudy's (1960) "Vibraflute" is a mechanically agitated tube with holes along it, permitting collection of fractions at intervals. The differences in size of various constituents of a residue cause the fractions to contain concentrated, "cleaned-up" samples of various palynomorphs. Hansen & Gudmundsson (1979) describe a technique for utilizing differential takeup of absolute ethyl alcohol by palynomorphs to permit their separation by specific gravity difference, from organic junk that does not take up the alcohol.

Batten & Morrison (1983) report that they have abandoned the use of the centrifuge entirely for palynological processing, using only dialysis, gradual decanting, and swirling. Unquestionably this procedure works, and we use it (excepting dialysis for neutralizing of acids) for very brittle or other easily damaged palynoflorules, but it really does not seem necessary to stop centrifuging altogether. Centrifuging is so *fast*!

About oxidation In addition to Schulze's mixture (see Fig. A.5), other oxidants, such as hydrogen peroxide or sodium hypochlorite (laundry bleach), may be used. Evitt (1984) suggests using HCl with the NaOCl to intensify the reaction. The reaction can be stopped by basifying the solution with 5% KOH.

Batten & Morrison (1983) point out that treatment with Schulze's mixture for the purpose of oxidation has the desirable side effect of eliminating pyrite (marcasite) crystals from the treated specimens. This is because of the solubility of FeS_2 in HNO_3.

Special techniques for megaspores, chitinozoans, and scolecodonts
Megaspores, from a botanical point of view, are any spores of a heterosporous plant from which the female gametophyte develops. Seed plants are heterosporous, but the megaspores are not released as such, remaining within the megasporangium. In this case, only one megaspore per sporangium is normally functional, and it develops into the megagametophyte ·(= embryo sac in angiosperms). Heterosporous free-sporing plants, however, release megaspores from the megasporangium, and they germinate to produce the megagametophyte, a much reduced but independent (gametophytic, 1N) plant. Free-sporing megaspores are now produced only by a few heterosporous pteridophytes such as *Selaginella* and *Isöetes*, and by a few heterosporous ferns such as *Marsilea* and *Azolla*. However, after heterospory first developed in the Devonian until seed plants "won out" in the Carboniferous, free-sporing megaspores were very common, and occur abundantly in sediments. They waned after the Carboniferous but occur frequently, sometimes relatively abundantly, in younger sediments. They usually exceed 200 µm in diameter, for which reason Guennel listed 200 µm as the upper limit for "miospores" (= "small spores"), and designated sporomorphs larger than 200 µm as "mega-spores". However, some pollen grains (microspore wall with included microgametophyte) are greater than 200 µm in diameter, and some functional megaspores are smaller than 200 µm. In any event, "large spores" are sporopolleninous, occur in the same sediments as miospores, mostly in non-marine sediments, and often are potentially useful for stratigraphy.

473

Chitinozoans are enigmatic marine microfossils with pseudochitinous walls which survive the vicissitudes of sedimentation and diagenesis as do palynomorphs generally, and they can be separated from sedimentary rocks in much the same way as sporomorphs. Their size range is mostly about 50–250 μm, which means that the smaller ones regularly occur in spores/pollen macerations of marine shales. (There are some smaller than 50 μm and some larger than 250 μm.)

Scolecodonts are chitinous mouth-lining parts of marine annelids. They range in size from about 100 μm to about 200 μm. Therefore, recognizable pieces of scolecodonts, as well as smaller kinds, occur in spores/pollen macerations, as do chitinozoans. Megaspores and many chitinozoans and scolecodonts have in common that they usually occur much less abundantly than spores/pollen in sediments, one or two orders of magnitude less abundantly. Furthermore, they are usually much larger than miospores and consequently visible as dots to the naked eye. Therefore different preparation techniques are required. First, the preliminary fine grinding used on miospores must be avoided; breaking into pieces about 1 cm across is as far as one should go. Secondly, strew slides are not acceptable because the concentration is too low. These large palynomorphs are best picked out of the maceration with a brush or a capillary tube and mounted on S.E.M stubs, slides, or on special microfossil (foraminifera-type) slides. Thirdly, all of these microfossils tend to be opaque because of the size and the thick walls, and are best studied by reflected light and by S.E.M. The following processing method, modified from Jenkins (1970), who developed it for chitinozoans, can be used also for scolecodonts and megaspores.

Preliminary treatment Fragment about 250 g of sediment into pieces about 1 cm in size (smaller fragments yield broken specimens, larger ones require too long for maceration). Discard fines from the fragmentation procedure.

Maceration

(1) Treat with 10% HCl until all reaction ceases. Wash and transfer to HF, as in conventional maceration procedure. Whenever time permits, use a gradual decantation process to go from one step to the next without centrifugation, as centrifuging damages large, ornate specimens. Thoroughly water wash.

(2) Sieve: use a brass sieve with openings appropriate to the category of fossil (ca. 125 μm for megaspores, 50 μm for chitinozoans, 100 μm for scolecodonts). Hold the sieve containing the residue in water to within 1 cm of the top of the sieve. Gently move the sieve up and down in the water without allowing the water to flow over the top of the sieve. Fines pass through. Continue this procedure, frequently changing the water, until no material passes through the sieve.

(3) Bleach (if necessary) in commercial laundry bleach (sodium hypochlorite solution), checking the course of the procedure under the microscope: transfer approximately one-fifth of the residue to a white-bottomed petri dish (about 10 cm in diameter), or a clear petri dish on a white card or tile. Adjust water depth to 0.5 cm. Add a few drops of the bleach solution, mix thoroughly, and observe under a low-power microscope. Stop the bleaching by adding an

excess of 5% KOH solution when the correct bleaching level is attained. The bleaching can be precisely controlled in this manner. Wash the residue by repeated addition of water and decanting until a drop of the residue in water leaves no trace of precipitate when evaporated on a microslide.

(4) "Picking" and mounting: spread the residue thinly over the bottom of a white-bottomed petri dish in about 0.5 cm of water. Search with a stereoscopic microscope. Pick up desired specimens with a capillary pipette. A pipette attached to a hypodermic syringe plunger by a plastic tube can be used, or a simple pipette drawn out at one end to about 1 mm in diameter (larger for some megaspores), closing the other end with a finger until the specimen is near the orifice of the tube. Release of the finger allows specimen to rush into the tube with water. It can then be blown out onto the desired surface. These are the same techniques as described for spores/pollen under single-grain mounts, except that the orifice of the pipette must be larger.

Specimens may be transferred, as picked, to a little distilled water in a second petri dish, e.g., to group similar specimens before mounting. When ready to mount, transfer the specimens to a watchglass containing a 2% aqueous solution of Cellosize. Take up specimens from the Cellosize solution with a pipette and put out as a small drop on a coverslip. (Up to 30 specimens can be grouped on one large coverslip.) Put the coverslip, face up, on a warming table at 50°C for 1h, or until dry. Invert the coverslip and carefully lower onto a drop of Canada balsam or other suitable permanent mountant on a slide. Alternatively, megaspores, chitinozoans, and scolecodonts can be mounted in glycerin jelly on slides, or they can be put (dry) on S.E.M. stubs or on microfossil slides of the sort used for foraminifera studies. If it is desired to study and photograph the whole specimens by reflected light, or to study by S.E.M., this is necessary. In this case, a very small camel hair brush is used for picking up the specimens, and, in the case of S.E.M. studies, the Cellosize must be washed off before mounting. Paris (1981) describes a useful method for mounting the specimens on a coverslip, which can be mounted on an S.E.M. stub, and then later removed for optical microscopy.

Collinson *et al.* (1985) suggest that, after maceration, a panning ("swirling") process in water can be used to concentrate megaspores. Residues are stirred, allowed to settle for a few seconds, then the supernatant is passed through a 125 µm sieve. The process is repeated until no more plant material is brought up by stirring. If the original maceration was not done with HF, an HF treatment is applied now to remove adhering minerals.

Field methods

Sample collecting   As stressed elsewhere, palynomorphs are sedimentary particles in the silt to very fine sand particle size range with a characteristic set of chemical and physical properties. It is here necessary only to mention those aspects of collecting samples which long experience has taught me are not always obvious to field geologists.

(1) Palynologists, or geologists well-instructed by experienced palynologists, should whenever possible collect their own samples, whether in the field or from cores in a core storage area.

(2) Cores or fresh outcrops are better than weathered outcrops. Palynomorphs are very sensitive to oxidation and much time is wasted macerating weathered rock. In hot, dry climates weathering often extends so deeply into rock outcrops that blasting or bulldozing is necessary to get down to acceptable samples. Where this is not possible, look for places where the target layers are exposed along a stream cut or actually in the stream bed. Avoid collecting on exposed ridges.

(3) The best rock type is relatively unconsolidated "fudgy" siltstone in the gray to light brown to green color range. Black shales are not always good because the black color often comes from minerals, or the organic matter may be predominantly non-palynomorph as in a sapropel. Coals are often highly vitrinitic, hard to process, and usually contain a flora typical of the original swamp and thus are less satisfactory for stratigraphy than shales. Sporonite-rich coals, however, may yield beautiful palynofloras! Red (and even dark brown) shales usually are oxidized and barren, an occasional exception being shales reddish from included red minerals where the rock itself is not oxidized. Because the siltstones we seek are soft, even a comparatively fresh outcrop, say a highway cut, will often show the best rock for our purposes only in the soft layers between consolidated, unsatisfactory shales with much recrystallization, or hard sandstones. Because they are soft and often thin and hence easily eroded away, it is frequently necessary to get at these layers by excavating deeply into the rock face between the hard layers, deeper the longer the outcrop has been exposed. Avoid rock that shows evidence of post-depositional alteration as evidenced by slickensides, cleat and cleavage, or obvious secondary cementation, or proximity to a lava flow or volcanic intrusion such as a dike.

(4) Test a very small piece of the rock between your incisor teeth for sediment size. If it is "creamy", it is no good – clay. If it is really gritty, it is no good – sand. If it is very slightly fine–granular, at least the particle size is acceptable. Remember that a "fudgy" silty rock is what you are looking for.

(5) Limestones are seldom good, though marls and calcareous shales are often productive, and there are exceptional cases of beautiful preservation in limestone. Dolomites are always barren, as previously noted. Look for stringers of siltstones associated with limestones/dolomites. These are sometimes productive.

(6) A moderately productive siltstone will yield a very good maceration from 10 g; a richly productive siltstone requires only 2–3 g. It is not necessary to collect large samples, although it is my custom where space or weight is not a problem to collect about 250 g (about half a cup) in labeled cloth bags, in case a future mass maceration might be desired. Later, after maceration of the first increment in the laboratory, I discard barren samples and consolidate all others to small plastic jars holding about 50 g. When visiting foreign countries where shipping rock samples is for various reasons a problem, it is possible to send perfectly good palynological samples home, folded flat into pieces of paper, in ordinary letter envelopes. Each sample will require only ordinary letter postage (1 oz = 29 g!). It is not necessary to take any special pains such as wrapping in plastic. I have found it useful when forced to have other geologists collect for me to mount a series of pieces of lithologically "super"

shale on a small board for the collector to use in the field for comparison purposes.

(7) Patient searching of outcrops in a sedimentary basin will almost always yield at least stringers of suitable siltstone. (An exception in my experience has been the Fundy Basin in Nova Scotia, where the prevailing rock type is red, oxidized shale and sandstone, the promisingly green horizons being apparently secondarily reduced, and the few grayish siltstones that occur are too close under the North Mountain Basalt and are carbonized.)

Keeping very good records in the field and later in the laboratory pays off! See Figure A.7 for some information about this. Despite our precautions, we sometimes have trouble when other geologists give us samples for which the field data are insufficient or seem to conflict. Where possible, palynologists should collect their own samples!

Field processing It is possible to produce studiable slides from rock samples by processing in the field, and, of course, to study them also, if a microscope is available. I have done this by using plastic milk bottles for HF digestion and working in the open, well away from buildings and people. A hand-operated centrifuge or a portable electric centrifuge run from a generator can be used, also working in the open. Final processing steps, slide making, and microscopic study can be accomplished in a motel room, or in a tent, provided that there is power for the microscope illuminator or that a microscope with a mirror for use of daylight is available. Kaars & Smit (1985) describe a method for field maceration based on use of dispersing agents, sieving, and decanting, without use of strong acids.

Manipulation of spores/pollen

Single-grain manipulations and mounts

Summary A century ago when microscopy was a popular hobby for the well-to-do, there were even folk who made beautiful *objets d'art* that could be viewed only with a microscope, by picking up and glueing down the variously colored scales from lepidopteran wings and diatoms. Manipulating individual spores/pollen and dinocysts would not have surprised such folk. However, modern-day students usually are appalled in elementary palynology when they discover that they will be expected to pick up and mount individual pollen grains for study. There are several reasons for teaching these techniques.

(1) Nothing so dramatically teaches the meaning of the size range of palynomorphs as making single-grain mounts. Until then, when they are only untouched objects under the lens, the actual size of palynomorphs is not really grasped. Furthermore spores/pollen seem to "come alive" for students when they have actually handled them.

(2) For reference purposes single-grain mounts are very helpful, e.g., as a reference collection of the constituents of a palynoflora to have near at hand during the course of the study. They are ideal for type specimens for new taxa. They are perfect for sending to another palynologist for discussion, because there is no question as to identity. (For this purpose I advocate mounting

PALEOPALYNOLOGICAL RESIDUE COLLECTION

NO.	STRATIGRAPHY	LOCATION	PREPARATOR	PREP'S NO.	PROCESS	DATE	COMMENTS
PRC 2872	Devonian (Frasnian?)	Otego, NY	V. Ediger		std. with Ediger's sieving technique	11/83	J. Bridge 83-35 (=T-796)
PRC 2873	"	"	"		"	"	J. Bridge 83-36 (=T-797)
PRC 2874	"	I-88 (NY)	"		"	"	J. Bridge 83-37 (=T-798)
PRC 2875	"	"	"		"	"	J. Bridge 83-43 (=T-799)
PRC 2876	Cretaceous, Upper (Senonian)	Cliffwood, NJ	B. Traverse	T-615 (Juniata dam)	std.	2/84	=T-615 Magothy Fm. (see PRC 2657 for earlier maceration)
PRC 2877	Jurassic, Lower (Sinemurian)	Dorset Coast, U.K.	D. Freudenrich	unk. #5	"	12/83	=T-802 (Robertson Research DC-23)
PRC 2878	Devonian	Centre Co., PA	R. Zerr	unk. #17	"	"	=T-581 Catskill Fm.
PRC 2879	Cretaceous, Upper	Cliffwood, NJ	J. Seitz	unk. #15	"	"	=T-615 Magothy Fm.
PRC 2880	Triassic, Latest	Warwickshire, England	D. Salabsky	unk. #3	"	"	=T-801 Rhaetian - Westbury Fm.
PRC 2881	Oligocene, Latest	San Juan Mtz, Colorado	A. Schuyler	unk. II 11	"	"	=T-776 Creede Fm.
PRC 2882	Pennsylvanian	Western Kentucky	J. Hitchens	unk. # 10	"	"	=T-804
PRC 2883	Oligocene	Forestdale, VT	L. McPherson	unk. # 8	"	"	=T-762 Brandon lignite
PRC 2884	Rhaetian-Triassic	Upper Basin, VA	A. Traverse	—	HCl- HF- In erg Sabo?(:?)	Nov.'83 -Feb84	Pamela Gore G5A1 = T-806
PRC 2885	"	"			"	"	Pamela Gore G1A1 = T-807
PRC 2886	Oligocene	Florissant, Colorado	A. Ahrens	unk. #18	"	2/84 (??)	=T-258 #5

a

b

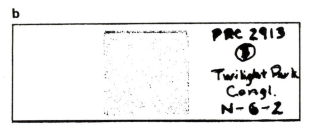

PRC 2913
Ⓑ
Twilight Park
Congl.
N-6-2

Figure A.7 Record keeping for paleopalynological maceration residues at Penn State. (a) Data pertaining to the PRC collection are keyed into a serial PRC number assigned each sample at the time of preparation. Field collecting notebooks present detailed information about the original locality. In line 1, "83–35" for PRC 2872 is a field number keying into the appropriate field notebook. There is also a permanent collection of the productive *rock* samples, and numbers from this collection are also recorded in the PRC book where applicable. For example, PRC 2872 is represented by rock sample T-796, and the rock sample collection book presents various information about lithology, etc., for this sample. (b) Photoduplicate of a representative slide from the PRC collection. PRC 2913 keys into the PRC book (see (a)); "B" (circled) refers to the particular slide (there is at least one more slide, "A"); Twilight Park Conglomerate is the field locality; and N-6-2 a field number keying into a field book. The slides are filed stratigraphically and geographically (Upper Devonian of New York in this case). Not shown, but filed by PRC number in cabinets, are vials containing the surplus residue. The vials fit into holes drilled into pieces of pine board.

478

in a "coverslip sandwich", so that viewing from both sides is facilitated.)

(3) For S.E.M. and T.E.M. work it is often best to isolate single specimens. One can, for example, make light and S.E.M., or even also T.E.M. (see Walker & Walker 1982), studies of the same specimen (see Fig. A.9).

Glycerin jelly–particle technique This is the "quick-and-dirty" technique (see Fig. A.8).

(1) Get some of the residue into glycerin and alcohol (50/50) or glycerin and water (25/75) in a tube or vial. Put a very small drop of this mixture on a slide and spread it out over the surface of the slide with a needle.

(2) Using the 10× objective of the scope locate a specimen of the desired taxon. Verify the identification at higher power (20× or 24×) if necessary, but return to 10× and clear away mountant and other debris completely from the area of the subject grain.

(3) Have ready a slide, a circular coverslip, and a slide map holder on which a "target" is ruled, in which the slide is placed. Also have ready a jar of fragmented paraffin, a jar with solid glycerin jelly, and an alcohol lamp.

(4) Take a needle holder equipped with a fine needle and use the point to cut out a very tiny piece of glycerin jelly (about 200 μm across) and get this firmly onto the very tip of the needle. While watching the process under the 10× objective, bring the needle with the tiny piece of jelly into the field and delicately and lightly touch it to the specimen and remove the needle from the field (check to be sure the grain is gone). Put the tiny piece of glycerin jelly on the slide in the target area.

(5) With a knife point arrange a small quantity of fragmented paraffin (melting point about 50°C) around the glycerin jelly blob. Heat slide gently over an alcohol flame until jelly and paraffin melt. They will appear to have run together. Hold slide level. Put slide on table and carefully lower coverslip. When paraffin solidifies, the glycerin jelly with enclosed specimen will appear as a clear window in the "ropey", whitish paraffin. If the slide is properly made the specimen will be almost precisely in the target area (so that all single grains are easily found using the microscope), and the paraffin will flow only to the edge of the coverslip. (If it flows beyond it is easy to clean up with a little xylene on a tissue.) The "window" should be tiny, but if it is too small, there is a tendency for the specimen to be enmeshed in the fibrous paraffin and therefore hard to see.

Capillary-tube methods Many palynologists favor a capillary-tube technique for single-grain mounts. The simple version Faegri taught me, which I still use to good advantage, is as follows:

(1) Draw out a supply of capillary tubes for the purpose. I use 3 mm tubing in pieces about 15–20 cm long. If one starts with a piece of tubing about 20 cm long and heats the center to melting and pulls carefully, one almost always gets two useful tubes that may carefully be broken apart with a triangular file.

(2) Spread out residue in 50/50 glycerin–water (or water alone, if water is added occasionally; or glycerin alone, but it is quite viscous). With needles, clear away debris and unwanted palynomorphs from around the specimen desired.

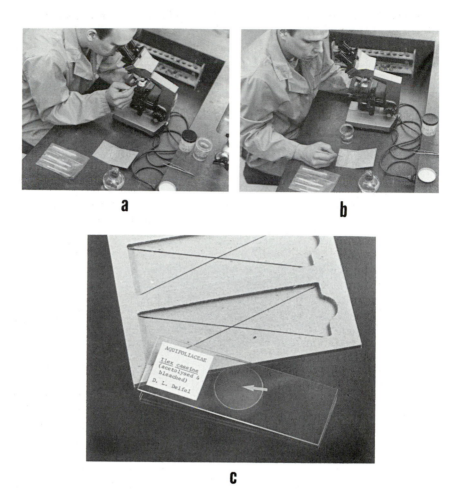

Figure A.8 Single-grain mounts by the glycerin jelly–needle method. (a) Under the microscope at low power, needles are used to clean debris away from the vicinity of the desired palynomorph, which is in a glycerin–water smear on a slide. (b) The palynomorph is picked up with a tiny chunk of solid glycerin jelly on the point of a needle. The block of glycerin jelly is put on a microslide, using the center of an "×" on a slide mailer as target. Pieces of paraffin wax are put around the glycerin jelly, and the wax and jelly are carefully melted over an alcohol lamp. (c) The wax hardens in about 30 s, leaving the glycerin jelly containing the palynomorph in the "window" (arrow).

(3) Use 10× objective (about 100× magnification with a stereo dissecting microscope is also good) to observe the capillary as it is brought to the vicinity of the desired specimen. Carefully touch the tip of the capillary tube to the liquid around the specimen. The specimen and some liquid will rush up into the tube, which must be quickly lifted to prevent too much liquid from entering.

(4) Blow the droplet of material from the capillary tube onto a small piece of glycerin jelly which can be arranged on a target area per the previous method, or blow the specimen or droplet onto a drop of glycerin on a slide or coverslip and cover, if mobility of the specimen is desired.

Variants on capillary-tube technique

(1) Pressure-tube technique: Instead of using capillary tubes alone, it is more elegant to use capillary tubes attached to a rubber or rubbery plastic (such as Tygon) tube. If the other end of the tube is attached to a hypodermic-type syringe, some glycerin may be sucked into the capillary and a negative pressure carefully controlled in the opening of the capillary, so that only a small amount of specimen plus liquid rushes into the tube when it is touched to the area of the specimen. The specimens may be ejected by depressing the syringe plunger. Also this control of the pressure makes it possible to put the capillary tube point into the specimen liquid without drawing up the liquid until desired. Another advantage for some purposes is that it is easy to pick up multiple (say 10) specimens to make a mount displaying specimens in different orientations. Evitt (1984) has described this sort of device in detail.

(2) Coverglass-sandwich technique (for S.E.M., etc.): If the specimen is put onto a coverglass in glycerin, and covered with another coverglass, the "sandwich" can be temporarily fastened to a slide with small pieces of transparent tape and the specimen photographed from both sides. The sandwich can then be removed from the slide, carefully pulled apart (inserting a razor blade is the best way) and the coverslips placed face up on a microslide. Find the specimen. Discard the empty coverslip. While watching the procedure under the scope, flush the specimen with water, or better with t-butyl alcohol, several times, wiping up the flush liquid, away from the specimen. This removes the glycerin, to allow proper gold coating. The coverglass may now be mounted on an S.E.M. stub with appropriate adhesives, gold-coated, and studied by S.E.M. The specimen may be turned over by squirting alcohol on it from a small hypodermic needle and maneuvering the grain with a small hair on a stick (see Artüz & Traverse 1980). In this manner S.E.M. pictures can also be made of both sides of the specimen, to match the light pictures (see Fig. A.9). (It is also possible to recover the specimen after S.E.M., remove the gold with aqua regia, and make thin sections for T.E.M. (see Walker & Walker 1982).)

Leffingwell & Hodgkin (1971) have published thorough descriptions of various other techniques for manipulation of palynomorphs in preparation for S.E.M. Leffingwell's laboratory used a micromanipulator attached to a microscope objective, for "picking" the fossils. Leffingwell also advocates polyurethane adhesive to stick the palynomorph to the S.E.M. mount surface. In the Penn State laboratory we have usually used no adhesives at all and find that only very rarely do the fossils come off the coverslip, which we mount on S.E.M. stubs. We store the preparation (stub plus coverslip) in covered plastic boxes, to assure that air currents are not a problem.

Manipulation and processing of single specimens Sometimes it is desirable to treat chemically a single specimen, in order to isolate the

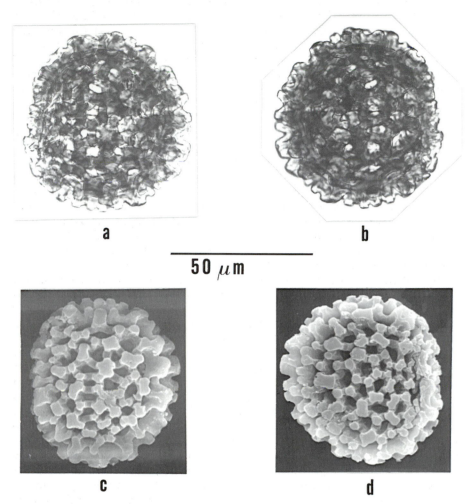

Figure A.9 S.E.M. micrograph and light microscope photomicrographs of a single specimen of *Reticulatisporites karadenizensis* Artüz, Carboniferous, Turkey. The specimen was picked with a capillary tube from a glycerin–water preparation, mounted in a coverglass "sandwich", and photographed "front" (a) and "back" (b). The coverglasses were then separated, the water evaporated and glycerin removed by repeated washings with water and t-butyl alcohol. The coverslip with specimen was then mounted on an S.E.M. stub and gold-coated. S.E.M. micrographs were made from both sides of the specimen. Ethyl alcohol applied by hypodermic needle was used to tip the spore over after the first S.E.M. micrograph, and gold coating was repeated. Note that the light pictures are upside-down mirror images of the S.E.M. pictures. Therefore individual sculptural elements are reversed as to up and down from (a) to (c) and from (b) to (d). The negatives of one set could be printed upside down to achieve exact correspondence.

procedure from other influences and to observe the results carefully. Johnson (1985), for example, did this in a study in our laboratory of the effect of oxidation on Silurian palynomorphs. Evitt (1984) has described a method for acetolyzing a single specimen, involving manipulation of the specimen in a syringe-type capillary tube (see above) and treating the specimen in the depressions of a glass cavity slide. The specimen is passed through two changes of glacial acetic acid, then heated in acetolysis mix, and finally washed, transferring the specimen with a capillary tube each time.

Storage of bulk palynomorph residues   I have always stored my residues in glycerin jelly. As noted elsewhere, glycerin jelly slides have a maximum life of about 30 years, much less in many cases, and the reason is both autoxidation of specimens (probably catalyzed by the glass of coverslip and slide) and dehydration and hardening of the glycerin jelly. Long after the slides have disintegrated, however, the original glycerin jelly residues can be remelted and new slides made. Chapman (1985) and others have reported glycerin jelly residues becoming so dehydrated that they cannot be reclaimed. All of mine, even those now almost 40 years old, can either still be used or are (so far) reclaimable by cooking in hydrochloric acid. However, clearly a better storage method should be found. Chapman recommends glycerin with phenol (to prevent fungal and bacterial growth), whereas Phipps & Playford (1984) recommend a 50/50 mixture of glycerin and 3% Cu_2SO_4 and a small amount of thimerosal ($C_9H_9HgNaO_2S$), a crystalline substance used in antiseptics such as "Merthiolate". Slides are made from these glycerin residues by addition of glycerin jelly. Other laboratories use water plus phenol or alcohol for storage. With either glycerin or water storage, inspection of the storage vials at regular intervals for loss of liquid is essential, as evaporation occurs over time even with very tight vials. If final mounting is to be with a resin substance, residues should be stored in alcohol–water or in the resin's solvent, so that the storage liquid can either be evaporated on a slide or coverslip prior to adding the drop of mountant, or be mixed easily with the resin (see "double mounting"). Some laboratories (see Felix & Burbridge 1985) have reported very satisfactory results from dry storage for years of macerated residues, after using the procedures outlined here for maceration of rocks. Felix & Burbridge advocate careful washing of the residue before dry storage, and re-wetting of the residue by treatment with KOH solution and washing before subsequent mounting.

Location of palynomorphs, photomicrography, and related matters

Location   Most palynologists make permanent (really, semipermanent) strew slides of macerations, e.g., in glycerin jelly, though some favor temporary glycerin or silicone oil mounts because the palynomorphs can be moved by tapping the coverglass, e.g., with a lead pencil. Actually, this can also be done with glycerin jelly slides by applying a very warm instrument to the coverglass in the near-vicinity of the palynomorph. Evitt (personal communication) uses a nail mounted on a wooden handle. The nail is connected to an electric source which makes it very warm (not hot), in the manner of a soldering iron.

APPENDIX: LABORATORY TECHNIQUES

The finding (refinding) of particular palynomorphs on a strew slide is a recurring problem. As long as one works with the same microscope and nobody has changed the settings, the mechanical stage vernier-coordinated readings work fine. Also, if other microscopes of the same manufacture are used, in the same laboratory or elsewhere, conversions can readily be calculated by carrying with one a slide on which a reference point is marked, for which mark the location readings are known on the original microscope. However, some brands of microscopes have mechanical stages on which the numbers run in opposite directions, making direct conversion difficult. In this case, it is possible to find a given palynomorph by marking the slide with a reference point, from which distances are given in millimeters and tenths of a millimeter above or below, to right or left. As mechanical stages all read in millimeters and tenths of a millimeter, palynomorphs can be located in this manner, but it is often difficult. Another approach is to use the England Finder (see Fig. A.10), a microslide on which a grid with reference numbers and letters appears. After a palynomorph is located (record mechanical stage location, in case the stage is accidentally moved), the slide is carefully removed, and the England Finder placed on the stage. A reading is taken from the England Finder and recorded. If the England Finder is then moved to another microscope, and the desired reference point found, the specimen slide can then be put on the microscope, and the fossil located. (In my experience, a small correction factor usually must be used with the England Finder if the second microscope's slide-holding device works differently from that of the first microscope or if the specimen slide is of slightly different dimensions.) In our laboratory, when moving between Leitz microscopes in the research microscopy room to Olympus microscopes in the teaching laboratory, we find the England Finder to be an ideal locating device.

Various techniques can be used to mark the surface of the coverslip with pen or brush, to help relocate palynomorphs. The coverslip can be marked with a circle of ink, and this ink protected with a little colorless fingernail polish. In our laboratory the favorite technique is to cut a tiny pointer from a gummed label, pick this up with a pair of iris forceps, moisten with the tongue and fasten down to the coverslip with the forceps points, while observing under low power. While still moist, the tiny pointer can be moved with a fine needle and finally labeled with a fine pen. (Devices are available that can be screwed onto the nosepiece of the microscope in place of one objective, and which have an inking device to make a small "o" around a desired palynomorph. In my experience these devices are a nuisance, hard to keep in adjustment and properly inked, and may accidentally break a coverslip.)

(Note: Much ink has been wasted in the paleopalynological literature by informing the reader what the mechanical stage coordinates are for illustrated or described specimens. Naturally, these readings are only good on the original microscope, so one would have to visit the subject laboratory to be able to use the readings. Furthermore, it is a simple matter to reset the stage of any microscope with a screwdriver, so that none of the readings would agree with those previously recorded. The best ways to go, say for type specimens of new taxa, are to use the England Finder, or mark the specimens with ink, protected by varnish, on the coverslip. Gummed label pointers as described

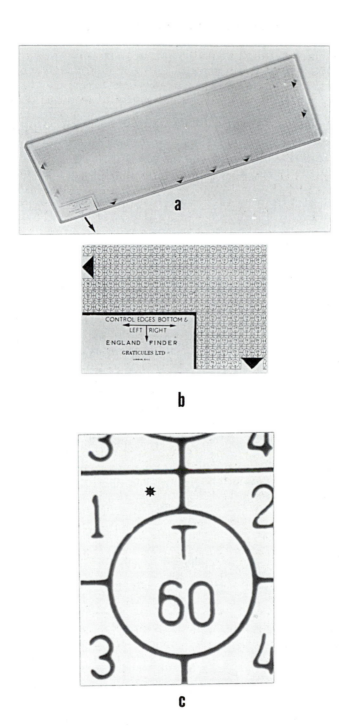

Figure A.10 England Finder (E.F.), the most convenient method for stating locations of specimens on microslides. (a) The England Finder is a glass 3 inch × 1 inch slide on

above would also be good. Measurements up or down, to right or left, from a reference point such as a scratched "×", in millimeters and tenths of a millimeter, also work well. These measurements can be made easily with the microscope's mechanical stage, which is always so graduated.)

Photomicrography   For routine purposes, I favor 35 mm photomicrography with an automatic camera or with an attachment camera. One problem with automatic cameras in comparison with simple attachment cameras is cleanliness; some authors even publish photos that show the same dust flecks on every picture! With automatic cameras it is difficult to cope with this because there are prism surfaces inside the camera that are in the plane of focus and cannot be cleaned by the user. With attachment cameras all surfaces that are in focus are available for operator cleaning. A small piece of fine chamois attached to a small stick is good for cleaning the surface of the offending prism in the attachment camera.

Magnification   Magnifications of 1,000×, 750×, 500× or 250× should be standard for most paleopalynological photomicrographs (100× or 150× for megaspores and other large palynomorphs). One reason for use of 1,000× is that a preponderance of palynomorphs are in the 30 μm range, and this produces a picture 30 mm across, which is about right as a compromise between the largest possible picture and the usual space restrictions. Another great advantage of 1,000× is that the size of the original fossil in micrometers can be read by measuring the photograph in millimeters. Within one paper an effort should be made to keep the number of different magnifications to a minimum. A bar should appear on plates of photomicrographs showing the size of specimens in micrometers, because enlargement or reduction in printing will not affect the relationship shown.

It is easy to calculate the magnification from the measured size given in descriptions. The ratio between the photo in millimeters and the measurement in micrometers × 1,000 is the magnification. With a millimeter scale one can

---

which a grid of letters and numbers is presented. The pattern of letters and numbers is upside down and backwards, to compensate for the fact that most light, compound microscopes present their images that way, and the E.F. images therefore appear in correct orientation under the scope, as shown in (c). A palynomorph is located and carefully centered, and the E.F. is put on the scope in the same position (if the stage moves easily, the mechanical stage coordinates must be used to control position). The reading of the center of the field is then recorded, e.g., "upper right of T-60-1" (see asterisk). Reversing the procedure locates the palynomorph on another microscope, providing that the specimen slide is also a 3 inch × 1 inch slide. When slide size differences or (very rarely) the design of the slide holders on the two scopes is radically different, conversions can always be worked out. The corners of England Finders are easily damaged by the spring clips of slide holders. If an England Finder is not available, a specimen can be located by giving its location with reference to a reference point such as a scratched "×" on the slide, e.g., "up 1.6 mm and to the left 2.9 mm", and this location can easily be found with any mechanical stage, as the stages are always calibrated in millimeters and tenths thereof.

then read off the fossil's size in micrometers (μm) if the magnification is known:

at 500×, 1 mm on photo =	2.00 μm on specimen
600×	1.67
750×	1.33
1,000×	1.00
1,200×	0.83

Unfortunately, application of this calculation to published figures frequently shows that the magnification is not as stated!

Microscopic techniques and photomicrography This is not the place for detailed instruction in the use of the microscope, though years of experience in teaching palynology have convinced me that many students have problems in this area. First of all, the student needs to learn how to use the condenser. Close down the field diaphragm as far as possible and then close down the iris diaphragm too. The leaves of the field diaphragm should appear as a clear silhouette. If this circle is not clear, raise or lower the condenser until it is. Then center this circle, using the centering knobs. Then open the field diaphragm until the leaves exceed the edge of the field, and open the iris diaphragm to achieve an amount of light that produces neither too much contrast nor too little. If the condenser cannot be centered, the student must still learn to use it at the right level for the various objectives, and to use the iris diaphragm efficiently. A blue filter is the most pleasing and restful for microscopy, but green filters produce the best black-and-white photo-micrographs of most fossil palynomorphs. Color films usually require no filter at all. However, sets of filters to insert in the light path, in order to alter the color if necessary, are available in camera shops. Do *not* leave the ordinary blue or green filter in place for color photomicrography.

Most palynomorphs are more or less three dimensional, though most are more or less flattened. Therefore the student must learn to interpret the three-dimensional structure by focusing up and down. This is apparently very difficult for some people to accomplish. Especially difficult is the interpretation of sculpture. Many students have problems distinguishing between positive and negative sculpture. As pointed out by Erdtman, this is possible by focusing up and down on the surface (LO analysis). Holes in the surface, for example, appear dark at high focus and lighter at lower focus. I have found it useful to instruct students to examine the outer edge of the grain in mid-focus to check their observations (edge analysis). Spines or verrucae will stick out! If there is time, making clay models of a form being studied is very helpful. I have found that beginning the instruction using a television camera in a demonstration microscope with a 12 inch black-and-white monitor is very helpful.

Interference contrast Ordinary transmitted-light microscopy is bright field. Another of the many varieties of light microscopy is interference contrast or phase contrast. Special condensers and objectives make the examined specimen stand out against the darkened background almost as if in S.E.M. The effect is especially helpful with specimens that are thin and/or colorless, such as some thin-walled pollen and various sorts of cysts. No special preparation of the specimens is required. Interference contrast photomicrographs have the advantage that, while they show surface features in S.E.M.-like contrast, internal features are shown also, which is not the case with S.E.M.

Fluorescence microscopy   This form of microscopy requires special light and filter equipment that can be used, however, with a regular light microscope. Specimens are illuminated with intense blue or ultraviolet light (either transmitted or incident) and fluoresce in different colors: that is, specimens on U.V. irradiation emit light of various intensity and wavelength. For example, different states of carbonization level yield different colors on fluorescing. The technique therefore has applications in study of coalification level (thermal alteration) and is discussed also in the section on that subject. Some specimens reveal structure in fluorescence microscopy that is not otherwise observable.

S.E.M. and T.E.M.   Scanning electron microscopy (S.E.M) provides a three-dimensional appearing view of the surface of palynomorphs. Indeed, as the depth of forms is great, S.E.M. pictures are ideal for stereopairs, and these have been published for megaspores (Higgs & Scott 1982, Plate 1 & Figs. 4&5) and other palynomorphs. Most people need a viewer to get the three-dimensional effect. Individual palynomorphs can be "picked" and mounted for S.E.M. work, as described in the coverglass-sandwich technique above, or a strew preparation can be gold-coated and studied, even counted, by S.E.M. However, because S.E.M. equipment is expensive, and should be in the care of a skilled technician, light microscopes will continue to be the overwhelmingly more common instruments of choice for most aspects of palynology. Even more specialized is transmission electron microscopy (T.E.M.). For a close look at internal structure of palynomorph walls, T.E.M. is indispensable. Preparing palynomorphs for this purpose involves special embedding, usually in various plastics, sometimes special staining of the sections, which are made with an ultramicrotome. Naturally, this technique is practised by only a few palynologists. Walker & Walker (1984) have pioneered the elegant study of single palynomorphs by light microscopy, S.E.M., and T.E.M. Such investigations are very important but are not likely to become routine.

Counting: how, and how many?

Nearly all paleopalynological projects sooner or later involve making a count of palynomorphs. As explained earlier, calculation of fossils per gram of sediment requires counting all or a fixed proportion (usually half) of all fossils on a slide. More commonly, a percentage is required, to provide an estimate of relative abundance. The percentage may be of all fossils or of a "pollen sum" that excludes some forms, or a ratio between the number of one kind of fossil and others may be calculated, e.g., in calculating steppe/forest index and other ratios. In any event, a strew slide is prepared, and the count is made by traversing the slide from side to side at a magnification sufficient to recognize the forms. The slide should not be too densely filled, or too many fossils will appear in each field for accurate counting. If only a percentage is sought, successive traverses should be counted far enough apart to prevent overlapping and double counting (especially of large) specimens. Where the whole slide is to be counted, double counting can be minimized by counting specimens that are only partly in the field at the top or bottom but not at both. In any event, large forms are more likely to be counted at the edge of the field (see Faegri 1951). A more serious problem really is that folding, squashing, and so forth make many specimens unrecognizable, and this is more true for some forms than for others. It is for this reason that many palynologists, even though they may make permanent preparations, as a rule prefer temporary, mobile preparations in glycerin or silicone oil for counting, because the grains can be manipulated for better inspection by tapping with a needle or pencil point.

The question of how many specimens must be counted for statistical reliability of later calculations has been given far too little attention. Early in the history of Quaternary pollen analysis, it was held that, given the numbers of taxa involved and their relative abundance, counts of about 200 were satisfactory for reliable percentage calculations (see Barkley 1934, Westenberg 1947). The statement that "200 specimens per slide were counted" as a satisfactory requirement has ever since been a bit of the palynological folklore, echoed in hundreds of papers, and for most analyses it is based on a solid mathematical foundation. Gordon Rittenhouse of Shell Development Company, who had a few years before made calculations relative to mineralogical counting, showed me in about 1957 how to calculate the number of specimens to count in order to get a reliable reading on the percentage of a certain species in a sample. He was convinced that many percentage-based palynological analyses being used for biostratigraphy were not statistically valid.

The Rittenhouse curve as applied to palynological counts is shown in Figure A.11. The gist of the chart is very simple: the number of specimens that must be counted to achieve a desired standard deviation depends on the abundance of the least abundant critical form. For example (not shown), in a total count of 200 (bottom axis), for a taxon that comprises about 60% of the total palynoflora (right-hand side of chart), a standard deviation of 6% (of 60%, i.e. 3.6%) is achieved (project to the left the intersection of the vertical 200 count line with the sloping 60% line). This means that two-thirds of all analyses of such a sample would fall in the range of 60±3.6%, a quite acceptable reliability; that is, a calculation of 58% for this form is meaningful. On the other hand, another taxon comprising about 2% of the palynoflora would have 50% standard deviation in a count of 200, meaning that two-thirds of the analyses of such a sample would be in the range 1–3%; so that many analyses would miss the form altogether and others would pick up four palynomorphs. Even a count of 1,000 would give a standard deviation of 25%, or two-thirds of analyses would be in the range 1.5–2.5%. To achieve a 6% standard deviation as for the form that comprises 50% of the palynoflora would require a count of several thousands (off the scale). The reliability in counts of 200 is satisfactory for abundant forms only, or if only a qualitative statement is required. To put it another way, a decrease of a taxon from 2.4% to 1.6% looks big, especially if plotted logarithmically, but actually cannot be

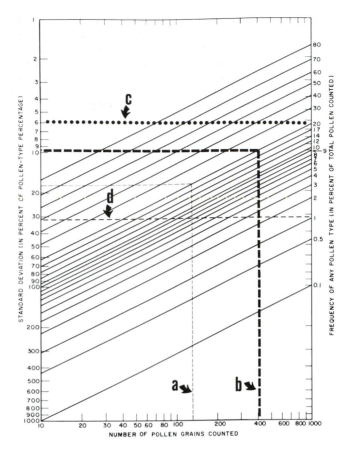

Figure A.11 Determination of standard deviation for pollen counts of various sizes. This graph provides a rough guide for estimating the size of count required for various levels of significance. We illustrate four examples. (a) If a certain pollen type represents roughly 20% of total pollen (based on preliminary counts), this percentage is shown by the sloping line labeled 20 on the right vertical axis. If 130 grains are counted, a vertical line from the base at 130 will intersect the sloping 20% line. Now project a horizontal line from the point of intersection to the left vertical axis to read the standard deviation of 17.4% (of 20%, i.e. 3.5%). This means that about two-thirds of all analyses may be expected to fall in the range of 20 ±3.5%. (b) If 400 grains of the same form are counted (about three times the number counted in (a)), using the same method of calculation shows a standard deviation of about 10%, meaning that two-thirds of all analyses would be expected to fall in the range of 20±2%. For most purposes, the extra labor is probably not justified. (c) Even with a count of 1,000, the standard deviation can only be reduced to 6%, demonstrating that for abundant forms large counts are not worth the extra work. (d) With a form present only at the 1% level, a count of 1,000 yields a standard deviation of about 30%, whereas a count of 100 yields 100%: two-thirds of samples would then be expected to be between 0 and 2%. It is clear that forms present in amounts less than 1% will yield very large standard deviations, even with counts of 1,000. In other words, counts of such forms are intrinsically unreliable from a statistical point of view. The number of specimens that need to be counted for statistical

490

depended on to mean anything at all, except presence/absence. Unfortunately, many published palynological data are in this category. Maher (1972) presents nomograms for calculation of the confidence limits of pollen analytical data, based on calculations by Mosimann (1965). The approach is considerably more sophisticated mathematically than that presented above, but the gist is the same, that large total counts are necessary to achieve meaningful data for uncommon forms. Birks & Birks (1980) and Faegri & Iversen (1975) present discussions of this and related matters.

---

significance depends on the abundance of the least abundant form of importance to the result. This graph was prepared by multiplying probable-error figures from Rittenhouse (1940) by 1/0.6745.

Glossary

The author wrote the spores/pollen palynological definitions for the 1972 *AGI glossary of geology*. The following definitions derive partly from that project. Other sources I have used in compiling definitions include: Beug (1961); preliminary versions of a glossary of fossil fungal spores morphology by Elsik *et al*.; Erdtman (1952); Evitt (1985); Evitt *et al*. (1977); Grebe (1971); the glossary of scolecodont terms from Jansonius & Craig (1971); Kremp (1965); Smith & Butterworth (1967); Walker & Doyle (1975); and Williams *et al*. (1978).

aboral pole   The end of a flask-shaped **chitinozoan** that includes the **chamber** of the **body** and the base. *See* **oral pole**.

absolute pollen frequency   The estimate of the actual amount of **pollen** deposited per unit area in a given length of time, achieved by correcting the amount of pollen per gram of sediment by factors based on rate of sedimentation. Abbrev. APF. The expression has been mostly replaced by **pollen influx** (= **pollen accumulation rate**).

acalymmate   Of the adhesion mode of members of a **polyad**, characterized by lack of direct contact between **exines** of adjacent grains. Ant. **calymmate**.

acanthomorphitic acritarch   An **acritarch** with clear differentiation of the **central body** and radially oriented processes.

accessory archeopyle suture   An **archeopyle suture** that consists of a short cleft in the wall adjacent to the principal **suture**, or that may be more fully developed on the **operculum** of the **dinoflagellate cyst**, dividing that structure into two or more separate pieces.

accessory spore   A **spore** present in a rock sample only in small quantities. Accessory spores include types with a very restricted range, and they have been used for correlation and for zoning (as of Coal Measures).

accumulation body   (Also called "eyespot", and other terms.) Of **dinoflagellate cysts**, a small clump of resistant matter within the cyst, usually interpreted as waste metabolic matter.

acetolysis   A chemical reaction in which acetic anhydride lyzes plant tissues, a sort of **maceration** in which peat or **palyniferous** material (e.g., angiosperm flowers) is heated in a mixture of nine parts acetic anhydride and one part concentrated sulfuric acid. The reaction breaks down cellulose and other organic compounds and thus concentrates **sporopollenin**.

acme zone   In **palynostratigraphy**, a zone which is of value for correlation based on dominance in percent of a **palynomorph** taxon or a related group of taxa, e.g., certain *Corollina* spp. for the lowest Jurassic.

acolpate   Of **pollen grains** without **furrows** (**colpi**). In practice, such pollen grains are sometimes difficult to distinguish from **alete spores**. Code P00. *See* **inaperturate**.

acritarch   A unicellular, or apparently unicellular, resistant-walled microscopic organic body of unknown or uncertain biologic relationship and characterized by varied **sculpture**, some being spiny and others smooth. Most acritarchs are of algal affinity, but the group is artificial. They range from Precambrian to Holocene, but are especially abundant in late Precambrian and early Paleozoic. The term was proposed by Evitt (1963, pp. 300–1) as "an informal, utilitarian, 'catch-all' category without status as a class, order, or other suprageneric unit" consisting of "small microfossils of unknown and probably varied biological affinities consisting of a central cavity enclosed by a wall of single or multiple layers and of chiefly organic composition". *See also* **hystrichosphaerid**.

actuopalynology   Study of extant **spores** and **pollen** and their distribution in atmosphere, hydrosphere and sediments, and related matters. *See* **pollen analysis** and **aerobiology**.

aerobiology   Study of organisms and parts of organisms in the atmosphere. **Spores** and **pollen** are a major concern of aerobiologists, especially in their role as pollutants and causes of **pollinosis**. *See* **iatropalynology**.

air sac   *See* **saccus**.

akinete   Modified, thick-walled vegetative green algal cell, e.g., in **Zygnema**. Some **acritarchs** are probably akinetes.

alete   Of a **spore** without a **laesura**. In practice, such spores are sometimes difficult to distinguish from **acolpate pollen**. Code S00. *See* **inaperturate**.

allochthonous   In **palynology**, **palynomorphs** that are produced at considerable distance from the site of their deposition. Ant. **autochthonous**.

amb   The contour or outline of a **pollen grain** (less commonly of a **spore**) as viewed from directly above one of the **poles**. Also called **equatorial limb**. (It is possible but uncommon to describe also a **polar limb** or **profile**, the outline of a grain from a "side view", at right angles to the equatorial limb. However amb is used almost exclusively for the outline from the pole.) *See also* **profile**.

ambitus   A less favored syn. of **amb**.

amphiesma   Of **dinoflagellates**, the cell covering in general.

ana-   A combining form used to indicate position on the **distal** surface of a **pollen grain** or **spore**, as anasulcate, having the **sulcus** on the distal surface. *See* **cata-**.

anemophily   A term for **pollination** by wind. Adj. anemophilous. *See* **entomophily** and **zoophily**.

annulus   A ring bordering a **pore** of a **pollen grain** and in which the **ektexine** is modified (usually thickened). *See* **margo** and **endannulus**.

antapex   Of **dinoflagellates**, the **area** at the posterior end of **theca** or **cyst**.

antapical series   The series of **plates** forming the germinal group anterior to the **postcingular series** in **dinoflagellate theca** or **cyst**. *See* **apical series**.

anteturma   One of the two major groupings in which **turmae** are classified in the **turmal system**: Sporites (for **spores**) and Pollenites (for **pollen**).

AP   Abbrev. for **arboreal pollen**.

aperture   Any of the various modifications in the exine of **spores** and **pollen** that can be a locus for exit of the contents; e.g., **laesura**, **colpus**, **pore**. *See also* **germinal aperture**.

APF   Abbrev. for **absolute pollen frequency**. *See* **pollen influx** and **pollen accumulation rate**.

apical archeopyle   An **archeopyle** formed in a **dinoflagellate cyst** by the loss of the entire **apical series** of **plates**. *See also* **haplotabular archeopyle** and **tetratabular archeopyle**.

apical area   Of an **embryophytic spore**, the **proximal** area associated with the **laesura**. *See* **contact area**.

apical papilla   A dot-like thickening of an interradial area of a **spore**. When present, there is generally one **apical papilla** per interradial area, hence three per spore.

apical prominence   In **megaspores**, mostly Paleozoic, variously constructed **proximal** projections formed by the intersection of the expanded contact areas.

apical series   The series of **plates** forming an apical cluster in the **epitheca** or **epicyst** of a **dinoflagellate**. *See* **antapical series**.

apocolpium   Area at **pole** of **pollen grain** delimited by a line joining the ends of the **colpi**. Syn. **polar area**.

apoporium   Area including the **pole** of **porate pollen grain**, delimited by the poleward limit of a line connecting the poleward limit of the pores. Cf. **mesoporium**.

aporate   Of **fungal spores** without **pores**.

appendage   An elongated **sculptural element**; a **process**.

arboreal (arborescent) pollen   The **pollen** of trees. Abbrev. AP. Syn. tree pollen. *See* **non-arboreal pollen**.

Archeophytic   An informal division of geologic time, before the regular appearance of robust-walled acritarchs (about 1.0×10^9 years ago = beginning of **Proterophytic**).

archeopyle   An opening in the wall of a **dinoflagellate cyst** by means of which the motile thecal stage emerges from the **cyst**. It is usually more or less polygonal in shape, and the **plate** (or plates) that drops out is the **operculum**. *See also* **apical archeopyle, cingular archeopyle, precingular archeopyle, combination archeopyle**, and **operculum**.

archeopyle suture   A line of dehiscence on the **dinoflagellate cyst** that more or less completely separates a part of the **cyst** wall to form an **operculum**. *See also* **accessory archeopyle suture**.

arcus   A band-like thickening in the **exine** of a **pollen grain** (as in *Alnus*), running from one **pore** apparatus to another.

armored   Of **dinoflagellates** (such as those of the order *Peridiniales*) possessing a cellulose envelope or cell wall that is subdivided into articulated **plates**. Ant. **unarmored**.

ascospore   A **fungal spore** produced in an ascus. Attachment scars usually lacking. May be chitinized and presumably occur as **palynomorphs**, but are not usually identifiable as such.

aseptate   Of **fungal spores**, lacking a **septum**.

aspidate   (Also spelled aspidote.) Having the **apertures** on dome-like protrusions (example *Betula*). Grains with aspidate external form are often **vestibulate** internally.

atectate   Of an outer **exine** lacking **columellae** and hence also lacking a **tectum**. But compare **intectate**.

atrium   A space between the external opening (**pore** or **exopore**) and a much larger internal opening (**endopore** or **os**) in the **endexine** of a **pollen grain** with a complex **porate structure**. The internal opening is so large that the endexine is missing in the endopore area. Atrium is usually reserved for the space immediately around the endopore, and **vestibulum** for lateral space between pore and os. (Thomson & Pflug (1953) define atrium and give *Myrica* as an example. *Betula*, by contrast, is **vestibulate**.) **Pl. atria**. *See* **pororate**.

attached operculum   The **operculum** of a **dinoflagellate cyst**, not completely surrounded by **archeopyle** sutures and hence remaining joined to the main part of the **cyst** where the suture is not developed. Syn. attached opercular piece. *See* **free operculum**.

auricula   One of the thickened "ears" of auriculate **spores**. Pl. auriculae. *See* **zone**.

auriculate   Of **zonate spores** having exine thickenings in the **equatorial** region that project like "ears", generally from the area of the ends of the radii of the **laesura**. *See* **valvate**.

auto-   In **dinoflagellate cysts**, single, such as autophragm (single wall).

autochthonous   In **palynology**, **palynomorphs** that are more or less locally produced in or near the site of deposition. Ant. **allochthonous**.

autocyst   Of **dinoflagellates**, **cysts** with a single wall (the **autophragm**).

autophragm   *See* **autocyst**.

azonate   Of **spores** without a **zone** or a similar (usually **equatorial**) extension.

baculate   Of **sculpture** of **pollen** and **spores** consisting of bacula (sing. **baculum**).

baculum   One of the tiny rods (not thickened or thinned at either end), varying widely in size and either isolated or clustered, that make up the **ektexine sculpture** of certain **pollen** or **spores**. Also used for baculate-like, internal structures of Normapolles **germinals**. Pl. bacula.

basal plate   In **scolecodonts**, a small to medium-sized right-hand **jaw** closely fitting into a posterior concavity of a simple jaw ("MId"). *See* **maxilla**.

basidiospore   A **fungal spore** produced by the basidium of a basidiomycete. Such **spores** with chitinous walls presumably occur as microfossils in some palynologic preparations but are usually not identifiable.

biform process   A **sculptural element** with two different thicknesses. Term is usually restricted to elements with a broader base and abruptly constricted, spiny tip.

bisaccate   Of **pollen** with two **sacci**. Usually occurs in conifers but also found in other gymnosperms, such as Caytoniales and seed ferns. Code Pv2. Syn. **bivesiculate** and **disaccate**.

bivesiculate   *See* **bisaccate**. Syn. **bisaccate** and **disaccate**.

bladder   Syn. of **saccus**, **vesicle**, and **wing**.

body   Of saccate pollen, the **corpus**; also loosely the **central body** of the saccate pollen of various **spores** and **pollen grains**. Of **chitinozoans**, the main, larger part of the unit, which lies below the **neck**.

B.P.   Before present, "present" conventionally taken to be 1950. Written as "yr B.P." or "years B.P."

Brevaxones   A group of mid-Cretaceous and younger dicot angiosperm **pollen** in which the **polar** axis is shorter than the **equatorial** diameter, representing an evolutionary advance over **Longaxones**, and including such forms as **Normapolles**.

brochate   *See* **heterobrochate** and **homobrochate**.

callose   A carbohydrate component of cell walls in certain plants; e.g., the amorphous cell wall substance that envelops the **pollen mother cell** during **pollen grain** development and acts as a barrier between **mother cells**, and subsequently between developing members of the **tetrad**, but that disappears as the **ektexine structure** is completed and impregnated with **sporopollenin**.

calymmate   Adhesion mode of members of a **polyad**, characterized by contact between all or portions of **exine** of adjacent grains. Ant. **acalymmate**.

camerate   As originally defined (Neves & Owens 1966) describes **spores** in which outer and inner **exine** are separated to various degrees by a chamber (= camera), but which lack infrareticulate **structure**. Includes **pseudosaccate**, which refers to spores with extensive separation of exine layers. In common usage, a syn. of **cavate**. Camerate (or cavate) is also used by some authors for spores with barely detectable or partial separation of wall layers.

capillate   A term for **sculpture** consisting of capillae, i.e., projections which are branched or jagged.

cappa   The thick-walled **proximal** side of the **corpus** of a **saccate pollen grain**. *See* **cappula**.

cappula   The **distal**, thin-walled region of the **corpus** of a **saccate pollen grain**. Unfortunately, the term is easy to confuse with **cappa**, the thick-walled **proximal** area.

cata-   Combining form indicating position on the **proximal** surface of a **pollen grain** or **spore**, as catasulcate, having the **sulcus** on the proximal surface, as for example in some Annonaceae (Walker 1971).

cavate   (a) Descriptive of **pollen** (or, less precisely, of **spores**) whose **exine** layers are separated by a cavea or cavum (= cavity). The separation may be rather slight or more extensive, or eventually producing a bladder-like protuberance approaching the **pseudosaccate** or **saccate** condition. (In practice, these three terms are difficult

495

to separate.) This term and **camerate** are also used of hollow **processes**. *See* **camerate**. (b) Of a **dinoflagellate cyst** with space or spaces of notable size between the **periphragm** and **endophragm** (as in *Deflandrea phosporitica*). *See* **chorate** and **proximate cysts**.

cella   For **fungal spores** and hyphae, one chamber within the structure. Pl. cellae.

Cenophytic   An informal division of geologic time, based on the abundant occurrence of angiosperms in the fossil record, therefore equals approximately Aptian–Albian to present. The next older such division is the **Mesophytic**.

central body   The main part of a **pollen grain** or **spore**; e.g., the **corpus** of a **saccate pollen grain**, as distinct from the **sacci**; or the central part of a **zonate** or **camerate spore**, exclusive of the **zona, pseudosaccus**, etc.; or the compact central part of a **dinoflagellate cyst** from which the projecting structures extend. Syn. **body**.

chagrenate   Smooth and translucent **sculpture** of **pollen** and **spores** (a variant of **psilate**).

chamber   Of a **chitinozoan**, the central cavity of the chitinozoid unit.

cheilocardioid   Of lipped, heart-shaped **spores**, as *Microbacilispora*. (The heart shape results from lateral compression.)

chevron   Describing the **laesura** of a **dilete spore**.

chiropterophily   A term for **pollination** by bats.

chitin   A long-chain nitrogen-containing polysaccharide, structurally similar to cellulose, having, however, the carbon–2 –OH groups replaced by acetylamine groups ($-NH-CO-CH_3$). The compound is resistant to acid attack and bio-degradation in a manner very similar to *sporopollenin*. The robust-walled **spores**, **fruit-bodies** and **hyphae** of some fungi consist of chitin, as do the walls of **microforaminifera** and **scolecodonts**. Compare **pseudochitin**.

chitinozoan   A pseudochitinous marine microfossil of the extinct group Chitinozoa, having uncertain affinity (generally assumed to represent animal remains), shaped in general like a flask, occurring individually or in chains. Stratigraphic range primarily from uppermost Cambrian to Devonian. Chitinozoans have thin, usually black, structureless, opaque walls, but they may be brown and translucent. Named by Eisenack (1931), who noted the resemblance of the walls to **chitin**.

chlamydospore   A thick-walled, non-deciduous **spore**, such as a unicellular **resting spore** in certain fungi, usually borne terminally on a **hypha** and rich in stored reserves. This **fungal spore** may have a chitinous wall and therefore occur as a microfossil in palynologic preparations.

chorate cyst   Encysted, unicellular alga with **processes**: especially a **dinoflagellate cyst** bearing little morphological resemblance to the motile **theca**. The ratio of the diameter of the main **body** to the total diameter of the **cyst** is 0.6 or less. Examples **marginate chorate cyst, membranate chorate cyst, pterate chorate cyst, trabeculate chorate cyst**. *See also* **proximochorate cyst** and **proximate cyst**.

cicatricose   Marked with scars; especially said of **sculpture** of **pollen** and **spores** consisting of more or less parallel **ridges** in the manner of a fingerprint.

cingular archeopyle   An **archeopyle** formed in a **dinoflagellate cyst** by breakage along and within the **cingulum** (= **girdle**).

cingulate   Having a **girdle**; especially said of a **spore** possessing a **cingulum**.

cingulizonate   Of **spores** with a **cingulum** to which an outer, thinner **equatorial** extension is appended.

cingulum   (a) An annular, more or less **equatorial** extension of a **spore** in which the wall is thicker than that of the main body of the spore. Pl. cingula. See **zone** and **crassitude**. (b) In a **dinoflagellate theca** or **cyst**, the **girdle**, a slightly depressed band of small **plates** separating **hypotract** from **epitract**.

clavate   Of **sculpture** consisting of clavae, i.e., rods with enlarged, club-like ends. *See* **pilate**.

clavus Of **scolecodonts**, a lateral rudder-shaped **plate** or ledge projecting from and more or less perpendicular to the inner face, located at the dorsal part of the **jaw**, in some types of MI. *See* **maxilla**.

-coel Of **dinoflagellate cysts**, it means cavity, as mesocoel, the middle cavity.

coenobium An (algal) colony with the number of cells fixed at its origin, as in *Pediastrum*.

collar Of **chitinozoans**, the lip-like edge, surrounding the **mouth**.

colpa A non-recommended synonym of **colpus**. Pl. colpae.

colpate Of **pollen grains** having more or less elongated, longitudinal **furrows** (**colpi**) in the **exine**.

colpi Pl. of **colpus**.

colpoid Erdtmanian term for a **colpus**-like **furrow** not situated either as **colpus**, **sulcus** or **sulculus**.

colporate Of **pollen grains** having **colpi** in which there is a **pore** or some other organized thinning of the **exine** (such as a **transverse furrow**), usually at the **equator**. *See* **colporoidate**.

colporoidate Of **pollen grains** having **colpi** in which there is no **pore** or other clearly recognizable thinning, but modifications of the **exine**, usually **equatorial**, are present: in other words, **colporate** but with a weakly developed pore.

colpus A longitudinal **furrow**- or groove-like modification in the **exine** of **pollen grains**, associated with germination (either enclosing a **germ pore** or serving directly as a place of emergence for the **pollen tube**) and often also important for **harmomegathic** swelling and shrinking. When the term is used strictly, a colpus must be meridional and will ordinarily cross the **equator**. In this sense the term is hence practically restricted to dicot angiosperms. In a looser sense, commonly used as synonymous with **sulcus** ("**monocolpate**" **pollen** are as a rule actually **monosulcate**).

colpus transversalis *See* **transverse furrow**.

columella One of the rodlets of **ektexine** that may branch and/or fuse distally to produce a **tectum** on **pollen grains** with complex **exine structure**. Pl. columellae.

columellate Possessing **columellae**; of **pollen grains** with a complex **ektexine structure** consisting of columellae.

combination archeopyle In **dinoflagellate cysts**, an **archeopyle** formed by the release of a part of the **cyst** wall that corresponds to **plates** of more than one **thecal plate series** (such as combining the plates of the **apical series** and of the **precingular series**).

commissure The groove more or less on the center line of a **laesura**, along which line an **embryophytic spore** usually germinates. It is essentially equivalent to **suture**.

compound operculum A **dinoflagellate cyst operculum** that is divided into two or more pieces that are completely separable from one another. *See* **simple operculum**.

conate Of **sculpture** of **pollen** and **spores** consisting of **coni**.

coni Pl. of **conus**.

conidiophore A structure that bears **conidia**; specifically a specialized, typically erect **hypha** that produces successive conidia in certain fungi.

conidiospore A **fungal spore** produced on a **conidiophore**. May have chitinous walls and therefore occur as a microfossil in palynologic preparations. Syn. **conidium**.

conidium An asexual **fungal spore** produced from the tip of a **conidiophore**; broadly, any asexual spore not borne within an enclosing structure, such as one not produced in a **sporangium**. Pl. conidia. Syn. **conidiospore**.

contact area One of the areas of the **proximal** side of a **spore** or **pollen grain** that contacted the other members of the **tetrad**. Contact areas are seldom visible in mature pollen grains but are frequently apparent in spores. **Trilete spores** have three contact areas; **monolete spores** have two contact areas.

conus   One of the small pointed projections making up the (**echinate**) **sculpture** of certain **pollen** and **spores**, being more or less rounded at the base and less than twice as high as the basal diameter. Pl. **coni**.

copula   Of **chitinozoans**, the stalk–like basal part of some species; it may terminate in a foot-like appendage. Both **copula** and **mucron** seemingly have to do with linkage of the chitinozoal units into chains.

corona   A more or less **equatorial** extension of a **spore**, similar in disposition to a **zone** but divided in fringe-like fashion, such as in *Reinschospora*. Adj. **coronate**.

corpus   That part of a **saccate pollen grain** or pseudosaccate **spore** exclusive of the **sacci** or **pseudosaccus**. *See* **body**.

costa   One of the rib-like thickenings in the **endexine** of **pollen**, associated with colpi. Costae are most often meridional and border colpi in pairs, but they may be transverse in association with **transverse furrows**. They are best seen in **equatorial** view. Example *Melia* (Fig. 5.7ak). Adj. costal.

crassinexinous   Having thick **nexine**. The usual limit is nexine at least twice as thick as **sexine**.

crassitude   A more or less local, usually **equatorial exine** thickening of a **spore**. *See* **cingulum** and **zone**.

crista   One of the elevations making up the **sculpture** of certain **pollen** and **spores**, characterized by long, curved bases (sometimes irregularly fused) and variously bumpy apices. Pl. cristae.

cristate   Crested, or having a crest; especially said of **sculpture** of **pollen** and **spores** consisting of **cristae**.

cryptarch   According to proposal of Diver & Peat (1979), those of what others call **acritarchs**, which lack **spines**, **plates** or other features suggesting algal affinity. Cryptarchs thus would include sphaeromorph acritarchs, organic filaments, etc., whether or not sporopolleninous.

cryptopore   Term used by some for **distal tenuitas** of *Corollina* (= *Classopollis*), and similar **pollen grains**. Some call the same structure a **pseudopore**.

cryptospore   Term proposed by Richardson *et al.* (1984) for mid-Paleozoic spore-like bodies. **Monad** is a more general term for such entities whether spore-like or not.

curvatura   A visible line of some (mostly mid-Paleozoic) **trilete spores**, connecting the extremities of the ends of the **laesura** and thus outlining the **contact areas**; a "curvatura perfecta" has three lines complete all around the spore's **proximal** face; a "curvatura imperfecta" has fork-like projections from the radial ends of the laesura, not joining with their neighbors (also called "reduced curvatura"). Pl. curvaturae.

cusp   Of **scolecodonts**, the largest tooth in a series of **denticles**, especially in **basal plate**, MII, MIII, MIV.

cyst   A microscopic resting body with a resistant wall, formed in algae by the breaking up of parts of filaments or by the enclosing of a cell or cell group and investment by a sheath or envelope. **Dinoflagellate cysts** form within **thecae**, as part of the normal life cycle; they are often sporopolleninous and hence occur abundantly as fossils. *See also* **statospore** and **hystrichosphaerid**.

-cyst   Of **dinoflagellate cysts**, combining form meaning **body**, as **endocyst**, the inner body.

decussate   Of **tetrads** of **pollen grains**, in which two pairs of elongated grains lie across one another, the pairs at right angles to each other.

demicolpate   Having **apertures** resembling what would happen if ordinary **colpi** were interrupted in the **equatorial** area to make sets of two, in-line colpi, not crossing the **equator**. A "3-demicolpate" form would therefore actually have six colpi-like apertures.

demicolporate Like **demicolpate**, but the demicolpi have **pores** or pore-like interruptions. A 3-demicolporate form can therefore have six pores (fossil example *Sindorapollis*).

densospore A **trilete spore**, chiefly Paleozoic, with a pronounced **cingulum** which has a tendency to be "doubled", with a thicker part toward the center of the **spore**, and a thinner more external part; e.g., the genus *Densosporites* and other similar genera such as *Cristatisporites*.

dentary Of **scolecodonts**, a series of **denticles** along the inner dorsal margin.

denticles Of **scolecodonts**, the individual elements or teeth on the dorsal margin of the **jaw**.

desmid Single-celled green algae in which the cell consists of two semicells. The **zygospores** are sometimes resistant-walled (probably sporopolleninous) and occur rarely as **palynomorphs**.

diad Alternate spelling for **dyad**.

diatom Usually single-celled algae of class Bacillariophyceae, the siliceous frustules of which occur in paleopalynological preparations if they have not been digested by HF or have been inadequately so digested.

dicolpate Of **pollen grains** having two **colpi**. Code Pb0.

dicolporate Of **pollen grains** having two **colpi**, with at least one colpus provided with a **pore** or **transverse furrow**. Dicolporate pollen are rare. Code Pb2.

dictyospore A multicellular **fungal spore** that has both cross septa and longitudinal walls. When chitinous, such **spores** may occur as microfossils in palynologic preparations.

dilete Of a **laesura** with only two **radii**. This morphological type is rare. *See* **chevron**.

dinocyst Contraction of **dinoflagellate cyst**, more or less synonymous for fossil **dinoflagellate**. The term is not popular with some dinoflagellate experts.

dinoflagellate A one-celled, microscopic, chiefly marine, usually solitary flagellated protist organism with resemblances to both animals (motility, ingestion of food) and plants (photosynthesis), characterized by one **transverse flagellum** encircling the **body** and usually lodged in the **cingulum** or **girdle**, and one posterior flagellum extending out from a similar median groove, the **sulcus**. Certain dinoflagellates have a **theca** (**test**; usually not resistant to decay) and that may be simple and smooth or variously sculptured and divided into characteristic **plates** and grooves; some produce a resting stage or cyst with a resistant, sometimes complex organic wall (e.g., spiny) and may differ markedly from the theca of the same species. **Dinoflagellate cysts** exist abundantly as fossils, and have a range primarily Triassic to present. Dinoflagellate cysts are known from the Paleozoic, but are important **palynomorphs** only from Jurassic to present. Dinoflagellates inhabit all water types and are capable of extensive diurnal vertical migrations in response to light. They constitute a significant element in marine plankton, including certain brilliantly luminescent forms and those that cause red tide. *See also* **hystrichosphaerid**.

diploxylonoid Of **bisaccate pollen**, in which the outline of the **sacci** in **distal–proximal** view is discontinuous with the **body** outline so that the grain appears to consist of three distinct, more or less oval figures. *See* **haploxylonoid**. (Terms come from *Diploxylon* and *Haploxylon*, sections of the genus *Pinus*.) The terms have unfortunately been used in different senses from this definition and are best avoided.

diporate Of **pollen grains** having two **pores**. Code P02.

disaccate *See* **bisaccate**.

distal The parts of **pollen grains** or **spores** away from the center of the original **tetrad**; e.g., of the side of a **monosulcate** pollen **grain** upon which the **sulcus** is borne, or of the side of a spore opposite the **laesura**. Ant. **proximal**.

distal pole The center of the **distal** surface of a **spore** or **pollen grain**, thus the very center of the **sulcus** of a **monosulcate grain**.

dyad An uncommon grouping in which mature **pollen grains**, **spores** or **cysts** occur as fused pairs. Code Pdy. *See* **tetrad** and **polyad**.

echinate Of **sculpture** consisting of **spines** (echinae).

ectexine Var. of **ektexine**.

ecto- Of **dinoflagellate cysts**, more or less extreme outer, as in **ectophragm**, extreme outer wall.

ectoexine Var. of **sexine**.

ectonexine Outer part of **nexine**.

ectophragm A thin membrane lying between **distal** ends of **processes** on a **dinoflagellate cyst**. *See* **endophragm** and **periphragm**.

ectosexine Outer part of **sexine**.

ektexine In Faegri & Iversen's (1975) scheme, the outer layer of the two layers of the **exine** of **spores** and **pollen**, normally more densely or deeply staining than the **endexine**, and characterized by richly detailed external **sculpture** and often by complex internal **structure** of granules, **columellae**, and other elements. In contrast to Erdtman's "geographically" based **sexine**, the ektexine must be set off from the underneath (endexine) layer by demonstrable staining difference. *See* **ectexine, ectoexine** and **sexine**.

elater The ribbon–like, usually hygroscopic, filamentous appendage of certain **spores** (as of *Equisetum*), consisting of more or less coiled strips of **exine**, perhaps homologous with **perine**. It aids in spore dispersal.

embryophytic Of plants producing a 2N (diploid) embryo as part of a 1N–2N life cycle – usually restricted to bryophytes and tracheophytes (pteridophytes, gymnosperms, angiosperms). Noun embryophyte. Essentially equivalent to archegoniate.

endannulus An **annulus** formed by the **endexine** of a **pollen grain**. (Thomson & Pflug (1953) illustrate it as characterizing some **Normapolles** pollen.)

endexine The inner, usually homogeneous layer of the two layers of the **exine** of **spores** and **pollen** in the Faegri & Iversen (1975) scheme, normally less deeply staining than the **ektexine**. *See* **intexine** and **nexine**.

endo- In **dinoflagellate cysts**, means inner, as **endophragm**, inner wall.

endoblast Of two-walled **dinoflagellate cysts**, the **endophragm** and its contents.

endocoel The cavity formed by the **endophragm** in a **cavate dinoflagellate cyst**. *See* **pericoel**.

endocyst Of more or less **cavate dinoflagellate cysts**, the separate inner body, of which the **endophragm** comprises the wall.

endogerminal Especially used for **Normapolles**. Essentially equivalent to **os**.

endonexine Inner layer of **nexine**.

endophragm The wall of the inner **body** of a **cavate dinoflagellate cyst**. *See* **endoblast**.

endopore The internal opening in the **endexine** of a **pollen grain** with a complex **porate** (= **pororate**) structure. Syn. **os**. *See* **exopore**.

endosexine Inner layer of the **sexine**.

endospore (a) Syn. of **intine**. The term is mostly applied to the **sporoderm** of **spores**, rather than to **pollen**. Syn. endosporium. *See* **exospore**. (b) Some **palynologists** have used endospore for the inner **exine** body of, e.g., **camerate spores**.

entomophily A term for **pollination** by insects. Adj. entomophilous. *See* **anemophily**.

epicyst The part of a **dinoflagellate cyst** anterior to the **cingulum** (**girdle**) region. *See* **hypocyst** and **epitheca**.

episome The anterior part of the cell **body** above the **cingulum** (**girdle**) of an **unarmored dinoflagellate**. *See* **hyposome**.

epitheca In a motile **dinoflagellate**, the portion of the **theca** anterior to the **cingulum**. *See* **epicyst**.

epitract Syn. of **epicyst**.

epityche An **excystment** (emergence) **aperture** in the **acritarch** genus *Veryhachium*. Originating as an arched slit between two **processes**, rupture allows the folding back of a relatively large flap.

equator An imaginary line connecting points midway between the **poles** of a **spore** or **pollen grain**.

equatorial extension Any equatorial extension of the **spore wall**, a less encumbered term for the general sense of **zone**.

equatorial limb A less satisfactory synonym for the term sometimes applied to the **amb** of a **pollen grain** or **spore**, as seen in **polar view**. *See* **limb** and **amb**.

equatorial view The lateral view of a **spore** or **pollen grain** from an aspect more or less midway between the **poles** and perpendicular to the **polar** axis.

excystment Having to do with the emergence of the contents of a **cyst**. The excystment site of a **dinoflagellate cyst** is the **archeopyle**.

exine The outer, very resistant layer of the two major layers forming the wall (**sporoderm**) of **spores** and **pollen**, consisting principally of **sporopollenin**, and situated immediately outside the **intine**. It is divided into two layers either on the basis of staining characteristics (**ektexine** and **endexine**), or somewhat arbitrarily on the basis of being related to **sculpture** (**sexine**) or not so related (**nexine**). Syn. **extine** and **exospore**. *See also* **perisporium**.

exinous Consisting of **exine**.

exoexine A less satisfactory synonym of **ektexine**.

exogerminal Especially used for **Normapolles**. Essentially equivalent to **pore**.

exopore The external opening in the **ektexine** of a **pollen grain** with a complex **porate structure**. *See* **endopore**.

exospore A synonym of **exine**, mostly applied to the non-perinous portion of the **sporoderm** of **spores**. Syn. exosporium.

extine Var. of **exine**. The term is not in good or current usage in **palynology**.

exuviation The removal of the **theca** of a **dinoflagellate**, either **plate** by plate, or as small groups of plates.

falx Of **scolecodonts**, a sickle-shaped extension of the anterior part of the **jaw**, often forming a hook.

fang Of **scolecodonts**, a poorly developed falcal hook.

flagellar pore One of the **pores** in a **dinoflagellate** for extrusion of flagella, usually located at the anterior or the posterior junction of **girdle** (**cingulum**) and **sulcus**.

Flandrian Western European term for what is variously called elsewhere "post-glacial", "recent", "Holocene", i.e., approximately the last 10,000 years, the present interglacial.

flange An **equatorial** extension of a **spore**; less precisely defined than **zone** or **cingulum**.

foot layer A downward extension of **ektexine**, partly surrounded by **endexine**.

forb A non-cultivated, dicotyledonous, herbaceous plant; a herb other than grass; a broad-leaved weed. The term appears in some palynologic literature dealing with Quaternary sediments.

fossulate Of **sculpture** of **pollen** and **spores** consisting of grooves that anastomose.
foveolate Means **pitted**; e.g., of **sculpture** of **pollen** and **spores** consisting of pits in the **ektexine**. *See* **scrobiculate**.
free operculum Part of a **dinoflagellate cyst** that is completely surrounded by **archeopyle** sutures, with no unsutured connection to the rest of the **cyst**. Syn. free opercular piece. *See* **attached operculum**.
fruit body Of fungi, a fructification, i.e., a spore-bearing organ. Chitinous fruit bodies occur as **palynomorphs**. Fossil fruit bodies characteristically lack **spores**. Microthyriaceous fruit bodies are more or less flattened and radially symmetrical. Syn. fruiting body.
fungal spore A **spore** of the usually multicellular, non-vascular, heterotrophic organisms belonging to kingdom Fungi. Such spores include a wide variety of types, from simple unicellular to multicellular sclerotia; they have a range of Precambrian to Holocene. Those fungal spores preserved in sediments and surviving **maceration** are chitinous, and such fungal spores range primarily from late Jurassic to present. Examples **basidiospore, chlamydospore, conidiospore, dictyospore, phragmospore, teleutospore,** and **urediospore**.
furrow A **colpus** or **sulcus**.

gametophyte The sexual generation of a plant that produces gametes or an individual of this generation; e.g., the haploid generation of an embryophytic plant, produced by germination of the **spores**. In lower vascular plants and bryophytes, the gametophyte is a separate plant, but in seed plants, it is confined to the cells (several to many) of the **microgametophyte** in the **pollen grain** and multicellular **megagametophyte** in the ovule (the seed consists of the fertilized ovule and investing tissues). *See* **prothallus** and **sporophyte**.
gemmate Of **sculpture** of **pollen** and **spores** consisting of more or less spherical projections (= gemma, pl. gemmae).
germ pore A membranous **pore** or thin area in the **exine** of a **pollen grain** through which the **pollen tube** emerges on germination. As ordinarily defined, a pore cannot have one dimension greater than or equal to twice another dimension. Where one dimension is greater than or equal to twice another, the thin area is a **colpus** or **sulcus**.
germinal Of **fungal spores**, a secondary opening in the **spore** wall at some point other than the principal **pore** or **furrow**. The germinating plasm may emerge here rather than at primary pores. Often used, especially for **Normapolles**, for both pore and **os** (**endopore**) and their associated structures.
germinal aperture An aperture (such as a **colpus, sulcus,** or **germ pore**) of a **pollen grain** through which the **pollen tube** emerges on germination of the grain. The term is sometimes used to include also the **laesura** of **spores** and **prepollen**.
germinal furrow A **colpus** or **sulcus**.
girdle *See* **cingulum**.
gonal spine A spine situated only at **plate** corners on a **dinoflagellate cyst**.
grain Syn. of **pollen grain**.
gula A projecting, neck-like, rather ornate extension of the **trilete laesura** of certain fossil **megaspores**. Pl. gulae. *See* **trifolium**.
gulate Of **megaspores**, possessing a **gula**.
gymnospore A naked **spore**, or one not developing in a **sporangium**. The term is not in good or current usage in **palynology**.

hamulate Of **sculpture** consisting of small, irregularly arranged hooks. The term is not widely used.

haplotabular archeopyle   An apical **archeopyle** in a **dinoflagellate cyst**, consisting of a single **plate**.

haploxylonoid   Of **bisaccate pollen**, in which the outline of the **sacci** in **distal–proximal** view is more or less continuous with the outline of the **body**, the **sacci** appearing more or less crescent-shaped, and the outline of the whole **grain** presenting a more or less smooth ellipsoidal form. *See also* **diploxylonoid**. (Both terms have been used in other senses and are now best avoided. Pleistocene **palynologists**, for example, usually define haploxylonoid as having, and diploxylonoid as lacking, **verrucae** on the **distal** face or **cappula**.)

haptotypic character   A feature of a **spore** or **pollen grain** that is a product of contact with other members of the **tetrad** in which it was formed; e.g., the **laesura** and contact areas of spores.

harmomegathus   The membrane of a **pore**, **colpus**, **leptoma**, etc. (of a **pollen grain**) when it serves to accommodate, by expansion and contraction, changes in volume of the grain, which usually results from the taking up or loss of water. Pl. harmomegathi. Adj. harmomegathic. Noun, for the phenomenon of accommodation to moisture-content change, harmomegathy.

heterobrochate   A not much used term for **reticulate sculpture** in which the **lumina** (and their enclosing **muri**) are of different sizes. In some sculpture described as heterobrochate, there are small lumina or perforations in the surface of the muri that surround the first order lumina. *See* **homobrochate**.

heterocolpate   Of **pollen grains** having **pores** in some **colpi**, not in others. This term is also used in other senses by some; e.g., for pollen with both colpi and (unassociated) pores.

heteropolar   Of **pollen grains** and **spores** with marked difference between the two **poles**, as monosulcate **pollen**. *See* **isopolar**.

heterosporous   Characterized by **heterospory**; specifically of plants that produce both **microspores** and **megaspores**.

heterospory   The condition in **embryophytic** plants in which **spores** are of two types: **microspores** and **megaspores**. *See* **homospory**.

hilate   Of a **spore** or **pollen grain** possessing a **hilum**.

hilum   An irregular **germinal aperture** of a **spore** or **pollen grain**, formed by the breakdown of the **exine** in the vicinity of one of the **poles**. The hilum in the spore *Vestispora* is associated with an **operculum** that may become separated from the spore.

holdfast   Of fungi, a pad-like **process** from the hyphal body for attachment to substrate. When chitinous and dispersed, may mimic a **spore**.

homobrochate   A not much used term for **reticulate sculpture** in which the **lumina** (plus their enclosing **muri**) are of about the same size. *See* **heterobrochate**.

homospore   One of the **spores** of an **embryophytic** plant which reproduces by **homospory**. Range is Silurian to Holocene. Syn. **isospore**. *See* **microspore**.

homosporous   Characterized by **homospory**.

homospory   The condition in **embryophytic** plants in which all **spores** produced are of the same kind; the production by various plants of **homospores**. Syn. **isospory**. *See* **heterospory**.

hypha   Of fungi, an individual, usually septate filament of the **mycelium**. If chitinous, pieces of hyphae may occur as **palynomorphs**.

hypnozygote   Algal cell resulting from gamete fusion, often with a thick, sculptured wall. **Dinoflagellate cysts** and perhaps some **acritarchs** are probably hypnozygotes.

hypocyst   The part of a **dinoflagellate cyst** posterior to the **cingulum**. *See* **epicyst**.

hyposome   The posterior part of the cell below the **cingulum** of an unarmored dinoflagellate. *See* **episome**.

hypotheca   In a motile **dinoflagellate**, the portion of the **theca** posterior to the **cingulum**.

hypotract   Syn. of **hypocyst**.

hystrichosphaerid   A general term formerly used for a great variety of resistant-walled organic microfossils, ranging from Precambrian to Holocene, and characterized by spherical to ellipsoidal, usually more or less spinose remains found among fossil microplankton. These are now divided among the **acritarchs** and **dinoflagellate cysts**. The term has no formal taxonomic status. Syn. hystrichosphere.

hystrichosphere   *See* **hystrichosphaerid**.

iatropalynology   Study of **spores** and **pollen** as applied to human health problems, especially **pollinosis**. *See* **aerobiology**.

inaperturate   Of **pollen** and **spores** having no **germinal**, **harmomegathic**, or other openings. *See* **acolpate** and **alete**.

infrasculpture   The **structure** of **spores** and **pollen** consisting of organized internal modifications of **exine**. *See* **structure**.

inner body   In **dinoflagellates**, the **endoblast**. In **spores**, the preferred term for the separate, inner **exine** layer of, e.g., **camerate spores**. *See* **endospore**.

intectate   Of an outer **exine** lacking a **tectum** but having **ektexinous** elements such as separated **bacula** that indicate **columellate structure**. Cf. **atectate**. (The difference between the two terms is not very useful.)

intercalary plate   Of a **dinoflagellate**, a **plate** whose position lies between two major **plate series**.

intercolpium   The area of a **colp(or)ate pollen exine** between the **colpi**, therefore delimited by the colpi and the **polar** areas. Syn. **mesocolpium**.

interloculum   Laterally extensive space between the **sexine** and **nexine**. Term especially used for **Normapolles**. Interloculum connects with the **vestibulum**, if such is present.

interporium   The area of a **porate pollen exine** delimited by lines tangential to the **pores**, plus the pores themselves; thus a comparatively narrow band of **exine** as wide as the pores, extending from pore to pore. Syn. **mesoporium**.

interradial   Pertaining to areas of the **proximal** face, or of the **amb** of **trilete spores**, lying between the arms of the **laesura**. Also sometimes applied to corresponding areas on the **distal** surface onto which the laesura can be projected. *See* **radial**.

intexine   A less satisfactory syn. of **endexine**. Also spelled intextine.

intine   The thin, inner layer of the two major layers forming the wall (**sporoderm**) of **spores** and **pollen**, composed of cellulose and pectates, and situated inside the **exine**, surrounding the living cytoplasm. The intine is not normally present in fossil **sporomorphs**. Syn. **endospore**.

intratabular   Of features of a **dinoflagellate cyst** that approximately correspond to the central parts of **thecal plates** rather than to the lines of separation between them. *See* **non-tabular** and **peritabular**.

isopolar   Of **pollen grains** with more or less radial symmetry and no marked difference between the two **poles**, as **tricolpate pollen**.

isopoll   A line on a map connecting locations with samples having the same percentage or other measure of amount of **pollen** of a given kind. Syn. isopollen (not much used).

isospore   A **spore** of plants producing only one kind of spore. Syn. **homospore**.

isospory   The quality or state of having or producing **isospores**. Syn. **homospory**.

jaw   Of **scolecodonts**, syn. for jaw piece, jaw plate, an individual major element of maxillary apparatus.

kyrtome   A more or less arcuate fold or band in the **interradial** area outside the **laesura** of a **trilete spore**. Some **palynologists** prefer to use **torus** for separate interradial bands, and kyrtome for a connected figure. A wide variety of other terms are used for modifications of the borders of laesurae. *See* **torus**, **labrum**, and **margo**.

labrum   Lip-like thickening of the edges of a **laesura**. Pl. labra. *See* **kyrtome**, **torus** and **margo**.

lacuna   A rarely used term for a depressed space, pit, or hole on the outer surface of a **pollen grain**.

laesura   The trace or scar on the **proximal** face of an **embryophytic spore**, that marks the original contact with other members of the **tetrad**. It may be **trilete**, **monolete**, or rarely **dilete**. Pl. laesurae. (Note: Some palynologists speak of each ray of the trilete laesura as "a laesura". In this usage, a trilete spore has three laesurae.) *See* **suture**, **tetrad scar**, and **Y mark**.

laevigate   Syn. of **psilate**. The term is more often applied to **spores** than to **pollen**.

lateral tooth   Of **scolecodonts**, a slender element usually formed by a simple single large **denticle**, placed immediately in front of any or all of the regular **jaws** of an apparatus.

leiosphaerid   A thin-walled, more or less spherical body of probable algal relationship, lacking **processes**, septa, etc., characterized by the genus *Leiosphaeridia*, and usually referred to the **acritarchs**. Mostly Ordovician to Silurian in age.

leiosphere   Syn. of **leiosphaerid**.

leptoma   A thin region of **exine** situated at the **distal pole** of a **pollen grain** and usually functioning as the point of emergence of the **pollen tube**. Example *Corollina*. *See also* **pseudopore** and **tenuitas**.

limb   Syn. of **amb** or **equatorial limb**.

limbus   A crease at the edge of the **saccus** or **pseudosaccus** where outer and inner **exine** layers are more or less fused.

lip   The **labrum**.

LO analysis   From Latin *lux–obscurus* (= light–dark), a microscopic technique depending on projecting elements appearing bright in high focus, dark in low focus, whereas holes appear dark in high focus, etc.

Longaxones   A group of primitive, usually lightly **sculptured**, **tricolpate**, lower Cretaceous and younger angiosperm **pollen**, in which the **polar** axis is as long as, or longer than, the **equatorial** diameter. *See* **Brevaxones**.

longitudinal flagellum   A thread-shaped flagellum in a **dinoflagellate**, trailing after the **body** and arising from the posterior **pore** in the **sulcus**. This flagellum propels the organism.

lumina   The depressions between **muri** of **reticulate sculpture**, and similar depressions. Sing. **lumen**. *See* **muri**.

maceration   The act or process of disintegrating sedimentary rocks by various chemical and physical techniques, in order to extract and concentrate acid-insoluble microfossils (including **palynomorphs**) from them. It includes mainly chemical treatment by halogen acids, oxidants, and alkalis, and use of other separating techniques that will remove extraneous mineral and organic constituents.

macrospore   Unsatisfactory syn. of **megaspore**.

main body   Of **dinoflagellates**, the **central body**.

mandible   Of **scolecodonts**, a single fixed pair of **jaw** pieces on the ventral side (not in the pharynx) of the animal, with long posterior shafts, often with an anterior calcareous cap. The mandibles are non-eversible (cannot be everted).

marginate chorate cyst   A **dinoflagellate chorate cyst**, whose outgrowths are characteristically localized on the lateral margins, leaving the dorsal and more often the ventral surfaces free of outgrowths.

margo   (a) Modified margin of the **colpus**, **sulcus** or **pore** of a **pollen grain**, consisting of a thickening or (less commonly) thinning of the **exine**. *See* **annulus**. (b) A term sometimes used for marginal features associated with the **laesura** of **spores**. *See* **krytome** and **labrum**.

marine influence   An expression for the proportion of a **palynoflora** composed of **palynomorphs** of marine origin. Abbrev. *MI*.

massula   (a) A more or less irregular, coherent mass of many fused **pollen grains** shed from the anther as a unit. *See* **pollinium**. (b) A term sometimes applied to a structure associated with the **laesura** and the attached non-functional spores of certain **megaspores**. Pl. massulae.

maxilla   Of **scolecodonts**, any major **jaw** piece of the mouth located in the pharynx. From posterior to anterior they are numbered MI, MII, etc.

megagametophyte   The female **gametophyte** or haploid generation that develops from the **megaspore** of a **heterosporous embryophytic** plant. In lower vascular plants, it is a small free-living plant bearing archegonia, but in seed plants it (e.g., embryo sac of angiosperms) is contained within the ovule, and the egg is produced in it. The embryo which develops from fertilization of the egg plus other gametophytic products and the enveloping tissues of the ovule comprise the seed. *See* **microgametophyte**.

megasporangium   A **sporangium** that develops or bears **megaspores**.

megaspore   (a) One of the **spores** of a **heterosporous embryophytic** plant that germinates to produce a **megagametophyte** (multicellular female **gametophyte**). It is ordinarily larger than the **microspore**. Range mid-Devonian to Holocene. Free megaspores were common in the late Paleozoic but are produced only by a few genera of pteridophytes today. Syn. **macrospore**. (b) A term arbitrarily defined in **paleopalynology** as a **spore** (or **pollen grain**!) greater than 200 μm in diameter (although it may not have been biologically a **megaspore** in function). *See* **miospore**.

melitopalynology   The study of **pollen** in honey and in connection with apiculture generally.

membranate chorate cyst   A **dinoflagellate chorate cyst** with a prominent membrane; e.g., *Membranilarnacia*.

meso-   Of **dinoflagellate cysts**, the middle, as in mesophragm, middle wall.

mesocolpium, mesoporium   (The "mesocolpus" and "mesoporus" of some authors.) Erdtmanian terms equivalent to **intercolpium** and **interporium**.

Mesophytic   An informal division of geologic time, extending from the coming to dominance of coniferophytes, ginkgophytes, cycadophytes, and other gymnosperms in mid-Permian time, to the **Cenophytic**.

MI   Abbreviation for marine influence.

micro-   A combining form used with **sculpture**, the elements of which are too small (less than 1 μm) for Faegri & Iversen's (1975) terms to apply: **micropitted**, etc.

microflora   An unsatisfactory syn. of **palynoflora**. The term properly applied to an assemblage of microscopic organisms such as the bacteria of an animal gut.

microforaminifera   The chitinous inner tests of certain, almost always spiral foraminifers, frequently found in palynologic preparations of marine sediments, and generally much smaller than "normal" whole foraminifers, but displaying recognizable characteristics of "normal" species.

microgametophyte   The male **gametophyte** or haploid generation that develops from the **microspore** of a **heterosporous embryophytic** plant. In lower vascular plants, a multicellular microgametophyte, as well as the sperm cells, develop within the microspore wall; in seed plants, the microgametophyte plus the surrounding microspore wall is the **pollen grain**, in which the microgametophyte is further reduced, being only 3-nucleate in the angiosperms. *See* **megagametophyte**.

micropitted Faegri & Iversen (1975) **sculpturing** terms used herein do not include a term for "holes" in the **exine** less than 1 µm. However, these can often be observed. I call such holes micropits, and **sculpture** consisting of them is micropitted.

microplankton In **paleopalynology**, a term often used for **acritarchs** and **dinoflagellate cysts**, to distinguish them from **spores** and **pollen**.

microreticulate Having **reticulate sculpture** in which the **muri** are so tiny that they can only be observed at high magnifications under oil.

microsporangium A **sporangium** that develops or bears **microspores**; e.g., the anther in an angiosperm or the **pollen sac** in all other seed plants. *See* **megasporangium**.

microspore One of the **spores** of a **heterosporous embryophytic** plant that germinates to produce a **microgametophyte** (male **gametophyte**). Ordinarily smaller than the **megaspore** of the same species. In seed plants, **pollen grains** consist of a **microspore** wall or **exine** with a microgametophyte contained inside. *See also* **miospore**. (Note that "microspore" should never be used as a general term for "small spore!". Use of "microspore" means that an author is sure the producing plants were heterosporous, not **homosporous**. Most Mesozoic and Cenozoic ferns, for example, produced homospores, not microspores. When in doubt, say miospore!)

miospore A term arbitrarily defined in **paleopalynology** as a **spore** or **pollen grain** less than 200 µm in diameter, regardless of biological function. The word is unfortunately pronounced the same as meiospore, a cell stage in meiosis. If "miospore" is used, all spores or pollen grains greater than 200 µm in diameter are called **megaspores**, regardless of biological function. *See* **megaspore**, **microspore**, and **small spore**.

monad Term used to describe single **sporopolleninous**-walled units, in contrast to **dyads**, **tetrads**, and **polyads**. Term is especially useful in studies of Silurian rocks, where all of the above-mentioned types occur, and it is desired to use the non-committal monad rather than **spore** or **acritarch**.

monocolpate Of **pollen grains** having a single, normally **distal colpus**. **Monosulcate** is preferred in most instances. Code Pa0.

monolete Of an **embryophytic spore** having a **laesura** consisting of a single line or mark. Code Sa0. *See* **trilete**.

monoporate Of **pollen grains** provided with a single **pore**, as in grasses.

monosaccate Of **pollen** with a single **saccus**, usually extending all around the **pollen grain** more or less at the **equator**. The monosaccate and **bisaccate** conditions are not sharply distinct. Many grains that appear bisaccate in one view can be shown to be actually monosaccate. Code Pv1.

monosulcate A term essentially equivalent to **monocolpate** in ordinary usage. Because the **germinal furrow** in such cases is practically always a **sulcus**, this is the preferred term. Code Pa0.

morphon As defined by Van der Zwan (1979), a group of palynological species (form-species) united by continuous variation of morphological characteristics. Others use the term "complex" in a very similar way. *See also* **palynodeme**.

mother cell A cell from which new cells are formed; e.g., **spore mother cell** and **pollen mother cell**.

mouth Of **chitinozoans**, the opening at the **oral pole** (see Fig. 6.11).

mucron Of **chitinozoans**, the nipple–like extension at the center of the base of some species. Both mucron and **copula** apparently have to do with the linkage of chitinozoal units into chains.

multisaccate   Of **pollen** with more than two **vesicles**. Code Pv3, etc.

muri   The more or less vertical walls which form positive **reticulate sculpture** in **pollen** and **spores**. Sing. murus. *See* **lumina**.

mycelium   The tissue of the vegetative structure of a fungus, consisting of **hyphae**.

nannofossil   Very small (mostly 5 µm or less) calcareous fossils, by definition not **palynomorphs**. The term is usually restricted to platelets of the walls of Coccolithophoridae, marine single-celled green algae (platelets are called coccoliths and discoasters). The term is unfortunately used by a few writers for all tiny fossils, including, e.g., **acritarchs** and **cryptarchs**.

NAP   Abbrev. for **non-arboreal pollen**.

neck   Of **chitinozoans**, the narrowed region between **collar** and main **body**.

negative sculpture   A term for **sculpture** engraved into rather than standing upon the outer surface of the **exine**; e.g., negative **reticulum**.

nexine   The inner layer of the **exine** of **pollen** in Erdtman's scheme. The nexine is purely "geographic" in definition: the lower part of the exine, unsculptured. *See* **endexine**.

non-arboreal (non-arborescent) pollen   The **pollen** of herbs and shrubs. Abbrev. NAP. Syn. non-tree pollen.

non-tabular   Of projecting surface features of a **dinoflagellate cyst** that are neither sutural nor **intratabular** and that have a random arrangement, or show no apparent relation to a tabulate scheme. *See* **peritabular**.

non-tree pollen   Syn. of **non-arboreal pollen**. Abbrev. NTP.

Normapolles   A group of Cretaceous and lower Paleogene **porate** (usually **triporate**) **pollen** with a complex **pore** apparatus (e.g., **oculus**) and sometimes other peculiarities such as double **Y marks**. *See* **Postnormapolles**.

NTP   Abbrev. for **non-tree pollen**. *See* **non-arboreal pollen**.

oblate   Of **pollen**, flattened (foreshortened) along the **polar** axis; e.g., pollen whose **equatorial** diameter is much longer than the dimension from **pole** to pole. Ant. **prolate**.

oculate   A group designation (*Oculata*) for *Wodehouseia* and similar Cretaceous **pollen**.

oculus   A much-enlarged part of the **pore** structure of (usually **triporate**) **pollen**, consisting of a bulging, very thick protrusion of **ektexine**. Pl. oculi.

operculate   Having an **operculum**; e.g., of **dinoflagellates**, possessing an **archeopyle** associated with an operculum, or of a **pollen grain** having **pore** membranes with an operculum.

operculum   (a) A lid consisting of the **plate** or plates that originally filled the **archeopyle** of a **dinoflagellate cyst** or the **pylome** of an **acritarch**. Also, the lid-like closure of the **mouth** of a **chitinozoan**. (b) A thicker central part of a **pore** membrane of a **pollen grain**, or a large section or cap of **exine** completely surrounded by a single **colpus**. For certain **hilate spores** and **pollen**, the operculum is a less well-defined lid of exine associated with the formation of the **hilum**.

ora   Pl. of **os**.

oral pole   That end of a flask-shaped **chitinozoan** that includes the **neck** and the **mouth**. *See* **aboral pole**.

orate   Of a **porate pollen grain** having an internal opening (**endopore** or **os**) in the **endexine**.

orbicle   Syn. for **orbicule**. *See* **ubisch bodies**.

orbicule   *See* **ubisch bodies**.

ornament, ornamentation   Less satisfactory syn. of **sculpture**.

ornithophily   A term for **pollination** by birds (mostly hummingbirds).

os   Syn. for **endopore**, an inner **aperture** of a complex structure. Pl. **ora**.

ostiole   Of fungi, the opening in the **neck** of a flask-shaped **fruit body**, or a rounded opening in any fruit body, through which the **spores** are released.

paleopalynology   A division of **palynology** concerned with the study of fossil **spores** and **pollen**. It is now interpreted broadly by most to include study of a wide range of fossil microscopic, usually organic bodies, in addition to spores and pollen: animal remains such as **chitinozoans**, as well as **fungal spores**, **dinoflagellates**, **acritarchs**, and other organisms resistant to acids and found in sedimentary rocks of all ages (**nannofossils** and **diatoms** should therefore not be included). The usual criteria for inclusion are that the bodies be microscopic in size (from about 5 μm to about 500 μm) and composed of a resistant organic substance (usually **sporopollenin**, **chitin**, or **pseudochitin**), resulting in the bodies being preserved in sedimentary rocks and available for separation by **maceration** from such rocks.

Paleophytic   An informal division of geologic time, extending from the first appearance of land plant **spores** (Ordovician/Silurian boundary) to the beginning of the **Mesophytic**.

palyniferous   Bearing **pollen**. The term in **palynology** usually refers to rocks or sediment samples that yield pollen, **spores**, or other **palynomorphs** on **maceration**.

palynodebris   The **palynomorph**-sized particles in a sediment other than palynomorphs, as wood fragments, etc. *See* **phytoclast**.

palynodeme   An expression for a group of **palynomorph** species that intergrade and probably represent ". . . the palynological reflection of a known or hypothetical plant species" (Visscher 1971). As originally used, the concept was phylogenetic and referred to characters changing in time. In practice, the term is used by many as if synonymous with **morphon** or the less formal term, "complex".

palynofacies   A term used in **paleopalynology** for: (a) the assemblage of **palynomorph** taxa in a portion of a sediment, representing local environmental conditions and not typical of the regional **palynoflora**; or (b) the assemblage of **phytoclasts** found in a certain kind of sediment, as palynomorphs, wood fragments, cuticles, etc.

palynoflora   The whole suite of **palynomorphs** from a given rock unit. The term **microflora** is sometimes used as a synonym, but should be avoided as it better applies to assemblages of extant microscopic algae, fungi, and bacteria. As palynomorphs do not really constitute a "flora", a term such as "palynofloral assemblage" is better. **Palynoflorule** refers to the palynomorph assemblage from a single sample or level.

palynology   That branch of science concerned with the study of **pollen** of seed plants, **spores** of other **embryophytic** plants (in a broad sense, also of other **palynomorphs**), whether living or fossil, including their dispersal and applications in stratigraphy and paleoecology. Term suggested by Hyde & Williams (1944, p. 6). Etymol. Greek παλυνω, "to strew or sprinkle", suggestive of παλη, "fine meal", cognate with Latin *pollen*, "fine flour, dust". *See also* **paleopalynology** and **pollen analysis**.

palynomorph   A microscopic, resistant-walled organic body found in palynologic **maceration** residues; a palynologic study object. Palynomorphs include **pollen**, **spores** of many sorts, **acritarchs**, **dinoflagellate thecae** and **cysts**, certain colonial algae, **scolecodonts**, **chitinozoans** and other acid-insoluble microfossils. *See* **sporomorph**.

palynostratigraphy   The stratigraphic application of palynologic methods.

pantocolpate, pantoporate   *See* **pericolpate** and **periporate**.

PAR   Abbrev. for **pollen accumulation rate**. *See* **pollen influx**.

para-   In **dinoflagellate** studies, a prefix sometimes assigned to **thecal** terms when these are applied to **cysts**, e.g., paracingulum, paratabulation, paraplate, etc.

patina   A thickening of the **exine** of **spores** that extends over approximately half of the surface, i.e., over the entire surface of one hemisphere. Adj. patinate.

PDR   Abbrev. for **pollen deposition rate**. *See* **pollen influx**.

peri-   Of **dinoflagellate cysts**, means outer, as **periphragm**, outer wall.

pericoel   The space between the **periphragm** and **endophragm** in a **cavate dinoflagellate cyst**. *See* **endocoel**.

pericolpate   Of **pollen grains** having more than three **colpi**, not meridionally arranged (*see* **stephanocolpate**). In Erdtman's terms, this is called **polyrugate**. Code Px0. Syn. **pantocolpate** (Pantocolpate means the colpi are "evenly distributed", and pericolpate is not so restricted.)

pericolporate   Of **pollen grains** having more than three **colpi**, not meridionally arranged, with at least part of the colpi provided with **pores** or **transverse furrows**. Code Pxx.

pericyst   Of more or less **cavate dinoflagellate cysts**, the separate outer body, of which the **periphragm** comprises the wall.

perine   Syn. for **perinium** and **perisporium**.

perinium   A sometimes present, more or less sculptured outer coat of a **pollen grain**. Pl. perinia. Adj. perinate. *See* **perisporium**.

periphragm   The outer layer of a **dinoflagellate cyst**, usually carrying extensions in the form of spines, and projecting to the position of former thecal wall. It may have served as a support during the period of cyst formation. *See* **ectophragm** and **endophragm**.

periporate   Of **pollen grains** having many **pores** scattered over the surface. In Erdtman's terminology, when there are more than 12 pores, the term is **polyforate**. Code P0x. Syn. **polyporate** and **pantoporate**. (Pantoporate means that the pores are "evenly distributed", whereas periporate is not so restricted. Polyporate is a less favored equivalent of periporate, both meaning more than three pores, not more or less on the **equator**, and either evenly or irregularly spaced.)

perispore   Syn. for **perisporium**.

perisporium   An additional wall layer external to the **exine** in certain **spores** and **pollen**. It is composed of thin and loosely attached **sporopollenin** and is therefore not usually encountered in dispersed fossil **sporomorphs**. Syn. **perine**, **perinium**, and **perispore**.

peritabular   Of the surface features of a **dinoflagellate cyst** that originate immediately interior to the margins of reflected **plate** areas (as in *Areoligera* and *Eisenackia*). *See* **intratabular** and **non-tabular**.

peroblate   Of the shape of a **spore** or **pollen grain**, having the **pole**-to-pole axis very much shortened, producing a discus-like shape. *See* **oblate** and **prolate**.

perprolate   Of the shape of a **spore** or **pollen grain**, having the **pole**-to-pole axis very much elongated, producing a cigar-like shape. *See* **prolate** and **oblate**.

-phragm   Of **dinoflagellate cysts**, means wall, as **autophragm**, single wall. As a synonym for **wall**, it is written **phragma**.

phragmospore   A **spore** having two or more septa; e.g., a septate **fungal spore** that may have a chitinous wall and therefore is encountered as a **palynomorph**.

phytoclast   Plant-derived, more or less resistant-walled particles in a sediment, including **palynomorphs**, cuticular and wood fragments, etc. *See* **palynodebris**.

PI   Abbrev. for **pollen-influx**. More or less the same as **PDR** and **PAR**.

pila Pl. of **pilum**.

pilate Of **spores** and **pollen** having **sculpture** that is similar to that of **clavate** forms but that consists of smaller hair-like **processes** (**pila**) with more or less spherical knobs. Syn. **piliferous**.

piliferous Bearing or producing hairs; e.g., said of **pollen grains** bearing **pila**. The synonymous term **pilate** is preferred by some **palynologists**.

pilum One of the small, spine-like rods comprising **sculpture** of the **exine** of certain **pollen** and **spores**. The rodlets are characterized by rounded or swollen knob-like ends. Pl. **pila**. *See* **clavate**.

pitted Syn. of **foveolate**.

plate Of **dinoflagellates**, the individual elements of the thecal wall numbered in Kofoid or Taylor systems. The corresponding **plates** of the **cyst** are sometimes called paraplates. Of **scolecodonts**, an alternative term for **jaw**.

plate equivalent Of the part of the **dinoflagellate cyst** wall judged to occupy a position equivalent to that occupied by a certain **plate** of the corresponding **theca**. *See* **para-**.

plate formula In **dinoflagellate** studies, a numerical expression for the arrangement and number of **plates**. There are several schemes for such enumeration, of which the most popular is Kofoid's (1907).

plica One of the ridge-like folds comprising most of the surface **exine** of *Ephedra* and some other fossil **pollen**. Also thickened fold-like area in the exine of certain **pollen grains**; in **Normapolles**, usually Y-shaped, centered over the **pole**. Adj. **plicate**.

polar area The part of a **pollen grain** poleward from the ends of the **colpi** and their associated structures. Syn. **apocolpium**.

polar area index The ratio between the diameter of the **polar area** of a **pollen grain** and the diameter of the grain.

polar limb A seldom used term to express the outline of a **pollen grain** as observed laterally (equatorially). Syn. **profile**. *See* **amb**.

polar view The view of a **spore** or **pollen grain** from directly above one of the **poles**. *See also* **amb**.

pole Either termination of the axis of a **pollen grain** or **spore** running from the center of the original **tetrad** to the center of the **distal** side of the grain; hence, the center of both distal and **proximal** surfaces. The term is especially useful for angiosperm pollen grains such as **tricolpate**, in which it is not apparent which is the proximal and which the distal surface. Also of **chitinozoans**, the two ends of a chitinozoal unit. *See* **aboral pole** and **oral pole**.

pollen The several-celled **microgametophyte** of seed plants, enclosed in the **microspore** wall. Fossil pollen consists entirely of the microspore wall or **exine**, from which the microgametophyte and **intine** were removed during or before lithification. The term "pollen" is a collective plural noun, and it is incorrect to say "a pollen"; the correct singular form is "pollen grain"; the correct plural is "pollen," not "pollens".

pollen accumulation rate Syn. for **pollen influx**, preferred by many.

pollen analysis (a) A branch of **palynology** dealing with the study of Quaternary (especially late Pleistocene and Holocene) sediments by employing **pollen diagrams**, **isopoll** maps, and other graphic displays to show the relative abundance of various **pollen** types in space and time; e.g., the identification and calculation of frequency of **pollen grains** of forest trees in peat bogs and lake beds as a means of reconstructing past plant communities and vegetation, thus of paleoclimates. It is used as a geochronologic and paleoecologic tool, sometimes in collaboration with archeology. The former application to dating is mostly superseded by radiocarbon and other absolute methods. Syn. **pollen statistics**. (b) A term used prior to the acceptance of the expression **palynology**, in a manner very similar to the present use of that word.

pollen deposition rate Syn. for **pollen influx**.

pollen diagram Any diagram of **pollen** abundance showing the fluctuations in time of concentration of various pollen types, as revealed from studies of cores and other samples of sediment; strictly, the graphical presentation of abundances of various genera of pollen and **spores** at successive levels of cores of late Neogene sediment studied in **pollen analysis**. (The expression is rarely used for studies of older rock.) Syn. **pollen profile**.

pollen grain Singular form for **pollen**.

pollen influx A mathematical expression for the amount of **pollen** (and/or **spores**, or other **palynomorphs**, as specified) sedimented/accumulated/deposited per year per square centimeter of a surface of deposition. In effect, pollen concentration per weight or volume is corrected for the rate of sedimentation to produce pollen influx. Syn. **pollen accumulation rate** and **pollen deposition rate**.

pollen mother cell A **mother cell** in the **pollen sac** of a seed plant giving rise by meiosis to four cells, each of which develops into a **pollen grain**. *See also* **spore mother cell**.

pollen profile A vertical section of an organic deposit (such as a peat bog) showing the sequence of buried or fossil **pollen** (essentially synonymous with **pollen diagram**). **Profile** is also used for the outline of **pollen grain** or **spore** as seen in **lateral** (= **equatorial**) view.

pollen rain The total deposit of **spores** and **pollen** in a given area and period of time, as estimated by study of sediment samples and by pollen-trapping devices.

pollen sac One of the pouch-like organs of a seed plant, containing the **pollen**; e.g., each of the pollen chambers in the anther of an angiosperm.

pollen spectrum A horizontal line in a **pollen diagram**, showing the relative abundances (percentages) of the various sorts of **pollen** and **spores** in a single sample from a single given level.

pollen statistics Syn. for **pollen analysis**.

pollen sum A portion of the total **pollen** count (in **pollen analysis**), from which certain sorts of pollen are excluded by definition, and which is used as the denominator for calculation of percentages. The most usual pollen sum excludes all **non-arborescent pollen** and sometimes **arborescent pollen** likely to be over-represented as well. Where pollen sums are used, pollen abundances are calculated as ratios of given sorts of pollen to the pollen sum, rather than as percentages of total count.

pollen symbol An arbitrary sign formerly much used in Quaternary **pollen diagrams**, representing a genus or other group of plants, and serving as an internationally understood identification for a line in the pollen diagram (see Erdtman 1943).

pollen tube A more or less cylindric extension that emerges from the wall of a **pollen grain** and protrudes through one of its apertures when the grain germinates on contact with the stigmatic surface of flowering plants or the **megasporangium** of gymnosperms. The tube acts primarily as a haustorial (absorptive) organ to nourish the **microgametophyte** in lower seed plants (such as cycads), but in flowering plants it also conducts the male nuclei to the vicinity of the female **gametophyte** (embryo sac) to effect fertilization.

pollenkitt A viscous complex lipid derived from the **tapetum**, which when occurring on the surface of animal-pollinated angiosperm **pollen** helps the pollen adhere to an animal **pollination** vector.

pollination In general, the fertilization of a seed plant; specifically the transfer of **pollen** from a stamen (anther) or **microsporangium** to an ovule or **megasporangium**.

pollinium   A large, coherent mass of **pollen**, usually the contents of a whole locule of an anther, shed in the mature stage as a unit (as in *Asclepias*). Pl. pollinia. Code Ppd (not distinguished from **polyad**). *See* **polyad** and **massula**.

pollinosis   The allergenic disease caused in some persons by adverse reaction to **spores** and **pollen** in the air. Syn. "hay fever". Subject matter of **iatropalynology**. One of the practical applications of **aerobiology** has to do with investigation and alleviation of this disease.

polster   A small cushion of plant material topped by layers of moss and/or lichen in which silt, sand and **palynomorphs** are frequently trapped.

polyad   A group of more than four mature **pollen grains** shed from the anther as a unit (as in *Acacia*). The number of grains within the polyad is usually a multiple of 4. Code Ppd. *See* **pollinium**, **dyad** and **tetrad**.

polyannulate   Refers especially to **germinal** structures of **Normapolles**, in which the **sexine** of the outer **germinal** has multiple layers of thickenings.

polyforate   *See* **periporate**.

polymorphic   (Dimorphic, etc.) Having **pollen** regularly of more than one size and/or morphological type, presumably as an anti-self-pollination device. The Rubiaceae are especially characterized by this phenomenon.

polyplicate   Of **pollen grains** (such as those of *Ephedra*) with multiple, longitudinal, linear thinnings in the **exine** that resemble, but are not, **colpi**. Code Pp1. *See* **plica**, **taeniate**, and **striate**.

polyporate   *See* **periporate**.

polyrugate   *See* **pericolpate**.

porate   Of **pollen grains** having a **pore** or pores in the **exine**.

pore   (a) One of the external, more or less circular or slightly oval thinnings or openings in the **exine** of **pollen grains**, having dimensions in ratios of less than 2:1. Pores may occur by themselves or in association with **colpi**. *See* **colpus**. *See also* **germ pore**. (b) In fungi, a primary, structural rounded opening in the **spore** wall. Some "pores" in fungi occur at the point of attachment of the spore, or they may be terminal or otherwise disposed.

pore canal   An elongated opening in **porate pollen** such as **Normapolles**, connecting the **pore** proper to the openings beneath, such as **atrium** and **vestibulum**.

pororate   Of **pollen grains** with complex **pore** structures having both external (pore) and internal (**os**) openings.

positive sculpture   A term for **sculpture** consisting of elements projecting from the surface, such as **scabrae**, **muri**, **spines**, etc. *See* **negative sculpture**.

postcingular series   The series of **plates** immediately below the **cingulum** of **dinoflagellate thecae**, usually fewer in number and often larger in size than those of the **precingular series**.

Postnormapolles   A group of Cretaceous–Cenozoic (usually **triporate**) **pollen** without the usual **pore** apparatus or other features of the **Normapolles** group, from which it presumably derived.

precingular archeopyle   An **archeopyle** formed in a **dinoflagellate cyst** by loss of the middorsal **plate** of the **precingular series**.

precingular series   The series of **plates** between the **apical series** and the **cingulum** in **dinoflagellate thecae**. *See* **postcingular series**.

prepollen   Functional **pollen grains** that have haptotypic characters like those of **spores**, e.g., a **trilete mark**, and germinated **proximally**. May also have a **sulcus** and/or such other pollen-like features as **sacci**. Prepollen is typical of extinct primitive gymnosperms (mostly Mississippian to Permian).

process   A longish element extending from the surface of a **spore**, **pollen grain**, **dinoflagellate cyst** or **acritarch**. They are much longer than ordinary sculptural

513

elements such as **spines**. Some have suggested a fixed limit such as 5 μm for minimum length of **processes** (shorter elements would be sculptural), but this has not been widely adapted.

profile   *See also* **pollen profile**. Sometimes used term for the outline of a **spore** or **pollen grain** as seen in **lateral** (= **equatorial**) view.

prolate   Extended or elongated in the direction of a line joining the **poles**; e.g., "prolate pollen" whose **equatorial** diameters are much shorter than the dimensions from pole to pole. Ant. **oblate**.

prosome   Of **chitinozoans**, an enigmatic, sometimes complex internal layer extending for some distance from the **neck** downward, either as a sort of plug or as an annulate tube.

Proterophytic   An informal division of geologic time, from the first regular appearance of robust-walled **acritarchs** (about 1.0×10^9 years ago) to the first appearance of **spore**-like bodies (beginning of the **Paleophytic**, about 440×10^6 years ago).

prothallus   The **gametophyte** of a fern or other pteridophyte; usually a flattened, thallus-like structure living on the soil. Pl. prothalli. Syn. prothallium.

proximal   Of the parts of **pollen grains** or **spores** nearest or toward the center of the original **tetrad**; e.g., of the side of a **monosulcate pollen grain** opposite the **sulcus**, or of the side of a **trilete spore** provided with **contact areas**. Ant. **distal**.

proximal pole   The center of the **proximal** surface of a **spore** or **pollen grain**, thus the midpoint of a **monolete laesura** or the point of a **trilete laesura** from which the **radii** originate.

proximate cyst   A **dinoflagellate cyst** of nearly the same size as, and closely resembling, the motile **theca** of the same species. The ratio of the diameter of the main **body** to the total diameter of the **cyst** exceeds 0.8. The term refers to the supposed proximity of the main cyst wall to the theca at the time of encystment. *See* **chorate cyst** and **proximochorate cyst**.

proximochorate cyst   A **dinoflagellate cyst** which is not as condensed (condensation of 60–80%) from the thecal cell as are **chorate cysts**, and which show some sutural evidence of **tabulation**.

pseudochitin   Resistant, C–H–O–N compound of uncertain structure, occurring in the walls of **chitinozoans** (*see* **chitin**). Behaves like chitin but does not give chitin test, e.g., on staining.

pseudocolpus   A **colpus**-like modification of the **exine** of **pollen grains**, differing from a true colpus in that it is never a site of **pollen tube** emergence. *See* **colpus**.

pseudopore   An especially thin area in the **leptoma** of certain coniferous **pollen** (as in the families Cupressaceae and Taxaceae, and the fossil family Cheirolepidiaceae). *See* **cryptopore** and **tenuitas**.

pseudopylome   A prominent thickening of the wall at the **antapical** end of the **vesicle** in some **acritarchs**, resembling the rim of a **pylome**.

pseudosaccus   An **ektexinous saccus**-like outer part of a fossil **spore** or **pollen grain** (such as *Endosporites*), resembling the true saccus of some pollen grains but not showing internal **structure** characteristic of sacci. The distinction between a pseudosaccus on the one hand and a saccus or the **cavate** condition on the other hand is often rather slight. Some would limit use of term pseudosaccus to spores only (e.g., Richardson (1965) in proposing the Subturma Pseudosaccititriletes).

psilate   Of the relatively smooth walls of **pollen** and **spores** lacking prominent **sculpture**. The term is usually also applied to **exines** with **pits** or **reticular** openings, **mural** units, etc., less than 1 μm in diameter. Syn. **laevigate**.

pterate   Of **chorate dinoflagellate cysts** which have **processes** linked distally in a mesh-like fashion.

pylome Of **acritarchs**, a more or less circular opening, presumably for excystment but lacking clues of **plate** arrangement present in **dinoflagellate archeopyles**.

radial Pertaining to **trilete spore** features associated closely with the arms of the **laesura**. *See* **interradial**.

radius One arm of a **trilete laesura**. Also called **ray**.

ramus Of **scolecodonts**, any arm–like lateral extension of the face of a **jaw**, usually pointing posteriorly.

ray One arm of **trilete laesura**. Also called **radius**.

resting spore A **spore** that remains dormant for a period before germination; e.g., a **chlamydospore**, or a desmid **zygospore**, having thick cell walls and able to withstand adverse conditions such as heat, cold, or drying out; some are apparently **sporopolleninous** and can occur as **palynomorphs**. *See* **statospore** and **cyst**.

reticulate A term for **sculpture** of **pollen** and **spores** consisting of a more or less regular network of ridges (**muri**). Such sculpture is a **positive reticulum**. *See* **negative reticulum** and **muri**.

retusoid Of **spores**, mostly Paleozoic, with prominent **contact areas** and **curvaturae**.

ridge Of **scolecodonts**, the main criterion for recognizing the constituent elements of compound **jaws**. The ridges indicate the existence of the elements.

rimula Of circumpolloid **pollen** (*Corollina* = *Classopollis, et al.*), the **furrow** of thinned **exine** encircling the **grain** "sub-equatorially". (Some sources say that the rimula is really pre-equatorial, that is, where its position can be accurately determined, the rimula is **distal** to the **equator**.)

ring furrow A sometimes used expression for the continuous encircling **sulculus** of **zonisulculate pollen**.

R_o Syn. of R_{oil}. A measurement made microscopically under oil of the reflectance of organic matter, usually vitrinite. The numbers, e.g. 0.2 for certain peaty organic matter, are the percent reflectance of vertical incident light, in terms of the reflectance of glass standards.

ruga In Erdtmanian terms, regularly disposed but not meridional germinal **furrows**. Because of confusion with **rugulate**, it is better to use **colpus**; thus **pericolpate**, not **polyrugate**.

rugulate Of **sculpture**, consisting of wrinkle-like **ridges** that irregularly anastomose. Often approaches **reticulate**.

saccate Of **pollen**, possessing **sacci**. Code Pv1, Pv2, etc.

saccus A wing-like extension, or **vesicle** of the **exine** in gymnosperms, especially (but not exclusively) in coniferous **pollen**. The saccus is an expanded, bladdery projection of **ektexine** extending beyond the main **body** of a **pollen grain** and typically displaying more or less complex internal **structure**. Pl. sacci. *See* **air sac, bladder, vesicle, wing**, and **pseudosaccus**.

scabrate Of **sculpture**, consisting of more or less isodiametric projections (**scabrae**) less than 1 μm in diameter.

Schulze's reagent An oxidizing mixture very commonly used in palynologic **maceration**, consisting of a saturated aqueous solution of $KClO_4$ and varying amounts of concentrated HNO_3. Named after Franz F. Schulze (1815–73). The term "Schulze's solution" is very commonly used for this mixture, but in biological microtechniques this expression means $ZnCl_2$–KCl–I mixture used for staining.

sclerine A term for **exine** and **perine** collectively – useful term for **spore wall** when it is not certain whether a fossil **spore** has a perine or not. *See* **sculptine**.

sclerotium Of fungi, a more or less rounded mass of hyphae with thickened walls. If chitinous, may occur as a **palynomorph**.

scolecodont   Any **jaw** piece of a polychaete annelid worm; originally intended only for dispersed, chitinous fossil elements. This glossary includes a few of the most important **scolecodont** terms. For more complete information, see Jansonius & Craig (1971).

scrobiculate   Of **sculpture** having **scrobiculi** (sing. scrobiculus). Scrobiculi have **lumina** which are too small, and the **muri** too wide, for the sculpture to be regarded as **reticulate**. Grades into **foveolate** (Erdtman (1952): "Very small foveolae are termed scrobiculi . . .").

sculptine   A term for **sexine** and **perine** collectively. Sometimes used if doubt exists whether the outermost **sporoderm** layer may include perine. *See* **sclerine**.

sculptural element   An individual unit of **sculpture**, such as a **spine**, a **clava**, a **baculum**, etc. Contrast **process**.

sculpture   The external textural modifications (such as spines, verrucae, grana, pila, pits, grooves, reticulations, etc.) of the **exine** of **pollen grains** and **spores**. It is usually a feature of the **ektexine** but may be a **perine** character. Syn. **ornamentation** and **process**. *See* **sculptine**, **sclerine**, and **structure**.

secondary pollen   As usually applied, e.g., Faegri & Iversen (1975), this means recycled or reworked **pollen**, e.g., pre-Pleistocene pollen from boulder clays in Holocene sediments. (Some authors, e.g., Bryant & Holloway (1985), speak of secondary pollen counts, meaning counts from which certain overrepresented forms are excluded.)

seed megaspore   Refers to the very large (half a centimeter or more across) **megaspores** of some Paleozoic lycopsids which although true free-sporing **megaspores** were in the size range of seeds and as disseminules presumably acted like seeds.

septal pore   Of **fungal spores** with one or more **septa**, **pore**(s) found in the center of the septum or septa.

septate   Of **fungal spores** having **septa**. *See* **septum**.

septum   Of **fungal spores**, a cross-wall partitioning the inner space of a **fungal spore**. Septa may be transverse or longitudinal.

sexine   The more or less arbitrarily delimited outer division of the **exine** of **pollen**. *See* **nexine** and **ektexine**.

shadow band   In **fungal spores**, incomplete **septa** showing as a shadowy line across the **spore**.

shaft   Of **scolecodonts**, a posterior extension of a **jaw** of proportionately large dimensions.

shank   Of **scolecodonts**, a backward extension (without teeth) of the posterior part of the inner face of a **jaw**.

shape class   The general group (**peroblate** to **perprolate**) to which a **pollen grain** belongs in terms of the ratio between the **equatorial** diameter and the **pole**-to-pole dimension.

skolochorate   Pertaining to **chorate dinoflagellate cysts** that have **processes** only, not **processes** and high **septa**. Low **ridges** and **septa** may be present.

small spore   A term that was formerly used as if synonymous with **microspore**, to contrast with "large spore" (= **megaspore**). However, the term has included **pollen** and **prepollen**, as well as spores other than true megaspores. The term is therefore essentially synonymous with the now more common **miospore**, but lacks its precise size definition.

spectrum   Syn. for **pollen spectrum**.

spine   One element in **echinate sculpture**.

spiraperturate   Of **pollen** with one or several spiral (sinuous or winding) **apertures**, such as those of *Anemone*, *Coffea* or *Thunbergia* (see Furness 1985).

516

Sporae dispersae The **spores** and **pollen** obtained by **maceration** of rocks, in contrast with those that have been found within the **sporangia** that bore them (**Sporae *in situ***).

Sporae *in situ* The **spores** and **pollen** obtained from fossil **sporangia** of **megafossil** plants.

sporal Being, pertaining to, or having the special characteristics of a **spore**. The term is not in good or current usage in **palynology**.

sporangium An organ within which **spores** are usually produced or borne; e.g., an organ in **embryophytic** plants in which spores are produced, such as a **pollen sac** of a gymnosperm or each chamber of the **anther** of an angiosperm. Pl. sporangia. *See also* **microsporangium** and **megasporangium**.

spore Any of a wide variety of minute, typically unicellular reproductive bodies or cells, often adapted to survive unfavorable environmental conditions. **Embryophytic spores** develop into **gametophytes**. Various fungal and algal **spores** develop into a number of different phases of the complex life cycles of these organisms. As usually used in **paleopalynology**, one of the haploid, dispersed reproductive bodies of **embryophytic** plants, having a very resistant outer wall (**exine**), and frequently occurring as fossils from Silurian to Holocene. *See* **sporomorph**.

spore case A **sporangium**. The term is not in good or current usage in **palynology**.

spore coat The **sporoderm**.

spore mother cell The **mother cell** in the **sporangium** of a spore-bearing plant, which, by reduction division, produces the **tetrad** of haploid **spores**. Syn. **sporocyte**. *See also* **pollen mother cell**.

spore wall The **sporoderm**.

sporocyte The **spore mother cell**.

sporoderm The entire **wall** of a **spore** or **pollen grain** collectively, consisting of an outer layer (**exine/exospore**) and an inner layer (**intine**), and, when present, an extra third layer (**perine**) outside of the **exine**. Syn. **spore coat** and **spore wall**.

sporogenous Producing or adapted to the production of **spores**, or reproducing by spores, e.g., "sporogenous tissue" in a **sporangium**, from which **spore mother cells** originate.

sporologic Of or pertaining to **palynology**. The term is rarely used as a synonym of the more current term "palynologic".

sporomorph A fossil dispersed **pollen grain** or **spore**. Some palynologists disapprove of the use of this term, but it is more specific than **palynomorphs** and often useful. *See* **palynomorph**.

sporophyte The asexual generation of a plant (or an individual of that generation), producing **spores**: therefore, the diploid generation of an **embryophytic** plant, produced by fusion of egg and spermatozoid in lower vascular plants, or by fusion of egg nucleus and the sperm nucleus (produced by the **pollen**) of seed plants. *See* **gametophyte**.

sporopollenin The very resistant and refractory organic substance of which the **exine** and **perine** of **spores** and **pollen** (and apparently also the walls of **dinoflagellates** and **acritarchs**) are composed. It is this substance which gives the **sporomorph** its extreme durability during geologic time, being readily destroyed only by oxidation or prolonged high temperature. It is a high-molecular-weight polymer of C–H–O, perhaps a carotenoid-like substance, but the exact structural composition has not yet been established. Adj. sporopolleninous.

statospore A **resting spore**, e.g., the siliceous, thick-walled resistant **cyst** formed within the frustules of various chiefly marine centric diatoms. The term is also used for certain other 2-partite algal **cysts** in the division Chrysophyta.

stephanocolpate Of **pollen grains** having more than three **colpi**, meridionally arranged. Code Pn0.

517

stephanocolporate Of **pollen grains** having more than three **colpi**, meridionally arranged and provided with **pores**. Code Pnn.

stephanoporate Of **pollen grains** having more than three **pores**, disposed on the **equator**. Code P0n.

striate "Streaked" **sculpture** characterized by multiple, more or less parallel grooves and ribs in the **exine**; also used in a general sense to describe **polyplicate** or **taeniate structure**, as of the *Striatiti*. It would be better to use polyplicate or taeniate, as appropriate, for these morphological features, and to restrict striate to description of streaked sculpture.

Striatiti Abundant upper Paleozoic and lower Mesozoic **pollen** with very characteristic **taeniate** or **plicate structure** in the **exine** of the **body** of the **pollen grain**, the grooves and "ribs" (taeniae) usually (but not always) oriented perpendicular to the axes of the **sacci** (if these are present). They are presumably pollen of conifers, gnetaleans, glossopterids, and other gymnosperms.

stroma Of fungi, a mass of unpatterned **hyphae**, generally of thickened cells, thus **sclerotium**-like or **fruit body**-like, but lacking the radial symmetry of the latter. If chitinous, may occur as **palynomorphs**.

structure (a) The internal makeup of the **ektexine** of **pollen grains** and **spores**, usually consisting of rodlets (**columellae**) that may be branched and more or less fused laterally. (When the fusion creates a coherent outer surface layer, this is the **tectum**.) (b) A term that is sometimes, but less desirably, used to describe major morphologic characteristics of spores, especially those of the Paleozoic.

subsaccate Syn. of **pseudosaccate**.

sulcal plate One of the **plates** of the **ventral furrow** region in **dinoflagellate thecae**. The plates are subdivided as to left or right, and anterior or posterior position.

sulculus Elongate **aperture** (**furrow**), more or less parallel to **equator**, either at the equator or displaced toward a **pole**, generally **distal** to the equator but not centered at the distal pole, as is a **sulcus**. Sulculi may be joined to form a ring. *See* **zonisulculate**.

sulcus (a) An elongate **aperture** (**furrow**) in the **exine** of **pollen grains**. The term is usually restricted to a **distal furrow** of pollen grains with only one such aperture, when this furrow has the **distal pole** in its center. *See* **colpus**. (b) A longitudinal groove on the **ventral** surface of **dinoflagellate thecae**. One of the two **flagella** runs posteriorly in it and trails behind the organism.

suture The line along which the **laesura** of an **embryophytic spore** opens on germination; loosely, the **laesura**. More precisely, the **commissure**.

syncolpate Of **pollen grains** in which the **colpi** join, normally near the **pole**. Code Pns. *See* **zonocolpate**.

tabulation Of **dinoflagellate thecae**, the pattern according to which the constituent **plates** are arranged. Reflected tabulation is the evidence in **dinoflagellate cysts** of the arrangement of the plates in the theca from which the cyst was derived.

taenia Strap-like, more or less elongated and parallel strips of **exine** characteristic of many upper Paleozoic and lower Mesozoic **pollen grains**. They occur on one or both sides of the corpus of **saccate** grains (e.g., *Lunatisporites*) or much less commonly on non-saccate grains such as *Vittatina*. Pl. taeniae. Adj. taeniate. *See* **striate**, **polyplicate** and **striatiti**.

tapetum Tissue of nutritive cells in the **sporangium** of **embryophytic** plants, largely used up during development of the **spores**. In angiosperms, it is the inner wall of the anther locules and provides nutritive substances for the developing **pollen**. **Pollenkitt** is a sticky tapetal residue often found on and in **exines**. *See* **ubisch bodies**.

tasmanitid An informal term for members of the genus *Tasmanites* and related forms, large, spherical **palynomorphs** with thick perforate walls presumably representing the resting **body** of certain green algae (Prasinophyceae). These fossils range from Ordovician to Cenozoic, and are usually classed with the **acritarchs**; certain organic-rich shales (tasmanite) contain enormous numbers of tasmanatids.

tectate Of a **pollen grain** whose **ektexine** has an outer surface supported by more or less complicated inner **structure** usually consisting of **columellae** supporting the **tectum**.

tectum (a) The surface of **tectate pollen grains**. (b) A term which should now be avoided, formerly used to designate thickened, upward projecting **exospore** associated with the **laesura**, usually of **megaspores**.

tegillum Erdtman distinguished between the **columellae**-supported outer surface of angiosperm **pollen** with less than 80% coverage as a **tegillum** and those with more than 80% coverage as a **tectum**. Most palynologists now call both tectum.

teleutospore A **fungal spore** developed in the final stage of the life cycle of rust fungi. When the thick walls are composed of chitin they may occur as microfossils in **palynologic** preparations. *See* **urediospore**.

teliospore Syn. of **teleutospore**.

tenuitas A thin area in the **exine** of a **pollen grain** or **spore**, as the **distal germinal** area of *Corollina* (= *Classopollis*) pollen grains. More regular in form than an **ulcus**, less regular than a **colpus**. Syn. **leptoma**.

test Syn. for **dinoflagellate cyst**. *See* **tract**.

tetrad A usually symmetric grouping of four **embryophytic spores** (or **pollen grains**) that result from meiotic division of one **mother cell**. Such tetrads may be **tetrahedral** (most common) or **tetragonal**, rarely of other types. A number of pollen types regularly remain in united tetrads as mature pollen when shed by the **pollen sacs** (as in some fossil circumpolloid forms). If the grains are always in such tetrads (as in Ericaceae) the tetrads are "obligate tetrads"). *See* **dyad** and **polyad**.

tetrad scar Syn. for **laesura**.

tetragonal tetrad A **tetrad** of **spores** or **pollen grains** in which the centers of the individual grains lie more or less in one plane.

tetrahedral tetrad A **tetrad** of **spores** or **pollen grains** in which each grain rests atop three others, so that the centers of the grains define a tetrahedron.

tetratabular archeopyle An **apical archeopyle** formed in a **dinoflagellate cyst** by the loss of four **plates**.

theca The outer covering or "shell" of a motile **dinoflagellate**. *See* **amphiesma**.

torus An arcuate invagination or protuberance of **exine** outside of and more or less paralleling the **laesura** of a **spore** in the **interradial** area. Some **palynologists** use torus for separate arcs, and **kyrtome** for a completely connected figure. Pl. tori. *See* **kyrtome** and **labrum**.

TP Abbrev. for **tree pollen**. *See* **arboreal pollen**.

tract Syn. for **test**, as in epitract, etc.

transverse flagellum A flagellum, often more or less ribbon-like, encircling a motile **dinoflagellate**, usually in a deep encircling groove or **cingulum**, arising from a **pore** in the **sulcus**.

transverse furrow A **colpus**-like thinning in the **exine** of a dicotyledonous **pollen grain**, usually occurring at the **equator** in association with and running perpendicular to a colpus. Syn. **colpus transversalis**.

tree pollen Syn. of **arboreal** (**arborescent**) **pollen**. Abbrev. TP.

trichotomocolpate Syn. of **trichotomosulcate**.

trichotomosulcate Of **monosulcate pollen grains** in which the **sulcus** is more or less triangular, often simulating a **trilete laesura**. Code Pac.

tricolpate   Of **pollen grains** having three meridionally arranged (120° apart) **colpi** which are not provided with **pores**, **transverse furrows** or other such modifications. Tricolpate pollen are produced by dicotyledonous plants, and they first appear in the fossil record in rocks of early Cretaceous age. Code Pc0. *See* **tricolporate**.

tricolporate   Of **pollen grains** having three **colpi** which are provided with **pores** or other, usually **equatorial**, modifications. Code Pc3.

tricolporoidate   An intermediate state between Pc0 and Pc3, in which some modification of the **colpus** is present equatorially, but not a **pore**, transverse colpus, or other organized thinning. Code Pc0.

trifolium   Of **megaspores**, a **proximal** figure similar to a **gula** but less massive, with three blade-like divisions and no broad base.

trilete   Of **embryophytic spores** and some **pollen grains** having a **laesura** consisting of a three-pronged mark somewhat resembling an upper-case "Y". The usage of this term as a noun ("a trilete") is improper. Code Sc0. *See* **monolete** and **Y mark**.

triplan   Refers to **trilete spores** with the **radii** of the **laesura** on deeply indented radial lobes. When seen laterally, such spores have a characteristic flapped appearance.

triporate   Of **pollen grains** having three **pores**, usually disposed at 120° from each other, on the **equator**. Code P03.

triprojectate   A group designation (Triprojectacites) for *Aquilapollenites* and similar, presumably related, forms of late Cretaceous–early Cenozoic angiosperm **pollen**, in which the three **colpi** are borne on the projecting ends of colpal arms, so that typically the three colpal arms (hence, "triprojectate") and two **polar** projections give the pollen a five-pronged appearance.

triradiate crest or **ridge**   Of **trilete spores**, the three-rayed figure on the **proximal** surface caused by intersection of the **contact areas**.

Turma   An artificial suprageneric grouping of form-genera of fossil **spores** and **pollen** (mostly in use for pre-Cenozoic forms), based on morphology. Turmae are grouped under two large headings, **Anteturmae** Sporites and Pollenites. Pollen and spores are functional in meaning, not morphological, creating some problems for classification. Turmae are subdivided into groups such as "Subturmae" and "Infraturmae". The system is not governed by the International Code of Botanical Nomenclature. *See* **Anteturma**.

ubisch bodies   Small (about 2–5 μm) pieces of **sporopollenin** formed from the **tapetum** after the rest of the sporopollenin available has been built into **spore** or **pollen sporoderm**. Ubisch bodies are sometimes rather abundant in palynological preparations. Also called **orbicules**. The name stems from description in 1927 of the particles by Ubisch (see Rowley 1963). However, others apparently described them earlier, which is one reason orbicules is favored by many.

ulcerate   Of **pollen** with **ulci**. Code Pu1.

ulcus   A thin place in the **exine**, more or less **pore**-like, but irregular in outline, and often broken up into patches, as in some Restionaceae. Pl. ulci. Cf. leptoma, tenuitas.

unarmored   *See* **armored**.

urediospore   A **fungal spore** of brief vitality, whose thin walls may be composed of chitin. Such spores may occur as **microfossils** in palynologic preparations. *See* **teleutospore**.

valva   One of the **radial** thickenings of a **valvate spore**.

valvate   Of **trilete, zonate spores** with the **equatorial** thickenings more pronounced at the "corners", that is in the areas beyond the ends of the **laesural radii. Auriculate** can thus be regarded as an extreme valvate condition.

ventral Of **dinoflagellates**, the side on which the **sulcus** and the ends of the **cingulum** occur.

vermiculate A **sculptural** pattern formed by elongate, irregularly placed depressions.

verrucate Warty, or covered with wart-like knobs or elevations; of **spores** and **pollen** having **sculpture** consisting of wart-like projections. Syn. verrucose.

vesicle Syn. of **saccus.**

vesiculate Syn. of **saccate.**

vestibulate Of **porate pollen** having a **vestibulum.**

vestibulum The space between the external opening (**exopore** or **pore**) in the **ektexine** and the internal opening (**endopore** or **os**) in the **endexine** of a **pollen grain** with a complex **porate structure**. The openings in ektexine and endexine are of similar or dissimilar size.

viscin threads Despite the name, non-viscous, **sporopolleninous** threads originating in the polar **ektexine** of a relatively limited number of angiosperms (Onagraceae, some legumes), functioning as attachment organs for dispersal of animal pollinators. One end is attached to the polar ektexine, the other end is free. Threads on **spores**, e.g., the **megaspore** *Balmeisporites*, not serving for attachment in **pollination**, and non-sporopolleninous threads (such as in some orchids), or threads not arising from the polar ektexine, or with both ends attached should not be called viscin threads (Patel *et al.* 1985). Viscin threads have been encountered in fossil onagraceous **pollen**.

wing Syn. of **saccus.**

Y mark A **trilete laesura** on **embryophytic spores, prepollen,** and some **pollen,** consisting of a three-pronged mark somewhat resembling an upper-case "Y". It commonly also marks a **commissure** or **suture** along which the spore germinates. The term is also applied to similar marks, which are not laesurae, on some pollen grains.

zona Same as **zone.**

zonasulculate Best regarded as a syn. for **zonisulculate.**

zonate Of **spores** or **pollen grains,** possessing a **zone** or other similar **equatorial** extension. Note that some authors would restrict "zonate" to spores with a zone, whereas others use it as here defined, to include other equatorial features as well. (Some specialists in angiosperm pollen use zonate to denote pollen grains with a **ring furrow.**)

zone A more or less **equatorial extension** of a **spore** or **pollen grain,** having varying equatorial width and being as thick as or thinner than the spore wall. It is much thinner than a **cingulum**. The term is also used, however, in a general sense for any equatorial extension of the spore wall. Because zone is used in a specific sense, simply "equatorial extension" would be better for the general term. Syn. zona. (Zone is also used stratigraphically, as in biozones, acme zones, etc. (see Hedberg 1976).) *See* **flange, corona, equatorial extension, auricula,** and **crassitude.**

zonisulculate Here advocated as a general term for **pollen** having a **sulculus** that encircles the **grain** forming a ring (**ring furrow**). Some palynologists use zonizonasulculate(!) for pollen where the ring furrow lies on the **equator,** and **zonasulculate** for the general case of pollen with ring furrows parallel to or on the equator. Zonisulculate (= zonasulculate) can be used for instances such as *Nypa* (Palmae), in which the ring furrow runs around the grain through both **poles.**

zonocolpate Of **pollen** with **colpal apertures** merging with each other at places other than the **poles**. *See* **syncolpate**.

zonotrilete Of a **trilete spore** characterized by an **equatorial zone** or other thickening.

zoophily A term for **pollination** by animals. *See* **entomophily** and **anemophily**.

zygospore A resting **spore** of various non-vascular plants (such as desmids), produced by sexual fusion of two protoplasts. The wall is often thick, apparently **sporopolleninous**, resistant, and can therefore occur as a **palynomorph**.

General Bibliography

(See also annotated bibliography, Chapter 1, pp. 17–27).

Abadie, M., G. Lachkar, E. Masure & J. Taugourdeau-Lantz 1977. Observations nouvelles sur le G. Schopfipollenites R. Potonié et Kremp 1954. In *Apport des techniques récentes en palynologie*, Y. Somers & M. Streel (eds.), 125–32. Inst. Nat. Ind. Extractives & Univ. Liège, Serv. Paléobot. Paléopalynol.

Adam, D. P. 1974. Palynological applications of principal component and cluster analyses. *J. Res. U.S. Geol. Surv.* **2**(6), 727–41.

Ager, T. A. & L. Brubaker 1985. Quaternary palynology and vegetational history of Alaska. In *Pollen records of late-Quaternary North American sediments*, V. M. Bryant, Jr. & R. G. Holloway (eds.), 39–70. Am. Assoc. Strat. Palynol. Found.

Akyol, E. 1975. Palynologie du Permien inférieur de Sariz (Kayseri) et de Pamucak Yaylasi (Antalya-Turquie) et contamination jurassique observée, due aux ruisseaux "Pamucak" et "Goynuk". *Pollen et Spores* **17**(1), 141–79.

Allen, K. C. 1980. A review of *in situ* late Silurian and Devonian spores. *Rev. Palaeobot. Palynol.* **29**, 253–70.

Alley, N. F. 1979. Middle Wisconsin stratigraphy and climatic reconstruction, southern Vancouver Island, British Columbia. *Quat. Res.* **11**, 213–37.

Ammann, B. 1977. A pollen morphological distinction between *Pinus banksiana* Lamb. and *P. resinosa* Ait. *Pollen et Spores* **19**(4), 521–9.

Anderson, J. M. & H. M. Anderson 1983. *Palaeoflora of southern Africa*. Vol. 1: *Molteno Formation (Triassic)*. Rotterdam: Balkema.

Anderson, R. S., O. K. Davis & P. L. Fall 1985. *Late glacial and Holocene vegetation and climate in the Sierra Nevada of California, with particular reference to the Balsam Meadow site*. Am. Assoc. Strat. Palynol. Contrib. Ser. 16, 127–40.

Anderson, T. W. 1974. The chestnut pollen decline as a time horizon in lake sediments in eastern North America. *Can. J. Earth Sci.* **11**, 678–85.

Anderson, T. W. 1980. Holocene vegetation and climatic history of Prince Edward Island, Canada. *Can. J. Earth Sci.* **17**, 1152–65.

Andrews, H. N., P. G. Gensel & A. E. Kaspar 1975. A new fossil plant of probable intermediate affinities (Trimerophyte–Progymnosperm). *Can. J. Bot.* **57**, 1719–28.

Andrews, H. N. & T. L. Phillips 1968. *Rhacophyton* from the Upper Devonian of West Virginia. *J. Linn. Soc.* **61**, 37–64.

Archangelsky, S. 1977. *Balmeopsis*, nuevo nombre generico para el palinomorfo *Inaperturopollenites limbatus* Balme, 1957. *Ameghiniana* **14**, 122–6.

Archangelsky, S. & J. C. Gamerro 1967. Pollen grains found in coniferous cones from the Lower Cretaceous of Patagonia (Argentina). *Rev. Palaeobot. Palynol.* **5**, 179–82.

Arnold, C. A. 1950. *Megaspores from the Michigan coal basin*. Univ. Mich. Mus. Paleont. Contrib. 8, 59–111.

Artüz, S. & A. Traverse 1980. SEM and light microscopy of single specimens of Namurian Turkish coal spores. *5. Int. Palynol. Conf.* Abstracts, **20**.

Ash, S. R. 1969. *Ferns from the Chinle Formation (Upper Triassic) in the Fort Wingate area, New Mexico*. U.S. Geol. Surv. Prof. Pap. 613D.

Ash, S. R. 1972. Late Triassic plants from the Chinle formation in north-eastern Arizona. *Palaeontology* **15**(4), 598–618.

Ash, S. R. 1979. *Skilliostrobus* gen. nov., a new lycopsid cone from the Early Triassic of Australia. *Alcheringa* **3**, 73–89.

Ash, S. R. 1986. The early Mesozoic land flora of the Northern Hemisphere. In *Land plants: notes for a short course organized by R. A. Gastaldo*, T. W. Broadhead (ed.) Univ. of Tennessee Dept. of Geological Sciences Studies in Geology **15**, 143–61.

Ash, S., R. J. Litwin & A. Traverse 1982. The Upper Triassic fern *Phlebopteris smithii* (Daugherty) Arnold and its spores. *Palynology* **6**, 203–19.

Aubran, J.–C. 1977. Présentation de quelques transformations structurales et texturales des exines de Cycadales traitees par l'acétolyse, puis fixées par le permanganate de potassium. In *Apport des techniques récentes en palynologie*, Y. Somers & M. Streel (eds.), 133–41. Inst. Nat. Ind. Extractives & Univ. Liège, Serv. Paléobot. Paléopalynol.

Axelrod, D. I. 1958. Evolution of the Madro-Tertiary geoflora. *Bot. Rev.* **24**(7), 433–509.

Axelrod, D. I. & P. H. Raven 1985. Origins of the Cordilleran flora. *J. Biogeog.* **12**, 21–47.

Azcuy, D. L. 1975a. *Las asocianes palinologices del Paleozoico superior de Argentine y sus relaciones.* Actas 1st Congr. Argentino Paleontol. Bioestrat. 1, 455–77.

Azcuy, D. L. 1975b. Palinologia estratografica de la Cuenca Paganzo. *Rev. Asoc. Geol. Argentina* **30**, 104–9.

Bacci, M. & M. R. Palandri 1985. Microwave drying of herbarium specimens. *Taxon* **34**(4), 649–53.

Baker, H. G. 1983. An outline of the history of anthecology, or pollination biology. In *Pollination biology*, L. Real (ed.), 7–28. Orlando: Academic Press.

Baker, R. G. 1986. Sangamonian (?) and Wisconsinan paleoenvironments in Yellowstone National Park. *Geol. Soc. Am. Bull.* **97**, 717–36.

Baker, R. G. & G. M. Richmond 1978. Geology, palynology, and climatic significance of two pre-Pinedale lake sediment sequences in and near Yellowstone National Park. *Quat. Res.* **10**, 226–40.

Baker, R. G. & K. A. Waln 1985. Quaternary pollen records from the Great Plains and central United States. In *Pollen records of late-Quaternary North American sediments*, V. M. Bryant, Jr. and R. G. Holloway (eds.), 191–203. Am. Assoc. Strat. Palynol. Found.

Balbach, M. K. 1966. Paleozoic lycopsid fructifications. II. *Lepidostrobus takhtajanii* in North America and Great Britain. *Am. J. Bot.* **53**, 275–83.

Balbach, M. K. 1967. Paleozoic lycopsid fructifications. III. Conspecificity of British and North American *Lepidostrobus* petrifactions. *Am. J. Bot.* **54**, 867–75.

Baldoni, A. M. & T. N. Taylor 1982. The ultrastructure of *Trisaccites* pollen from the Cretaceous of southern Argentina. *Rev. Palaeobot. Palynol.* **38**, 23–33.

Balme, B. E. 1964. The palynological record of Australian pre-Tertiary floras. In *Ancient Pacific floras*, L. M. Cranwell (ed.), 49–80. Honolulu: University of Hawaii Press.

Balme, B. E. 1970. Palynology of Permian and Triassic strata in the Salt Range and Surghar Range, West Pakistan. In *Stratigraphic problems: Permian and Triassic of West Pakistan*, B. Kummel & C. Teichert (eds.), 305–453. Univ. Kansas Dept. Geol. Spec. Publ. 4.

Banks, H. P. 1968. The early history of land plants. In *Evolution and environment*, E. T. Deake (ed.), 73–107. New Haven, Conn.: Yale University Press.

Barghoorn, E. S. 1948. Sodium chlorite as an aid in paleobotanical and anatomical study of plant tissues. *Science* **107** (2784), 480–1.

Barghoorn, E. S. 1951. Age and environment: a survey of North American Tertiary floras in relation to paleoecology. *J. Paleont.* **25**, 736–44.

Barkley, F. A. 1934. The statistical theory of pollen analysis. *Ecology* **15**(3), 283–9.

Barnard, P. D. W. 1968. A new species of *Masculostrobus* Seward producing *Classopollis* pollen from the Jurassic of Iran. *J. Linn. Soc. (Bot.)* **61**(384), 167–76.

Barnosky, C. W. 1984. Late Miocene vegetational and climatic variations inferred from a pollen record in northwest Wyoming. *Science* **223**, 49–51.

Barss, M. S. 1967. *Illustrations of Canadian fossils: Carboniferous and Permian spores of Canada.* Geol. Surv. Can. Pap. 67–11.

Barss, M. S., E. H. Davies & G. L. Williams 1984. Palynological data management at the Atlantic Geoscience Centre. *VI Int. Palynol. Conf.* Abstracts, 7.

Bartlett, H. H. 1929. Fossils of the Carboniferous coal pebbles of the glacial drift at Ann Arbor. *Pap. Mich. Acad. Sci. Arts Lett.* **9**, 11–28.

Basinger, J. F. & D. L. Dilcher 1984. Ancient bisexual flowers. *Science* **224**, 511–13.

Basso, K. H. & N. Anderson 1973. A Western Apache writing system: the symbols of Silas John. *Science* **180**, 1013–22.

Batten, D. J. 1969. Some British Wealden megaspores and their facies distribution. *Palaeontology*, **12**(2), 333–50.

Batten, D. J. 1973. Use of palynologic assemblage-types in Wealden correlation. *Palaeont* **16**(1), 1–40.

Batten, D. J. 1974. Wealdon palaeoecology from the distribution of plant fossils. *Proc. Geol. Assoc.* **85**(4), 433–58.

Batten, D. J. 1980. Use of transmitted light microscopy of sedimentary organic matter for evaluation of hydrocarbon source potential. *IV Int. Palynol. Conf. Proc.* 2, 589–94.

Batten, D. J. 1981a. Palynofacies, organic maturation and source potential for petroleum. In *Organic maturation studies and fossil fuel exploration*, J. Brooks (ed.), 201–23. London: Academic Press.

Batten, D. J. 1981b. Stratigraphic, palaeogeographic and evolutionary significance of Late Cretaceous and Early Tertiary Normapolles pollen. *Rev. Palaeobot. Palynol.* **35**, 125–37.

Batten, D. J. 1982. Palynofacies, palaeoenvironments and petroleum. *J. Micropalaeont.* **1**, 107–14.

Batten, D. J. 1984. Palynology, climate and the development of Late Cretaceous floral provinces in the Northern Hemisphere; a review. In *Fossils and climate*, P. Brenchly (ed.), 127–64. New York: Wiley.

Batten, D. J. 1985. Coccolith moulds in sedimentary organic matter and their use in palynofacies analysis. *J. Micropalaeont.* **4**(2), 111–16.

Batten, D. J. & R. A. Christopher 1981. Key to the recognition of Normapolles and some morphologically similar pollen genera. *Rev. Palaeobot. Palynol.* **35**, 359–83.

Batten, D. J. & L. Morrison 1983. Methods of palynological preparation for palaeoenvironmental, source potential and organic maturation studies. *Norwegian Petroleum Directorate Bull.* **2**, 35–53.

Baxter, R. W. 1971. *Carinostrobus foresmani*, a new lycopod genus from the middle Pennsylvanian of Kansas. *Palaeontographica B* **134**, 124–30.

Baxter, R. W. & G. A. Leisman 1967. A Pennsylvanian calamitean cone with *Elaterites triferens* spores. *Am. J. Bot.* **54**, 748–54.

Beck, C. B. 1981. *Archaeopteris* and its role in vascular plant evolution. In *Paleobotany, paleoecology and evolution*, Vol. 1, K. J. Niklas (ed.), 193–230. New York: Praeger.

Benda, L. 1971. Grundzüge einer pollenanalytischen Gliederung des türkischen Jungtertiärs. *Beih. Geol. Jahrb.* **113**, 1–45.

Benda, L., H. A. Jonkers, J. E. Meulenkamp & P. Steffens 1979. Biostratigraphic correlations in the eastern Mediterranean Neogene. *Newsl. Strat.* **8**, 61–9.

Benda, L. & J. E. Meulenkamp 1979. Biostratigraphic correlations in the eastern Mediterranean Neogene. *Ann. Géol. Pays Hellén.* **1** 61–70 (7th Int. Congr. Mediterranean Neogene).

Bennie, J. & R. Kidston 1886. On the occurrence of spores in the Carboniferous Formation of Scotland. *Proc. Phys. Soc.* **9**, 82–117.

Benninghoff, W. S. & C. W. Hibbard 1961. Fossil pollen associated with a late-glacial woodland musk ox in Michigan. *Pap. Mich. Acad. Sci.* **46**, 155–9.

Benson, L. V. 1978. Fluctuation in the level of pluvial lake Lahontan during the last 40,000 years. *Quat. Res.* **9**, 300–18.

Berggren, W. A. 1971. Tertiary boundaries and correlations. In *The micropaleontology of oceans*, B. M. Funnel & W. R. Riedel (eds.), 693–809. Cambridge: Cambridge University Press.

Berggren, W. A. & J. A. van Couvering 1974. The Late Neogene: biostratigraphy, geochronology and paleoclimatology of the last 15 million years in marine and continental sequences. *Developments in Palaeontology and Stratigraphy* Vol. 2 Amsterdam: Elsevier.

Berggren, W. A. *et al.* 1980. Towards a Quaternary time scale. *Quat. Res.* **13**, 277–302.

Berggren, W. A. *et al.* 1985. Cenozoic geochronology. *Geol. Soc. Am. Bull.* **96**, 1407–18.

Berglund, B. E. (ed.) 1986. *Handbook of Holocene palaeoecology and palaeohydrology*. Chichester: Wiley.

Berglund, B. E., G. Erdtman & J. Praglowski 1959. On the index of refraction of embedding media and its importance in palynological investigations. *Svensk Botanisk Tidskrift* **53**(4), 462–8.

Bernabo, J. C. & T. Webb, III 1977. Changing patterns in the Holocene pollen record of northeastern North America: a mapped summary. *Quat. Res.* **8**, 64–96.

Bertolani Marchetti, D. 1975. Preliminary palynological data on the proposed Plio-Pleistocene boundary type-section of Le Castella. *L'Ateneo Parmense Acta Naturalia* **11**, 467–85.

Bertolani Marchetti, D., C. A. Accorsi, G. Pelosio & S. Raffi 1979. Palynology and stratigraphy of the Plio-Pleistocene sequence of the Stirone River (northern Italy). *Pollen et Spores* **21**, 149–67.

Bertsch, K. 1942. *Lehrbuch der Pollenanalyse: Handbücher der praktischen Vorgeschichts-forschung*, Bd 3. Stuttgart: Ferdinand Enke.

Beug, H.-J. 1961. *Leitfaden der Pollenbestimmung für Mitteleuropa und angrenzende Gebiete*. Stuttgart: Gustav Fischer.

Bharadwaj, D. C. 1964. *Potonieisporites* Bhard., ihre Morphologie, Systematik und Stratigraphie. *Fortschr. Geol. Rheinld. Westf.* **12**, 45–54.

Bharadwaj, D. C. & B. S. Venkatachala 1968. Suggestions for a morphological classification of sporae dispersae. *Rev. Palaeobot. Palynol.* **6**, 41–59.

Birks, H. J. B. 1973. *Past and present vegetation of the Isle of Skye: a palaeoecological study*. Cambridge: Cambridge University Press.

Birks, H. J. B. & H. H. Birks 1980. *Quaternary palaeoecology*. London: Edward Arnold.

Birks, H. J. B. & A. D. Gordon 1985. *Numerical methods in Quaternary pollen analysis*. London: Academic Press.

Birks, H. J. B., T. Webb III & A. A. Berti 1975. Numerical analysis of pollen samples from central Canada: a comparison of methods. *Rev. Palaeobot. Palynol.* **20**, 133–69.

Blatt, H., G. Middleton & R. Murray 1972. *Origin of sedimentary rocks*. Englewood Cliffs, NJ: Prentice-Hall.

Blome, C. D. & N. R. Albert 1985. Carbonate concretions: an ideal sedimentary host for microfossils. *Geology* **13**, 212–15.

Boenigk, W., G. Von der Brelie, K. Brunnacker, *et al.* 1977. Jungtertiär und Quartär im Hofloff-Graben/Vogelsberg. *Geol. Abh. Hessen* **75**, 1–80.

Bolick, M. R. *et al.* 1984. On cavities in spines of Compositae pollen – a taxonomic perspective. *Taxon* **33**(2), 289–93.

Bonamo, P. M. 1977. *Rellimia thomsonii* (Progymnospermopsida) from the Middle Devonian of New York State. *Am. J. Bot.* **64**, 1271–85.

Bonamo, P. M. & H. P. Banks 1966a. A study of the fertile branches of *Tetraxylopteris*. *Am. J. Bot.* **53**, 628.

Bonamo, P. M. & H. P. Banks 1966b. *Calamophyton* in the Middle Devonian of New York State. *Am. J. Bot.* **53**, 778–91.

Bonamo, P. M. & H. P. Banks 1967. *Tetraxylopteris schmidtii*: its fertile parts and its relationships within the Aneurophytales. *Am. J. Bot.* **54**, 755–68.

Bonny, A. P. 1976. Recruitment of pollen to the seston and sediment of some English Lake District lakes. *J. Ecol.* **64**, 859–87.

Bonny, A. P. 1978. The effect of pollen recruitment processes on pollen distribution over the sediment surface of a small lake in Cumbria. *J. Ecol.* **66**, 385–416.

Bortolami, G. C., J. C. Fontes, V. Markgraf & J. F. Saliege 1977. Land, sea and climate in the northern Adriatic region during late Pleistocene and Holocene. *Palaeogeog. Palaeoclimatol. Palaeoecol.* **21**, 139–56.

Bose, M. N. & S. C. Srivastava 1973. *Nidistrobus* gen. nov., a pollen-bearing fructification from the Lower Triassic of Gopad River Valley, Nidpur. *Geophytology* **2**(2), 211–12.

Bostick, N. H. 1971. Thermal alteration of clastic organic particles as an indicator of contact and burial metamorphism in sedimentary rocks. *Geosci. & Man* **3**, 83–92.

Boulter, M. C. 1971a. A palynological study of two of the Neogene plant beds in Derbyshire: *Bull. Br. Mus.* (*Nat. Hist.*) *Geol.* **19**(7), 359–410.

Boulter, M. C. 1971b. A survey of the Neogene flora from two Derbyshire pocket deposits. *Mercian Geol.* **4**(1), 45–62.

Boulter, M. C. 1979. Taxonomy and nomenclature of fossil pollen from the Tertiary. *Taxon* **28**, 337–44.

Boulter, M. C. & D. L. Craig 1979. A middle Oligocene pollen and spore assemblage from the Bristol Channel. *Rev. Palaeobot. Palynol.* **28**, 259–92.

Boulter, M. C. & R. N. L. B. Hubbard 1982. Objective paleoecological and biostratigraphic interpretation of Tertiary palynological data by multivariate statistical analysis. *Palynology* **6**, 55–68.

Boulter, M. C. & G. C. Wilkinson 1977. A system of group names for some Tertiary pollen. *Palaeontology* **20**(3), 559–79.

Brack-Hanes, S. D. 1978. On the megagametophytes of two lepidodendraceaean cones. *Bot. Gaz.* **139**(1), 140–6.

Brack-Hanes, S. D. 1981. On a lycopsid cone with winged spores. *Bot. Gaz.* **141**(2), 294–304.

Brack-Hanes, S. D. & T. N. Taylor 1972. The ultrastructure and organization of *Endosporites*. *Micropaleontology* **18**(1), 101–9.

Brack-Hanes, S. D. & B. A. Thomas 1983. A re-examination of *Lepidostrobus* Brongniart. *Bot. J. Linn. Soc.* **86**, 125–33.

Brady, H. & H. Martin 1978. Ross Sea region in the middle Miocene: a glimpse into the past. *Science* **203**, 437–8.

Brakenridge, G. R. 1978. Evidence for a cold, dry full-glacial climate in the American Southwest. *Quat. Res.* **9**, 22–40.

Braman, D. R. & L. V. Hills 1985. The spore genus *Archaeoperisaccus* and its occurrence within the Upper Devonian Imperial Formation, District of McKenzie, Canada. *Can. J. Earth Sci.* **22**(8), 1118–32.

Brenner, G. J. 1963. *The spores and pollen of the Potomac Group of Maryland*. Maryland Dept. Geol. Mines & Water Res. Bull. 27.

Brenner, G. J. 1976. Middle Cretaceous floral provinces and early migrations of angiosperms. In *Origin and early evolution of angiosperms*, C. B. Beck (ed.), 23–47. New York: Columbia University Press.

Brenner, G. J. 1984. Late Hauterivian angiosperm pollen from the Heloz Formation, Israel. *VI Int. Palynol. Conf.* Abstracts, 15.

Brooks, J., *et al.* (eds.) 1971. *Sporopollenin*. London: Academic Press.

Brown, R. C. & B. E. Lemmon 1984. Spore wall development in *Andreaea* (Musci: Andreaeopsida). *Am. J. Bot.* **71**(3), 412–20.

527

Brugman, W. A. 1983. *Permian–Triassic palynology*. Lab. Palaeobot. Palynol., State Univ. Utrecht, Netherlands, informal publ.

Bryant, V. M., Jr. 1974a. The role of coprolite analysis in archeology. *Bull. Texas Arch. Soc.* **45**, 1–28.

Bryant, V. M., Jr. 1974b. Pollen analysis of prehistoric human feces from Mammoth Cave. In *Archeology of the Mammoth Cave area*. 203–9. New York: Academic Press.

Bryant, V. M. Jr. & R. G. Holloway 1985. A late-Quaternary paleoenvironmental record of Texas: an overview of the pollen evidence. In *Pollen records of late-Quaternary North American sediments*, V. M. Bryant, Jr. and R. G. Holloway (eds.), 39–70. Am. Assoc. Strat. Palynol. Found.

Bujak, J. P., M. S. Barss & G. L. Williams 1977. Offshore East Canada's organic type and color and hydrocarbon potential. *Oil & Gas J.* April 4, 1977, 198–202; April 11, 1977, 96–100.

Bujak, J. P. & G. L. Williams 1979. Dinoflagellate diversity through time. *Marine Micropaleont.* **4**(1), 1–12.

Burga, C. A. 1980. Pollenanalytische Untersuchungen zur Vegetationsgeschichte des Schams und des San Bernardino-Passgebiete (Graubünden, Schweiz). *Dissertationes Botanicae* Vol. 56, 165. Vaduz: J. Cramer.

Burger, D. 1981. Observations on the earliest angiosperm development with special reference to Australia. *4th Int. Palynol. Conf.* Lucknow (1976–77), Vol. 3, 418–28.

Butterworth, M. A. 1966. The distribution of densospores. *Palaeobotanist* **15**(1 & 2), 16–28.

Cain, S. A. & L. G. Cain 1944. Size–frequency studies of *Pinus palustris* pollen. *Ecology* **25**, 229–32.

Candilier, A. M., R. Coquel & S. Loboziak 1982. Mégaspores du Dévonien terminal et du Carbonifère inférieur des bassins d'illizi (Sahara Algérien) et de Rhadames (Libye occidentale). *Palaeontographica B* **183**, 83–107.

Caratini, C. 1980. Ultrasonic sieving to improve palynological processing of sediments: a new device. *Int. Comm. Palynol. Newsl.* **3**(1), 4.

Caratini, C., J. Bellet & C. Tissot 1983. Les palynofaciès: représentation graphique, intérêt de leur étude pour les reconstitutions paléogéographiques. *Géochem. organique des sédiments marins*, 327–52. D'Orgon a Misedor: C.N.R.S.

Chadwick, A. V. 1983. *Cappasporites*: a common middle Pennsylvanian palynomorph. *Palynology* **7**, 205–10.

Chaloner, W. G. 1953a. A new species of *Lepidostrobus* containing unusual spores. *Geol. Mag.* **90**(2), 97–110

Chaloner, W. G. 1953b. On the megaspores of *Sigillaria*. *Ann. Mag. Nat. Hist. Ser.* **12**(6), 881–97.

Chaloner, W. G. 1954. Notes on the spores of two British Carboniferous lycopods. *Ann. Mag. Nat. Hist. Ser.* **12**(7), 81–91.

Chaloner, W. G. 1958a. *Polysporia mirabilis* Newberry, a fossil lycopod cone. *J. Paleont.* **32**, 199–209.

Chaloner, W. .G. 1958b. A Carboniferous *Selaginellites* with *Densosporites* microspores. *Palaeontology* **1**, 245–53.

Chaloner, W. G. 1961. Palaeo-ecological data from Carboniferous spores. *Recent Advances in Botany*, Sect. 10, 980–3. Toronto University of Toronto Press.

Chaloner, W. G. 1962. *Sporangiostrobus* with *Densoporites* microspores. *Palaeontology* **5**, 73–85.

Chaloner, W. G. 1967. Spores and land-plant evolution. *Rev. Palaeobot. Palynol.* **1**, 83–93.

Chaloner, W. G. 1968a. British pre-Quaternary palynology: a historical review. *Rev. Palaeobot. Palynol.* **6**, 21–40.

Chaloner, W. G. 1968b. The paleoecology of fossil spores. In *Evolution and environment*, E. T. Drake (ed.), 125–38. New Haven, Conn.: Yale University Press.

Chaloner, W. G. 1968c. The cone of *Cyclostigma kiltorkense* Haughton, from the Upper Devonian of Ireland. *J. Linn. Soc. (Bot.)* **61**, 25–36.

Chaloner, W. G. 1970a. The rise of the first land plants. *Biol. Rev.* **45**, 353–77.

Chaloner, W. G. 1970b. The evolution of miospore polarity. *Geosci. & Man* **1**, 47–56.

Chaloner, W. G. 1976. The evolution of adaptive features in fossil exines. In *The evolutionary significance of the exine*, I. K. Ferguson & J. Muller (eds.), 1–14. Linnean Soc. Symp. Ser. I. London: Academic Press.

Chaloner, W. G. 1984. The biology of spores. *17th Ann. Mtg. A.A.S.P.* Program & Abstracts, 3.

Chaloner, W. G. & W. S. Lacey 1973. *The distribution of late Paleozoic floras.* Spec. Pap. Palaeontol. **12**, 271–89.

Chaloner, W. G. & M. Muir 1968. Spores and floras. In *Coal and coal-bearing strata*, D. G. Murchison & T. S. Westall (eds.), 127–46. Edinburgh: Oliver & Boyd.

Chaloner, W. G. & A. Sheerin 1981. The evolution of reproductive strategies in early land plants. *Evolution today*, Proc. 2nd Int. Congr. Systematic Evolutionary Biol., 93–100.

Chapman, J. L. 1985. Preservation and durability of stored palynological specimens. *Pollen et Spores* **17**(1), 112–20.

Charlesworth, J. K. 1957. *The Quaternary era: with special reference to its glaciation.* Vol. 2. London: Edward Arnold.

Charpin, J., R. Surinyach & A. W. Frankland 1974. *Atlas of European allergenic pollens.* Paris: Sandoz Editions.

Chi, B. I. & L. V. Hills 1976. Biostratigraphy and taxonomy of Devonian megaspores, Arctic Canada. *Bull. Can. Petrolm. Geol.* **24**, 640–818.

Chinappa, C. C. & B. G. Warner 1982. Pollen morphology in the genus *Coffea* (Rubiaceae). II. Pollen polymorphism. *Grana* **21**, 29–37.

Choi, D. K. 1983. *Paleopalynology of the Upper Cretaceous–Paleogene Eureka Sound Formation of Ellesmere and Axel Heiberg Islands, Canadian Arctic Archipelago.* Unpubl. Ph.D. thesis, Pennsylvania State University.

Chowdhury, K. R. 1982. Distribution of recent and fossil palynomorphs in the south-eastern North Sea (German Bay). *Senckenbergiana Marit.* **14**(3/4), 79–145.

Christensen, B. B. 1954. New mounting media for pollen grains. *Denmarks Geol. Unders.* **2**(80), 7–11.

Christopher, R. A. 1978. Quantitative palynologic correlation of three Campanian and Maestrichtian sections (Upper Cretaceous) from the Atlantic coastal plain. *Palynology* **2**, 1–27.

Clayton, G., *et al.* 1974. Palynological correlations in the Cork Beds (Upper Devonian–?Upper Carboniferous) of southern Ireland. *Proc. R. Irish Acad.* **74B**(10), 145–55.

Clayton, G., *et al.* 1977. Carboniferous miospores of western Europe: illustration and zonation. *Meded. Rijks Geol. Dienst* **19**, 1–71.

Clement-Westerhof, J. A. 1974. *In situ* pollen from gymnosperm cones from the upper Permian of the Italian Alps – a preliminary account. *Rev. Palaeobot. Palynol.* **17**, 63–73.

Cloud, P. 1974. Evolution of ecosystems. *Am. Scientist* **62**, 54–66.

Cohen, A. D. & A. L. Guber 1968. Production of pollen-sized "microforaminifera" from "normal" foraminifera. *Micropaleontology* **14**(3), 361–2.

Cohen, A. D. & W. Spackman 1972. Methods in peat petrology and their application to reconstruction of paleoenvironments. *Geol. Soc. Am. Bull.* **83**, 129–42.

Colbath, G. K. & S. K. Larson 1980. On the chemical composition of fossil polychaete jaws. *J. Paleont.* **54**(2), 485–8.

Collinson, M. E. 1978. Dispersed fern sporangia from the British Tertiary. *Ann. Bot.* **42**, 233–50.

Collinson, M. E. 1980. A new multiple-floated *Azolla* from the Eocene of Britain with a brief review of the genus. *Palaeontology* **23**(1), 213–29.

Collinson, M. E. 1983. Palaeofloristic assemblages and palaeoecology of the Lower Oligocene Bembridge Marls, Hamstead Ledge, Isle of Wight. *Bot. J. Linn. Soc.* **86**, 177–225.

Collinson, M. E., D. J. Batten, A. C. Scott & S. N. Ayonghe 1985. Palaeozoic, Mesozoic and contemporaneous megaspores from the Tertiary of southern England: indicators of sedimentary provenance and ancient vegetation. *J. Geol. Soc. Lond.* **142**, 375–95.

Collinson, M. E., K. Fowler & M. C. Boulter 1981. Floristic changes indicate a cooling climate in the Eocene of southern England. *Nature* **291**, 315–17.

Combaz, A. 1964. Les palynofacies. *Rev. Micropaléont.* **7**, 205–18.

Combaz, A., *et al.* 1967. *Les chitinozoaires*, C.I.M.P. Microfossiles Organiques du Paléozoique, Part 2. Paris: C.N.R.S.

Cookson, I. C. 1947. Plant microfossils from the lignites of Kerguelen Archipelago. *B.A.N.Z. Antarctic Research Expedition 1929–1931, Reports. Series A* Vol. 2, 127–42.

Cookson, I. C. 1959. Fossil pollen grains of *Nothofagus* from Australia. *Proc. R. Soc. Victoria* **71**, 25–30.

Cornell, W. C. 1982. Acritarcha: ¿Quien sabe?. *Proc. 3rd Am. Paleont. Conv.* Vol. 1, 105–8.

Cornet, B. 1977a. *The palynology and age of the Newark Supergroup.* Unpubl. Ph.D. Thesis, Pennsylvania State University.

Cornet, B. 1977b. Preliminary investigation of two Late Triassic conifers from York County, Pennsylvania. In *Geobotany*, R. C. Romans (ed.), 165–72. New York: Plenum.

Cornet, B. & P. E. Olsen 1985. A summary of the biostratigraphy of the Newark Supergroup of eastern North America with comments on early Mesozoic provinciality. *III Congr. Latinoamer. Paleontol.* Mexico Simp. Sobre Floras Trias. Tard. Fitogeog. Paleoecol. Mem., 67–81.

Cornet, B. & A. Traverse 1975. Palynological contributions to the chronology and stratigraphy of the Hartford Basin in Connecticut and Massachusetts. *Geosci. & Man* **11**, 1–33.

Cornet, B., A. Traverse & N. G. McDonald 1973. Fossil spores, pollen, and fishes from Connecticut indicate early Jurassic age for part of the Newark Group. *Science* **182**, 1243–7.

Couper, R. A. 1956. Evidence of a possible gymnosperm origin for *Tricolpites troedsonii* Erdtman. *New Phytol.* **55**, 280–4.

Couper, R. A. 1958. British Mesozoic microspores and pollen grains: a systematic and stratigraphic study. *Palaeontographica B* **103**, 75–179.

Courvoisier, J. M. & T. L. Phillips 1975. Correlation of spores from Pennsylvanian coal-ball fructifications with dispersed spores. *Micropaleontology*, **21**, 45–59.

Cousminer, H. L. & W. Manspeizer 1977. Autunian and Carnian palynoflorules contributions to the chronologic and tectonic history of the Moroccan Pre-Atlantic borderland. In *Stratigraphic micropaleontology of Atlantic Basin and borderlands*, F. M. Swain (ed.), 185–204. New York: Elsevier.

Cramer, F. H. 1969. Possible implications for Silurian paleogeography from phytoplancton assemblages of the Rose Hill and Tuscarora Formations of Pennsylvania. *J. Paleont.* **43**(2), 485–91.

Cramer, F. H. 1970a. *Distribution of selected Silurian acritarchs.* Revista Española Micropaleontologia Num. Extraord.

Cramer, F. H. 1970b. Middle Silurian continental movement estimated from phytoplankton facies transgression. *Earth & Planet. Sci. Lett.* **10**, 87–93.

Cramer, F. H. & M. d. C. R. Diez 1972. North America Silurian palynofacies and their spatial arrangement: Acritarchs. *Palaeontographica B* **138**, 107–80.

Cramer, F. H. & M. d. C. R. Diez 1974a. Early Paleozoic palynomorph provinces and paleoclimate. In *Paleogeographic provinces and provinciality*, C. A. Ross (ed.), Soc. Econ. Pal. Mineral. Spec. Pub. 21, 177–88.

Cramer, F. H. & M. d. C. R. Diez 1974b. Silurian acritarchs, distribution and trends. *Rev. Palaeobot. Palynol.* **18**, 137–54.

Cramer, F. H. & M. d. C. R. Diez 1979. Lower Paleozoic acritarchs. *Palinologia* **1**, 17–160.

Crane, P. R. 1985. Phylogenetic analysis of seed plants and the origin of angiosperms. *Ann. Missouri Bot. Gard.* **72**, 716–93.

Crane, P. R., E. M. Friis & K. R. Pedersen 1986. Lower Cretaceous angiosperm flowers: fossil evidence on early radiation of dicotyledons. *Science* **232**, 852–4.

Crepet, W. L. 1979. Some aspects of the pollination biology of middle Eocene angiosperms. *Rev. Palaeobot. Palynol.* **27**, 213–38.

Crepet, W. L. 1983. The role of insect pollination in the evolution of the angiosperms. In *Pollination biology*, L. Real (ed.), 29–49. Orlando: Academic Press.

Crepet, W. L. & C. P. Daghlian 1980. Castaneoid inflorescences from the middle Eocene of Tennessee and the diagnostic value of pollen (at the subfamily level) in the Fagaceae. *Am. J. Bot.* **67**, 739–57.

Crepet, W. L. & C. P. Daghlian 1981. Lower Eocene and Paleocene Gentianaceae: floral and palynological evidence. *Science* **214**, 75–7.

Crepet, W. L., C. P. Daghlian & M. Zavada 1980. Investigations of angiosperms from the Eocene of North America: a new juglandaceous catkin. *Rev. Palaeobot. Palynol.* **30**, 361–70.

Crepet, W. L. & D. L. Dilcher 1977. Investigations of angiosperms from the Eocene of North America: a mimosoid inflorescence. *Am. J. Bot.* **64**, 714–25.

Crepet, W. L., D. L. Dilcher & F. W. Potter 1975. Investigations of angiosperms from the Eocene of North America: a catkin with juglandaceous affinities. *Am. J. Bot.* **62**, 813–23.

Cridland, A. A. 1966. *Biscalitheca kansana* sp. n. (Coenopteridales, Zygopteridaceae), a compression from the Lawrence shale (Upper Pennsylvania), Kansas, U.S.A. *Am. J. Bot.* **53**, 987–94.

Cross, A. T. 1984. Pollen and spore distribution in surface sediments of Baja California, Mexico, and the Gulf of California as indicators of source plant communities. *VI Int. Palynol. Conf. Calgary* Abstracts, **27**.

Cross, A. T., G. G. Thompson & J. P. Zaitzeff 1966. Source and distribution of palynomorphs in bottom sediments, southern part of Gulf of California. *Marine Geol.* **4**, 467–524.

Cushing, E. J. 1966. Evidence for differential pollen preservation in Late Quaternary sediments in Minnesota. *Rev. Palaeobot. Palynol.* **4**, 87–101.

Cwynar, L. C., E. Burden & J. H. McAndrews 1979. An inexpensive sieving method for concentrating pollen and spores from fine-grained sediments. *Can. J. Earth Sci.* **16**, 1115–20.

Darrell, J. H. II 1973. *Statistical evaluation of palynomorph distribution in the sedimentary environments of the modern Mississippi River delta.* Unpubl. Ph.D. Thesis, Louisiana State University.

Darrell, J. H. II & G. F. Hart 1970. Environmental determinations using absolute miospore frequency, Mississippi River delta. *Geol. Soc. Am. Bull.* **81**, 2513–18.

Davies, E. H., J. P. Bujak & G. L. Williams 1982. The application of dinoflagellates to paleoenvironmental problems. *3rd North Am. Paleont. Conv. Proc.* 1, 125–31.

Davis, M. B. 1963a. Estimation of absolute pollen rain from pollen frequencies in sediments of known accumulation rate. *Bull. Ecol. Soc. Am.* **44**, 81.

Davis, M. B. 1963b. On the theory of pollen analysis. *Am. J. Sci.* **261**, 897–912.

Davis, M. B. 1965. A method for determination of absolute pollen frequency. In *Handbook of paleontological techniques*, B. Kummel & D. M. Raup (eds.), 674–85. San Francisco: W. H. Freeman.

Davis, M. B. 1967. Late-glacial climate in northern United States: a comparison of New England and the Great Lakes region. In *Quaternary paleoecology*, E. J. Cushing & H. E. Wright, Jr. (eds.), Vol. 7, 11–43. Proc. 7th Congr. INQUA. New Haven, Conn.: Yale University Press.

Davis, M. B. 1968. Pollen grains in lake sediments: redeposition caused by seasonal water circulation. *Science* **162**, 796–9.

Davis, M. B. 1969. Climatic changes in southern Connecticut recorded by pollen deposition in Rogers Lake. *Ecology* **50**, 409–22.

Davis, M. B. 1973. Redeposition of pollen grains in lake sediments. *Limnol. Oceanog.* **18**, 44–52.

Davis, M. B. 1980. Was the mid-Holocene hemlock decline in North America caused by a pathogen? *5. Int. Palynol. Conf.* Abstracts 104.

Davis, M. B. & D. B. Botkin 1985. Sensitivity of cool–temperate forests and their fossil pollen record to rapid temperature change. *Quat. Res.* **23**, 327–40.

Davis, M. B., L. B. Brubaker & T. Webb III 1973. Calibration of absolute pollen influx. In *Quaternary plant ecology*, H. J. B. Birks & R. G. West (eds.), 9–25. London: Blackwell.

Davis, M. B., R. W. Spear & L. C. K. Shane 1980. Holocene climate of New England. *Quat. Res.* **14**, 240–50.

Deevey, E. S., Jr. 1939. Studies on Connecticut lake sediments, 1. A postglacial climatic chronology for southern New England. *Am. J. Sci.* **237**, 691–724.

Deevey, E. S., Jr. 1949. Biogeography of the Pleistocene. Part I: Europe and North America. *Bull. Geol. Soc. Am.* **60**, 1315–416.

Deevey, E. S., Jr. & R. F. Flint 1957. Postglacial hypsithermal interval. *Science* **125**, 182–4.

Delcourt, H. R. 1979. Late Quaternary vegetation history of the eastern highland rim and adjacent Cumberland plateau of Tennessee. *Ecol. Monographs* **49**(3), 255–280.

Delcourt, H. R. & P. A. Delcourt 1985a. Quaternary palynology and vegetational history of the southeastern United States. In *Pollen records of late-Quaternary North American sediments*, V. M. Bryant, Jr. & R. G. Holloway (eds.). Am. Assoc. Strat. Palynol.

Delcourt, H. R. & P. A. Delcourt 1985b. Comparison of taxon calibrations, modern analogue techniques, and forest-stand simulation models for the quantitative reconstruction of past vegetation. *Earth Surface Processes and Landforms* **10**, 293–304.

Delevoryas, T. & R. C. Hope 1973. Fertile coniferophyte remains from the late Triassic Deep River Basin, North Carolina. *Am. J. Bot.* **60**(8), 810–18.

Denizot, D. 1977. Au sujet de la solubilité de l'exine dans l'éthanolamine. In *Apport des techniques récentes en palynologie*, Y. Somers & M. Streel (eds.), 143–51. Inst. Nat. Ind. Extractives & Univ. Liège, Serv. Paléobot. Paléopalynol.

Dennis, R. L. & D. A. Eggert 1978. *Parasporotheca* Gen. Nov., and its bearing on the interpretation of the morphology of permineralized medullosan pollen organs. *Bot. Gaz.* **139**, 117–39.

Dettman, M. E. 1961. Lower Mesozoic megaspores from Tasmania and South Australia. *Micropaleontology* **7**(1), 71–86.

Dettmann, M. E. 1963. Upper Mesozoic microfloras from south-eastern Australia. *Proc. R. Soc. Victoria* **77**(1), 1–148.

Dibner, A. F. 1973. Morphology and classification of late Paleozoic monosaccate miospores. *Rev. Palaeobot. Palynol.* **16**, 263–70.

Dijkstra, S. J. & P. Piérart 1957. Lower Carboniferous megaspores from the Moscow Basin. *Mededel. Geol. Sticht.* (*N.S.*) **11**, 5–19.

Dilcher, D. L. 1979. Early angiosperm reproduction: an introductory report. *Rev. Palaeobot. Palynol.* **27**, 291–328.

Dimbleby, G. W. 1961. Soil pollen analysis. *J. Soil Sci.* **12**(1), 1–11.

Dimbleby, G. W. 1985. *The palynology of archaeological sites.* London: Academic Press.

DiMichele, W. A., J. F. Mahaffy & T. L. Phillips 1979. Lycopods of Pennsylvanian age coal: *Polysporia. Can. J. Bot.* **57**(16), 1740–53.

Diver, W. L. & C. J. Peat 1979. On the interpretation and classification of Precambrian organic-walled microfossils. *Geology* **7**, 401–4.

Dodson, J. 1977. Late Quaternary palaeoecology of Wyrie Swamp, southeastern South Australia. *Quat. Res.* **8**, 97–114.

Doher, L. I. 1980. *Palynomorph preparation procedures currently used in the paleontology and stratigraphy laboratories, U.S. Geological Survey.* U.S. Geol. Surv. Circ. 830.

Dolby, J. H. & B. E. Balme 1976. Triassic palynology of the Carnarvon Basin, Western Australia. *Rev. Palaeobot. Palynol.* **22**, 105–68.

Doran, J. B., P. G. Gensel & H. N. Andrews 1978. New occurrences of trimerophytes from the Devonian of eastern Canada. *Can. J. Bot.* **56**(24), 3052–68.

Dorning, K. J. 1981. Silurian acritarch distribution in the Ludlovian shelf sea of South Wales and the Welsh Borderland. In *Microfossils from recent and fossil shelf seas*, J. W. Neale & M. D. Brasier (eds.), 31–6. Chichester: Ellis Horwood, for British Micropalaeont. Soc.

Dorning, K. J. 1984. Acritarch, Chitinozoa, conodont, dinocyst, graptolite, miospore and scolecodont coloration: multiple indications of the thermal alteration index. *6th Int. Palynol. Conf.*, Calgary, Abstracts, 34.

Dorning, K. J. 1985. The acritarch zonation of the Llandovery, Wenlock, and Ludlow series of the Silurian system. *Am. Assoc. Strat. Palynol. 18th Ann. Mtg.* Abstracts, 9.

Dorning, K. J. 1986. Organic microfossil geothermal alteration and interpretation of regional tectonic provinces. *J. Geol. Soc. London* **143**, 219–20.

Doubinger, J. & L. Grauvogel-Stamm 1971. Présence de spores du genre *Thymospora* chez *Pecopteris hemitelioides* du Mont-Pelé (Stéphanien moyen du Bassin d'Autun). *Pollen et Spores* **13**, 597–607.

Douglas, J. G. 1973. *Spore–plant relationships in Victorian Mesozoic cryptogams.* Geol. Soc. Australia Spec. Publ. 4, 119–26.

Dow, W. G. 1977. Kerogen studies and geological interpretations. *J. Geochem. Explor.* **7**, 79–99.

Downie, C. 1967. The geological history of the microplankton. *Rev. Palaeobot. Palynol.* **1**, 269–81.

Downie, C. 1973. Observations on the nature of the acritarchs. *Palaeontology* **16**(2), 239–59.

Downie, C. 1979. *The acritarchs: exploration short course.* Multilithed notes. Baton Rouge: Louisiana State University.

Doyle, J. A. 1969. Cretaceous angiosperm pollen of the Atlantic Coastal Plain and its evolutionary significance. *J. Arnold Arboretum* **50**(1), 1–35.

Doyle, J. A. 1984. Evolutionary, geographic, and ecological aspects of the rise of angiosperms. *Proc. 27th Int. Geol. Congr.* **2**, 23–33.

Doyle, J. A., P. Biene, A. Doerenkamp & S. Jardiné 1977. Angiosperm pollen from the pre-Albian Lower Cretaceous of Equatorial Africa *Bull. Cent. Rech. Explor.-Prod. Elf-Aquitaine* **1**(2), 451–73.

Doyle, J. A. & L. J. Hickey 1976. Pollen and leaves from the Mid-Cretaceous Potomac group and their bearing on early angiosperm evolution. In *Origin and early evolution of angiosperm*, C. B. Beck (ed.), 129–206. New York: Columbia University Press.

Doyle, J. A., M. Van Campo & B. Lugardon 1975. Observations on exine structure of *Eucommiidites* and Lower Cretaceous angiosperm pollen. *Pollen et Spores* **17**(3), 429–86.

Dragastan, O., J. Petrescu & L. Olaru 1980. *Palinologie cu applicati in geologie*. Bucharest: Editura Didactica si Pedagogica.

Ducker, S. C. & R. B. Knox 1985. Pollen and pollination: a historical review. *Taxon* **34**(3), 401–49.

Dudgeon, M. J. 1985. *Palynology of the Yaamba Basin (Eocene), central Queensland.* Unpubl. Ph.D. Thesis, University of Queensland, Australia.

Duffield, S. L. & J. A. Legault 1982. Gradational morphological series in early Silurian acritarchs from Anticosti Island, Quebec. *3rd North Am. Paleont. Conv. Proc.* I, 137–41.

Durand, B. 1980 (ed.). *Kerogen: insoluble organic matter from sedimentary rocks.* Paris: Technip.

Dybova-Jachowicz, S., *et al.* 1982. Révision des megaspores à gula du Carbonifère (I). *Prace Inst. Geol. (Warszawa Wydawnictwa Geol.)* **107**, 1–44

Ecke, H.-H. 1984 Palynology of the Zechstein and lower Bundsandstein (*Germanic Basin*) – an example for a Permian–Triassic transition in evaporite facies. *6th Int. Palynol. Conf.* Calgary, Abstracts, 38

Ecke, H.-H. & T. Löffler 1985. Palynostratrigraphic data from an evaporite-solution breccia at the eastern margin of the Leinetal Graben near Northeim (Southern Lower Saxony, F. R.). *N. Jb. Geol. Palaeont. Abh* **170**(2), 167–82.

Edgar, D. R. 1984. Polychaetes of the lower and middle Paleozoic: a multi-element analysis and a phylogenetic outline. *6th Int. Palynol. Conf.* Calgary, Abstracts, 39.

Ediger, V. S. 1981. Fossil fungal and algal bodies from Thrace Basin, Turkey. *Palaeontographica B* **179**, 87–102.

Ediger, V. S. 1986a. Sieving techniques in palynological sample-processing with special reference to the "MRA" system. *Micropaleontology* **32**(3), 256–70.

Ediger, V. S. 1986b. *Paleopalynological biostratigraphy, organic matter deposition, and basin analysis of the Triassic–(?) Jurassic Richmond Rift Basin, Virginia.* Unpubl. Ph.D. Thesis, Pennsylvania State University.

Edwards, D., V. Fanning and J. B. Richardson 1986. Stomata and sterome in early land plants. *Nature* **323**, 438–40.

Edwards, L. E. 1978. Range charts and no-space graphs. *Comput. Geosci.* **4**, 247–55.

Edwards, L. E. 1980. Dinoflagellate biostratigraphy: a first look. In Reinhardt, Juergen, and Gibson, *Upper Cretaceous and lower Tertiary geology of the Chattahoochee River Valley, western Georgia and eastern Alabama*, in *Excursions in southeastern geology*, Vol. 2, R. W. Frey (ed.), Geol. Soc. Am. 93rd Ann. Mtg. Field Trip Guidebooks, 424–7, pl. 9, fig. 16.

Edwards, L. E. 1982a. Numerical and semi-objective biostratigraphy: review and predictions. *3rd North Am. Paleont. Conv. Proc.* I, 147–52.

Edwards, L. E. 1982b. Biostratigraphically important species of *Pentadinium* Gerlach, 1961, and a likely ancestor, *Hafniasphaera goodmanii* n. sp., from the Eocene of the Atlantic and Gulf Coastal Plains. *Palynology,* **6**, 105–17.

Edwards, L. E. 1984a. Miocene dinocysts from Deep Sea Drilling Project Leg 81, Rockall Plateau, eastern North Atlantic Ocean. In *Initial reports Deep Sea Drilling Project 81*, D. G. Roberts, D. Schnitker, *et al.* (eds.), 581–94, pl. 1–5.

Edwards, L. E. 1984b. Insights on why graphic correlation (Shaw's method) works. *J. Geol.* **92**, 583–97.

Edwards, L. E., D. K. Goodman & R.J. Witmer 1984. Lower Tertiary (Pamunkey Group) Dinoflagellate biostratigraphy, Potomac River area, Virginia and Maryland. In *Cretaceous and Tertiary stratigraphy, paleontology, and structure, southwestern Maryland and northeastern Virginia*, N.O. Frederiksen and K. Krafft (eds.). Am. Assoc. Strat. Palynol. Field Trip Vol. and Guidebook, 137–52.

Eggert, D. A. & T. Delevoryas 1967. Studies of Paleozoic ferns: *Sermaya*, gen. nov. and its bearing on filicalean evolution in the Paleozoic. *Palaeontographica B* **120**, 169–80.

Eggert, D. A. & R. W. Kryder 1969. A new species of *Aulacotheca* (Pteridospermales) from the middle Pennsylvanian of Iowa. *Palaeontology* **12**(3), 414–19.

Eggert, D. A. & G. W. Rothwell 1979. *Stewartiotheca* gen. N. and the nature and origin of complex permineralized medullosan pollen organs. *Am. J. Bot.* **66**(7), 851–66.

Eggert, D. A. & T. N. Taylor 1968. A *Telangium*-like pollen organ from the Upper Mississippian of Arkansas. *Am. J. Bot.* **55**, 726.

Eisenack, A. 1931. Neue Mikrofossilien des baltischen Silurs 1. *Paläontol. Zeitschr.* **13**, 74–118.

Eisenack, A., F. H. Cramer & M. d. C. R. Diez 1973. *Katalog der fossilen Dinoflagellaten, Hystrichospharen und verwandten Mikrofossilien*, Vol. III, *Acritarcha*, Part I. Stuttgart: E. Schweizerbart'sche.

Elsik, W. C. 1966. Biologic degradation of fossil pollen grains and spores. *Micropaleontology* **12**(4), 515–18.

Elsik, W. C. 1968a. Palynology of a Paleocene Rockdale lignite, Milam County, Texas. I. Morphology and taxonomy. *Pollen et Spores* **10**(2), 263–314.

Elsik, W. C. 1968b. Palynology of a Paleocene Rockdale lignite, Milam County, Texas. II. Morphology and taxonomy. *Pollen et Spores* **10**(3), 599–664.

Elsik, W. C. 1976. Microscopic fungal remains and Cenozoic palynostratigraphy. *Geosci. & Man* **15**, 115–20.

Elsik, W. C. 1979. *Definitions of fungal morphology*. Multilithed discussion-book for Kent State fungal spore working-group meeting, May 1979.

Elsik, W. C. (Chmn.) 1983. *Annotated glossary of fungal palynomorphs*. Am. Assoc. Strat. Palynol. Contrib. Ser. 11.

Elsik, W. C. & D. M. Jarzen 1984. Recent fungal palynomorphs from Zambia. *6th Int. Palynol. Conf.* Calgary, Abstracts, 43.

Englund, K. J. 1979. Virginia. In the Mississippian and Pennsylvanian (Carboniferous) systems in the United States (Anonymous). *U.S. Geol. Surv. Prof. Pap.* 1110 A–L, C1–C21.

Englund, K. J., H. H. Arndt & T. W. Henry (eds.) 1979. Proposed Pennsylvanian system stratotype, Virginia and West Virginia. *Am. Geol. Inst. Sel. Guidebook Ser.* 1.

Erdtman, G. 1943. *An introduction to pollen analysis*. Waltham, Mass.: Chronica Botanica. (2nd printing 1954.)

Erdtman, G. 1947. Suggestions for the classification of fossil and recent pollen grains and spores. *Svensk Botan. Tidskr.* **41**, 104–14.

Erdtman, G. 1952–71 (corrected reprint, 1971). *Pollen morphology and plant taxonomy: angiosperms – an introduction to palynology*, Vol. 1. Stockholm: Almqvist & Wiksell. (Vol. 2, 1957; Vol. 3, 1965; Vol. 4, 1971.)

Erdtman, G. 1954. *An introduction to pollen analysis*, 2nd printing. Waltham, Mass: Chronica Botanica.

Eshet, Y. & H. L. Cousminer 1986. Palynozonation and correlation of the Permo-Triassic succession in the Negev, Israel. *Micropaleontology* **32**(3) 193–214.

Evitt, W. R. 1961. Observations on the morphology of fossil dinoflagellates. *Micropaleontology* **7**(4), 385–420.

Evitt, W. R. 1963. A discussion and proposals concerning fossil dinoflagellates, hystrichospheres, and acritarchs. *Proc. Nat. Acad. Sci.* **49**(2–3), 158–64, 298–302.

Evitt, W. R. 1967. *Dinoflagellate studies II. The archeopyle.* Stanford Univ. Pub. Geol. Sci. **10**(3).

Evitt, W. R. 1984. Some techniques for preparing, manipulating and mounting dinoflagellates. *J. Micropaleont.* **3**(2), 11–18.

Evitt, W. R. 1985. *Sporopollenin dinoflagellate cysts: their morphology and interpretation.* Am. Assoc. Strat. Palynol. Found.

Evitt, W. R., J. K. Lentin, M. E. Millioud, L. E. Stover & G. L. Williams 1977. *Dinoflagellate cyst terminology.* Geol. Surv. Can. Pap. **76**(24).

Evreinova, T. N. *et al.* 1976. Role of coacervate systems in the evolution of matter and origins of life. In *Evolutionary biology*, V. J. A. Novák & B. Pacltová (eds.), 61–72. Prague: Czechoslovak Biological Soc.

Faegri, K. 1951. An unrecognized source of error in pollen analysis. *Geol. Fören. Förhandl.* **73**, 51–6.

Faegri, K. 1974. Quaternary pollen analysis – past, present and future. *Adv. Pollen Spore Res.* **1**, 62–9.

Faegri, K. 1981. Some pages of the history of pollen analysis. In *Florilegume Florinis Dedicatum*, L.-K. Konigsson & K. Paabo (eds.), *Striae* **14**, 42–7.

Faegri, K. & J. Iversen 1975. *Textbook of pollen analysis.* 3rd edn. New York: Hafner.

Faegri, K. & P. Ottestad 1949. Statistical problems in pollen analysis. *Univ. Bergen Årbok, Naturvidenskap*, **3**, 1–29.

Faegri, K. & L. van der Pijl 1979. *The principles of pollination ecology*, 3rd edn., Oxford: Pergamon.

Farley, M. B. 1982. *An assessment of the correlation between miospores and depositional environments of the Dakota Formation (Cretaceous), north-central Kansas and adjacent Nebraska.* M. A. Thesis, Indiana University.

Farley, M. B. 1987. Palynomorphs from surface water of the eastern and central Caribbean Sea. *Micropaleontology* **33**, 270–9.

Fay, R. O. *et al.* 1979. Oklahoma. In the Mississippian and Pennsylvanian (Carboniferous) systems in the United States (Anonymous), *U.S. Geol. Surv. Prof. Pap.* 1110 M–DD, R1–R35.

Federova, K. V. 1952. Dispersal of spores and pollen by water currents. *Trudy Inst. Geog. Akad. Nauk USSR* **52**, 46–72 (in Russian).

Federova, V. A. 1977. The significance of the combined use of microphytoplankton, spores and pollen for differentiation of multi-facies sediments. In *Questions of phytostratigraphy*, S. R. Samoilovich & N. A. Timoshina (eds.), 70–88. Trudy VNIGRI No. 398. (in Russian)

Felix, C. & P. Burbridge 1985. Reappraisal of a palynological storage technique. *Pollen et Spores* **27**, 491–2.

Ferguson, I. K. 1977. Technique utilisant le méthylate de sodium comme solvart de la résine époxy des bloes d'inclusion "Type MET" pour les observations d l'exine des grains de pollen. In *Apport des techniques récentes en palynologie*, Y. Somers & M. Streel (eds.), 153–7. Inst. Nat. Ind. Extractives & Univ. Liège, Serv. Paléobot. Paléopalynol.

Ferguson, I. K. 1984. Some observations on the variation and its significance in the pollen morphology of the Palmae. *6th Int. Palynol. Conf.* Abstracts, 46.

Ferguson, I. K. & J. J. Skvarla 1982. Pollen morphology in relation to pollinators in Papilionoideae (Leguminosae). *J. Linn. Soc. (Bot.)* **84**, 183–93.

Feuer, S., C. J. Niezgoda & L. I. Nevling 1985. Ultrastructure of *Parkia* polyads (Mimosoideae: Leguminosae). *Am. J. Bot.* **72**, 1871–90.

Filatoff, J. 1975. Jurassic palynology of the Perth Basin, western Australia. *Palaeonto-graphica B* **154**, 1–113.

Firbas, F. 1949. *Spät und nacheiszeitliche Waldgeschichte Mitteleuropas nördlich der Alpen*, Vol. 1. Jena: Gustav Fischer.

Firbas, F. 1952. *Spät und nacheiszeitliche Waldgeschichte Mitteleuropas nördlich der Alpen*, Vol. 2. Jena: Gustav Fischer.

Firbas, F., H. Losert and F. Broihan 1939. Untersuchungen zur jüngeren Vegetations-geschichte im Oberharz. *Planta* **30**, 422–56.

Fisher, M. J. & R. E. Dunay 1981. Palynology and the Triassic/Jurassic boundary. *Rev. Palaeobot. Palynol.* **34**, 129–35.

Fleming, R. F. 1984. Palynological observations of the Cretaceous–Tertiary boundary in the Raton Formation, New Mexico. *Am. Assoc. Strat. Palynol. 17th Ann. Mtg.* Abstracts, 8.

Flenley, J. 1979. *The equatorial rain forest: a geological history*. London: Butterworths.

Flohn, H. & S. Nicholson 1980. Climatic fluctuations in the arid belt of the "Old World" since the last glacial maximum; possible causes and future implications. *Palaeoecol. Africa* **12**, 3–21.

Flugel, E. 1979 (ed.). *Fossil algae: recent results and developments*. Berlin: Springer.

Follieri, M. 1979. Late Pleistocene evolution near Rome. *Pollen et Spores* **21**, 135–48.

Francis, J. E. 1983. The dominant conifer of the Jurassic Purbeck Formation, England. *Palaeontology* **26**(2), 277–94.

Francis, J. E. 1984. The seasonal environment of the Purbeck (Upper Jurassic) fossil forests. *Palaeogeog. Palaeoclimatol. Palaeoecol.* **48**, 285–307.

Frederiksen, N. O. 1972. The rise of the Mesophytic flora. *Geosci. & Man* **4**, 17–28.

Frederiksen, N. O. 1980a. Paleogene sporomorphs from South Carolina and quantitative correlation with the Gulf Coast. *Palynology* **4**, 125–79.

Frederiksen, N. O. 1980b. Significance of monosulcate pollen abundance in Mesozoic sediments. *Lethaia* **13**, 1–20.

Frederiksen, N. O. 1983. *Middle Eocene palynomorphs from San Diego, California*. Am. Assoc. Strat. Palynol. Contrib. Ser. 12.

Frederiksen, N. O. 1985. *Review of early Tertiary sporomorph paleoecology*. Am. Assoc. Strat. Palynol. Contrib. Ser. 15.

Frenzel, B. 1968. *Grundzüge der Pleistozänen Vegetationsgeschichte Nord-Eurasiens*. Wiesbaden: Franz Steiner.

Friis, E. M. 1981. Upper Cretaceous angiosperm flowers, fruits and seeds from Sweden. *13th Int. Bot. Congr. Prog.*, Sydney, Australia, Abstracts, 129.

Friis, E. M. 1983. Upper Cretaceous (Senonian) floral structures of juglandalean affinity containing Normapolles pollen. *Rev. Palaeobot. Palynol.* **39**, 161–88.

Friis, E. M. 1984. Preliminary report of Upper Cretaceous angiosperm reproductive organs from Sweden and their level of organization. *Ann. Missouri Bot. Gard.* **71**, 403–18.

Friis, E. M. 1985a. Structure and functions in Late Cretaceous angiosperm flowers. *Kongelige Dansk. Videnskab. Selsk. Biol. Skriften* **25**, 1–37.

Friis, E. M. 1985b. *Actinocalyx* gen. nov., sympetalous angiosperm flowers from the Upper Cretaceous of southern Sweden. *Rev. Palaeobot. Palynol.* **45**, 171–83.

Friis, E. M., P. R. Crane & K. R. Pedersen 1986. Floral evidence for Cretaceous chloranthoid angiosperms. *Nature* **320**, 163–4.

Friis, E. M. & A. Skarby 1982. *Scandianthus* gen. nov., angiosperm flowers of saxifragalean affinity from the Upper Cretaceous of southern Sweden. *Ann. Bot.* **50**, 569–83.

Funkhouser, J. W. & W. R. Evitt 1959. Preparation techniques for acid-insoluble microfossils. *Micropaleontology* **5**, 369–75.

Furness, C. A. 1985. A review of spiraperturate pollen. *Pollen et Spores* **27**(3/4), 307–20.

Galtier, J. & A. C. Scott 1979. Studies of Paleozoic ferns: on the genus *Corynepteris*: a redescription of the type and some other European species. *Palaeontographica B* **170**, 81–125.

Gamerro, J. C. 1968. Orbiculas (Corpusculos de Ubisch) y membranas tapetales cutinizadas en cuatro coniferas del Cretacico Inferior de Santa Cruz, Republica Argentina. *Ameghiniana* **5**, 271–8.

Gastaldo, R. A. 1981a. *Palaeostachya dircei* n. sp., an authigenically cemented equisetalean strobilus from the Middle Pennsylvanian of southern Illinois. *Am. J. Bot.* **68**(10), 1306–18.

Gastaldo, R. A. 1981b. An ultrastructural and taxonomic study of *Valvisporites auritus* (Zerndt) Bhardwaj, a lycopsid megaspore from the middle Pennsylvanian of southern Illinois. *Micropaleontology* **27**(1), 84–93.

Gensel, P. G. 1976. *Renalia hueberi*, a new plant from the Lower Devonian of Gaspe. *Rev. Palaeobot. Palynol.* **22**, 19–37.

Gensel, P. G. 1977. Morphologic and taxonomic relationships of the Psilotaceae relative to evolutionary lines in early land vascular plants. *Brittonia* **29**(1), 14–29.

Gensel, P. G. 1980a. On the spores of Psilophyton. *5th Int. Palyn. Conf.* Abstracts, 150.

Gensel, P. G. 1980b. Devonian *in situ* spores: a survey and discussion. *Rev. Palaeobot. Palynol.* **30**, 101–32.

Gensel, P. G. 1982a. A new species of *Zosterophyllum* from the Early Devonian of New Brunswick. *Am. J. Bot.* **69**(5), 651–69.

Gensel, P. G. 1982b. *Orcilla*, a new genus referable to the zosterophyllophytes from the late Early Devonian of northern New Brunswick. *Rev. Palaeobot. Palynol.* **37**, 345–59.

Gensel, P. G. & H. N. Andrews 1984. *Plant life in the Devonian*. New York: Praeger.

Gensel, P. G., H. N. Andrews & W. H. Forbes 1975. A new species of *Sawdonia* with notes on the origin of microphylls and lateral sporangia. *Bot. Gaz.* **136**, 50–62.

Gensel, P. G. & A. R. White 1983. The morphology and ultrastructure of spores of the early Devonian trimerophyte *Psilophyton* (Dawson) Hueber & Banks. *Palynology* **7**, 221–33.

Germeraad, J. H. 1979. Fossil remains of fungi, algae and other organisms from Jamaica. *Scr. Geol.* **52**, 1–39.

Germeraad, J. H. 1980. Dispersed scolecodonts from Cainozoic strata of Jamaica. *Scr. Geol.* **54**, 1–24.

Germeraad, J. H., C. A. Hopping & J. Muller 1968. Palynology of Tertiary sediments from tropical areas. *Rev. Palaeobot. Palynol.* **6**(3/4), 189–348.

Given, P. H. 1984. An essay on the organic geochemistry of coal. In *Coal science*, Vol. 3, M. L. Gorbaty, J. W. Larsen and I. Wender (eds.), 65–234. New York: Academic Press.

Given, P. H. *et al.* 1985. Aspects of the origins of some coal macerals. *Proc. 1983 Int. Conf. on Coal Science*, Pittsburgh, Pennsylvania, 389–92.

Givulescu, R. 1962. Die fossile Flora von Valea Neogra, Bezirk Crisana, Rumanien. *Palaeontographica B* **110**, 128–87.

Góczán, F. *et al.* 1967. Die Gattungen des "Stemma Normapolles Pflug 1953b" (Angiospermae). *Paläontologische Abh.* (*Berlin*) *B* **II**(3), 427–633.

Godwin, H. 1956. *The history of the British flora*, 1st edn. Cambridge: Cambridge University Press.

Godwin, H. 1975. *The history of the British flora: a factual basis for phytogeography*, 2nd edn. Cambridge: Cambridge University Press.

Good, B. H. & R. I. Chapman 1978. The ultrastructure of *Phycopeltis* (Chroolepidaceae; Chlorophyta). I. Sporopollenin in the cell walls. *Am. J. Bot.* **65**(1), 27–33.

Good, C. W. 1977. Taxonomic and stratigraphic significance of the dispersed spore genus *Calamospora*. In *Geobotany*, R. C. Romans (ed.), 43–64. New York: Plenum.

Good, C. W. 1978. Taxonomic characteristics of sphenophyllalean cones. *Am. J. Bot.* **65**, 86–97.

Good, C. W. 1979. *Botryopteris* pinnules with abaxial sporangia. *Am. J. Bot.* **66**(1), 19–25.

Good, C. W. & T. N. Taylor 1975. The morphology and systematic position of the calamitean elater-bearing spores. *Geosci. & Man* **11**, 133–9.

Gothan, W. & W. Remy 1957. *Steinkohlenpflanzen*. Essen: Glückauf.

Gothan, W. & H. Weyland 1964. *Lehrbuch der Paläobotanik*. 2nd edn. Berlin: Akademie Verlag.

Gould, R. 1981. Illustrations of petrified Glossopteris. *13th Int. Bot. Congr.* Sydney, Australia, Abstracts, 202.

Graham, A. & G. Barker 1981. Palynology and tribal classification in the Caesalpinioideae. In *Advances in legume systematics*, R. M. Polhill & P. H. Raven (eds.), 801–32. London: Brit. Mus. Nat. Hist.

Granoff, J. A., P. G. Gensel & H. N. Andrews 1976. A new species of *Pertica* from the Devonian of Eastern Canada. *Palaeontographica* (B) **155** (5–6), 119–28.

Grauvogel-Stamm, L. 1978, *La flore du Grès à Voltzia (Buntsandstein Supérieur) des Vosges du Nord (France)*. Sci. Géol. Mem. 50.

Grauvogel-Stamm, L. & P. Duringer 1983. *Annalepis zeilleri* FLICHE 1910 emend., un organe reproducteur de Lycophyte de la Lettenkohle de l'Est de la France. Morphologie, spores in situ et paléoécologie. *Geol. Rundschau* **72**(1), 23–51.

Grauvogel-Stamm, L. & L. Grauvogel 1973. *Masculostrobus acuminatus* nom. nov., un nouvel organe reproducteur male de Gymnosperme de Grès à Voltzia (Trias inférieur) des Vosges (France). *Geobios* **6**(2), 101–14.

Grauvogel-Stamm, L. & L. Grauvogel 1980. Morphologie et anatomie d'*Anomopteris mougetii* Brongniart (synonymie: *Pecopteris sulziana* Brongniart): une fougère du Buntsandstein Supérieur des Vosges (France). *Sci. Géol. Bull.* **33**, 53–66.

Gray, H. H. & G. K. Guennel 1961. Elementary statistics applied to palynologic identification of coal beds. *Micropaleontology* **7**(1), 101–6.

Gray, J. 1965. Extraction techniques. In *Handbook of paleontological techniques*, B. Kummel and D. Raup (eds.), 530–87. San Francisco: W. H. Freeman.

Gray, J., D. Massa & A. J. Boucot 1982. Carodocian land plant microfossils from Libya. *Geology* **10**, 197–201.

Grebe, H. 1971. A recommended terminology and descriptive method for spores. In *Les Spores*, C.I.M.P. Microfossiles Organiques du Paléozoique, Part 4, 11–34. Paris: C.N.R.S.

Green, D. G. 1983. Interactive pollen time series analysis. *Pollen et Spores* **25**(3–4), 531–40.

Gribbin, J. 1978. *Climatic change*. Cambridge: Cambridge University Press.

Grindrod, J. 1985. The palynology of mangroves on a prograded shore, Princess Charlotte Bay, North Queensland, Australia. *J. Biogeog.* **12**, 323–48.

Groot, J. J. 1966. Some observations on pollen grains in suspension in the estuary of the Delaware River. *Marine Geol.* **4**, 409–16.

Groot, J. J. & C. R. Groot 1966. Marine palynology: possibilities, limitations, problems. *Marine Geol.* **4**, 387–95.

Groot, J. J. & C. R. Groot 1971. Horizontal and vertical distribution of pollen and spores in Quaternary sequences. In *The micropaleontology of oceans*, B. M. Funnell & W. R. Riedel (eds.), 493–504. Cambridge: Cambridge University Press.

Guedes, M. 1982. Exine stratification, ectexine structure and angiosperm evolution. *Grana* **21**, 161–70.

Guennel, G. K. 1952. *Fossil spores of the Alleghenian coals in Indiana*. Indiana Geol. Surv. Rep. Prog. 4.

Gutjahr, C. C. M. 1966. Carbonization measurements of pollen-grains and spores and their application. *Leidse Geol. Mededel.* **38**, 1–29.

Habib, D. 1966. Distribution of spores and pollen assemblages in the Lower Kittanning Coal of western Pennsylvania. *Palaeontology* **9**(4), 629–66.

Habib, D. 1979. Sedimentary origin of North Atlantic Cretaceous palynofacies. In *Deep drilling results in the Atlantic Ocean: continental margins and paleoenvironment*, 420–37. Am. Geophys. Union.

Habib, D. 1982a. Sedimentation of black clay organic facies in a Mesozoic oxic North Atlantic. *3rd North Am. Paleont. Conv. Proc.* **1**, 217–20.

Habib, D. 1982b. Sedimentary supply origin of Cretaceous black shales. In *Nature and origin of Cretaceous carbon-rich facies*, S. O. Schlanger & M. B. Cita (eds.), 113–27. New York: Academic Press.

Habib, D. & S. D. Knapp 1982. Stratigraphic utility of Cretaceous small acritarchs. *Micropaleontology* **28**(4), 335–71.

Hafsten, J. 1956. Pollen-analytic investigations on the late Quaternary development in the inner Oslofjord area. *Univ. Bergen Årbok, Naturvidenskap.* **8**, 1–161.

Hafsten, J. 1959. Bleaching + HF + acetolysis: a hazardous preparation process. *Pollen et Spores* **1**(1), 77–9.

Hall, D. W. 1981. Microwave: a method to control herbarium insects. *Taxon* **30**(4), 818–19.

Hall, J. W. & B. M. Stidd 1971. Ontogeny of *Vesicaspora*, a Late Pennsylvanian pollen grain. *Palaeontology* **14**, 431–6.

Hall, S. A. 1985. Quaternary pollen analysis and vegetational history of the Southwest. In *Pollen records of late-Quaternary North America sediments*, V. M. Bryant, Jr. & R. H. Holloway (eds.), 95–123. Am. Assoc. Strat. Palynol. Found.

Hamer, J. J. & G. W. Rothwell 1983. *Phillipopteris* gen. nov. – anatomically preserved sporangial fructifications from the Upper Pennsylvanian of the Appalachian Basin. *Am. J. Bot.* **70**(9), 1378–85.

Hamilton, A. 1976. The significance of patterns of distribution shown by forest plants and animals in tropical Africa for the reconstruction of upper Pleistocene palaeoenvironments: a review. *Palaeoecol. Africa* **9**, 63–97.

Hansen, J. M. & L. Gudmundsson 1979. A method for separating acid-insoluble microfossils from organic debris. *Micropaleontology* **25**(2), 113–17.

Hantke, R. 1978. *Eiszeitalter: die jüngste Erdgeschichte der Schweiz und ihrer Nachbargebiete*, Bd. 1. Thun, Switzerland: Ott.

Hantke, R. 1983. *Eiszeitalter: die jüngste Erdgeschichte der Schweiz und ihrer Nachbargebiete*, Bd. 3. Thun, Switzerland: Ott.

Haq, B. V. & A. Boersma (eds.) 1978. *Introduction to marine micropaleontology*. New York: Elsevier.

Harland, R., S. J. Morbey & W. A. S. Sarjeant 1975. A revision of the Triassic to lowest Jurassic dinoflagellate *Rhaetogonyaulax*. *Palaeontology* **18**(4), 847–64.

Harris, T. M. 1973. Pollen from fossil cones. *The Botanique* **4**(1), 1–8.

Harris, T. M. 1974a. *Williamsoniella lignieri*: its pollen and the compression of spherical pollen grains. *Palaeontology* **17**(1), 125–48.

Harris, T. M. 1974b. Notes by A. Traverse on lecture at U.S. Museum of Natural History, Smithsonian Institution, Washington, DC.

Hart, G. F. 1971. The Gondwana Permian Palynofloras. *Ann. Acad. Brasil. Cienc.* **43** (Suppl.), 145–85.

Hart, G. F. 1974. Permian palynofloras and their bearing on continental drift. In *Paleogeographic provinces and provinciality*, C. A. Ross (ed.), 148–64. Soc. Econ. Paleontol. Mineralog. Spec. Pub. 21.

Hatcher, P. G. *et al.* 1982. Organic matter in a coal ball: peat or coal? *Science* **217**, 831–3.

Havinga, A. J. 1967. Palynology and pollen preservation. *Rev. Palaeobot. Palynol.* **2**, 81–98.

Havinga, A. J. 1971. An experimental investigation into the decay of pollen and spores in various soil types. In *Sporopollenin*, J. Brooks *et al.* (eds.), 446–79. London: Academic Press.

Havinga, A. J. 1984. A 20-year experimental investigation into the differential corrosion susceptibility of pollen and spores in various soil types. *Pollen et Spores* **26**(3–4), 541–58.

Hayes, J. M., I. R. Kaplan & K. W. Wedeking 1983. Precambrian organic geochemistry, preservation of the record. In *Earth's earliest biosphere: its origin and evolution*, J. W. Schopf (ed.), 93–134. Princeton, New Jersey: Princeton University Press.

Hays, J. D. 1969. Climatic record of late Cenozoic Antarctic ocean sediments related to the record of world climate. *Palaeoecol. Africa* **5**, 139–63.

Hays, J. D. & W. A. Berggren 1971. Quaternary boundaries and correlations. In *The micropaleontology of oceans*, B. M. Funnell & W. R. Riedel (eds.), 669–91. Cambridge: Cambridge University Press.

Hedberg, H. D. (ed.), 1976. *International stratigraphic guide.* New York: Wiley-Interscience.

Helby, R. & A. R. H. Martin 1965. *Cylostrobus* gen. nov., cones of lycopsidean plants from the Narrabean group (Triassic) of New South Wales. *Australian J. Bot.* **13**, 389–404.

Herngreen, G. F. W. & A. F. Chlonova 1981. Cretaceous microfloral provinces. *Pollen et Spores* **23**, 441–555.

Hesse, M. 1981. Pollenkitt and viscin threads: their role in cementing pollen grains. *Grana* **20**, 145–52.

Heusser, C. J. 1965. A Pleistocene phytogeographical sketch of the Pacific Northwest and Alaska. In *The Quaternary of the United States*, H. E. Wright, Jr. & D. G. Frey (eds.), 469–83. Rev. Vol. for 7th INQUA Congr. Princeton, N.J.: Princeton University Press.

Heusser, C. J. 1977a. Quaternary palynology of the Pacific slope of Washington. *Quat. Res.* **8**, 282–306.

Heusser, C. J. 1977b. *A survey of pollen types of North America.* Am. Assoc. Strat. Palynol. Contrib. Ser. 5A, 111–29.

Heusser, C. J. & L. E. Florer 1973. Correlation of marine and continental Quaternary pollen records from the northeast Pacific and western Washington. *Quat. Res.* **3**, 661–70.

Heusser, C. J., L. E. Heusser & S. S. Streeter 1980. Quaternary temperature and precipitation for the north-west coast of North America. *Nature* **286**, 702–4.

Heusser, L. E. 1978. Spores and pollen in the marine realm. In *Introduction to marine micropaleontology*, B. V. Haq & A. Boersma (eds.). New York: Elsevier.

Heusser, L. E. 1983. Pollen distribution in the bottom sediments of the western North Atlantic Ocean. *Marine Micropaleont.* **8**, 77–88.

Heusser, L. E. & W. L. Balsam 1977. Pollen distribution in the northeast Pacific Ocean. *Quat. Res.* **7**, 45–62.

Hickey, L. J. & J. A. Doyle 1977. Early Cretaceous fossil evidence for angiosperm evolution. *Bot. Rev.* **43**(1), 3–104.

Hickey, L. J. & S. L. Wing 1983. *Report to 5th Northeastern Paleobotanical Conf.*, Harvard Forest, Petersham, Mass.

Higgs, K. & A. P. Beese 1986. A Jurassic microflora from the Colbord clay of Cloyne, County Cork. *Irish J. Earth Sci.* **7**, 99–109.

Higgs, K. & A. C. Scott 1982. Megaspores from the uppermost Devonian (Strunian) of Hood Head, County Wexford, Ireland. *Palaeontographica B* **181**(4–6), 79–108.

Higgs, K. & M. Streel 1984. Spore stratigraphy at the Devonian–Carboniferous boundary in the northern "Rheinisches Schiefergebirge", Germany. *Cour. Forsch. Inst. Senckenberg* **67**, 157–79.

Hills, L. V. 1984. Devonian–Carboniferous megaspore zonation, northern Canada: a review. *6th Int. Palynol. Conf.* Abstracts, 63.

Hills, L. V. & A. R. Sweet 1972. The use of "Quaternary O" in megaspore palynological preparations. *Rev. Palaeobot. Palynol.* **13**, 229–31.

Hochuli, P. A. 1979. The palaeoclimatic evolution in the late Palaeogene and the early Neogene. *Ann. Géol. Pays Hellén.* **2**, 515–23.

Hochuli, P. A. 1984. Correlation of Middle and Late Tertiary sporomorph assemblages. *Paléobiol. continentale* **14**(2), 301–14.

Hoffmeister, W. S. 1954. *Microfossil prospecting for petroleum*. U. S. Patent 2,686,108.

Hoffmeister, W. S. 1959. Lower Silurian plant spores from Libya. *Micropaleontology* **5**(3), 331–4.

Holloway, R. G. & V. M. Bryant, Jr. 1984. *Picea glauca* pollen from late glacial deposits in central Texas. *Palynology* **8**, 21–32.

Holloway, R. G. & V. M. Bryant, Jr. 1985. Late-Quaternary pollen records and vegetational history of the Great Lakes region: United States and Canada. In *Pollen records of late-Quaternary North America sediments*, V. M. Bryant, Jr. & R. G. Holloway (eds.), 205–45. Am. Assoc. Strat. Palynol. Found.

Hooghiemstra, H. & C. O. C. Agwu 1986. Distribution of palynomorphs in marine sediments: a record for seasonal wind patterns over NW Africa and adjacent Atlantic. *Geol. Rundschau* **75**(1), 81–95.

Hopkins, D. M., J. V. Matthews, J. A. Wolfe & M. L. Silberman 1971. A Pliocene flora and insect fauna from the Bering Strait region. *Palaeogeog. Palaeoclimatol. Palaeoecol.* **9**, 211–31.

Hopping, C. A. 1967. Palynology and the oil industry. *Rev. Palaeobot. Palynol.* **2**, 23–48.

Horodyski, R. J. 1980. Middle Proterozoic shale-facies microbiota from the Lower Belt Supergroup, Little Belt Mountains, Montana. *J. Paleont.* **54**(4), 649–63.

Horowitz, A. 1973. Late Permian palynomorphs from southern Israel. *Pollen et Spores* **15**(2), 315–41.

Hsü, K. J. & F. Giovanoli 1979. Messinian event in the Black Sea. *Palaeogeog. Palaeoclimatol. Palaeoecol.* **29**, 75–93.

Hueber, F. M. 1982. *Megaspores and a palynomorph from the Lower Potomac Group in Virginia*. Smithsonian Contrib. Paleobiol. 49, 1–69.

Hughes, N. F. 1961. Further interpretation of *Eucommiidites* Erdtman 1948. *Palaeontology* **4**(2), 292–9.

Hughes, N. F. 1970. The need for agreed standards of recording in paleopalynology and palaeobotany. *Paläont. Abh. B.* **3**, 357–64.

Hughes, N. F. 1975. The challenge of abundance in palynomorphs. *Geosci. & Man* **11**, 141–4.

Hughes, N. F. 1976. *Palaeobiology of angiosperm origins: problems of Mesozoic seed-plant evolution*. Cambridge: Cambridge University Press.

Hughes, N. F. 1984. Earliest fossil pollen with angiospermid characters. *6th Int. Palynol. Conf.* Calgary, Abstracts, 67.

Hughes, N. F. & C. A. Croxton 1973. Palynologic correlation of the Dorset "Wealden". *Palaeontology* **16**(3), 567–601.

Hughes, N. F., G. E. Drewry & J. F. Laing 1979. Barremian earliest angiosperm pollen. *Palaeontology* **22**(3), 513–35.

Hughes, N. F. & J. Moody-Stuart 1966. Descriptions of schizaeaceous spores taken from early Cretaceous macrofossils. *Palaeontology* **9**(2), 274–89.

Hughes, N. F. & J. C. Moody-Stuart 1969. A method of stratigraphic correlation using early Cretaceous spores. *Palaeontology* **12**(1), 84–111.

Huntley, B. & H. J. B. Birks 1983. *An atlas of past and present pollen maps for Europe: 0–13,000 years ago.* Cambridge: Cambridge University Press.

Hyde, H. A. & D. W. Williams 1944. Right word. *Pollen Anal. Circ.* **8**, 6.

Ibrahim, A. C. 1933. *Sporenformen des Aegirhorizonts des Ruhr-Reviers.* Würzburg: Buchdruckerei Konrad Triltsch.

Iversen, J. 1936. Secondary pollen as a source of error. *Danmarks. Geol. Unders. IV* (2),**15**, 1–24.

Iversen, J. 1941. Land occupation in Denmark's Stone Age. A pollen-analytical study of the influence of farmer culture on the vegetational development. *Danmarks. Geol. Unders.* (2), **66**, 20–65.

Iversen, J. 1944. *Viscum, Hedera,* and *Ilex* as climate indicators. *Geol. Fören. Stockholm Förh.* **66**, 493.

Iversen, J. 1954. The late-glacial flora of Denmark and its relation to climate and soil. *Danmarks Geol. Unders.* (2) **80**, 87–119.

Iversen, J. & J. Troels-Smith 1950. Pollenmorfologiske definitionen og types. *Danmarks Geol. Unders. IV* (3), **8**.

Jacobson, G. L., Jr. & R. H. W. Bradshaw 1981. The selection of sites for paleovegetational studies. *Quat. Res.* **16**, 80–96.

Jacobson, S. R. & D. J. Nichols 1982. Palynological dating of syntectonic units in the Utah–Wyoming thrust belt: the Evanston Formation, Echo Canyon Conglomerate, and Little Muddy Creek Conglomerate. In *Geologic studies of the Cordilleran thrust belt,* Vol. 2, R. B. Powers (ed.), 735–750. Denver, Col.: Rocky Mountain Assoc. Geologists.

Jacobson, S. R., B. R. Wardlaw & J. D. Saxton 1982. Acritarchs from the Phosphoria and Park City Formations (Permian, northeastern Utah). *J. Paleont.* **56**(2), 449–58.

Jain, R. K. 1968. Middle Triassic pollen grains and spores from Mines de Petroleo beds of the Cacheuta Formation (Upper Gondwana) Argentina. *Palaeontographica B* **122**, 1–47.

Jan du Chêne, R. 1974. Etude palynologique du Néogene et du Pleistocène inférieur de Bresse. *Bull. Rech. Geol. Miniéres Ser. 2,* **1**, 209–35.

Jansen, J. H. F., A. Kuijpers & S. R. Troelstra 1986. A mid-Brunhes climatic event: long-term changes in global atmosphere and ocean circulation. *Science* **232**, 619–21.

Jansonius, J. 1970. Classification and stratigraphic application of Chitinozoa. *Proc. North Am. Paleontol. Conv.* G, 789–808.

Jansonius, J. 1978. A key to the genera of fossil angiosperm pollen. *Rev. Palaeobot. Palynol.* **26**, 143–72.

Jansonius, J. & J. H. Craig 1971. Scolecodonts: I. Descriptive terminology and revision of systematic nomenclature: II. Lectotypes, new names for homonyms, index of species. *Bull. Can. Petrolm. Geol.* **19**(1), 251–302.

Jansonius, J. & J. H. Craig 1974. Some scolecodonts in organic association from Devonian strata of western Canada. *Geosci. & Man* **9**, 15–26.

Jansonius, J. & L. V. Hills 1976–. *Genera file of fossil spores and pollen.* Dept. Geol., University of Calgary, Alberta, Spec. Publ.

Janssen, C. R. 1967. A comparison between the recent regional pollen rain and the sub-recent vegetation in four major vegetation types in Minnesota (U.S.A.). *Rev. Palaeobot. Palynol.* **2**, 331–42.

Janssen, C. R. 1974. *Verkenningen in de Palynologie*. Utrecht: Oosthoek, Scheltema & Holkema.

Jardiné, S. & L. Magloire 1963. Palynologie et stratigraphie du Crétacé des bassins du Sénégal et de Côte d'Ivoire. *Mém. Bur. Rech. Géol. Miniéres* 32, 187–245.

Jardiné, S. *et al.* 1974. Distribution stratigraphique des acritarches dans le Paléozoïque du Sahara Algérien. *Rev. Palaeobot. Palynol.* 18, 99–129.

Jenkins, W. A. M. 1967. Ordovician chitinozoa from Shropshire. *Palaeontology* 10(3), 436–88.

Jenkins, W. A. M. 1970. Chitinozoa. *Geosci. & Man* 1, 1–21.

Jennings, J. R. & M. A. Millay 1978. A new permineralized fern from the Pennsylvanian of Illinois. *Palaeontology* 21, 709–16.

de Jersey, N. J. 1965. *Plant microfossils in some Queensland crude oil samples.* Geol. Surv. Queensland Publ. 329.

de Jersey, N. J. 1982. An evolutionary sequence in *Aratrisporites* miospores from the Triassic of Queensland, Australia. *Palaeontology* 25(3), 665–72.

Jessen, K. 1920. Bog investigations in North-East Sjaelland. *Danmarks. Geol. Unders.* (2)34, (in Danish).

Jessen, K. 1935. The composition of the forests in Northern Europe in Epipalaeolithic time. K. Danske Videnske. Selske., *Biol. Medd.* XII:1.

Jiang, D. 1984. Pollen and spores from crude oil of Talimu Basin, China. Poster display at *Am. Assoc. Strat. Palynol.* 17th Ann. Mtg., Arlington, Va.

Jiang, D. & H. Yang 1980. Petroleum sporopollen assemblages and oil source rock of Yumen oil-bearing region in Gansu. *Acta Bot. Sinica* 22(3), 280–5.

Johnson, N. G. 1984. *Early Silurian palynomorphs from the Tuscarora Formation in central Pennsylvania and their paleobotanical and geological significance.* M.Sc. Thesis, Pennsylvania State University.

Johnson, N. G. 1985. Early Silurian palynomorphs from the Tuscarora Formation in central Pennsylvania and their paleobotanical and geological significance. *Rev. Palaeobot. Palynol.* 45, 307–60.

Jones, P. 1984. A note on D.P.X mountant. *Am. Assoc. Strat. Palynol. Newsl.* 17(4), 7–8.

Kaars, W. A. & J. Smit 1985. A palynological field preparation technique. *Pollen et Spores* 27, 493–6.

Kapp, R. O. 1969. *How to know pollen and spores.* Dubuque, Iowa: Wm. C. Brown.

Kedves, M. 1983. Development of the European Brevaxones pollen grains and the main stages of their evolution during the lower and middle Senonian. *Pollen et Spores* 25(3/4), 487–98.

Kedves, M., L. Endrédi & Z. Szeley 1966. Problémes palynologiques concernant le remaniemant des sédiments Paléo- et Mésozoiques dans des bassins du Pannonien supérieur de Hongrie. *Pollen et Spores* 8, 315–36.

Keegan, J. B. 1981. Palynological correlation of the Upper Devonian and Lower Carboniferous in central Ireland. *Rev. Palaeobot. Palynol.* 34, 99–105.

Kemp, E. M. (=Truswell, E. M.) 1968. Probable angiosperm pollen from the British Barremian to Albian strata. *Palaeontology* 11(3), 421–34.

Kemp, E. M. (=Truswell, E. M.) 1972. Reworked palynomorphs from the West Ice Shelf area, East Antarctica, and their possible geological and palaeoclimatological significance. *Marine Geol.* 13, 145–57.

Kemp, E. M. (=Truswell, E. M.) 1978. Microfossils of fungal origin from Tertiary sediments on the Ninety-east Ridge, Indian Ocean. *Bureau Min. Resources (Australia) Bull.* 192, 73–81.

Kemp, E. M. (=Truswell, E. M.) & W. K. Harris, 1977. *The palynology of Early Tertiary sediments Ninetyeast Ridge, Indian Ocean*. Palaeont. Assoc. Spec. Pap. in Palaeont. 19.

Kerr, R. A. 1982. New evidence fuels Antarctic ice debate. *Science* **216**, 973–4.

Klapper, G. & W. Ziegler 1979. Devonian conodont biostratigraphy In *The Devonian system*, M. R. House *et al.* (eds.), 199–224. Palaeont. Assoc. Spec. Pap. 23.

Klaus, W. 1955. Über die Sporendiagnose des deutschen Zechsteinsalzes und des alpinen Salzgebirges. *Z. Deutschen Geol. Gesellschaft* **105**, 776–88.

Knoll, A. H. & E. S. Barghoorn 1975. Precambrian eukaryotic organisms: a reassessment of the evidence. *Science* **190**, 52–4.

Kofoid, C. A. 1907. The plates of *Ceratium* with a note on the unity of the genus. *Zoolog. Anzeiger* **32**, 177–83.

Kofoid, C. A. 1909. On *Peridinium steini* Jörgensen, with a note on the nomenclature of the skeleton of the Peridinidae. *Archiv Protistenkunde* **16**, 25–47.

Koreneva, E. V. 1964. *Distribution and preservation of pollen in sediments in the western part of the Pacific Ocean*. Geol. Inst., Acad. Sci. U.S.S.R. Trudy 109 (in Russian).

Koreneva, E. V. 1966. Marine palynological researches in the U.S.S.R. *Marine Geol.* **4**, 565–74.

Koreneva, E. V. 1971. Spores and pollen in Mediterranean bottom sediments. In *The micropalaeontology of oceans*, B. M. Funnel & W. R. Riedel (eds.), 361–71. Cambridge: Cambridge University Press.

Kosanke, R. M. 1950. *Pennsylvanian spores of Illinois and their use in correlation*. Illinois St. Geol. Surv. Bull. 74.

Kovach, W. & D. L. Dilcher 1985. Morphology, ultrastructure and paleoecology of *Paxillitriletes vittatus* sp. nov. from the mid-Cretaceous (Cenomanian) of Kansas. *Palynology* **9**, 85–94.

Krassilov, V. 1978. Mesozoic lycopods and ferns from the Bureja Basin. *Palaeontographica B* **166**, 16–29.

Kremp, G. O. W. 1965. *Morphologic encyclopedia of palynology*. Tucson: University of Arizona Press.

Kress, W. J. & D. E. Stone 1982. Nature of the sporoderm in monocotyledons, with special reference to the pollen grains of *Canna* and *Heliconia*. *Grana* **21**, 129–48.

Krutzsch, W. 1970. Zur Kenntnis fossiler disperser Tetradenpollen. *Paläont. Abh. B* **3**(3/4), 399–426.

Kummel, B. & D. Raup (eds.) 1965. *Handbook of paleontological techniques* (see esp. 598–613). San Franciso: W. H. Freeman.

Kurmann, M. H. 1983. The ultrastructure of pollen extracted from *Boulaya fertilis*. *Am. J. Bot* **70**(5), 72–3.

Kuroda, T. 1980. *An exploratory study of evaluation of factors controlling the dispersal process of pollen and spores by the statistical factor analysis*. Doctoral dissertation. Dept. Geol., Kyushu University.

Kuyl, O. S., J. Muller & H. T. Waterbolk 1955. The application of palynology to oil geology with reference to western Venezuela. *Geol. en Mijnbouw*(N.S.) **17**(3), 49–76.

Laing, J. F. 1976. The stratigraphic setting of early angiosperm pollen. In *The evolutionary significance of the exine*, I. K. Ferguson & J. Muller (eds.), 15–26. London: Academic Press. Linn. Soc. Symp. Ser. 1.

Laveine, J.-P. 1969. Quelques pécoptéridinées houillères a la lumière de la palynologie. *Pollen et Spores* **11**, 619–68.

Leffingwell, H. A. & N. Hodgkin 1971. Techniques for preparing fossil palynomorphs for study with the scanning and transmission electron microscopes. *Rev. Palaeobot. Palynol.* **11**, 177–99.

Legault, J. A. & G. Norris 1982. Palynological evidence for recycling of Upper Devonian into Lower Cretaceous of the Moose River Basin, James Bay Lowland, Ontario. *Can. J. Earth Sci.* **19**(1), 1–7.

Leisman, G. A. 1970. A petrified *Sporangiostrobus* and its spores from the Middle Pennsylvanian of Kansas. *Palaeontographica B* **129**, 166–77.

Leisman, G. A. & J. S. Peters 1970. A new pteridosperm male fructification from the Middle Pennsylvanian of Illinois. *Am. J. Bot.* **57**, 867–73.

Leisman, G. A. & T. L. Phillips 1979. Megasporangiate and microsporangiate cones of *Achlamydocarpon varius* from the Middle Pennsylvanian. *Palaeontographica B* **168**, 100–28.

Leisman, G. A. & B. M. Stidd 1967. Further occurrences of *Spencerites* from the middle Pennsylvanian of Kansas and Illinois. *Am. J. Bot.* **54**(3), 316–23.

Lele, K. M., P. K. Maithy & J. Mandal 1981. *In situ* spores from Lower Gondwana ferns – their morphology and variation. *Palaeobotanist* **28–29**, 128–54.

Leopold, E. B. 1967. Late-Cenozoic patterns of plant extinction. In *Pleistocene extinctions: the search for a cause*, P. S. Martin & H. E. Wright, Jr. (eds.), 203–46. New Haven, Conn.: Yale University Press.

Leopold, E. B. 1969. Late Cenozoic palynology. In *Aspects of palynology*, R. H. Tschudy & R. A. Scott (eds.), 377–438. New York: Wiley-Interscience.

Leopold, E. B. 1984. Comparative age of steppe and grassland east and west of the rocky mountains, USA. *6th Int. Palynol. Conf. Calgary*, Abstracts, 88.

Leopold, E. B., R. Nickmann, J. I. Hedges & J. R. Ertel 1982. Pollen and lignin records of Late Quaternary vegetation, Lake Washington. *Science* **218**, 1305–7.

Leopold, E. B. & V. C. Wright 1985. Pollen profiles of the Plio- Pleistocene transition in the Snake River plain, Idaho. In *Late Cenozoic history of the Pacific Northwest*, C. J. Smiley (ed.), 323–48. Pacific Div., Am. Assoc. Adv. Sci.

Leroi-Gourhan, A. 1975. The flowers found with Shanidar IV, a Neanderthal burial in Iraq. *Science* **190**, 562–4.

Leschick, G. 1956. Die Entstehung der Braunkohle der Wetterau und ihre Mikro- und Makroflora. *Palaeontographica B* **100**, 26–64.

Lesnikowska, A. D. & M. A. Millay 1985. Studies of Paleozoic marattialeans: new species of *Scolecopteris* (Marattiales) from the Pennsylvanian of North America. *Am. J. Bot.* **72**(5), 649–58.

Le Thomas, A. 1980–81. Ultrastructural character of the pollen grains of African Annonaceae and their significance for the phylogeny of primitive angiosperms. (1st part) *Pollen et Spores* **22**(3/4), 267–342; (2nd part) *Pollen et Spores* **23**(1), 5–36.

Leuschner, R. M. & G. Boehm 1979. Investigations with the 'Individual Pollen Collector' and the 'Burkhard Trap' with reference to hay fever patients. *Clin. Allergy* **9**, 175–84.

Leuschner, R. M. & G. Boehm 1981. Pollen and inorganic particles in the air of climatically very different places in Switzerland. *Grana* **20**, 161–7.

Litwin, R. J. 1985. Fertile organs and *in situ* spores of ferns from the Late Triassic Chinle Formation of Arizona and New Mexico, with discussion of the associated dispersed spores. *Rev. Palaeobot. Palynol.* **44**, 101–46.

Livingstone, D. A. & T. Van der Hammen 1978. Palaeogeography and palaeoclimatology. In *Tropical forest ecosystems*, UNESCO/UNEP/FAO. Paris: Unesco. *Natural Resources Res.* **14**(3), 61–90.

Loeb, R. E. 1984. The accuracy and reliability of the pollen record in representing regional forest change in the past century. *6th Int. Palynol. Conf.* Calgary, Abstracts, 95.

Loewus, F. A., *et al.* 1985. Pollen sporoplasts: dissolution of pollen walls. *Plant Physiol.* **78**, 652–4.

Lona, F., R. Bertoldi & E. Ricciardi 1969. Plio-Pleistocene boundary in Italy based on the Leffian and Tiberian vegetational and climatological sequences. *8th INQUA Congr.* Paris, Vol. 2, 573–4.

Loquin, M. V. 1981. Affinités fongiques probables des Chitinozoaires devenant Chitinomycétes. *Cah. Micropaléont.* **1**, 29–36.

Lucas-Clark, J. 1984. Some friends of Cribroperidinium: Spongodinium and a new genus. *17th Ann. Mtg. A.A.S.P.*, Program & Abstracts, 13.

Lund, J. J. & K. R. Pedersen 1985. Palynology of the marine Jurassic formations in the Vardekloft ravine, Jameson Land, East Greenland. *Bull. Geol. Soc. Denmark* **33**, 371–99.

Mack, R. N. & V. M. Bryant, Jr. 1974. Modern pollen spectra from the Columbia Basin, Washington. *Northwest Sci.* **48**(3), 183–94.

Mack, R. N., V. M. Bryant, Jr. & W. Pell 1978. Modern forest pollen spectra from eastern Washington and northern Idaho. *Bot. Gaz.* **139**(2), 249–55.

Mack, R. N., N. W. Rutter & S. Valastro 1979. Holocene vegetation history of the Okanogan Valley, Washington. *Quat. Res.* **12**, 212–25.

Mädler, K. 1963. Die figurierten organischen Bestandteile der Posidonienschiefer. *Beih. Geol. Jahrb.* **58**, 287–406.

Madsen, D. B. & D. R. Currey 1979. Late Quaternary glacial and vegetation changes, Little Cottonwood Canyon Area, Wasatch Mountains, Utah. *Quat. Res.* **12**, 254–70.

Maher, L. J., Jr. 1964. *Ephedra* pollen in sediments of the Great Lakes region. *Ecology* **45**, 391–5.

Maher, L. J., Jr. 1972. Nomograms for computing 0.95 confidence limits of pollen data. *Rev. Palaeobot. Palynol.* **13**, 85–93.

Maheshwari, H. K. & S. V. Meyen 1975. *Cladostrobus* and the systematics of cordaitalean leaves. *Lethaia* **8**, 103–23.

Malyavkina, V. S. 1949. *Key to spores and pollen: Jurassic–Cretaceous.* Trudy VNIGRI N.S. no. 33 (in Russian).

Manchester, S. R. & P. R. Crane 1983. Attached leaves, inflorescences, and fruits of Fagopsis, an extinct genus of fagaceous affinity from the Oligocene Florissant flora of Colorado, U.S.A. *Am. J. Bot.* **70**(8), 1147–64.

Mandrioli, P., M. G. Negrin & A. L. Zanotti 1982. Airborne pollen from the Yugoslavian coast to the Po Valley (Italy). *Grana* **21**, 121–8.

Mangerud, J., E. Sønstegaard & H.-P. Sejrup 1979. Correlation of the Eemian (interglacial) stage and the deep-sea oxygen-isotope stratigraphy. *Nature* **277**, 189–92.

Manum, S. 1976a. Dinocysts in Tertiary Norwegian–Greenland sea sediments (Deep Sea Drilling Project Leg 38), with observations on palynomorphs and palynodebris in relation to environment. In *Initial reports Deep Sea Drilling Project 38*, M. Talwani, L. Udintsev *et al.* (eds.), 897–919.

Manum, S. 1976b. Individual site reports, D.S.D.P. Leg 38. In *Initial Reports Deep Sea Drilling Project 38*, M. Talwani, L. Udintsev *et al.* (eds.), II: 44–5, 196, 198, 203–4, 206, 399, 464, 535–8, 606–7, 662–3.

Manum, S. B. & T. Throndsen 1978. Dispersed organic matter (kerogen) in the Spitsbergen Tertiary. *Norsk Polarinst. Årbok* **1977**, 179–87.

Mapes, G. 1983. Permineralized *Lebachia* pollen cones. *Am. J. Bot.* **70**(5), 74.

Mapes, G. & J. T. Schabilion 1979a. A new species of *Acitheca* (Marattiales) from the Middle Pennsylvanian of Oklahoma. *J. Paleont.* **53**, 685–94.

Mapes, G. & J. T. Schabilion 1979b. *Millaya* gen. n., an upper Paleozoic genus of marattialean synangia. *Am. J. Bot.* **66** (10), 1164–72.

Marcinkiewicz, T. 1979. Megaspores. *Budowa geologiezna Polski*, Tom 3, *Atlas skamienialości Przewodnich i charakterystycznych*, czę'ść 2a, *Mesozoik, Trias*, 201–214 (in Polish).

Margulis, L. 1981. *Symbiosis in cell evolution*. San Francisco: W. H. Freeman.

Markgraf, V. 1974. Paleoclimatic evidence derived from timberline fluctuations. *Colloq. Int. C.N.R.S.* **219**, 67–77.

Markgraf, V. & L. Scott 1981. Lower timberline in central Colorado during the past 15,000 yr. *Geology* **9**(5), 231–4.

Markova, L. G. 1966. Plant spores of the family Schizaeaceae in Cretaceous deposits of the western-Siberian depression and their significance for stratigraphy. In *About the method of paleopalynological research*, N. A. Bolkhovitina, *et al.* (eds.). 2nd Int. Palynol. Conf., Leningrad. VSEGEI, 214–35 (in Russian).

Martin, P. S. & P. J. Mehringer, Jr. 1965. Pleistocene pollen analysis and biogeography of the Southwest. In *The Quaternary of the United States*, Rev. Vol. for 7th INQUA Congr., H. E. Wright, Jr. & D. G. Frey (eds.), 433–51. Princeton, N.J.: Princeton University Press.

Martin, P. S. & F. W. Sharrock 1964. Pollen analysis of prehistoric human feces: a new approach to ethnobotany. *Am. Antiquity* **30**, 168–80.

Masran, T. C. 1984. Sedimentary organic matter of Jurassic and Cretaceous samples from North Atlantic deep-sea drilling sites. *Bull. Can. Petrolm. Geol.* **32**(1), 52–73.

Masran, T. C. & S. A. J. Pocock 1981. The classification of plant-derived particulate organic matter in sedimentary rocks. In *Organic maturation studies and fossil fuel exploration*, J. Brooks (ed.). London: Academic Press.

Matsushita, M. 1982. Palynological researches of surface sediments in the Harima Nada, Seto Inland Sea – behaviour of pollen grains and spores. *Quat. Res. (Japan)* **21**(1), 15–22.

McAndrews, J. H. 1967. Pollen analysis and vegetational history of the Itasca region, Minnesota. In *Proc. 7th Cong. INQUA*, E. J. Cushing & H. E. Wright, Jr. (eds.), 219–36. New Haven, Conn.: Yale University Press.

McAndrews, J. H. 1984. Pollen analysis of the 1973 ice core from Devon Island glacier, Canada. *Quat. Res.* **22**, 68–76.

McAndrews, J. H. & D. M. Power 1973. Palynology of the Great Lakes: the surface sediments of Lake Ontario. *Can. J. Earth Sci.* **10**(5), 777–92.

McDowell, L. L., R. M. Dole, M. Howard, Jr. & R. A. Farrington 1971. Palynology and radio-carbon chronology of Bugbee wildflower sanctuary and natural area, Caledonia County, Vermont. *Pollen et Spores* **13**, 73–91.

McGregor, D. C. 1965. *Triassic, Jurassic, and Lower Cretaceous spores and pollen of Arctic Canada*. Geol. Surv. Can. Pap. 64–55.

McGregor, D. C. 1979a. *Spores in Devonian stratigraphical correlation*. Spec. Pap. Paleont. **23**, 163–84.

McGregor, D. C. 1979b. Devonian miospores of North America. *Palynology* **3**, 31–52.

McGregor, D. C. 1981. Spores and the Middle-Upper Devonian Boundary. *Rev. Palaeobot. Palynol.* **34**, 25–47.

McGregor, D. C. & G. M. Narbonne 1978. Upper Silurian trilete spores and other microfossils from the Read Bay Formation, Cornwallis Island, Canadian Arctic. *Can. J. Earth Sci.* **15**(8), 1292–303.

McIntire, D. J. 1972. Effect of experimental metamorphism on pollen in a lignite. *Geosci. & Man* **4**, 111–17.

Medus, J. 1970. Contribution à la classification des grains de pollen du groupe des *Circumpolles* (Pflug) Klaus. *Pollen et Spores* **12**(2), 205–16.

Medus, J. 1977. The ultrastructure of some Circumpolles. *Grana* **16**, 23–8.

Medus, J. 1982. Pollens de Restionales dans des sediments Paléocènes du Sénégal. *Cah. Micropaléont.* **2**, 75–80.

Medus, J. 1983. Quelques elements palynofloristiques de l'Hauterivien du Portugal. *Actas del IV Simposio de Palinologia*, APLE-Barcelona, Oct. 1982. 397–409.

Meeuse, B. & S. Morris 1984. *The sex life of flowers*. London: Faber & Faber.

Mehringer, P. J., Jr., *et al.* 1977. Pollen influx and volcanic ash. *Science* **198**(4314), 257–61.

Melia, M. B. 1982. The distribution and relationship between palynomorphs in aerosols and deep-sea sediments off the coast of northwest Africa. *Palynology* **6**, 285 (abstract).

Melia, M. B. 1984. The distribution and relationship between palynomorphs in aerosols and deep-sea sediments off the coast of northwest Africa. *Marine Geol.* **58**, 345–71.

Melville, R. 1981. Surface tension, diffusion and the evolution and morphogenesis of pollen aperture patterns. *Pollen et Spores* **23**(2), 179–203.

Menke, B. 1975. Vegetationsgeschichte und Florenstratigraphie Nordwestdeutschlands im Pliozän und Frühquartär mit einem Beitrag zur Biostratigraphie des Weichsel-frühglazials. *Geol. Jahrb. A*, **26**, 3–151.

Menke, B. 1976. Pliozäne und altestquartäre Sporen- und Pollenflora von Schleswig-Holstein. *Geol. Jahrb. A*, **32**, 3–197.

Meon-Vilain, H. 1970. *Palynologie des formations Miocènes supérieures et Pliocènes du Bassin du Rhone (France)*. Doc. Lab. Géol. Fac. Sci. Lyon 38.

Mercer, J. H. 1973. Cainozoic temperature trends in the southern hemisphere: Antarctic and Andean glacial evidence. *Palaeoecol. Africa* **8**, 85–114.

Meyer, K. J., L. Benda & P. Steffens 1978. Pollenanalytische Untersuchungen im plio-pleistozänen Grenzbereich Norditaliens. *Newsl. Strat.* **7**, 26–44.

Mickle, J. E. 1980. *Ankyropteris* from the Pennsylvanian of eastern Kentucky. *Bot. Gaz.* **141**(2), 230–43.

Mickle, J. E. & G. W. Rothwell 1986. Vegetative and fertile structures of *Cyathotheca ventilaria* from the Upper Pennsylvanian of the Appalachian Basin. *Am. J. Bot.* **73**, 1474–85.

Millay, M. A. 1977. *Acaulangium* Gen. Nov., a fertile marattialean from the Upper Pennsylvanian of Illinois. *Am. J. Bot.* **64**, 223–9.

Millay, M. A. 1978. Studies of Paleozoic marattialeans: the morphology and phylogenetic position of *Eoangiopteris goodii*, sp. n. *Am. J. Bot.* **65**, 577–83.

Millay, M. A. 1979. Studies of Paleozoic marattialeans: a monograph of the American species of *Scolecopteris. Palaeontographica B* **169**, 1–69.

Millay, M. A. 1982a. Studies of Paleozoic marattialeans: an Early Pennsylvanian species of the fertile fern *Scolecopteris. Am. J. Bot.* **69**(5), 728–33.

Millay, M. A. 1982b. Studies of Paleozoic marattialeans: the morphology and probable affinities of *Telangium pygmaeum* Graham. *Am. J. Bot.* **69**(10), 1566–73.

Millay, M. A. 1982c. Studies of Paleozoic marattialeans: an evaluation of the genus *Cyathotrachus* (Watson) Mamay. *Palaeontographica B* **180**, 65–81.

Millay, M. A., D. A. Eggert & R. L. Dennis 1978. Morphology and ultrastructure of four Pennsylvanian prepollen types. *Micropaleontology* **24**(3), 303–15.

Millay, M. A. & G. W. Rothwell 1983. Fertile pinnae of *Biscalitheca* (Zygopteridales) from the Upper Pennsylvanian of the Appalachian Basin. *Bot. Gaz.* **144**(4), 589–99.

Millay, M. A., G. W. Rothwell & D. A. Eggert 1980. Ultrastructure of *Stewartiotheca* (Medullosaceae) prepollen. *Am. J. Bot.* **67**(2), 223–7.

Millay, M. A. & T. N. Taylor 1970. Studies of living and fossil saccate pollen grains. *Micropaleontology* **16**, 463–70.

Millay, M. A. & T. N. Taylor 1974. Morphological studies of Paleozoic saccate pollen. *Palaeontographica B* **147**, 75–99.

Millay, M. A. & T. N. Taylor 1976. Evolutionary trends in fossil gymnosperm pollen. *Rev. Palaeobot. Palynol.* **21**, 65–91.

Millay, M. A. & T. N. Taylor 1977. *Feraxotheca* Gen. N., a lyginopterid pollen organ from the Pennsylvanian of North America. *Am. J. Bot.* **64**, 117–85.

Millay, M. A. & T. N. Taylor 1980. An unusual botryopterid sporangial aggregation from the Middle Pennsylvanian of North America. *Am. J. Bot.* **67**(5), 758–73.

Millay, M. A. & T. N. Taylor 1982. The ultrastructure of Paleozoic fern spores: I. *Botryopteris. Am. J. Bot.* **69**(7), 1148–55.

Millay, M. A. & T. N. Taylor 1984. The ultrastructure of Paleozoic fern spores: II. *Scolecopteris* (Marattiales). *Palaeontographica B* **194**, 1–13.

Miller, F. X. 1977. Biostratigraphic correlation of the Mesaverde Group in Southwestern Wyoming and Northwestern Colorado. *Rocky Mtn. Assoc. Geologists 1977 Symp.*, 117–137.

Misset, M.-T., J. P. Gourret & A. Huon 1982. Le pollen d'*Ulex* L. (Papilionoideae): morphologie des grains et structure de l'exine. *Pollen et Spores* **24**(3–4), 369–95.

Mosimann, J. E. 1965. Statistical methods for the pollen analyst: multinomial and negative multinomial techniques. In *Handbook of paleontological techniques*, B. Kummel & D. M. Raup (eds.), 636–73. San Francisco: W. H. Freeman.

Moy, C. J. 1987. Variations of fern spore ultrastructure as reflections of their evolution. *Grana.*

Mudie, P. J. 1982. Pollen distribution in recent marine sediments, eastern Canada. *Can. J. Earth Sci.* **19**, 729–47.

Muir, M. D. 1967. Reworking in Jurassic and Cretaceous spore assemblages. *Rev. Palaeobot. Palynol.* **5**, 145–54.

Muller, J. 1959. Palynology of Recent Orinoco Delta and shelf sediments. *Micropaleontology* **5**, 1–32.

Muller, J. 1970. Palynological evidence on early differentiation of angiosperms. *Biol. Rev.* **45**, 417–50.

Muller, J. 1981. Fossil pollen records of extant angiosperms. *Bot. Rev.* **47**(1).

Muller, J. 1984. Significance of fossil pollen for angiosperm history. *Ann. Missouri Bot. Gard.* **71**(2), 419–43.

Nagy, E. 1985. *Sporomorphs of the Neogene in Hungary.* Inst. Geol. Hungary. (In Hungarian and English.)

Naumova, S. N. 1953. *Complexes sporo-pollinique du Devonien supérieur de la plateforme russe et leur valeur stratigraphique.* Works on the Inst. Sci., Geol. 143, Geol. Ser. 60 (in Russian).

Nautiyal, A. C. 1977. The paleogeographic distribution of Devonian acritarchs and biofacies belts. *J. Geol. Soc. India* **18**, 53–64.

Needham, H. D., D. Habib & B. C. Heezen 1969. Upper Carboniferous palynomorphs as a tracer of red sediment dispersal patterns in the northwest Atlantic. *J. Geol.* **77**, 113–20.

Neves, R. & B. Dale 1963. A modified filtration system for palynological preparations. *Nature* **198**, 775–6.

Neves, R. & B. Owens 1966. Some Namurian camerate miospores from the English Pennines. *Pollen et Spores* **8**(2), 337–60.

Nichols, D. J. 1984. Latitudinal variation in terrestrial palynomorph assemblages from Maestrichtian deposits in western North America. *6th Int. Palynol. Conf.* Abstracts, 117.

Nichols, D. J. & S. R. Jacobson 1982. Palynostratigraphic framework for the Cretaceous (Albian–Maestrichtian) of the Overthrust Belt of Utah and Wyoming. *Palynology* **6**, 119–47.

Niklas, K. 1976. Chemical examination of some non-vascular Paleozoic plants. *Brittonia* **28**(1), 113–37.

Niklas, K. J. 1984. The motion of windborne pollen grains around ovulate cones: implications on wind pollination. *Am. J. Bot.* **71**(3), 356–74.

Niklas, K. J. 1985. Wind-pollination – a study in controlled chaos. *Am. Scientist* **73**, 462–70.

Nitsch, J. P. & C. Nitsch 1969. Haploid plants from pollen grains. *Science* **163**, 85–7.

Nowicke, J. W. & J. J. Skvarla 1977. *Pollen morphology and the relationship of the Plumbaginaceae, Polygonaceae, and Primulaceae to the order Centrospermae.* Smithsonian Contrib. Bot. 37.

Ogden, E. C., *et al.* 1974. *Manual for sampling airborne pollen.* New York: Hafner.

Olsen, P. E. 1978. On the use of the term Newark for Triassic and Early Jurassic rocks of eastern North America. *Newsl. Strat.* **7**(2), 90–5.

Orbell, G. 1973. Palynology of the British Rhaeto-Liassic. *Bull. Geol. Surv. Great Br.* **44**, 1–44.

Orłowska-Zwolińska, T. 1979. Microplankton and miospores. *Budowa geologiezna Polski,* Tom 3, *Atlas skamienialości Przewodnich i charakterystycznych,* część 2a, *Mesozoik, Trias,* 150–6, 159–201 (in Polish).

Orłowska-Zwolińska, T. 1983. Palynostratigraphy of the upper part of Triassic epicontinental sediments in Poland. *Prace Inst. Geol. (Warsaw)* **104**, 1–89.

Overbeck, F. 1975. *Botanisch–geologische Moorkunde* (see esp. tabs. 24–25). Neumünster: Karl Wackholtz.

Overbeck, F. & S. Schneider 1938. Mooruntersuchungen bei Lüneburg und bei Bremen und die Reliktnatur von *Betula nana* L. in Nordwestdeutschland. *Ztschr. Bot.* **33**, 1–54.

Overpeck, J. T., T. Webb III & I. C. Prentice 1985. Quantitative interpretation of fossil pollen spectra: dissimilarity coefficients and the method of modern analogs. *Quat. Res.* **23**, 87–108.

Pant, D. D. & N. Basu 1979. Some further remains of fructifications from the Triassic of Nidpur, India. *Palaeontographica B* **168**, 129–46.

Pant, D. D. & D. D. Nautiyal 1960. Some seeds and sporangia of Glossopteris flora from Raniganj coalfield, India. *Palaeontographica B* **107**, 41–64.

Paris, F. 1981. *Les Chitinozoaires dans le Paleozoïque du sud-ouest de l'Europe.* Mem. Soc. Géol. Min. Bretagne 26.

Paris, F. 1984. Arenigian chitinozoans from the Klabava Formation, Bohemia. *6th Int. Palynol. Conf.* Calgary, Abstracts, 124.

Parsons, R. W. & I. C. Prentice 1981. Statistical approaches to *R*-values and the pollen–vegetation relationship. *Rev. Palaeobot. Palynol.* **32**, 127–52.

Patel, V., *et al.* 1985. The nature of threadlike structures and other morphological characters in *Jacqueshuberia* pollen (Leguminosae: Caesalpinioideae). *Am. J. Bot.* **72**(3), 407–13.

Pearson, D. L. 1984. *Pollen/spore color "standard",* version 2. Phillips Petroleum Co., privately distributed.

Peck, R. M. 1973. Pollen budget studies in a small Yorkshire catchment. In *Quaternary plant ecology,* H. J. B. Birks & R. G. West (eds.), 43–60. Oxford: Blackwell Scientific.

Penck, A. & E. Brückner 1909. *Die Alpen in Eiszeitalter,* 3 Vols. Leipzig.

Peppers, R. A., in Phillips, T. L. & W. A. DiMichele 1981, 259–63.

Peppers, R. A. & H. W. Pfefferkorn 1970. A comparison of the floras of the Colchester (No. 2) coal and Francis Creek Shale. In *Depositional environments in parts of the Carbondale Formation – western and northern Illinois,* W. H. Smith *et al.* (eds.), 61–74. Illinois St. Geol. Surv. Guidebook Ser. 8.

Peterson, G. M., T. Webb III, J. E. Kutzback, T. van der Hammen, T. A. Wijmstra & F. A. Street 1979. The continental record of environmental conditions at 18,000 yr. B.P.: an initial evaluation. *Quat. Res.* **12**, 47–82.

Pettit, J. M. & W. G. Chaloner 1964. The ultrastructure of the Mesozoic pollen Classopollis. *Pollen et Spores* **6**(2), 611–20.

Pfefferkorn, H. W. *et al.* 1971. *Some fern-like fructifications and their spores from the Mazon Creek compression flora of Illinois (Pennsylvanian)*. Illinois St. Geol. Surv. Circ. 463.

Pflug, H. D. 1953. Zur Entstehung und Entwicklung des angiospermiden Pollens in der Erdgeschichte. *Palaeontographica B* **95**(4–6), 60–171.

Pflug, H. D. & E. Reitz 1985. Earliest phytoplankton of eukaryotic affinity. *Naturwissenschaften* **72**, 656–7.

Phillips, T. L. & W. A. DiMichele 1981. Paleoecology of middle Pennsylvanian age coal swamps in southern Illinois/Herrin coal member at Sahara Mine no. 6. In *Paleobotany, Paleoecology, and Evolution*, K. J. Niklas (ed.), Vol. 1, 231–84. New York: Praeger.

Phillips, T. L., R. A. Peppers, M. J. Avcin & P. F. Laughnan 1974. Fossil plants and coal: patterns of change in Pennsylvanian coal swamps of the Illinois Basin. *Science* **187**(4144), 1367–9.

Phillips, T. L., R. A. Peppers & W. A. DiMichele 1985. Stratigraphic and inter-regional changes in Pennsylvanian coal-swamp vegetation: environmental inferences. *Int. J. Coal Geol.* **5**(1–2), 43–109.

Phipps, D. & G. Playford 1984. Laboratory techniques for extraction of palynomorphs from sediments. *Pap. Dept. Geol. Univ. Queensland* **11**(1), 1–23.

Piérart, P. 1955. Les mégaspores contenues dans quelques couches de houille du Westphalien B et C aux charbonnages Limbourg Meuse. *Publ. Assoc. Étud. Paléont. (Bruxelles)* **21**(8), 124–42.

Piérart, P. 1980. *Evolution of sporopollenin during thermic treatments*. Univ. Etat Mons, Serv. Biol. Ecol. Palynol. (paper presented at 5th Int. Palynol. Conf., Cambridge).

Pigg, K. B. 1982. *Megasporangiate Mazocarpon from the Upper Pennsylvanian of the Appalachian Basin*. Bot. Soc. Am. Misc. Publ. 162.

Pigg, K. B. & G. W. Rothwell 1983. Chaloneria gen. nov.: heterosporous lycophytes from the Pennsylvanian of North America. *Bot. Gaz.* **144**, 132–47.

Planderová, E. 1972. Pliocene sporomorphs from the West Carpathian Mountains and their stratigraphic interpretation. *Geol. Prace* **59**, 209–83.

Playford, G. 1976. Plant microfossils from the Upper Devonian and Lower Carboniferous of the Canning Basin, Western Australia. *Palaeontographica B* **158**, 1–71.

Playford, G. & R. S. Dring 1981. *Late Devonian acritarchs from the Carnarvon Basin, Western Australia*. Spec. Pap. Palaeont. 27.

Pocock, S. A. J. 1973. *Jurassic short course*. Louisiana State University multilithed and privately circulated.

Pohl, F. 1937a. Die Pollenerzeugung der Windblütler. *Beih. Bot. Centralblatt* **56A**, 365–470.

Pohl, F. 1937b. Die Pollenkorngewichte einiger windblütigen Pflanzen und ihre ökologische Bedeutung. *Beih. Bot. Centralblatt* **57A**, 112–72.

Potonié, R. 1934. *Zur Mikrobotanik des eocänen Humodils des Geiseltals*. Preuss. Geol. Landesanstalt. Arbeiten Inst. Paläobot. Petrog. Brennsteine, 25–125.

Potonié, R. 1951. Revision stratigraphisch wichtiger Sporomorphen der mittel-europäischen Tertiärs. *Palaeontographica B* **91**, 131–51.

Potonié, R. 1956. Die stratigraphische Inkongruität der Organe des Pflanzenkörpers. *Paläont. Z.* **30**, 88–94.

Potonié, R. 1956–. *Synopsis der Gattungen der Sporae Dispersae*, 7 vols: I–VI, Beihefte zum geologischen Jahrbuch 23(1956), 31(1958), 39(1960), 72(1966), 87(1970), and 94(1970); VII, Fortschr. Geol. Rheinld. Westf. 25(1975).

Potonié, R. 1958. Views on spore nomenclature. *Geol. Mag.* **95**, 491–6.

Potonié, R. 1960. See "Synopsis", part III.

Potonié, R. 1962. *Synopsis der Sporae in situ*. Beih. Geol. Jahrb. 52.

Potonié, R. 1967. *Versuch der Einordnung der fossilen Sporen dispersae in das phylogenetische System der Pflanzenfamilien*. Forschungsber. Landes Nordrhein-Westfalen 1761.

Potonié, R. 1975. Artbegriff und Palaeobotanik. *Rev. Palaeobot. Palynol.* **19**, 161–72.

Potonié, R. & G. Kremp 1954. Die Gattungen der paläozoischen *Sporae dispersae* und ihre Stratigraphie. *Geol. Jahrb.* **69**, 111–94.

Potonié, R. & G. Kremp 1955. Die *Sporae dispersae* des Ruhrkarbons, Teil I. *Palaeontographica B* **98**, 1–136.

Potonié, R. & G. Kremp 1956a. Die *Sporae dispersae* des Ruhrkarbons, Teil II. *Palaeontographica B* **99**, 85–191.

Potonié, R. & G. Kremp 1956b. Die *Sporae dispersae* des Ruhrkarbons, Teil III. *Palaeontographica B* **100**, 65–121.

Potonié, R. & K. Rehnelt 1971. Aspects of sporin: on the aromatisation of sporin and the hydrogen density of the sporin of Carboniferous lycopsids. In *Sporopollenin*, J. Brooks *et al.* (eds.), 295–304. London: Academic Press.

Potonié, R., P. W. Thomson & F. Thiergart 1950. Zur Nomenklatur und Klassifikation der neogenen Sporomorphae (Pollen und Sporen). *Geol. Jahrb.* **65**, 35–70.

Powis, G. D. 1981. Comparison of plant microfossil assemblages in Late Palaeozoic Gondwana glacial sequences. *Palaeobotanist* **28–29**, 93–98.

Pratt, L. M., T. L. Phillips & J. M. Dennison 1978. Evidence of non-vascular land plants from the early Silurian (Llandoverian) of Virginia, U.S.A. *Rev. Palaeobot. Palynol.* **25**(2), 121–49.

Punt, W. (ed.) 1976. *The northwest European pollen flora I*. Amsterdam: Elsevier.

Ravn, R. L. 1983. Paleobotanical relationships and stratigraphic importance of the Carboniferous miospore genus *Vestispora* and questionably allied genera. *J. Paleont.* **57**(3), 568–80.

Raymond, A. 1985. Floral diversity, phytogeography, and climatic amelioration during the Early Carboniferous (Dinantian). *Paleobiology* **11**(3), 293–309.

Raymond, A., W. C. Parker & S. F. Barrett 1985a. Early Devonian phytogeography. In *Geological factors and the evolution of plants*, B. H. Tiffney (ed.), 129–67. New Haven, Conn.: Yale University Press.

Raymond, A., W. C. Parker & J. T. Parrish 1985b. Phytogeography and paleoclimate of the Early Carboniferous. In *Geological factors and the evolution of plants*, B. H. Tiffney (ed.), 169–222. New Haven, Conn.: Yale University Press.

Reid, E. M. 1920. A comparative review of Pliocene floras, based on the study of fossil seeds. *Q. J. Geol. Soc. London* **76**, 145–59.

Remy, W. & R. Remy 1955. Mitteilungen über Sporen, die aus inkohlten Fruktifikationen von echten Farnen des Karbon gewonnen wurden, Teil 1. *Abh. Deutschen Akad. Wiss. Berlin*, 41–47.

Remy, W. & R. Remy 1957. Durch Mazeration fertiler Farne des Paläozoikums gewonnene Sporen. *Paläont. Z.* **31**, 55–65.

Retallack, G. 1980. Late Carboniferous to Middle Triassic megafossil floras from the Sydney Basin. In *A guide to the Sydney Basin*, C. Herbert & R. J. Helby (eds.), 384–430. Geol. Surv. New South Wales Bull. 26.

Retallack, G. & D. L. Dilcher 1981a. A coastal hypothesis for the dispersal and rise to dominance of flowering plants. In *Paleobotany, paleoecology and evolution*, Vol. 2, K. J. Niklas (ed.), 27–77. New York: Praeger.

Retallack, G. & D. L. Dilcher 1981b. Early angiosperm reproduction: *Prisca reynoldsii*, gen. et sp. nov. from mid-Cretaceous coastal deposits in Kansas, U.S.A. *Palaeontographica B* **179**, 103–37.

Reymanówna, M. 1968. On seeds containing *Eucommiidites troedsonii* pollen from the Jurassic of Grojec, Poland. In *Studies on fossil plants*, K. P. Alvin *et al.* (eds.). *Bot. J. Linn. Soc.* **61**(384), 147–52.

Reyre, Y. 1970. Stereoscan observations on the pollen genus *Classopollis* Pflug 1953. *Palaeontology* **13**(2), 303–22.

Richardson, J. B. 1965. Middle Old Red Sandstone spore assemblages from the Orcadian Basin north-east Scotland. *Palaeontology* **7**(4), 559–605.

Richardson, J. B., J. H. Ford & F. Parker 1984. Miospores, correlation and age of some Scottish Lower Old Red Sandstone sediments from the Strathmore region (Fife and Angus). *J. Micropaleont.* **3**(2), 109–24.

Richardson, J. B. & N. Ioannides 1973. Silurian palynomorphs from the Tanezzuft and Acacus Formations, Tripolitania, North Africa. *Micropaleontology* **19**(3), 257–307.

Richardson, J. B. & D. C. McGregor 1986. *Silurian and Devonian spore zones of the Old Red Sandstone continent and adjacent regions.* Geol. Surv. Can. Bull. 364.

Richardson, J. B. & S. M. Rasul 1978. Palynological evidence for the age and provenance of the lower Old Red Sandstone from the Apley Barn borehole, Witney, Oxfordshire. *Proc. Geol. Assoc.* **90**(1), 27–42.

Richardson, J. B., S. M. Rasul & T. Al-ameri 1981. Acritarchs, miospores and correlation of the Ludlovian–Downtonian and Silurian–Devonian boundaries. *Rev. Palaeobot. Palynol.* **34**, 209–24.

Richmond, G. M. 1970. Comparison of the Quaternary stratigraphy of the Alps and Rocky Mountains. *Quat. Res.* **1**, 3–28.

Ride, W. D. L. *et al.* (eds.) 1985. *International Code of Zoological Nomenclature.* Int. Trust Zool. Nomencl. London: Br. Mus. Nat. Hist.

Riegel, W. 1974. Phytoplankton from the upper Emsian and Eifelian of the Rhineland, Germany – a preliminary report. *Rev. Palaeobot. Palynol.* **18**, 29–39.

Riggs, S. D. & G. W. Rothwell 1985. *Sentistrobus goodii* n. gen. and sp., a permineralized sphenophyllalean cone from the Upper Pennsylvanian of the Appalachian Basin. *J. Paleont.* **59**(5), 1194–202.

Ritchie, J. C. 1985. Quaternary pollen records from the western interior and the Arctic of Canada. In *Pollen records of late-Quaternary North American sediments*, V. M. Bryant, Jr. & R. G. Holloway (eds.), 327–52. Am. Assoc. Strat. Palynol. Found.

Rittenhouse, G. 1940. Exhibit F: curves for determining probable errors in heavy mineral studies. *Nat. Res. Council Rept. Comm. on Sedimentation*, 1939–40, 97–101.

Robbins, E. I. 1982. *"Fossil Lake Danville": the paleoecology of a Late Triassic ecosystem on the North Carolina–Virginia border.* Unpubl. Ph.D. Thesis, Pennsylvania State University.

Rogers, C. M. & B. D. Harris 1969. Pollen exine deposition: a clue to its control. *Am. J. Bot.* **56**(10), 1209–11.

Rossignol-Strick, M. 1973. Pollen analysis of some sapropel layers from the deep-sea floor of the eastern Mediterranean. In *Initial reports Deep Sea Drilling Project 13*, 971–91.

Rossignol-Strick, M. & D. Duzer 1979a. A late Quaternary continuous climatic record from palynology of three marine cores off Senegal. *Palaeoecol. Africa* **11**, 185–8.

Rossignol-Strick, M. & D. Duzer 1979b. West African vegetation and climate since 22,500 B.P. from deep-sea cores palynology. *Pollen et Spores* **21**, 105–34.

Rotai, A. P. 1978. Carboniferous stratigraphy of the USSR; proposal for an international classification. In The Carboniferous of the U.S.S.R., *Geol. Soc. Oceas. Publ.* **4**, 225–47.

Rothwell, G. W. 1972. Evidence of pollen tubes in Paleozoic pteridosperms. *Science* **175**, 772–4.

Rothwell, G. W. 1978. *Donneggia complura* gen. et sp. nov., a filicalean fern from the Upper Pennsylvanian of Ohio. *Can. J. Bot.* **56**(24), 3096–104.

Rothwell, G. W. 1982. New interpretations of the earliest conifers. *Rev. Palaeobot. Palynol.* **37**, 7–28.

Rothwell, G. W. & J. E. Mickle 1982. *Rhetinotheca patens* n. sp., a medullosan pollen organ from the Upper Pennsylvanian of North America. *Rev. Palaeobot. Palynol.* **36**, 361–74.

Rouse, G. E. & S. K. Srivastava 1970. Detailed morphology, taxonomy, and distribution of *Pistillipollenites macgregorii*. *Can. J. Bot.* **48**, 287–92.

Rowley, J. R. 1963. Ubisch body development in *Poa annua*. *Grana Palynol.* **4**(1), 25–36.

Rowley, J. R. 1976. Dynamic changes in pollen wall morphology. In *The evolutionary significance of the exine*, I. K. Ferguson & J. Muller (eds.), 39–66. Linn. Soc. Symp. Ser. 1.

Rowley, J. R. 1978. The origin, ontogeny, and evolution of the exine. *4th Int. Palynol. Conf.* Lucknow (1976–77), Vol. 1, 126–36.

Rowley, J. R. 1981. Pollen wall characters with emphasis upon applicability. *Nord. J. Bot.* **1**, 357–80.

Rue, D. J. 1982. *Floral Park: migration, invasion or trade*. Unpubl. M. A. Paper, Pennsylvania State University, Dept of Anthropology.

Rue, D. J. 1986. *A palynological analysis of pre-Hispanic human impact in the Copan Valley, Honduras*. Unpubl. Ph.D. Thesis, Pennsylvania State University.

Salgado-Labouriau, M. L. 1984. Reply to "On cavities in spines of Compositae pollen – a taxonomic perspective". *Taxon* **33**(2), 293–5.

Sarjeant, W. A. S. 1970. Acritarchs and tasmanitids from the Chhidru Formation, Uppermost Permian of West Pakistan. In *Stratigraphic boundary problems: Permian and Triassic of West Pakistan*, B. Kummel & C. Teichert (eds.), 277–304. Dept. Geol., Univ. Kansas, Spec. Publ. 4.

Sarjeant, W. A. S. 1973. *Acritarchs and Tasmanitids from the Mianwali and Tredian Formations (Triassic) of the Salt and Surghar Ranges, West Pakistan*. Can. Soc. Petrolm. Geol. Mem. 2, 35–73.

Sarjeant, W. A. S. 1978. Christian Gottfried Ehrenberg, 1795–1876. *Palynology* **2**, 209–11.

Sarjeant, W. A. S. 1984. Charles Downie and the early days of palynological research at the University of Sheffield. *J. Micropaleont.* **3**(2), 1–6.

Sarro, T. J., L. E. Heusser & J. J. Morley 1984. Quaternary paleoclimatic evidence from the Sea of Japan: pollen and radiolaria records from the last 60,000 years. *17th Ann. Mtg. Am. Assoc. Strat. Palynol.*, Program & Abstracts, 19.

Sauvage, J. 1979. The palynological subdivisions in Hellenic Cainozoic and their stratigraphical correlations. *Ann. Géol. Pays Hellén.* **3**, 1091–5. (7th Int. Congr. Medit. Neogene).

Sauvage, J. & M. Sebrier 1977. Données palynologiques et stratigraphiques sur le passage plio-pléistocène: coupe de la Trapeza (Corinthie, Grèce). *C.R. Acad. Sci. Paris B*, **284**, 1963–6.

Scagel, R. F., *et al.* 1965. *An evolutionary survey of the plant kingdom*. Belmont, Cal.: Wadsworth.

Schaarschmidt, F. 1963. Sporen und Hystrichosphaerideen aus dem Zechstein von Büdingen in der Wetterau. *Palaeontographica B* **113**, 38–91.

Schaarschmidt, F. & V. Wilde 1986. Palmenblüten und -blätter aus dem Eozän von Messel. *Cour.-Forsch. Inst. Senckenberg* **86**, 177–202.

Schäfer, W. 1972. *Ecology and paleoecology of marine environments*. Chicago: University of Chicago Press.

Scheihing, M. H. & H. W. Pfefferkorn 1984. The taphonomy of land plants in the Orinoco Delta: a model for the incorporation of plant parts in clastic sediments of Late Carboniferous age of Euramerica. *Rev. Palaeobot. Palynol.* **41**, 205–40.

Schlanker, C. M. & G. A. Leisman 1969. The herbaceous Carboniferous lycopod *Selaginella fraiponti* comb. nov. *Bot. Gaz.* **130**, 35–41.

Schopf, J. M., L. R. Wilson & R. Bentall 1944. *An annotated synopsis of Paleozoic fossil spores and the definition of generic groups*. Illinois St. Geol. Surv. Rep. Invest. 91.

Schopf, J. W. 1968. Microflora of the Bitter Springs Formation, late Precambrian, central Australia. *J. Paleont.* **42**(3), 651–88.

Schopf, J. W. 1977. Biostratigraphic usefulness of stromatolitic Precambrian microbiotas: a preliminary analysis. *Precambrian Res.* **5**, 143–73.

Schopf, J. W. & J. M. Blacic 1971. New microorganisms from the Bitter Springs Formation (Late Precambrian) of the north-central Amadeus Basin, Australia. *J. Paleont.* **45**(6), 925–60.

Scott, A. C. 1978. Sedimentological and ecological control of Westphalian B plant assemblages from West Yorkshire. *Proc. Yorkshire Geol. Soc.* **41**, 461–508.

Scott, A. C. 1979. The ecology of Coal Measure floras from northern Britain. *Proc. Geol. Assoc.* **90**(3), 97–116.

Scott, A. C. 1984. The early history of life on land. *J. Biol. Educ.* **18**(3), 207–19.

Scott, A. C., W. G. Chaloner & S. Paterson 1985a. Evidence of pteridophyte–arthropod interactions in the fossil record. *Proc. R. Soc. Edinburgh* **86B**, 133–40.

Scott, A. C. & M. E. Collinson 1978. Organic sedimentary particles: results from scanning electron microscope studies of fragmentary plant material. In *Scanning electron microscopy in the study of sediments*, W. B. Whalley (ed.), 137–67. Norwich: Geo. Abstracts.

Scott, A. C., J. Galtier & G. Clayton 1985b. A new late Tournaisian (Lower Carboniferous) flora from the Kilpatrick Hills, Scotland. *Rev. Palaeobot. Palynol.* **44**, 81–9.

Scott, A. C. & G.R. King 1981. Megaspores and coal facies: an example from the Westphalian A of Leicestershire, England. *Rev. Palaeobot. Palynol.* **34**, 107–13.

Scott, A. C. & B. Meyer-Berthaud 1985. Plants from the Dinantian of Foulden, Berwickshire, Scotland. *Trans. R. Soc. Edinburgh: Earth Sci.* **76**, 13–20.

Scott, A. C. & G. Playford 1985. Early Triassic megaspores from the Rewan Group, Bowen Basin, Queensland. *Alcheringa* **9**, 297–323.

Scott, L. & E. M. Van Zinderen Bakker, Sr. 1985. Exotic pollen and long-distance wind dispersal at a sub-Antarctic island. *Grana* **24**, 45–54.

Scott, R. W. 1982. Palynomorphs are neglected elements in paleocommunities. *3rd North Am. Paleont. Conv. Proc.* Vol. 2, 471–6.

Shackleton, N. J. & J. P. Kennett 1975. Paleotemperature history of the Cenozoic and the initiation of Antarctic glaciation: oxygen and carbon isotope analyses in D.S.D.P. sites 277, 279, and 281. *Initial reports Deep Sea Drilling Project 39*, 743–55.

Shackleton, N. J. & R. K. Matthews 1977. Oxygen isotope stratigraphy of Late Pleistocene coral terraces in Barbados. *Nature* **268**, 618–20.

Shackleton, N. J. & N. D. Opdyke 1977. Oxygen isotope and paleomagnetic evidence for early Northern Hemisphere glaciation. *Nature* **270**, 216–19.

Shaw, A. B. 1964. *Time in stratigraphy*. New York: McGraw-Hill.

Sherwood-Pike, M. A. & J. Gray 1985. Silurian fungal remains: probable records of the Class Ascomycetes. *Lethaia* **18**, 1–20.

Short, S. K. 1985. Palynology of Holocene sediments, Colorado Front Range: vegetation and treeline changes in the subalpine forest. In *Pollen records of late-*

Quaternary North American sediments, V. M. Bryant, Jr. & R. H. Holloway (eds.), 7–30. Am. Assoc. Strat. Palynol. Found.

Singh, C. 1983. *Cenomanian microfloras of the Peace River area, northwestern Alberta.* Alberta Res. Council Bull. 44.

Sirkin, L. A. 1971. Surficial glacial deposits and postglacial pollen stratigraphy in central Long Island, New York. *Pollen et Spores* **13**, 93–100.

Skarby, A. 1968. *Extratriporopollenites (Pflug) emend. from the Upper Cretaceous of Scania, Sweden.* Stockholm Contrib. Geol. 16.

Skarby, A. 1981. Upper Cretaceous Normapolles anthers, S. Sweden. *13th Int. Bot. Congr.* Sydney, Australia, Abstracts, 193.

Skog, J. E. 1980. *The genus Ruffordia (Schizaeaceae) from the Lower Cretaceous Potomac Group.* Bot. Soc. Am. Misc. Ser. Publ. 158, 106.

Skog, J. E. 1982. *Pelletixia amelguita*, a new species of fossil fern in the Potomac Group (Lower Cretaceous). *Am. Fern J.* **72**(4), 115–21.

Smith, A. H. V. 1962. The palaeoecology of Carboniferous peats based on miospores and petrography of bituminous coals. *Proc. Yorkshire Geol. Soc.* **33**, 423–74.

Smith, A. H. V. & M. A. Butterworth 1967. *Miospores in the coal seams of the Carboniferous of Great Britain.* Spec. Pap. Palaeont. 1.

Smith, P. H. 1978. Fungal spores of the genus *Ctenosporites* from the early Tertiary of southern England. *Palaeontology* **21**(3), 717–22.

Solomon, A. M. 1983. Pollen morphology and plant taxonomy of white oaks in eastern North America. *Am. J. Bot.* **70**(4), 481–94.

Solomon, A. M. & T. Webb III 1985. Computer-aided reconstruction of Late-Quaternary landscape dynamics. *Ann. Rev. Ecol. Syst.* **16**, 63–84.

Southworth, D. 1974. Solubility of pollen exines. *Am. J. Bot.* **61**(1), 36–44.

Sowunmi, M. A. 1981. Late Quaternary environmental changes in Nigeria. *Pollen et Spores* **23**(1), 125–48.

Spaulding, W. G., E. B. Leopold & T. R. Van Devender 1983. Late Wisconsin paleoecology of the American Southwest. In *The Late Pleistocene*, S. C. Porter (ed.), in *Late-Quaternary environments of the United States*, Vol. 1, H. E. Wright, Jr. (ed.), 259–93.

Spector, D. L. 1984. *Dinoflagellates.* Orlando: Academic Press.

Srivastava, S. C. 1984. *Lelestrobus*: a new microsporangiate organ from the Triassic of Nidpur, India. *Palaeobotanist* **32**(1), 86–90.

Srivastava, S. K. 1976a. Biogenic infection in Jurassic spores and pollen. *Geosci. & Man* **15**, 95–100.

Srivastava, S. K. 1976b. The fossil pollen genus *Classopollis*. *Lethaia* **9**, 437–57.

Srivastava, S. K. 1981. Evolution of the Upper Cretaceous phytogeoprovinces and their pollen flora. *Rev. Palaeobot. Palynol.* **35**, 155–73.

Srivastava, S. K. 1983. Cretaceous phytogeoprovinces and paleogeography of the Indian plate based on palynological data. *Proc. Symp. on Cretaceous of India*, 141–57. Indian Assoc. Palynostrat.

Srivastava, S. K. & P. L. Binda 1984. Siliceous and silicified microfossils from the Maastrichtian Battle Formation of southern Alberta, Canada. *Paléobiol. Continentale* **14**(1), 1–24.

Stach, E. 1957. Die Anschliff-Sporendiagnose des Ruhrkohlenflözes Baldur. *Palaeontographica B* **102**(4–6), 71–95.

Stach, E. 1964. Zur Untersuchung des Sporinits in Kohlen-Anschliffen. *Fortschr. Geol. Rheinld. Westf.* **12**, 403–20.

Stanley, E. A. 1965. Abundance of pollen and spores off the eastern coast of the U.S. *Southeastern Geol.* **7**, 25–33.

Stanley, E. A. 1966. The problem of reworked pollen and spores in marine sediments. *Marine Geol.* **4**, 397–408.

Stanley, E. A. 1967. Palynology of six ocean-bottom cores from the southwestern Atlantic Ocean. *Rev. Palaeobot. Palynol.* **2**, 195–203.

Stanley, E. A. 1969. The occurrence and distribution of pollen and spores in marine sediments. *Proc. 1st Int. Conf. Planktonic Microfossils*, Geneva, 1967, 640–3. Leiden: E. J. Brill.

Stanton, M. L., A. A. Snow & S. N. Handel 1986. Floral evolution: attractiveness to pollinators increases male fitness. *Science* **232**, 1625–7.

Staplin, F. L., *et al.* 1982. *How to assess maturation and paleotemperatures.* Soc. Econ. Paleont. Mineral. Short Course 7.

Stewart, W. N. 1983. *Paleobotany and the evolution of plants.* Cambridge: Cambridge University Press.

Stidd, B. M. 1978. An anatomically preserved *Potoniea* with *in situ* spores from the Pennsylvanian of Illinois. *Am. J. Bot.* **65**, 677–83.

Stidd, B. M., G. A. Leisman & T. L. Phillips 1977. *Sullitheca dactylifera* gen. et sp. n.: a new medullosan pollen organ and its evolutionary significance. *Am. J. Bot.* **64**, 994–1002.

Stidd, B. M., M. O. Rischbieter & T. L. Phillips 1985. A new lyginopterid pollen organ with alveolate pollen exines. *Am. J. Bot.* **72**(4), 501–8.

Stockmarr, J. 1971. Tablets with spores used in absolute pollen analysis. *Pollen et Spores* **13**, 615–21.

Stolar, J. 1978. *Megaspores and the Devonian–Mississippian boundary along Route 322, Centre County, Pennsylvania.* Unpubl. D. Ed. Thesis, Pennsylvania State University.

Stover, L. E. & G. L. Williams 1982. Dinoflagellates. *3rd North Am. Paleont. Conv. Proc.* **2**, 525–33.

Straka, H. 1975. *Pollen- und Sporenkunde.* Stuttgart: Gustav Fischer.

Streel, M. 1972. Dispersed spores associated with *Leclercqia complexa* Banks, Bonamo & Grierson from the late middle Devonian of Eastern New York state (U.S.A.). *Rev. Palaeobot. Palynol.* **14**, 205–15.

Streel, M. & A. Traverse 1978. Spores from the Devonian/Mississippian transition near the Horseshoe Curve Section, Altoona, Pennsylvania, U.S.A. *Rev. Palaeobot. Palynol.* **26**, 21–39.

Street, F. A. & A. T. Grove 1979. Global maps of lake-level fluctuations since 30,000 yr. B.P. *Quat. Res.* **12**, 83–118.

Strother, P. K. & A. Traverse 1979. Plant microfossils from Llandoverian and Wenlockian rocks of Pennsylvania. *Palynology* **3**, 1–21.

Stubblefield, S. P. & T. N. Taylor 1984. Fungal remains in the lycopod megaspore *Triletes rugosus* (Loose) Schopf. *Rev. Palaeobot. Palynol.* **41**, 199–204.

Stubblefield, S. P., T. N. Taylor, C. E. Miller & G. T. Cole 1983. Studies of Carboniferous fungi II. The structure and organization of *Mycocarpon, Sporocarpon, Dubiocarpon*, and *Coleocarpon* (Ascomycotina). *Am. J. Bot.* **70**(10), 1482–98.

Suc, J.-P. 1980. *Contribution à la connaissance du Pliocène et du Pléistocène inférieur des régions méditerraneennes d'Europe occidentale par l'analyse palynologique des dépôts du Languedoc-Rousillon (Sud de la France) et de la Catalogne (Nord-est de l'Espagne).* Univ. Sci. & Tech. Languedoc, These Doct., Tome 1 (text), Tome 2 (plates & figs.)

Sukh-Dev, 1980. Evaluation of *in situ* spores and pollen grains from the Jurassic–Cretaceous fructifications. *4th Int. Palynol. Conf.* Lucknow (1976–77), **2**, 753–68.

Sullivan, H. J. 1967. Regional differences in Mississippian spore assemblages. *Rev. Palaeobot. Palynol.* **1**, 185–92.

Sweet, A. R. 1979. *Jurassic and Cretaceous megaspores.* Am. Assoc. Strat. Palynol. Contrib. Ser. 5B, 1–30.

Takahashi, K. 1984. Stratigraphic significance of three important pollen groups in the late Upper Cretaceous and early Palaeogene. *J. Palynol. (Japan)* **30**(1), 15–24.

Takahashi, K. & H. Shimono 1982. Maestrichtian microflora of the Miyadani-gawa formation in the Hida district, central Japan. *Bull. Fac. Liberal Arts, Nagasaki Univ., Nat. Sci.* **22**(2), 11–188.

Tappan, H. 1980. *The paleobiology of plant protists.* San Francisco: W. H. Freeman.

Tappan, H. & A. R. Loeblich, Jr. 1965. Foraminiferal remains in palynological preparations. *Rev. Micropaleont.* **8**(2), 61–3.

Tauber, H. 1965. Differential pollen dispersion and the interpretation of pollen diagrams. *Danmarks. Geol. Unders. II,* **89**.

Tauber, H. 1967. Differential pollen dispersion and filtration. In *Quaternary paleoecology*, E. J. Cushing & H. E. Wright, Jr. (eds.), 131–41. *Proc. 7th Congr. INQUA*, Vol. 7. New Haven, Conn.: Yale University Press.

Tavera, C. M. 1982. *Light and electron microscope studies of spore structure in selected tropical ferns.* Unpubl. Ph.D. Thesis, Pennsylvania State University.

Taylor, T. N. 1972. A new sporangial aggregation. *Rev. Palaeobot. Palynol.* **14**, 309–18.

Taylor, T. N. 1976a. Fossil ubisch bodies. *Trans. Am. Microsc. Soc.* **95**(1), 133–6.

Taylor, T. N. 1976b. The ultrastructure of *Schopfipollenites*: orbicules and tapetal membranes. *Am. J. Bot,* **63**(6), 857–62.

Taylor, T. N. 1981. *Paleobotany: an introduction to fossil plant biology.* New York: McGraw-Hill.

Taylor, T. N. 1982. Ultrastructural studies of Paleozoic seed fern pollen: sporoderm development, *Rev. Palaeobot. Palynol.* **37**, 29–53.

Taylor, T. N. & K. L. Alvin 1984. Ultrastructure and development of Mesozoic pollen: *Classopollis. Am. J. Bot.* **71**(4), 575–602.

Taylor, T. N. & D. F. Brauer, 1983. Ultrastructural studies of *in situ* Devonian spores: *Barinophyton citrulliforme. Am. J. Bot.* **70**(1), 106–12.

Taylor, T. N., M. A. Cichan & A. M. Baldoni 1984. The ultrastructure of Mesozoic pollen: *Pteruchus dubius* (Thomas) Townrow. *Rev. Palaeobot. Palynol.* **4**, 319–27.

Taylor, T. N. & C. P. Daghlian 1980. The morphology and ultrastructure of *Gothania* (Cordaitales) pollen. *Rev. Palaeobot. Palynol.* **29**, 1–14.

Taylor, W. A. 1986. Ultrastructure of sphenophyllalean spores. *Rev. Palaeobot. Palynol.* **47**, 105–28.

Teichmüller, M. & B. Durand 1983. Fluorescence microscopical rank studies on liptinites and vitrinites in peat and coals, and comparison with results of the Rock-Eval pyrolysis. *Int. J. Coal Geol.* **2**, 197–230.

Teichmüller, M. & K. Ottenjahn 1977. Liptinite und lipoide Stoffe in einem Erdölmuttergestein. *Erdöl Kohle* **30**, 387–98.

Terasmae, J. 1973. Notes on late Wisconsin and early Holocene history of vegetation in Canada. *Arctic Alpine Res.* **5**, 201–22.

Thomas, B. A. 1970. A new specimen of *Lepidostrobus binneyanus* from the Westphalian B of Yorkshire. *Pollen et Spores* **12**, 217–34.

Thomas, B. A. & D. M. Crampton 1971. A fertile *Zeilleria avoldensis* from the British Upper Carboniferous. *Rev. Palaeobot. Palynol.* **11**(3/4), 283–95.

Thomson, P. W. & H. Pflug 1953. Pollen und Sporen des mitteleuropäischen Tertiärs. *Palaeontographica B* **94**, 1–138.

Thusu, B. 1973. Acritarchs of the middle Silurian Rochester Formation of southern Ontario. *Palaeontology* **16**(4), 799–826.

Ting, W. S. 1966. *Determination of Pinus species by pollen statistics.* Univ. California Publ. Geol. Sci. 58.

Tiwari, R. S. 1984. The fundamental form deviation in Permo-Triassic palynology: Gondwanaland vis-à-vis northern continents. *6th Int. Palynol. Conf.* Abstracts, 165.

Townrow, J. A. 1962. On some disaccate pollen grains of Permian to middle Jurassic age. *Grana Palynol.* **3**(2), 13–44.

Townrow, J. A. 1965. A new member of the Corystospermaceae Thomas. *Ann. Bot.* (*N.S.*) **29**, 495–511.

Traverse, A. 1951. *The pollen and spores of the Brandon Lignite, a coal in Vermont of lower Tertiary age.* Unpubl. Ph.D. Thesis, Harvard University.

Traverse, A. 1955. *Pollen analysis of the Brandon Lignite of Vermont.* U.S. Bur. Mines Rep. Invest. 5151.

Traverse, A. 1965. Preparation of modern pollen and spores for palynological reference collections. In *Handbook of paleontological techniques*, B. Kummel & D. Raup (eds.), 598–613. San Francisco: W. H. Freeman.

Traverse, A. 1972. A case of marginal palynology: a study of the Franciscan melanges. *Geosci. & Man* **4**, 87–90.

Traverse, A. 1974a. Paleopalynology, 1947–1972. *Ann. Missouri Bot. Gard.* **61**, 203–36.

Traverse, A. 1974b. *Palynologic investigation of two Black Sea cores.* Am. Assoc. Petrol. Geol. Mem. **20**, 381–8.

Traverse, A. 1978a. Palynological analysis of D.S.D.P. Leg 42B (1975) cores from the Black Sea. In *Initial reports Deep Sea Drilling Project 42*, D. A. Ross & Y. P. Yeprochnov (eds.), 2: 28, 41, 48, 147, 159, 161, 199, 305, 993–1015.

Traverse, A. 1978b. Application of simple arithmetic ratios to study of D.S.D.P. Black Sea cores. *Palaeobotanist* **25**, 525–8.

Traverse, A. 1982. Response of world vegetation to Neogene tectonic and climatic events. *Alcheringa* **6**, 197–209.

Traverse, A. 1986. Closing the semicircle: palynofloras of the Eagle Mills Formation and South Georgia Basin date the initial rifting-precursors of the Gulf of Mexico and Atlantic Ocean. *Am. Assoc. Strat. Palynol.* Program & Abstracts, 36.

Traverse, A., J. S. Bridge, M. E. Bowers & A. Schuyler 1984. Palynostratigraphic zonation and paleoecology of part of the Late Devonian Catskill magnafacies, south-central New York. *Palynology* **9**, 255.

Traverse, A., K. H. Clisby & F. Foreman 1961. Pollen in drilling-mud "thinners", a source of palynological contamination. *Micropaleontology* **7**, 375–7.

Traverse, A. & R. N. Ginsburg 1966. Palynology of the surface sediments of Great Bahama Bank, as related to water movement and sedimentation. *Marine Geol.* **4**, 417–59.

Trivett, M. L. 1983. Pollen cones of a mesarch cordaitean from the Upper Pennsylvanian of the Appalachian Basin. *Am. J. Bot.* **70**(5), 81.

Trivett, M. L. & G. W. Rothwell 1985. Morphology, systematics, and paleoecology of Paleozoic fossil plants: *Mesoxylon priapi*, sp. nov. (Cordaitales). *Systematic Bot.* **10**(2), 205–23.

Truswell, E. M. 1980. Permo-Carboniferous palynology of Gondwanaland: progress and problems in the decade to 1980. *BMR J. Australian Geol. Geophys.* **5**, 95–111.

Truswell, E. M. 1981. Pre-Cenozoic palynology and continental movements. *Paleo-reconstruction of the Continents, Geodynamics Ser.* **2**, 13–25.

Truswell, E. M. 1983. Recycled Cretaceous and Tertiary pollen and spores in Antarctic marine sediments: a catalogue. *Palaeontographica B* **186**, 121–74.

Truswell, E. M. & W. K. Harris 1982. The Cainozoic paleobotanical record in arid Australia: fossil evidence for the origins of an arid-adapted flora. In *Evolution of the flora and fauna of Australia.* W. R. Barker & P. J. M. Greenslade (eds.), 67–75. Adelaide: Peacock.

Tschudy, R. H. 1960. "Vibraflute". *Micropaleontology* **6**(3), 325–6.

Tschudy, R. H. 1970. *Two new pollen genera (Late Cretaceous and Paleocene) with possible affinity to the Illiciaceae.* U.S. Geol. Surv. Prof. Pap. 643–F.

Tschudy, R. H. 1973. *Complexiopollis pollen lineage in Mississippi embayment rocks.* U.S. Geol. Surv. Prof. Pap. 743–C.

Tschudy, R. H. 1975. *Normapolles pollen from the Mississippi embayment.* U.S. Geol. Surv. Prof. Pap. 865.

Tschudy, R. H. & B. D. Tschudy 1984. The "fern spike" at the Cretaceous–Tertiary boundary, western interior, United States. *Am. Assoc. Strat. Palynol. 17th Ann. Mtg. Abstracts,* 22.

Tsukada, M. 1957. Pollen analytical studies of postglacial age in Japan. I. Hyotan-ike ponds on Shiga Heights, Nagano Prefecture. *J. Inst. Polytech., Osaka City Univ. D* **8**, 203–16.

Tsukada, M. 1958. Pollen analytical studies of postglacial age in Japan. II. Northern region of Japan North-Alps. *J. Inst. Polytech., Osaka City Univ. D* **9**, 235–49.

Tsukada, M. 1982. *Pseudotsuga menziesii* (Mirb.) Franco: its pollen dispersal and late Quaternary history in the Pacific Northwest. *Jap. J. Ecol.* **32**, 159–87.

Turekian, K. K. 1971. *The Late Cenozoic glacial ages.* New Haven, Conn.: Yale University Press.

Urban, J. B. 1971. Palynology and the Independence Shale of Iowa. *Bull. Am. Paleont.* **60**, 103–89.

Van de Weerd, A. 1979. Palynology of some upper Miocene and Lower Pliocene sections in Greece: preliminary results; biostratigraphic implications. *Ann. Géol. Pays Hellén.* **3**, 1253–61. (7th Int. Congr. Medit. Neogene).

Van der Eem, J. G. L. A. 1983. Aspects of Middle and Late Triassic palynology. 6. Palynological investigations in the Ladinian and lower Karnian of the western Dolomites, Italy. *Rev. Palaeobot. Palynol.* **39**, 189–300.

Van der Hammen, T., J. H. Werner & H. Van Dommelen 1973. Palynological record of the upheaval of the northern Andes: a study of the Pliocene and lower Quaternary of the Columbian eastern Cordillera and the early evolution of its high-Andean biota. *Rev. Palaeobot. Palynol.* **16**, 1–122.

Van der Hammen, T., T. A. Wijmstra & W. H. Zagwijn 1971. The floral record of the Late Cenozoic of Europe. In *The late Cenozoic glacial ages,* K. K. Turekian (ed.), 391–424. New Haven, Conn.: Yale University Press.

Van der Zwan, C. J. 1979. Aspects of Late Devonian and Early Carboniferous palynology of southern Ireland. I. The *Cyrtospora cristifer* morphon. *Rev. Palaeobot. Palynol.* **28**, 1–20.

Van der Zwan, C. J. 1981. Palynology, phytogeography and climate of the Lower Carboniferous. *Palaeogeog. Palaeoclimatol. Palaeoecol.* **33**, 279–310.

Van der Zwan, C. J., M. C. Boulter & R. N. L. B. Hubbard 1985. Climatic change during the Lower Carboniferous in Euramerica, based on multivariate statistical analysis of palynological data. *Palaeogeog. Palaeoclimatol. Palaeoecol.* **52**, 1–20.

Van Geel, B. 1972. Palynology of a section from the raised peat bog "Wietmarscher Moor", with special reference to fungal remains. *Acta Bot. Neerl.* **21**, 261–84.

Van Gijzel, P. 1975. Polychromatic UV-fluorescence microphotometry of fresh and fossil plant substances, with special reference to the location and identification of dispersed organic material in rocks. *Colloq. Int. C.N.R.S.,* Sept. 1973, 67–91.

Van Gijzel, P. 1981. Applications of the geomicrophotometry of kerogen, solid hydrocarbons and crude oil to petroleum exploration. In *Organic maturation studies and fossil fuel exploration,* J. Brooks (ed.), 351–77. London: Academic Press.

Van Konijnenburg-Van Cittert, J. H. A. 1978. Osmundaceous spores in situ from the Jurassic of Yorkshire, England. *Rev. Palaeobot. Palynol.* **26**, 125–41.

Van Uffelen, G. 1984. *Recent Polypodiaceae s.s. (Filicales) and a quest for fossil ones* (handout at 6th Int. Palynol. Conf., Calgary). Leiden: Rijksherbarium.

Van Zant, K. 1979. Late glacial and postglacial pollen and plant microfossils from Lake West Okoboji, northwestern Iowa. *Quat. Res.* **12**, 358–80.

Van Zinderen Bakker, E. M. 1976. The evolution of late Quaternary palaeoclimates of southern Africa. *Palaeoecol. Africa* **9**, 160–202.

Vavrdova, M. 1974. Geographical differentiation of Ordovician acritarch assemblages in Europe. *Rev. Palaeobot. Palynol.* **18**, 71–175.

Vidal, G. & A. H. Knoll 1983. Proterozoic plankton. *Geol. Soc. Am. Mem.* **161**, 265–77.

Vincèns, A. 1984. Environment végétal et sédimentation pollinique lacustre actuelle dans le Bassin du Lac Turkana (Kenya). *Rev. Paléobiol.* Spec. Vol., 235–42.

Vishnu-Mittre 1956. *Masculostrobus sahnii* sp. nov., a petrified conifer male cone producing three-winged and one- and four-winged abnormal pollen grains from the Jurassic of the Rajmahal Hills, Bihar. *Grana Palynol.* (*N.S.*) **1**, 99–107.

Visscher, H. 1971. *The Permian and Triassic of the Kingscourt outlier, Ireland.* Geol. Surv. Ireland Spec. Pap. 1.

Visscher, H. 1980. Die Stellung des Rhät aus palynologischer Sicht. *Cour. Forsch.-Inst. Senckenberg* **42**, 56–63.

Visscher, H. & C. J. Van der Zwan 1981. Palynology of the circum-Mediterranean Triassic: phytogeographical and paleoclimatological implications. *Geol. Rundschau* **70**(2), 625–34.

Von Post, L. 1916. Om skogsträdpollen i sydsvenska torfmossclagerfölyder. *Geol. Fören. Stockholm Förh.* **38**, 384–90.

Von Post, L. 1944. The prospect for pollen analysis in the study of the earth's climatic history. *New Phytologist* **45**, 193–217.

Voss, E. G. *et al.* (eds.) 1983. *International Code of Botanical Nomenclature.* Regnum vegetabile III. Int. Assoc. Plant. Taxon.

Vozzhennikova, T. F. & B. V. Timofeev 1973. Microfossils of the oldest deposits. *Proc. 3rd Int. Palyn. Conf.* Moscow: Nauka.

Waldman, M. & W. S. Hopkins, Jr. 1970. Coprolites from the Upper Cretaceous of Alberta, Canada, with a description of their microflora. *Can. J. Earth Sci.* **7**, 1295–303.

Walker, A. G. & J. W. Walker 1982. *Same grain combined light, scanning electron, and transmission electron microscopy of Lower Cretaceous angiosperm pollen.* Bot. Soc. Am. Misc. Publ. 162, 66.

Walker, J. W. 1971. *Pollen morphology, phytogeography, and phylogeny of the Annonaceae.* Contrib. Gray Herbarium 202.

Walker, J. W. 1976. Evolutionary significance of the exine in the pollen of primitive angiosperms. In *The evolutionary significance of the exine,* I. K. Ferguson & J. Muller (eds.), 251–308. Linn. Soc. Symp. Ser. 1. London: Academic Press.

Walker, J. W. 1986. Classification and evolution of the monocotyledons. *Am. J. Bot.* **73**(5), 746.

Walker, J. W. & J. A. Doyle 1975. The bases of angiosperm phylogeny: palynology. *Ann. Missouri Bot. Gard.* **62**(3), 664–723.

Walker, J. W. & A. G. Walker 1984. Ultrastructure of Lower Cretaceous angiosperm pollen and the origin and early evolution of flowering plants. *Ann. Missouri Bot. Gard.* **71**, 464–521.

Wall, D. 1962. Evidence from recent plankton regarding the biological affinities of *Tasmanites* Newton 1875 and *Leiosphaeridia* Eisenack 1958. *Geol. Mag.* **94**(4), 353–62.

Wall, D. 1965. Modern hystrichospheres and dinoflagellate cysts from the Woods Hole region. *Grana Palynol.* **6**(2), 297–314.

Wall, D. & B. Dale 1967. The resting cysts of modern marine dinoflagellates and their paleontological significance. *Rev. Palaeobot. Palynol.* **2**, 349–54.

Wall, D., B. Dale, G. P. Lohmann & W. K. Smith 1977. The environmental and climatic distribution of dinoflagellate cysts in modern marine sediments from regions in the North and South Atlantic Oceans and adjacent seas. *Marine Micropaleont.* **2**, 121–200.

Wang, K., Y. Zhang & Y. Sun 1982. The spore-pollen and algae assemblages from the surface layer sediments of the Yangtze river delta. *Acta Geog. Sinica* **37**(3), 261–71.

Watts, W. A. 1979. Late Quaternary vegetation of central Appalachia and the New Jersey coastal plan. *Ecol. Monographs* **49**, 427–69.

Watts, W. A. 1980. Late-Quaternary vegetation history at White Pond on the inner Coastal Plain of South Carolina. *Quat. Res.* **13**, 187–99.

Webb, T., III, E. J. Cushing & H. E. Wright, Jr. 1983. Holocene changes in the vegetation of the Midwest. In *Late-Quaternary environments of the United States*, Vol. 2: *The Holocene*, H. E. Wright, Jr. (ed.) 142–65. Minneapolis: University of Minnesota Press.

Webb, T., III, S. E. Howe, R. H. W. Bradshaw & K. M. Heide 1981. Estimating plant abundances from pollen data: the use of regression analysis. *Rev. Palaeobot. Palynol.* **34**, 269–300.

Wells, P. V. 1979. An equable glaciopluvial in the west: pleniglacial evidence of increased precipitation on a gradient from the Great Basin to the Sonoran and Chihuahuan deserts. *Quat. Res.* **12**, 311–25.

West, R. G. 1968. *Pleistocene geology and biology, with especial reference to the British Isles.* London: Longman.

West, R. G. 1977. *Pleistocene geology and biology, with especial reference to the British Isles*, 2nd edn. London: Longman.

West, R. G, 1985. Climatic change in the Quaternary – evidence and ideas. *J. Geol. Soc. London* **142**, 413–16.

Westenberg, J. 1947. Mathematics of pollen diagrams. *Koninkl. Ned. Akad. Wetenschap. Proc. Ser. A* **50**, 509–20, 640–8.

White, D. 1915. Some relations in origin between coal and petroleum. *J. Washington Acad. Sci.* **5**, 189–212.

White, D. 1935. Metamorphism of organic sediments and derived oils. *Bull. Am. Assoc. Petrolm. Geol.* **19**(5), 589–617.

White, D. & R. Thiessen 1913. The origin of coal. *U.S. Bur. Mines Bull.* **38**, 1–390.

Whitehead, D. R., S. T. Jackson, M. C. Sheehan & B. W. Leyden 1982. Late-glacial vegetation associated with caribou and mastodon in central Indiana. *Quat. Res.* **17**, 241–57.

Wicander, R. 1983. *A catalog and biostratigraphic distribution of North American Devonian acritarchs.* Am. Assoc. Strat. Palynol. Contrib. Ser. 10.

Wicander, R. 1984. Middle Devonian acritarch biostratigraphy of North America. *J. Micropaleont.* **3**(2), 19–24.

Wicander, R. & G. D. Wood 1981. *Systematics and biostratigraphy of the organic-walled microphytoplankton from the Middle Devonian (Givetian) Silica Formation, Ohio, U.S.A.* Am. Assoc. Strat. Palynol. Contrib. Ser. 8.

Wiggins, V. D. 1976. Fossil oculata pollen from Alaska. *Geosci. & Man* **15**, 51–76.

Wijmstra, T. A. 1969. Palynology of the first 30 metres of a 120 m deep section in northern Greece. *Acta. Bot. Neerl.* **18**(4), 511–17.

Wijmstra, T. A. 1978. Paleobotany and climatic change. In *Climatic change*, J. Gribbin (ed.), 25–45. Cambridge: Cambridge University Press.

Williams, D. B. 1971. The occurrence of dinoflagellates in marine sediments. In *The micropaleontology of oceans*, B. M. Funnell & W. R. Riedel (eds.), 231–43. Cambridge: Cambridge University Press.

563

Williams, G. L. 1975. *Dinoflagellate and spore stratigraphy of the Mesozoic–Cenozoic offshore eastern Canada*. Geol. Surv. Can. Pap. 74–30, 107–61.

Williams, G. L. 1977. Dinocysts: their classification, biostratigraphy and palaeoecology. In *Oceanic micropalaeontology*, A. T. S. Ramsay (ed.), Vol. 2, 1231–1325.

Williams, G. L., W. A. S. Sarjeant & E. J. Kidson 1978. *A glossary of the terminology applied to dinoflagellate amphiesmae and cysts and acritarchs*. Am. Assoc. Strat. Palynol. Contrib. Ser. 2A.

Willman, H. B., E. Atherton & T. C. Buschbach, *et al.* 1975. Handbook of Illinois stratigraphy. *Illinois Geol. Surv. Bull.* **95**.

Willmann, R. 1980. Die Alterstellung kontinentaler Neogenablagerungen in der südostlichen Agäis (Rhodos und Kos/Dodekanes, Datça/Südwestanatolien). *Newsl. Strat.* **9**, 1–18.

Wilson, G. J. & C. D. Clowes 1980. *A concise catalogue of organic-walled fossil dinoflagellate genera*. New Zealand Geol. Surv. Rep. 92.

Wilson, L. R. 1964. Recycling, stratigraphic leakage, and faulty techniques in palynology. *Grana Palynol.* **5**(3), 425–36.

Wilson, L. R. 1967. Technique for illustrating palynological succession in sedimentary deposits. *Oklahoma Geol. Notes* **27**, 9–13.

Wilson, L. R. 1976. *A study of Paleozoic rocks in Arbuckle and western Ouachita Mountains of southern Oklahoma*. Shreveport (Louisiana) Geol. Soc. Field Trip Guidebook, 83–9.

Wilson, L. R. & B. S. Venkatachala 1963. *Thymospora*, a new name for *Verrucosporites*. *Oklahoma Geol. Notes* **23**(3), 75–9.

Wilson, L. R. & R. M. Webster 1946. Plant microfossils from a Fort Union coal of Montana. *Am. J. Bot.* **33**, 271–8.

Wing, S. L. 1981. *A study of paleoecology and paleobotany in the Willwood Formation (early Eocene, Wyoming)*. Unpubl. Ph.D. Thesis, Yale University.

Wodehouse, R. P. 1932. Tertiary pollen. I. Pollen of the living representatives of the Green River flora. *Bull. Torrey Bot. Club* **59**, 313–40.

Wodehouse, R. P. 1933. Tertiary pollen II. The oil shales of the Eocene Green River formation. *Bull. Torrey Bot. Club* **60**, 479–524.

Wodehouse, R. P. 1935. *Pollen grains*. New York: McGraw-Hill.

Woillard, G. M. 1978. Grande Pile peat bog: a continuous pollen record for the last 140,000 years. *Quat. Res.* **9**, 2–21. (See also Gruger 1979.)

Wolfe, J. A. 1971. Tertiary climatic fluctuations and methods of analysis of Tertiary floras. *Palaeogeog. Palaeoclimatol. Palaeoecol.* **9**, 27–57.

Wolfe, J. A. 1977. *Paleogene floras from the Gulf of Alaska region*. U.S. Geol. Surv. Prof. Pap. 997.

Wolfe, J. A. 1980. Tertiary climates and floristic relationships at high latitudes in the Northern Hemisphere. *Palaeogeog. Palaeoclimatol. Palaeoecol.* **30**, 313–23.

Wolfe, J. A. 1986. Tertiary floras and paleoclimates of the Northern Hemisphere. In *Land plants: notes for a short course organized by R. A. Gastaldo*, T. W. Broadhead (ed.), 182–226. Univ. of Tennessee Dept. of Geological Sciences Studies in Geology 15.

Wood, G. D. 1978. Silurian trilete spores and plant fragments from northern Indiana, and their paleobotanical implication. *Micropaleontology* **24**(3), 327–31.

Wood, G. D. 1984. A stratigraphic, paleoecologic, and paleobiogeographic review of the acritarchs *Umbellasphaeridium deflandrei* and *Umbellasphaeridium saharicum*. In *Biostratigraphy*, P. K. Sutherland & W. L. Manger (eds.), Compte Rendu, Vol. 2, 191–211. 9th Int. Congr. on Carboniferous Strat. & Geol. Southern Illinois Press.

Wrenn, J. H., S. L. Duffield & J. A. Stein (eds.) 1986. *Papers from the first symposium on dinoflagellate cyst biostratigraphy*. Am. Assoc. Strat. Palynol. Contrib. Ser. 17.

Wright, H. E., Jr. 1972. Interglacial and postglacial climates: the pollen record. *Quat. Res.* **2**, 274–82.

Wright, H. E., Jr. 1976. The dynamic nature of Holocene vegetation: a problem in paleoclimatology, biogeography, and stratigraphic nomenclature. *Quat. Res.* **6**, 581–96.

Yang, H. & D. Jiang 1981. Pollen and spores extracted from petroleum of Liaohe oilfield and their significance. *Acta Bot. Sinica* **23**(1), 52–7.

Zagwijn, W. H. 1975. Variations in climate as shown by pollen analysis, especially in the Lower Pleistocene of Europe. In *Ice Ages: ancient and modern*, H. E. Wright & F. Moseley (eds.), 137–52. Liverpool: Seel House.

Zauer, V. V. 1977. Pollen of Mesozoic conifers: *Protoabietipites*, Mal., *Minutisaccus* Mädl., *Leiosaccus* Sauer Gen. Nov. *Trudy VNIGRI* **398**, 100–13 (in Russian).

Zavada, M. S. 1983. Pollen wall development of *Zamia floridana*. *Pollen et Spores* **25**(3/4), 287–304.

Zetsche, F. & K. Huggler 1928. Untersuchungen über die Membran der Sporen und Pollen. I. 1. *Lycopodium clavatum*. *Justus Liebigs Ann. Chem.* **461**, 89–107.

Zetsche, F. & O. Kalin 1931. Untersuchungen der Sporen und Pollen. V. 4. Zur Autooxydation der Sporopollenine. *Helv. Chim. Acta* **14**, 517–19.

Zetsche, F. & H. Vicari 1931. Untersuchungen über die Membran der Sporen und Pollen. II. 2. *Lycopodium clavatum*. *Helv. Chim. Acta* **14**, 58–78.

Zhang, L.-J. 1980. Late Triassic spores and pollen from regions to the south and north of Qinling Range. Paper for *5th Int. Palyn. Conf.*, Nanjing Inst. Geol. Palaeontol. Acad. Sinica, 1–15.

Index

Note: I believe that books of this sort are best served by a single index, in which names of taxa are indexed along with other key words. Multiple indices have always seemed an unnecessary nuisance to me. Main entries in the Glossary are not separately indexed, as the Glossary is its own index. The reader may therefore wish to check both Index and Glossary for more information on a particular item. Figures are referenced by italic numbers, as *10.5*. References to the figures that illustrate a taxon are indicated by an asterisk, as "*1.1★, 8.11, 8.13★–15*" for *Retispora lepidophyta*; this means that the taxon is mentioned in the five figures cited but is illustrated only in *1.1* and *8.13*. References in the text to individuals are *not* indexed, but their cited publications are of course listed in the bibliography, and those responsible for information in figures or tables are listed under Acknowledgements. Journals, books and other publications are not indexed, but are listed in the annotated bibliography of such items in Chapter 1.